CAMBRIDGE
UNIVERSITY PRESS

Shaftesbury Road, Cambridge CB2 8EA, United Kingdom

One Liberty Plaza, 20th Floor, New York, NY 10006, USA

477 Williamstown Road, Port Melbourne, VIC 3207, Australia

314–321, 3rd Floor, Plot 3, Splendor Forum, Jasola District Centre, New Delhi – 110025, India

103 Penang Road, #05–06/07, Visioncrest Commercial, Singapore 238467

Cambridge University Press is part of Cambridge University Press & Assessment, a department of the University of Cambridge.

We share the University's mission to contribute to society through the pursuit of education, learning and research at the highest international levels of excellence.

www.cambridge.org
Information on this title: www.cambridge.org/9781107061750

First published 2015

A catalogue record for this publication is available from the British Library

ISBN 978-1-107-06175-0 Hardback

Contents

Color plates follow page 206.

About the Authors

Daniel R. Lynch is MacLean Professor of Engineering at Dartmouth College and Adjunct Scientist at the Woods Hole Oceanographic Institution (WHOI). He holds degrees from the Massachusetts Institute of Technology and Princeton University. He has published extensively on simulation methods in coastal oceanography and is coeditor of the American Geophysical Union (AGU) volume *Quantitative Skill Assessment for Coastal Ocean Models* (CES 47, 1995) and a companion volume, *Skill Assessment for Coupled Physical-Biological Models of Marine Systems* (JMS 76, 2009). He co-founded the Gordon Research Conference in Coastal Ocean Modeling; was Executive Director of the Regional Association for Research on the Gulf of Maine; and served on the Executive Committee of the US Global Ocean Ecosystem Dynamics (GLOBEC) Northwest Atlantic Program (National Oceanic and Atmospheric Administration [NOAA] and National Science Foundation [NSF]). Sabbatical and other visiting leaves include WHOI, the Institut de Mecanique de Grenoble, the Institute of Ocean Sciences, and the University of Bergen.

Lynch is a NSF Presidential Young Investigator, received the B.H. Ketchum Award for interdisciplinary estuarine studies, and is a Fellow of the American Society of Civil Engineers. He developed the Numerical Methods Laboratory at Dartmouth, around the theme of interdisciplinary computational engineering, and authored the graduate text *Numerical Solution of Partial Differential Equations for Environmental Scientists and Engineers* (Springer, 2005).

Recent investigations focus on Sustainability, Natural Resources, and Professional Education. Included are visiting appointments at the University of Notre Dame (Melchor Visiting Chair); the School of Philosophy at the Catholic University of America; and the Woodrow Wilson School of Public and International Affairs at Princeton. He is coeditor of the special issue *Professions and the Common Good* (ACCU 2006) and a contributor to the *Civil Engineering Body of Knowledge for the 21st Century* (ASCE 2008). He authored the undergraduate text *Sustainable Natural Resource Management for Scientists and Engineers* (Cambridge University Press, 2009).

David A. Greenberg has been working in Canadian Marine Science since his sessions as a summer student with the Canadian Department of Mines and Technical Surveys in the late 1960s. He received bachelor's and master's degrees in Mathematics from the University of Waterloo (1969 and 1970) and a PhD from Liverpool University (1975). His thesis on modeling Bay of Fundy tides and tidal power set the stage for much of his later work, which includes continued work on tidal power, model development, and application to many theoretical and practical problems. His early work with finite difference models included mesh nesting, transport algorithms, and open boundary limitations. With finite element models he has looked at simulating wetting and drying of intertidal areas, and adapting models from Cartesian coordinates to geographic coordinates for larger domains and constraints on resolution in all ocean models. Processes examined include tidal, wind, density, and boundary-driven circulation; mixing; and sea level rise. Many model configurations have been created for studies in the Bay of Fundy and the adjacent Gulf of Maine, looking at tidal power, biological productivity, sediment transport, paleo tides, and aquaculture parasite and disease transport. Studies outside the Fundy–Maine domain include the tsunami generated from the 1917 explosion in Halifax Harbour, the circulation on the Newfoundland shelf, and the freshwater flow through the Canadian Arctic Archipelago. He continues to work actively in model development and application.

Ata Bilgili received the BSc from the Istanbul Technical University's Faculty of Naval Architecture and Ocean Engineering. He received the MSc and PhD degrees from the University of New Hampshire (UNH) in Ocean and Mechanical Engineering, respectively. His master's and PhD dissertations on Eulerian bed-load transport of coarse sediments and finite element modeling of the tides in the Great Bay Estuarine System in New Hampshire shaped the foundation for his later work. During his postdoctoral studies at UNH and later at Dartmouth College, he worked on improving the tidal models for well-mixed coastal systems with special emphasis on the N2 and S2 tidal components and their residual effects, using both Eulerian and Lagrangian methods. During his residency at Dartmouth College as a Research Assistant Professor from 2001 to 2005, Dr. Bilgili's work concentrated on coding and testing particle tracking methods for the characterization of estuary–ocean exchange and mixing using a statistically significant number of particles on parallel processors. Studies outside this domain included unstructured grid generation for finite element circulation modeling, modeling concentration and dilution of pollutant particles originating from wastewater treatment facilities, and interdisciplinary transfer of modeling products to researchers in the environmental science and education fields. In 2005, Dr. Bilgili started working for the Istanbul Technical University, where he obtained tenure as an Associate Professor of Coastal and Ocean Engineering. His work at this institution includes the application of the techniques developed earlier to wind-dominated coastal systems and the optimization of particle tracking methods on high-performance computing platforms. His additional current interests include the advancement of marine engineering education and environmental contingency planning for ports and harbors.

Dennis J. McGillicuddy, Jr., is a Senior Scientist in the Department of Applied Ocean Physics and Engineering at the Woods Hole Oceanographic Institution. His primary research interest is the interface between the fluid dynamics and the biology of the sea. He has pursued physical–biological–chemical interactions in four contexts: (1) the role of eddies in biogeochemical cycling of the open ocean, (2) impacts of coastal circulation on zooplankton dynamics, (3) the dynamics of harmful algal blooms, and (4) larval dispersal in deep-sea vent communities.

Dr. McGillicuddy received a PhD from Harvard in 1993 and joined the Woods Hole Oceanographic Institution in 1993 as a Postdoctoral Scholar. He was appointed to the scientific staff in 1996. He was a recipient of the Office of Naval Research Young Investigator Award (1998), the American Society of Limnology and Oceanography's Lindeman Award (2000), and the University of Miami's Rosenstiel Award (2008). He is author or coauthor of more than one hundred refereed publications.

Dr. McGillicuddy has been very active in the oversight of large interdisciplinary oceanographic programs on both national and international levels, having served on the scientific steering committees of the Joint Global Ocean Flux Study, the Global Ecosystem Dynamics Program, and the Global Ecology and Oceanography of Harmful Algal Blooms. Dr. McGillicuddy currently serves as Deputy Director of the Woods Hole Center for Oceans and Human Health.

James P. Manning After receiving an undergraduate degree in mathematics (University of Maine, 1979) and a master's degree in oceanography (University of Rhode Island, 1987), James P. Manning has been at NOAA's Northeast Fisheries Science Center for more than 25 years. Given a career largely devoted to building low-cost observing systems with the help of New England lobstermen, there is now an extensive network of individuals who deploy both moored and drifting instrumentation throughout the Gulf of Maine and the southern New England shelf. He also works closely with local ocean circulation modelers, providing them with data for assimilation and validation purposes. At the time of this writing, his web-served archive includes more than 5 million hourly values of moored bottom temperatures and more than a million kilometers of surface drifter tracks. The motivation behind most of these efforts is to describe the physical oceanographic environments of critical offshore spawning grounds. With a melding of data and model simulations, he hopes to help biologists explain the variability in recruitment of larvae and its relationship to the physical conditions.

Alfredo L. Aretxabaleta is a physical oceanographer working at the USGS Science Center in Woods Hole, MA. He received a BS in Marine Sciences from the University of Las Palmas de Gran Canaria in Spain in 1998 and a PhD in Marine Sciences from the University of North Carolina at Chapel Hill (UNC-CH) in 2005. While at UNC-CH, he was part of the USGLOBEC real-time forecasting team, and he worked on the development of a coastal observing system for the Southeast United States. He then worked as a postdoctoral investigator at the Woods Hole Oceanographic Institution on the physical processes affecting harmful algal blooms ("red tides") in the Gulf of Maine and Bay of Fundy, as part of the US EcoHab and BoHab projects. Then he worked at the Institut de Ciencias del Mar (Barcelona, Spain) on the development and improvement of the first satellite measuring salinity from space (SMOS, Soil Moisture and Ocean Salinity of the European Space Agency). Currently, Aretxabaleta is working with the US Geological Survey in Woods Hole on sediment transport problems in the coastal ocean, ranging from cohesive sediment behavior to resuspension of material from the bed. His main interest is the implications of the physical processes in the ocean for problems of interdisciplinary nature such as the behavior of biological organisms or the transport of pollutants and/or sediment. Aretxabaleta has collaborated in teaching introductory-level classes on physical oceanography and the modeling of marine and Earth systems. He is also one of the authors of the book *Project Earth Science: Physical Oceanography* of the National Science Teachers Association.

Preface

This is a book about the coastal ocean; specifically, about simulating particle motion within it. It unifies essential ideas, for a multidisciplinary group of scientific professionals.

We take as given realistic computer fields for the primary variables: velocity, pressure, temperature, salinity, and turbulence. These fields are becoming available for wide use, in site-specific realizations that account for detailed forcing and topography. Emerging are some biogeochemical fields in some cases: nutrients, light, plankton. This is our *virtual environment*. Though imperfect, it has reached a high level of realism, and it is continuously improving.

Our subject is the transport, evolution, and fate of particles in these environments. For simple particles, physical motion is of fundamental concern. For chemically active particles, one adds reaction, transformations, and interaction with bounding surfaces. For living particles, one adds suspension, growth, death, transformation, behavior, and reproduction. Nonlinear physical–biological interaction is implied and natural.

The material is culled from 20 years of research projects. A diverse collection of ideas has emerged, variously titled *Lagrangian*, *ensemble*, *kinetic*, *Monte Carlo*, *site-specific*, or *individual-based* methods. They share the common theme of simulating many small units (particles) individually, allowing variability to develop among them without prior aggregation. Current computing power makes broad operational use possible.

To date there is no single source that unifies these ideas for the audiences intended. Our book fills that gap. Included are basic ideas about applied probability; random walk; interpolating and navigating general computational meshes and data products; and Monte Carlo simulation methods. Applications include drifters, sediment motion, oil spills, and larval ecology.

Overall, our intention is to facilitate the analysis of coastal ocean phenomena using simulation of *particle dynamics*. The notion of *trajectory* is general – the physical and biochemical history experienced by an individual *particle in motion*. The work is configured as a teaching and reference source, suitable for year-4 and beyond university students and for working professionals.

Acknowledgments

We have benefited over the years from collaboration with many individuals. In particular, we are grateful for the enormous contributions of Francisco Werner, Charles Hannah, Christopher Naimie, Charlie Miller, Craig Lewis, Jeff Proehl, John Loder, Fred Page, Brian Blanton, Keston Smith, Justin Ip, and Ruoying He in developing the ideas covered here.

Special thanks go to Malcolm Spaulding for generous assistance with Chapter 10.

Additional thanks to Rick Lumpkin, Jon Hare, and Jeff Pessutti, who provided valuable reviews, and to Vitalii Shermet for providing particle tracking code.

We thank our sponsors: NSF, NSERC, NIEHS, NOAA (OAR, SeaGrant, CICEET, CSCOR), NASA, ONR, the special interagency programs including GLOBEC, ECOHAB, MERHAB, GOMTOX, and NOPP; and others who supported these developments. DJM gratefully acknowledges additional internal WHOI support from the Holger W. Jannasch Senior Scientist Chair.

We thank the National Center for High Performance Computing at the Istanbul Technical University (http://www.uybhm.itu.edu.tr/index.php) for the use of its parallel computer clusters for some of the simulations in this book, under contract number 1001192011.

Acronyms

ACCU	Association of Catholic Colleges and Universities
ADCP	Acoustic Doppler Current Profiler
AGU	American Geophysical Union
ASCE	American Society of Civil Engineers
ASLO	American Society of Limnology and Oceanography
CICEET	Cooperative Institute for Coastal and Estuarine Environmental Technology
CSCOR	Center for Sponsored Coastal Ocean Research
ECOHAB	Ecology of Harmful Algal Blooms
GEOHAB	Global Ecology and Oceanography of Harmful Algal Blooms
GLOBEC	Global Ocean Ecosystem Dynamics
GOMTOX	Gulf of Maine Toxicity Project
IOOS	Integrated Ocean Observing System
JGOFS	Joint Global Ocean Flux Study
LAPCOD	Lagrangian Analysis and Prediction of Coastal and Ocean Dynamics
MERHAB	Monitoring and Event Response for Harmful Algal Blooms
NASA	National Aeronautics and Space Administration
NCCOS	National Centers for Coastal Ocean Science
NEFSC	Northeast Fisheries Science Center
NIEHS	National Institute of Environmental Health Sciences
NSERC	Natural Sciences and Engineering Research Council
NSF	National Science Foundation
NOAA	National Oceanic and Atmospheric Administration
NOPP	National Ocean Partnership Program
OAR	Office of Oceanic and Atmospheric Research
ONR	Office of Naval Research
SG	Sea Grant
SMOS	Soil Moisture and Ocean Salinity
USGS	United States Geological Survey

Definitions and Notation

Vectors and Scalars

Bold indicates vector (e.g., $\mathbf{V}$)
Nonbold indicates a scalar (e.g., ρ)

Eulerian Fields

$\mathbf{x}$ is a geographic coordinate. It can be 2-D or 3-D. Implied is a Cartesian system with scalar parts (x, y, z); if other projections are used, they will be explicitly identified.
Unit vectors are indicated by underlining, for example, $\underline{\mathbf{x}}$.
$\mathbf{v}(\mathbf{x})$ and $\zeta(\mathbf{x})$ are Eulerian fields, respectively a vector field $\mathbf{v}$ and a scalar field ζ.
$\frac{\partial}{\partial x}$, $\frac{\partial}{\partial t}$, ∇, and so forth indicate conventional Eulerian partial derivatives in the $(\mathbf{x}, t)$ domain.

Lagrangian Particle Trajectory and Properties

$\frac{dA_j}{dt} \equiv \dot{A}_j$ indicates a Lagrangian time derivative of A following particle j.
$\mathbf{X}_j(t)$ is a specific location in $\mathbf{x}$, the position of the jth particle at time t.
$\mathbf{V}_j = \dot{\mathbf{X}}_j$.
$\mathbf{V}_j$ is the velocity of particle j: $\mathbf{V}_j(t) = \mathbf{v}(\mathbf{X}_j(t))$.
$M_j(t)$ is the scalar mass of an individual particle j at time t.

Numerical Analysis

$\mathcal{O}(\alpha)$ indicates "small terms of order α," that is, the leading or largest term scales as α.
ΔQ^k indicates the backward difference $Q^k - Q^{k-1}$.

Elements

$\mathbf{x}_i$ is the value of the coordinate $\mathbf{x}$ at node i.
$\phi_i(\mathbf{x})$ is the ith basis function, centered at node i.
ϕ_{ik} is the location of particle k on an element (the value of ϕ_i at point k).
ρ_i is the value of a function ρ at node i.

The continuous function $\rho(\mathbf{x})$ is interpolated with the bases:

$$\rho(\mathbf{x}) = \sum \rho_i \phi_i(\mathbf{x}) \tag{1}$$

A_e is the area of element e.

Singularities

$\delta(x)$ is the Dirac delta function. It is zero everywhere but has a removable singularity at the origin, $x=0$. Specifically, $\int \delta(x)\,dx = 1$ if the domain of integration includes the origin; otherwise, the integral is zero.

$\delta_j \equiv \delta(x - x_j)$ is the Dirac delta function, singular at position $x = x_j$.

$\int f(x) \cdot \delta_j \, dx = f(x_j)$. This integral samples the function $f(x)$ at the point x_j.

Random Numbers

Several symbols are reserved for random numbers:

ϵ indicates a random number.
$\mathcal{R}$ indicates a random number.
η indicates a random number.

Generally, an overbar indicates a mean or average, for example, $\overline{F}$. A departure from the mean is indicated by either of two markers, $\widehat{F}$ or $\widetilde{F}$.

Some familiar distributions:

$\mathcal{U}$ is the uniform distribution.
$\mathcal{G}$ is the Gaussian distribution.
$\mathcal{N}$ is the standard Normal distribution: Gaussian with zero mean, unit variance.

The random variable z as generated from the distribution f is denoted $z \sim f$.

Example: $z \sim \mathcal{U}[0,1]$: z is a deviate uniformly distributed on the interval $[0,1]$.
Example: $z \sim \mathcal{G}(a, b^2)$: z is a Gaussian deviate with mean a and variance b^2.
Example: $n \sim Pois(\lambda)$: n is a Poisson deviate with rate λ.

Parameters and Properties of Distributions

f is the probability density function (PDF).
F is the cumulative distribution function (CDF); $F(x) = \int f\,dx$.
E is the Expectation operator: $E(x) = \int x f(x)\,dx$.
μ indicates the expected value (mean) of a distribution: $\mu_\epsilon = E(\epsilon)$.
Var indicates the variance of a distribution; equivalently, its second moment about the mean:

$$Var(\epsilon) = E(\epsilon - \mu_\epsilon)^2 \tag{2}$$

σ indicates standard deviation; $\sigma^2 \equiv Var$.
ρ indicates a (dimensionless) autocorrelation coefficient in a random time series.
$\nu(\mathbf{x})$ is the particle *intensity*, number of particles per unit area.
$P[event]$ is the probability of that event occurring. Example: $P[x \geq 5] = .1$.

Covariance

Cov_{xy} is the expected value of the product of x and y: $Cov_{xy} = E[(x - \mu_x)(y - mu_y)]$.

$C_m(\epsilon)$ is the autocovariance of the discrete, demeaned series $\widehat{\epsilon}$, at points separated by lag m:

$$C_m(\epsilon) \equiv E\left[\widehat{\epsilon}^k\,\widehat{\epsilon}^{k-m}\right] \tag{3}$$

C is symmetric: $C_i(\,) = C_{-i}(\,)$.

$C_0(\epsilon)$ is the same as σ_ϵ^2.

ρ_m is the autocorrelation coefficient, at lag m – the normalized autocovariance: $\rho_m = \frac{C_m}{C_0}$.

Time Series

$\mathcal{R}$ generally indicates a random displacement over Δt. Units: length

$\mathcal{R}_1$ generally indicates a random velocity change over Δt. Units: length/time

$\mathcal{R}_2$ generally indicates a random acceleration change over Δt. Units: length/time/time

$d\mathcal{W}$ is the Wiener increment.

$\Delta\mathcal{W}$ is the Wiener integral $= \int_t^{t+\Delta t} d\mathcal{W} \sim \sqrt{\Delta t} \cdot \mathcal{G}(0,1)$.

$\mathcal{R}_v$ is the Wiener noise input to the AR1 model.

S_v is its size.

$\mathcal{R}_v\,dt = S_v\,d\mathcal{W}$ is the scaled Wiener increment.

$\mathcal{R}_a$ is the Wiener noise input to the AR2 model.

S_a is its size.

$\mathcal{R}_a\,dt = S_a\,d\mathcal{W}$ is the scaled Wiener increment.

Discrete Convolution

Z_i are convolution coefficients. For a discrete (linear) time series Q forced by the series F

$$Q^N = \sum_{j=1}^{N} F^j Z_{N-j} \tag{4}$$

The series Z is the response to unit forcing. The convolution is the superposition of responses to all past influences F.

B^k is the partial sum of convolution coefficients,

$$B^k = \sum_{i=0}^{k} Z_i \tag{5}$$

Introduction and Scope

The Coastal Ocean

The *coastal ocean* comprises the semi-enclosed seas on the continental shelf, including estuaries and extending to the shelf break. Geographically the scope is approximately that of the Exclusive Economic Zone for which nation-states are responsible.

The coastal ocean is the focus of many serious concerns, including coastal inundation by tide, storm, or sea level change; fisheries and aquaculture management; water quality, health, and pollution; harmful algae blooms; facilities siting and planning; ports and harbor development and maintenance; spills and other accidents. All fall within the scope of science, engineering, and development. All demand professional competence in promoting wise and healthy use of this critical region, and in anticipating, monitoring, and responding to events therein.

The Virtual Environment

Computational Circulation Products. A realistic simulated or virtual environment (herein "the environment") is becoming available for scientific and engineering use. This takes the form of site-specific computational products for the baseline physical fields. Simulated particles in these fields offer the possibility of high realism, facilitating advanced analysis.

Equations and Variables. Well-defined equations and variables characterize the coastal ocean. For our purpose, the bundle of "physical" variables includes pressure, velocity, temperature, salinity, light, nutrients, and subgrid or unresolved mixing rates (typically in the form of turbulent energy, mixing length, a stability parameter, and eddy diffusivity). These all share very conventional formulations for their evolution and transport. This scientific foundation has advanced very far and supports the construction of site-specific environments.

Randomness. There will always be unresolved dynamics, and they must be accounted for, both in motion and in biotic development. At the simplest level, there is truncation error, equation error, and forcing uncertainty. These all contribute to a distribution of possibilities for the environment itself. In addition, there is uncertainty about basic biotic rates and their distribution across individuals in the same population. Perhaps most fundamentally, in setting out a gridded calculation and product, the real motion at the subgrid will have at best a statistical description. Particles live in the subgrid, and influences there can only be statistically described. A Monte Carlo approach, historically always anticipated, is now clearly possible for the coastal ocean.

User Groups and Regional Systems. Networked availability of site-specific environments are emerging in the form of regional observation/modeling systems. The Integrated Ocean Observing System (IOOS) provides an international network for these, and some standards. As a class these offer continuous, cooperative improvement. Specific products include site-specific climatologies and data archives, hindcast products, and forecasts. These can support many applications: scientific studies, infrastructure planning, response to emergencies, and so forth.

Technical user groups provide model-family support spanning applications globally. These share a laboratory/university research and demonstration focus. Further practical implementation is a professional matter, to be done by a proper blend of private-sector initiative and public-sector support.

Professional competence demands scientific bundling of the advanced supporting ideas. Accordingly, *our text serves this need.*

The Present Work

Our purpose is to present the basics of a Lagrangian particle-based approach to simulation. We do not seek to construct the virtual environment, but to describe phenomena within it. Required is a basic description of principles: the Lagrangian trajectory in physical–biological space and its statistical nature; the data file standards that are needed to import the detailed environment and to record the simulation results; and enough basic simulation mathematics to make ideas easy to implement in practical programs.

A fundamental assumption is that geometric realism in the coastal ocean demands an unstructured horizontal mesh – so we assume the simplest in this category, the linear mesh (triangles or quadrilaterals). This is not to constrain the sophistication of the fields themselves – neither their interpolation nor their generation. Rather, we seek to liberate the geometrical description of the coastal ocean from any constraints imposed by horizontal mesh. It is a truism that any published field on a structured horizontal mesh structure can be converted to one of triangles for the purpose of particle simulation. So, nothing is lost, but a vast degree of generalization is gained. This is the baseline. We extend the ideas to quadrilateral meshes for generality.

This book is *not* a replacement for conventional descriptions of the coastal ocean or how to simulate it – there are many excellent works in that category. Rather, it is about the *use* of these to test ideas and hypotheses using particle methods. There is an eager and prepared audience for this, a group of professionals who want to solve complex problems. This book is for them – for their education and as a handbook of practice and reference.

Audiences

The material presented reflects the experiences of the authors:

In teaching: We address upper-level (fourth-year university) undergraduate students with scientific/mathematical preparation and first-year graduate students. In either case the book supports courses, theses, project-based academic work, and scientific seminars and study groups.

In research: The material is backed up by two decades of scientific research. These research projects have been published at the frontiers of simulation, oceanography, ecology, and ocean engineering. This is an exciting interdisciplinary field that has emerged during our careers, and its continuation is clearly needed on a wider scale.

In practice: The examples given show relevance to many contemporary problems, including plant siting, water quality, tidal power, and storm surge. These are reflective of current engineering practice. The purpose of this book is to collect the apparatus necessary for wide use, without demanding that professionals first construct a new and competitive virtual environment – instead, one *uses* available products.

Preparation Necessary

The target audience has upper-level undergraduate preparation in science or science-based majors:

- Physics, chemistry, biology, ecology, oceanography, earth sciences
- Ocean, civil, chemical engineering
- Computer science and applied math

Involvement of each is needed in the coastal ocean; each should be able to engage this material and find it useful. The necessary baseline is standard for science and engineering students: calculus through ordinary differential equations (ODEs), linear algebra, and a first exposure to numerical analysis. Programming in a modern structured language is assumed; there are many paths to the necessary level, and most involve a healthy dose of experience.

Beyond this basic level, multivariable calculus and partial differential equations (PDEs) are the basis of the simulation products one is expected to use. Preparation in this area will be discipline-specific, so some instructor attention to the specific audience is important here. In engineering disciplines, this is likely to come via first courses in fluid mechanics and in transport theory. In the geosciences, it is likely to arise naturally in geophysical transport phenomena, in particular, physical oceanography and oceanographic ecosystems.

A first exposure to statistics and probability is assumed. There is a very wide variation in preparation in this area, yet getting to the level of a first university course in probability and statistics is fundamental. In some engineering disciplines, notably electrical engineering, communications, dynamics and control, and computer science/engineering, this is basic. In other disciplines there is more variety in preparation and readers need to make up any deficit in parallel. A useful bottom line: a mature understanding of ensembles of samples, underlying probability distributions, and Monte Carlo simulation. In this and many other areas of applied mathematical analysis, one benefits greatly from the very practical material in Press et al. ([362]).

The phenomena of interest are at the intersection of ecology, oceanography, and engineering. These are the features of the coastal ocean; if they are not familiar, then parallel study of them is recommended.

A practical "test" of adequate preparation might be: understand the processes underlying coastal phenomena, their description in PDEs, and their numerical simulation; be able to generate random sequences with known properties on modern computers; force simulated processes stochastically with them; and gather the simulated outcomes and evaluate the properties of the outcomes. This is one description of a Monte Carlo simulation approach. If one can understand it, carry it out on machines, and relate that to the coastal ocean, then the preparation cited above is likely adequate.

Numbered Problems are included throughout, located near the text material that they amplify. Generally these problems are easy, open-ended explorations or extensions of material just exposed. In most cases they are set up to encourage discussion of the ideas. An example: "Suppose the analysis above were redone with the following change: ... a) what do you expect, and why? b) simulate the effect; do the results agree with your expectations? why or why not?" All of these simulations are readily done with desktop packages, accessible by the intended audiences. (For example, Matlab has a comprehensive library of statistical and linear algebra routines.) Rather than having simple answers, these problems are intended to be starting points for further exploration. And they are readily tailored to a specific audience's strengths or weaknesses.

Related Books

There is no model for this book; the material is broadly distributed across many sources, but is neither coordinated nor focused on the intersection of particle methods and the coastal ocean.

Related texts covering physics include some classics, such as Csanady [95], and some current works such as Kantha and Clayson [223], Knauss [231], and Simpson and Sharples [419]. Parker [350] provides an excellent treatise on tides. Robinson and Brink [384, 54] give an excellent summary of the coastal ocean with processes and regional syntheses. There are many textbooks covering general oceanography, suitable for broad audiences; Townsend [452] is a good current example.

Many works are directed to oceanic transport, for example, Fischer [152, 153], Csanady [94]; these are typical in that the Eulerian or concentration-based approach is stressed. They extend more foundational work on diffusive processes as in the (non-oceanic) classics by Crank [91], Carslaw and Jaeger [69], and Finlayson [151]. Useful works that expand on turbulence closure include Kantha and Clayson [224] and Baumert et al. [22].

Numerical analysis is heavily covered in the literature. Gresho and Sani [184] provide a comprehensive study of advection–diffusion problems. Lynch [279] looks at PDEs generally; Glover et al. [172] focus on oceanography. Generally, these texts treat deterministic problems well but are short on statistical variation. And they focus on Eulerian descriptions.

In terms of Lagrangian dynamics, Bennett [24] gives a theoretical treatment of fluid dynamic theory. Griffa et al. (LAPCOD) [186] provide an excellent summary of recent blending of theory and observation with only limited coverage in the coastal seas and closely focused on physics.

Individual-based modeling is covered by several authors. Okubo and Levin [343] set the stage with early descriptions of individuals and probability theory. Grimm and Railsback [188] and DeAngelis et al. [108, 107] provide more contemporary summaries of this area.

Our book unifies the essentials needed to begin simulating particles in the coastal ocean. That synthesis does not exist in current books.

Part I

Background

1
The Coastal Ocean

Man's interface with the ocean begins at the coast. Just the mention of estuaries, harbors, bays, lagoons, open coastline, and shelf seas conjures up a myriad of processes and scales with complex interactions. It is at the coast where human influences are most concentrated – waste disposal, construction, transportation, agriculture, aquaculture, recreation, mineral and biological exploitation. Equally, it is at the coast that the ocean most impacts humans, dictating patterns of everyday life and long-term viability. Oceanic motion mediates all these interactions, at several scales: surface waves, tide, storm surge, seasonal, decadal and beyond.

Our scope extends from land across the continental shelf to roughly the shelf break. The ocean dynamics change abruptly near the shelf break, as do many practical concerns including resource management, pollution, and navigation. This large, semienclosed region has come to be identified as the *coastal ocean*. Geographically the scope is approximately that of the Exclusive Economic Zone. The transport of matter within and through the coastal ocean is of paramount importance.

The understanding of water motion and its simulation in the coastal ocean is quite advanced. Many modeling methods are in use today that will give a faithful representation of currents and sea level. The resolution possible is surprisingly high and continually improving. Model output in the form of hindcasts and even forecasts is growing in availability as well. These trends are likely to continue. Simultaneously the geographic coverage of models is expanding under national and international guidance.

It is the current fields from these models that are our starting point. We seek to use them, not produce them. Together with their companion fields (temperature, salinity, nutrients, turbulence) they constitute a sophisticated base from which to examine many important issues.

The common starting point is to examine the tracks that simple particles follow. It is the connections among time, place, and a particle's past that make this such an important matter. Immediate applications include the tracking of floating objects or predicting the path of a stricken vessel or marine mammal. This "Lagrangian" description rises rapidly in complexity as one addresses marine chemistry and biota. Oceanic chemical interactions are complex enough in themselves; the uncontrolled outdoor environment enhances the difficulty in measuring and understanding them. A living organism will add its own behavioral motion and biochemical dynamics, both of which will develop as it matures and moves through the environment.

An excellent introduction to coastal ocean phenomena can be found in the text by Simpson and Sharples [419] and in the earlier text by Csanady [95]. Townsend [452] provides an excellent general introduction to marine science, with special emphasis on contemporary issues in biological oceanography. These texts are recommended as basic source material and a gateway to the wider oceanographic literature. For more advanced texts, the two Robinson/Brink volumes for processes [54] and regional studies [384] are entirely focused on the coastal ocean.

1.1 Typical Motions and Scales

Water movement on the continental shelf is heavily modulated by topography and coastline. Motions are steered by topography; water banks up against the coast; onshore flows can be diverted vertically. Typical horizontal motions in the coastal ocean can be crudely distinguished by four characteristics: velocity magnitude, time scale, length scale of variation, and size of displacement (Table 1.1). The "seasonal" category includes mean seasonal wind and mean baroclinicity reflecting seasonal water mass formation. These can extend across the shelf from land to the shelf break and account for significant and persistent along-shore displacement.

These motions can be strongly affected by local scale variations and the geographic setting. A small harbor in a straight coastline on a smooth continental shelf will experience very different scales of motion than those seen offshore, but still be driven principally by the larger scale shelf motions. River motions will tend to be dominant along the axis of the river, with transverse motions strongly influenced by curvature and longitudinal variations in width and topography. Coastline variations and structures can modify motions well up and downstream. With sufficient tide or surge, shallow areas can dry and dry areas can flood. There is no substitute for local resolution in all these cases, but it cannot come at the expense of geographic coverage. One needs both if the relevant motions are to be properly represented.

Below we elaborate on some physical processes, emphasizing the characteristics that will be important to transport and dispersion. There is an abundance of literature on such processes.

Tides

In the deep sea, the tides are forced primarily by gravitational interactions of the rotating and orbiting Earth, moon, and sun. The tidal motion on the shelf is driven

Table 1.1. *Scales of Motion*

	Velocity	Time scale	Length scale	Displacement
Tide	1 m/s	Daily, periodic	200 km	10 km
Wind	0.5 m/s (surface)	3-day events	100 km	100 km
Seasonal shelf setup	0.01 – 0.10 m/s	Monthly, steady	400 km	0.86–8.6 km/d

by co-oscillation with the astronomically driven deep ocean tides, with the direct gravitational forcing having a minimal effect. This shelf oscillation can be amplified and damped by the structure of the coastline and bathymetry. For example, the large tides in the Bay of Fundy are often attributed to a resonant oscillation (Defant [109]). Tidal resonance is a global phenomena seen in studies of the Bristol Channel [156], the Gulf of California [173], Ungava Bay and Hudson Strait [7], the Coral Sea [470], and even in relatively low-tide environments like the Gulf of Gabes, Tunisia in the Mediterranean Sea [399].

In the presence of sloping topography, tidal motions can be rectified to produce a mean subtidal circulation (Huthnance [204]). For example, through an examination of Stokes drift, Loder [257] noted that there were significant differences between the Eulerian and Lagrangian mean currents around Georges Bank. These motions have major ecological importance; the processes can be expected to be operative over steep topography in tidally energetic waters.

Simpson and Hunter [418] called attention to the importance of vertical mixing fronts initiated by bottom stress during tidal flow. They found that at a particular value of the ratio h/u^3 (h is the mean water depth, and u is the depth-averaged current of the spring tide) the water would be vertically mixed. The frontal boundary between tidally mixed areas and the surrounding stratified waters tend to be associated with enhanced biological activity. "Bottom trapped" fronts have been identified in a great many coastal ocean locations.

Wetting and Drying

In many nearshore applications, explicit treatment of the moving boundary between land and water is a necessity. Accurately representing this process is a challenge for both modeling and observation. Drying schemes can be implemented in models at various levels of complexity [31, 70, 155, 179]. Greenberg [181] has demonstrated that in tidally energetic areas with extensive tidal flats, there can be significant dynamic influences hundreds of kilometers away. Bilgili Mclaughlin et al. [307] found that even on shorter length scales, the tidal prism can have significant contributions from wetting/drying areas, with important distortions in deeper currents.

Westerink et al. [480] found that properly simulating inundation in southern Louisiana required extreme resolution to reproduce the overtopping of small structures. This has been a critical concern and requires land use and other data not typically part of oceanography. A remarkable feature of this model is the coupling of these finely resolved local flooding dynamics to broad basin–scale coverage necessary to replicate hurricane processes.

Wind Response

The surface wind stress can strongly influence the circulation of coastal seas. Wright and others [487] computed the large-scale response to regional wind using a simple linearized two-dimensional (2-D) barotropic model of the Gulf of Maine and adjacent continental shelf. They showed that to get a reasonable representation it was necessary to account properly for upstream shelf influences (in the sense of trapped topographic waves) outside the model domain through appropriate specification of the open boundary (see also Hayashi et al. [191]). An improvement on this study was done in Greenberg et al. [177] in a three-dimensional (3-D) model where the domain boundaries were expanded outward along the shelf and into the deep sea. This demonstrated the importance of surface and bottom Ekman layers leading to vertical structure in the currents, giving rise to wind-driven upwelling and downwelling, distinct surface currents, and circulation patterns in deep channels and along steep topography.

Buoyancy-Driven Flow

Density gradients influence many ocean flows and processes. They are critical to many biological processes, as illustrated in Boicourt et al. [43]. The arrested topographic wave of Csanady [93] describes the basic motion of coastal freshwater trapped along the coast under the influence of friction and the Earth's rotation. The motion is very much dependent on topography [409]. Wright [486] showed how on a flat shelf the transport would migrate to the shelf break and Chapman and Lentz [77] demonstrated that the bottom boundary layer, even when thin, is a major component of the dynamics. The buoyancy flows can be driven along the coast by coastal trapped freshwater as described in Chapman and Lentz [77] and Lentz and Fewings [247] and by dense bottom water as seen in Shaw and Csanady [409] and Brooks and Townsend [57]. Wind and waves will change the circulation patterns associated with coastal river input [247], but the meteorological influence is very much dependent on the scale of the river outflow [246].

Buoyancy-driven circulation is ubiquitous in the coastal ocean, but as seen earlier, the motion is often heavily modified by other forcings. Although simple in concept, an accurate physical description of the compounded processes involved can be very complex.

Comprehensive Subtidal Circulation

The increasing complexity of circulation models is demonstrated by the incremental improvements in studies of the subtidal circulation in the Gulf of Maine. The two-dimensional structured approach of Greenberg [182] and Wright et al. [487] mentioned previously, showed the importance of tidal, wind, and boundary forcing. Lynch et al. [278] included baroclinic computations using an unstructured mesh and a linear harmonic model. Lynch and Naimie advanced this by expanding the domain and increasing the resolution over the outer banks of the Gulf of Maine [278, 286]. These two studies provided the physical fields for a major Lagrangian investigation into the fate of cod and haddock early life stages [478]. Further advances were made by Greenberg et al. [177] looking at detailed wind-driven solutions. Naimie et al. [326] included the major nonlinearities by iterating on the linear harmonic solutions. The work progressed to fully nonlinear prognostic modeling with the development of a comprehensive model by Lynch et al. [282]. Naimie [327], taking advantage of the flexibility of a triangular unstructured mesh, showed that locally adding high resolution around Georges Bank without changing the topography gave more accurate solutions.

Mixing, Residence Time, and Exchange Rates

The study of mixing, residence time and exchange rates relates to many practical problems. Historically box models and tidal prisms have been used to approximate the dynamic. Needless to say this can be a highly aggregated form of analysis, leaving many questions unaddressable. A good introduction can be found in Zimmerman's papers [504, 505, 506] and in Fisher et al. [153].

Bilgili et al. [31] assert "... residence and exchange are inherently Lagrangian concepts and are best treated as such." There are many complexities involved in making Lagrangian computations, part of the motivation for this book. Thompson et al. [447] demonstrate how in tidal computations, the work of particle tracking can be reduced by using a Markov chain approach linking specific subregions. Bilgili et al. [31] use the technique in a complex estuary system. Differences in retention due to the choice of particle release time (high or low water) were significant. This echoes results found by Ridderinkhof et al. [381], where exchange was very much dependent on the starting tidal phase. All of these findings rely on tidal-time particle tracking.

Zimmerman and Ridderinkhof used tidal-time particle tracking and found "Lagrangian chaos" in periodically forced tidal simulations. This "chaotic stirring" occurred in areas of very irregular topography and coastline [380, 507], despite the expectation of ordinary, periodic behavior. The variation from regular to chaotic can depend on the relative scales of the topography and tidal excursion [379]. Particle tracks can be interpreted as manifolds delineating fixed points that are either elliptic (centers of eddies) or hyperbolic (saddlepoints where streamlines converge and separate) [379, 256]. Saddlepoint analysis provides another method of describing the chaotic versus regular state of the regime.

Proehl et al. [366] demonstrated the importance of wind forcing over complex tidal–time flows. Significant horizontal dispersion was found in modeled and observed systems, effectively created through 3-D tidal time advection of particles and their 1-D vertical diffusion. No horizontal eddy diffusivity was used, the operative mechanism was shear dispersion; and it was very much event specific. The accurate modeling of the near-surface response to wind stress is of crucial importance.

With respect to coastal ocean motion, we can perhaps generalize Zimmerman's [507] parting words: "Thinking about the latter regime [Lagrangian chaos] particularly has prompted this paper's title [The Tidal Whirlpool ...], which stands as much for the complex dynamical reality of a tidal area, as it is a metaphor for the state of mind one enters in trying to get a grip on it." (p.152)

Summary: Motion, Processes, Scale

There are still challenges in the modeling of many coastal ocean processes. There will always be a need for site-specific knowledge and understanding; and there will always be processes unresolved. Particle tracking can help expose shortcomings in circulation models; it will certainly benefit from the constant progress being made in modeling methodology and application.

1.2 Particle Simulation

We simulate many individual oceanic particles. Each particle will have its own dynamics, subject to numerous, variable influences. All outcomes will be retained at the individual particle level. At any time, these outcomes can be summarized or aggregated without loss of any information. Rather than simulate the average particle, we simulate many particles and average the results when appropriate.

1.2.1 Motion

In terms of motion, we have to account for particle advection, dispersion, and behavior; all jointly determine the individual trajectory $\mathbf{X}_i(t)$. There will be a deterministic component – the advective fluid velocity $\mathbf{v}$ – that depends on a particle's location in space and time. In addition, the motion will have an individualized random displacement $\mathcal{R}$, representing unresolved or unknown motion – this will be parameterized with an eddy diffusion coefficient A. Both $\mathbf{v}$ and A are assumed available as gridded fields from a computational coastal ocean model. There may be other parameters influencing the random $\mathcal{R}$, including an individual particle's condition or memory. If the particle is finite in size or mass, then an additional law of motion is required, dealing with particle drag and inertia. Finally, there is the influence of the boundary layers encountered – air–sea, land–sea, benthic etc. – these may or may not be resolved in a coastal ocean model; but their effects need to be recognized.

Beyond these aspects of motion is behavior – something that develops as an individual develops, for example, buoyancy, swimming speed, preferred location, etc. This is a Lagrangian property of an individual particle; it is a prime example of coupling of Lagrangian biology with physical motion. It is critically important to include this behavioral portion of motion in the vertical; the scales of motion and resolution in most intended applications make the horizontal motion not subject to behavioral modification. Of course, if an individual changes its depth, it can dramatically alter its environment including the basic fluid motion and several of the environmental variables important to it.

The computational task is simple and well explored: integrate $\dot{\mathbf{X}}_i = \mathbf{V}_i(\mathbf{X})$ with particle velocity $\mathbf{V}_i$ dependent on the current particle location, history, and environment, plus some randomness.[1]

An individual is presumed to have embedded memory – memory of where it has been, how it reacted to the environment then/there, and how it has accumulated development. There are dynamical relations for random walk memory as well as biotic rates of change; these have to be embedded in a "particle." We thus imagine a particle's state to include location; sufficient to collect all environmental variables; plus a group of "onboard" relations and rates, sufficient to process the environment, generate vital rates and process random influences; plus a memory of its past and whatever it inherited from either initial conditions, launch or birth. A special possibility for biotic particles is to retain lineage through generations.

The *passive* particle is the fluid mechanical baseline; an inert particle follows the 3D fluid motion, including upwelling, surface and boundary activity, etc. At the other extreme, there is the *fixed depth* particle – this is familiar in analysis, and compatible with simple forms of drifting instruments (surface or drogued at fixed depth). This is not a passive situation, however – implied by perfect depth regulation is potentially vigorous vertical motion to maintain position. Examples include drogued drifters and larvae with assumed diel or seasonal migratory behavior.

Floating objects involve special considerations of motion associated with atmospheric or surface-wave forcing occurring in the surface boundary layer. Typical models do not have adequate resolution here, so some form of vertical subgrid closure is needed. These considerations pertain to search-and-rescue operations, as well as to the motion of floating plankton and other material.

[1] Throughout we use $\mathbf{v}$ for the gridded fluid velocity, $\mathbf{V}$ for the particle velocity.

1.2.2 Rates

In terms of biotic processes, an individual must carry with it a complete description of the relevant state and its rate of change, sufficient to integrate its particular dynamics. For each individual, there is a collection of ordinary differential equations (ODEs) with individual parameters, forced by deterministic and stochastic influences, coming from within (the individual) and without (the environment).

There is an important history of modeling many biotic and chemical processes in this way – and in identifying state variables and their rates of change. A fundamental form is

$$\frac{1}{B}\frac{dB}{dt} = r(B,E\,|\,\alpha) \tag{1.1}$$

r is the process rate, B the biological state, E the environment, and α the parameters.

If there is a useful reference state with rate r_o, for example, low abundance or an equilibrium, then this may be expressed as

$$\frac{1}{B}\frac{dB}{dt} = r_o \cdot L_o(B,E\,|\,\alpha) \tag{1.2}$$

where $L_o = r/r_o$ is a rate limiter, unity when in the reference state. For example, the popular Michaelis–Menten form has $L_o = \frac{B_h}{B_h+B}$, with parameter B_h, the "half saturation constant." $L = 1$ at low B, the reference state, and decreases monotonically through 0.5 at half-saturation to zero at large B. The reaction is saturated at high B; r is slowed to nothing. In Appendix D we summarize this and some other well-used rate expressions. We anticipate putting this level of realism to work onboard every particle.

These rates describe continuous changes, for example, growth, weight gain, etc. Some transformations however, are fundamentally *binary*. Examples include instrument failure or stranding; death; reproduction; morphological transition. In these situations every individual is binary, say with probability P it will die during the coming timestep; the aggregation of outcomes (number of positive versus negative outcomes) will obey the binomial distribution, there being many combinations of individuals for a given aggregate outcome.

For N identical particles, the total number of transitions will have mean NP, such that on average, $\frac{\Delta N}{N} = P$. This is intuitive. The outcome will be variable, however, with standard deviation $\sigma = \sqrt{NP(1-P)}$. A single realization will have

$$\frac{\Delta N}{N} = P + \epsilon \tag{1.3}$$

with the random number ϵ having zero mean and standard deviation scaling with $\frac{1}{\sqrt{N}}$. At high N we approach the limit of vanishing variability even though each individual

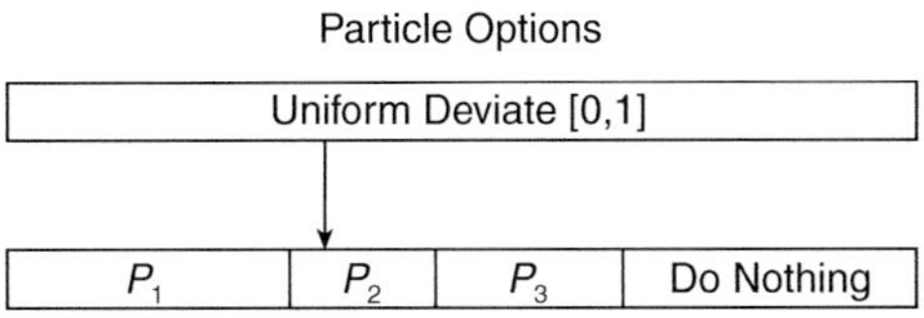

Figure 1.1. A single particle with three mutually exclusive options, with unequal probabilities P_1, P_2, P_3. A single random deviate from the uniform distribution allows selection from among them, outcome 2 in this case. The do-nothing alternative has probability $1-\sum P_i$.

is binary. By comparison with Equation 1.1, we have the equivalence

$$\frac{1}{B}\frac{dB}{dt} = r \iff P = r\Delta t \tag{1.4}$$

This is particularly important when bulk rates r describe aggregates of binary individual particles. (Often the aggregate is all that is easily observable.) In such cases, the bulk rate expression r is immediately transferable to individual $P = r\Delta t$; and the particle simulation will replicate the bulk relation, on average.

Naturally P must not exceed unity; so this puts a limit on the time step, $\Delta t < \frac{1}{r}$. Essentially, Δt must not exceed the natural response time embedded in the rate r.

This idea readily generalizes to discrete processes with more than two mutually exclusive outcomes in a time step. The multinomial distribution governs; it is a generalization of the binomial; and the conclusion is the same, $P = r\Delta t$ for each outcome separately[2]:

$$\frac{1}{B}\frac{dB}{dt} = \sum_i r_i \iff P_i = r_i\Delta t \tag{1.5}$$

with r_i the rate for process i. The time step requirement is

$$\Delta t < \frac{1}{\sum_i r_i} \tag{1.6}$$

Given P_i, a single random number, uniformly distributed on the interval [0, 1], is sufficient to select among mutually exclusive alternatives for a given particle (Figure 1.1).[3]

All of the distributions mentioned previously are available in high-level languages on current computers, making their use in particle simulation simple.

[2] It is useful to note that the do-nothing alternative always has probability $1-\sum P_i$; and rate $r = 0$. In describing the multinomial, this is often included as a separate outcome.

[3] Figure 1.1 implies that several comparisons be made, a consequence of using the uniform distribution. The categorical distribution would produce a unique integer without the comparisons.

We thus arrive at the coupled physical–biological simulation, with inputs and outcomes specific to a particular phenomenon of interest. Properly done, this allows improvements in physical realism (the coastal ocean circulation model) to proceed in parallel with improvements in biotic and chemical realism. The site-specific, individual-based representation blends the complexity of physical and biotic processes; the probabilistic approach proceeds to outcomes via simulation; the outcomes may be aggregated (e.g., abundance, distribution) following simulation, without loss of information.

Fundamental to this approach is that repeated simulations will be different in detail; the injection of randomness guarantees that. But, there can be significance in the outcomes in terms of mean and distribution of important metrics. The importance of ensemble size N is key. Simulation features that are robust to changes in N are likely to be reliable; those that are not robust require either model refinement, hypothesis simplification, or larger N.

1.2.3 Gather, Scatter

In many cases one needs the effective density of a set of particles; or in reverse, the generation of a set of particles from a density field. This pair of topics is treated in detail in Chapter 8.

In simulations that require interaction among particles, there will be N^2 interactions. This does not scale favorably and can easily discourage simulation experiments or force simplifications. Fields can be an efficient intermediary. If one generates a single field for all particles to interact with, one replaces N^2 interactions with N interactions with a single middleman. This is a common approach, for example, where there are density effects in living systems.

Environmental fields routinely affect particles; but in some cases, the fields themselves are affected in a two-way exchange. An example might be a nutrient field being affected by the presence of large aggregations of planktonic particles. An undesirable effect of particles grazing on a gridded field would be a set of holes in that field. A better option would be to generate a field from the particles, and have the two fields interact. That way, the singular effect of each particle is smoothed out. Again, the generated field is the middleman.

Display of particle locations can be awkward or ineffective when the ensemble is large, while quantitative display of an equivalent density field is well developed and easy to read and communicate.

We refer to this as the "gathering" problem. It can be formalized as the *projection* of particle positions $\mathbf{X}$ into a smooth field. We will present two forms of projection:

1. Projection of particles onto a fixed grid – a lattice of given nodes with predetermined interpolation among them

2. Projection of particles with kernel methods, where each particle is accompanied by its own smooth interpolant

The reverse is "scattering" – given a known density, generating an ensemble of particles. The scattering problem is one of random number generation at a spatially variable intensity. This can be implemented as a two-stage process: scatter among elements, then scatter within each element. Within an element, the most fundamental issue is obtaining a uniform distribution; the rejection method can always be used then, to produce variable intensity. The most obvious example is the generation of input particles from known or assumed source fields.

During simulation, one may need to consolidate or condense an unanticipated concentration of particles. Both operations are involved – first gather a given ensemble into a density field, and then re-scatter it with a smaller number of bigger particles. The opposite pertains to an unexpected sparsity of particles.

Both procedures need to be practical for unstructured grids, high resolution, and potentially large numbers of particles.

1.2.4 Simulation

Particle simulation was among the earliest uses of the "modern" computer.

During the period in which von Neumann and others were assembling the first prototype machines, the canonical problems included the forecasting and control of weather, and the performance of nuclear devices. In these two examples, one can see the early branching of geophysical mechanics toward partial differential equations (PDEs), data assimilation, and estimation theory, and subatomic physics toward particle methods. In the latter area, the name "Monte Carlo" was applied to probabilistic particle simulations, a metaphoric allusion to gambling and to game simulation generally. This form of simulation is our main subject, carried out in oceanographic contexts. The continued development of the geophysical, PDE branch has made possible our confident use of detailed coastal ocean model output.

Randomness was and remains a prominent concern in both branches, representative of vastly different scales. The basic problem of generating random sequences on finite machines had to be attacked early. In addition, the problem of random functions (continuous noise) had to be characterized in terms of discrete accumulations (integrals over finite time) and their properties. What is currently called "Wiener noise" emerged from these early roots. Fundamental properties include (1) the independence of nonoverlapping integrals and (2) the scaling of an integral with the square root of the duration. In a sense, Wiener noise is the continuous analog of the discrete random sequence, and fundamental to the analysis of random walk processes.

Some fascinating reading about this period is available, for example, Dyson [123] and Feynman [141].

Our contemporary approach is rooted in the Langevin-type formulation, that of deterministic processes driven by stochastic input,

$$\frac{dU}{dt} + g(U) = f(\mathbf{x}, t) + \epsilon \tag{1.7}$$

where f is deterministic, ϵ stochastic. U could be particle position $\mathbf{X}$, or more generally some state of a particle. There will be one such ODE per process and one set of ODEs per particle. g represents intrinsic or "onboard" processes for a particle (e.g., biotic rates). All environmental information is contained in gridded fields, keyed to location $(\mathbf{x}, t)$. The original Langevin equation addressed Brownian motion per se. [238].[4] Over time this has come to signify a more generic set of processes with a blend of deterministic and stochastic forcing.

Path integration for stochastic processes raises special issues; the Ito calculus provides suitable extensions and a basis for numerical integration. As we simulate arbitrarily small time steps, we imagine use of conventional ODE numerical methods (e.g., Runge-Kutta) with stochastic forcing (Wiener noise). We do not imply a choice here, and encourage experiments with more sophisticated methods. Some background is available in Kloeden and Platen [229], Oksendal [341], and Kloeden et al. [230].

We simulate finite ensembles of particles. Each has its own Lagrangian ODE. The limiting case of unlimited N makes particle intensity the outcome. The Fokker–Planck or Smoluschowsky equation describes this limiting case in terms of an Eulerian PDE for the evolution of the particle intensity field (unnormalized probability density function). That constitutes another serious frontier, an alternative to particle simulation and its equivalent in the limit as $N \to \infty$.

1.2.5 Aggregation and Identity

Statistical outcomes can be modeled in two ways. In the first, one seeks the statistics of outcomes by analytically processing the distributions of inputs. This of course requires serious simplifications, but can lead to powerful results. When situations are not easily simplified, the recourse is to the second – simulation.

The particle approach is to simulate outcomes (1) for each individual and (2) one step at a time. Both are sources of individual variation. Variability is built-in from the start among individuals, with no necessary aggregation. Subsequent variability emerges from their simulation, realized at the individual level. Aggregation may be done as needed, without loss of individual identity. The aggregate properties are summary outcomes of a simulation.

[4] Lemons and Gythiel [245] provide an English Translation.

A simple example is probably the "particles in a box" problem examined in Section 3.6. This example has wide applicability in residence times, stage completion, mortality, failure rate, etc. There are N individuals; each has binary outcomes per step. The aggregate performance over time can be analyzed; if that path is followed, individuals lose their identity to the aggregate, from the very start.

Alternatively, the same results would emerge from per-step simulation of individual particles. But simulation need not aggregate until the end; every individual retains its identity and history. Simulation of N particles this way would be a sampling from the full population represented by the analytic result.

In subsequent chapters, we explore many of these issues in greater detail. Chapters 2 through 6 discuss several individual aspects of particle simulations. Chapters 7 and 8 deal with handling unstructured grids – navigating, interpolating, gathering, and scattering. Chapters 9 through 11 describe how models have been assembled and applied for three different contemporary problems.

2

Drifters and Their Numerical Simulation

2.1 Introduction

Who has not heard a note-in-a-bottle story and marveled at the natural process? Because water parcels can travel from one coast to another, continents are physically connected on one large globe. It is a romantic notion that humanity is actually linked in this way across vast expanses of ocean.

The subject of this book is the tracks water parcels follow, particularly those near the coast. In this chapter we provide an introduction to both the observed and modeled tracks; in subsequent chapters we develop some details and extensions. By observed tracks we mean the trajectories of "drifters" and "floats," instruments deployed to document potential passive transport pathways. By modeled tracks we mean the trajectories simulated by numerical ocean models. In both cases, we do not assume these follow water parcels exactly but they are the best we have.

The quantity of these observed and modeled tracks has exploded in the last few decades. Figure 2.1, for example, shows the composite of observed drifter tracks off the northeast United States, where many oceanographers have been investigating the flow fields for many years.

The coverage is extensive, and though each individual track represents an orderly set of observations, the aggregated outcomes appear tangled, chaotic. But there is order in the aggregate. In Figure 2.2 we see the bin-averaged velocity field that results from these observations. The field recovers the general circulation in these waters, as reported early by observation and inference (Bigelow [30], Bumpus [59], Bumpus and Lauzier [60]) and later by modelers (Brooks [56], Lynch et al. [282], He et al. [251]).

Also shown in Figure 2.2 is the subtidal variability of the same binned data. This is information not readily available from circulation models, but is an essential complement to the data averages and combines all sources of subtidal variability as observed.

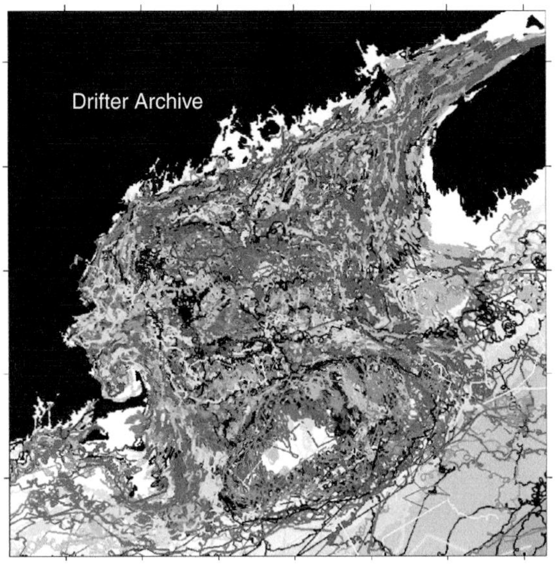

Figure 2.1. Observed drifter tracks off the northeast United States. (Manning et al. [299]). Reprinted with permission. For a color version of this figure, please see the color plate section.

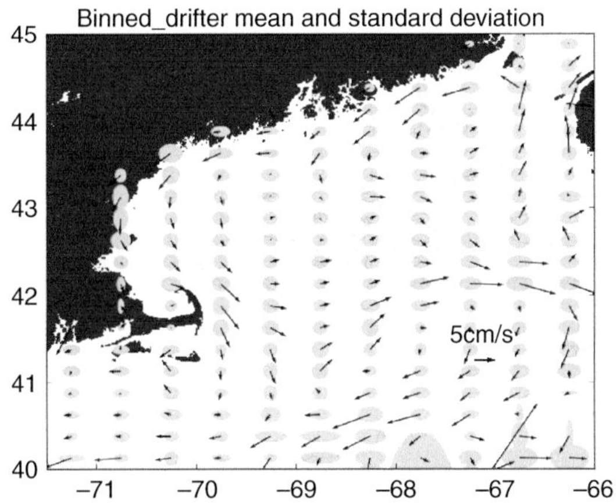

Figure 2.2. Bin-averaged mean velocity and the low-frequency, subtidal standard deviations for the Figure 2.1 data archive.

Among the primary purposes for deploying drifters is to help validate model simulations. Numerical modelers have made tremendous progress in the last few decades in being able to reproduce the flow fields off our coast. An example comparison for spring/summer bin averages appears in Figure 2.3. This is an Eulerian[1] comparison.

[1] Leonhard Euler, considered one of the greatest mathematicians ever and born in 1707, contributed much to the analysis of fluids and was an academic advisor to Lagrange.

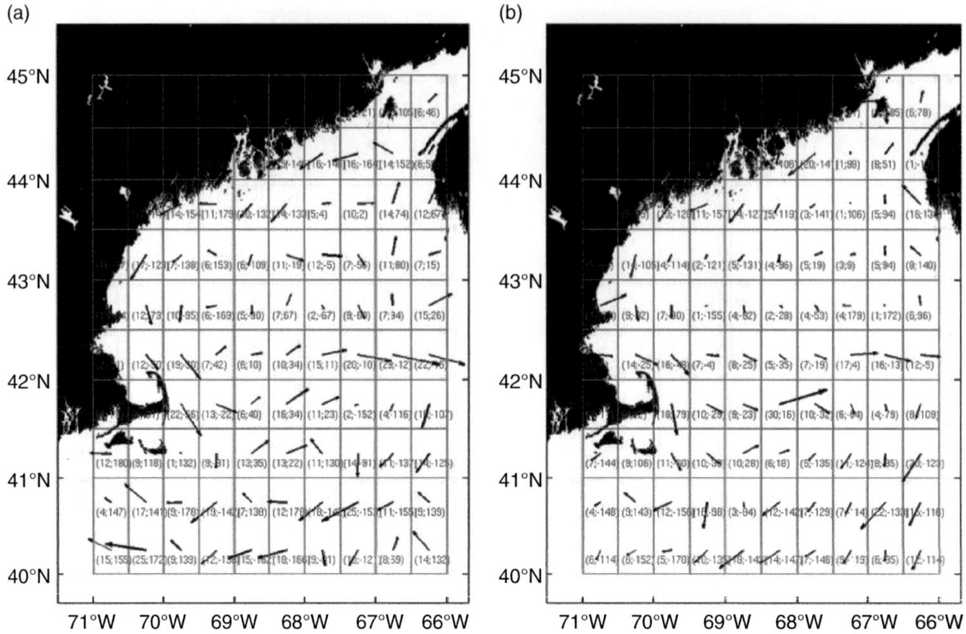

Figure 2.3. Example model-data comparison for bin-averaged drifter data for spring/summer 2004–2009. (From Li et al. [251], Reproduced by permission.)

As we discuss in later sections of this chapter, there are a variety of metrics we can use to quantify model performance both in a mean flow sense as well as in simulating all the various processes described in the previous chapter.

A great number of investigations are inspired by Lagrangian[2] considerations. Modeled Eulerian fields are of course the starting point – the instantaneous state of motion. But it is transport between places and activities that we typically wish to understand – Lagrangian displacement, destination, connection in space and time.

The location of branch points and/or hyperbolic points is critical in many practical situations; these are fundamentally Lagrangian properties. Similarly, trapping lines occur for fixed-depth drifters along convergence fronts. Figure 2.4 illustrates these. When these types of features vary on short length scales relative to tide and topography, trajectories can exhibit a chaotic mixing effect (Ridderinkhof and Zimmerman [258]).

A specific instance of two of these features appears in Figure 2.5: (1) A branch point is known to be operative near the launch point shown at Cape Ann; all three drifters followed the same southward path and avoided the alternate path toward the southwest. (2) A hyperbolic point is operative on the western end of Georges

[2] Giuseppe Luigi Lagrancia, a mathematician born in 1736, introduced another way of looking at the mechanics of moving particles.

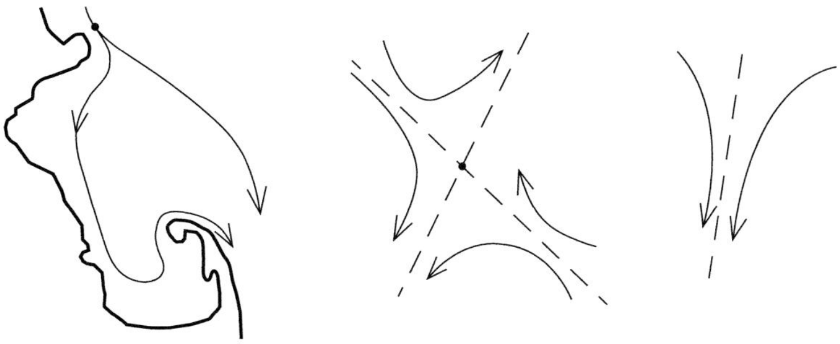

Figure 2.4. Some common coastal ocean features. Left: Branch point at Cape Ann [4]; Middle: Hyperbolic point in the Great South Channel [378]. Right: Surface convergence along a downwelling front in the coastal Gulf of Maine [193].

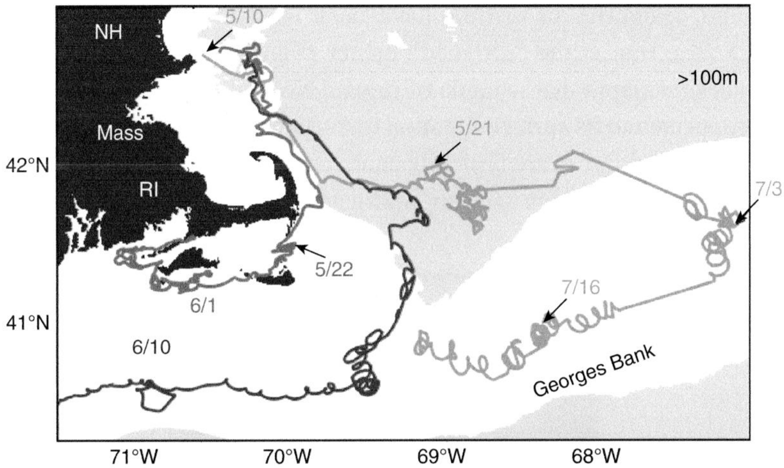

Figure 2.5. Observed branch and hyperbolic points in exit pathways from the Gulf of Maine. Three drifters launched in close proximity to Cape Ann followed very different pathways to Nantucket Sound, Great South Channel, and Georges Bank. Adapted from Anderson et al. [4]. (Reproduced by permission.) For a color version of this figure, please see the color plate section.

Bank; the three drifters took different paths there. These features are key in regulating Lagrangian connectivity in these waters; one particular instance of them is evident in the Figure 2.5 tracks.

After some discussion of drifter technology and the observational data in this chapter, we examine the basics for a Lagrangian particle model. We outline the idea of "particle tracking" through numerical fields and mention a variety of practical issues associated with this process. We discuss briefly the motivation behind some of this work and provide a brief preview of some applications.

2.2 Drifter Technology

2.2.1 Design

There are several commercial drifter manufacturers that work together with various governmental agencies (NOAA, NSF, NASA, etc.) to ensure drifter configurations conform to the oceanographic standards. Most importantly, the Global Drifter Program, set up originally by the World Ocean Circulation Experiment, has developed a set of standards and best practices to ensure that drifters have nearly the same size, shape, and therefore water-following characteristics (Lumpkin and Pazos [268]). Several hundred drifters have been deployed around the world's oceans each year for the past few decades and more than a thousand in each of the last several years but there has been very little change in the type of drifter deployed. An extensive body of literature has been published from these projects.

Two general categories of drifters have been used most extensively in the last few decades. The first is the "drogued" drifter (Figure 2.6), which has a cylindrical "holey-sock" at depth that is made of rugged cloth material supported by circular hoops and connected to its surface flotation (0.3–0.4 m sphere) with a thin wire tether. For most programs, drogues are typically centered at 15 m below the surface. Though the original drogue dimension was approximately 6 m by 0.9 m, it was reduced in

Figure 2.6. HoleySock drifter. Photo of Jack Barth (Oregon State University) preparing for the NEP GLOBEC project. (Reproduced with permission.) For a color version of this figure, please see the color plate section.

size beginning in about 2003 to 5 m by 0.6 m because the size of the surface flotation holding the electronics as well as the diameter of the tether was proportionally reduced. The objective has always been to maintain a 40-to-1 drag ratio and minimize the effect of wind and waves. The biggest problem with both the old style and the new "mini" drogued drifters is detecting drogue loss (Pazan and Niiler [352]; Lumpkin et al. [270]). When the drogue becomes detached from the surface float for whatever reason, the drifter track is driven primarily by the surface current, the wind, and the waves rather than the subsurface current, as it is intended. Devices are used to measure either the tension of the tether (strain sensor) or the conductivity in salt water near the surface float (submergence sensor) but the timing of drogue loss is not always obvious.

The second type of drifter often used marks the upper few meters of the water column. This type is typically referred to as the "Davis-style" or "CODE" drifter first developed by Russ Davis for the Coastal Ocean Dynamics Experiment in the early 1980s (Davis [103]) and pictured below from Poulain [359]. Though it is normally deployed very near the surface, the same configuration is sometimes hung subsurface via a tether as, for example, that sketched and described in Thomson [448] (Figure 2.8).

While this CODE surface drifter is the most common, there are a variety of other types of drifters such as the Tri-Star drifter as shown in Figure 2.7.

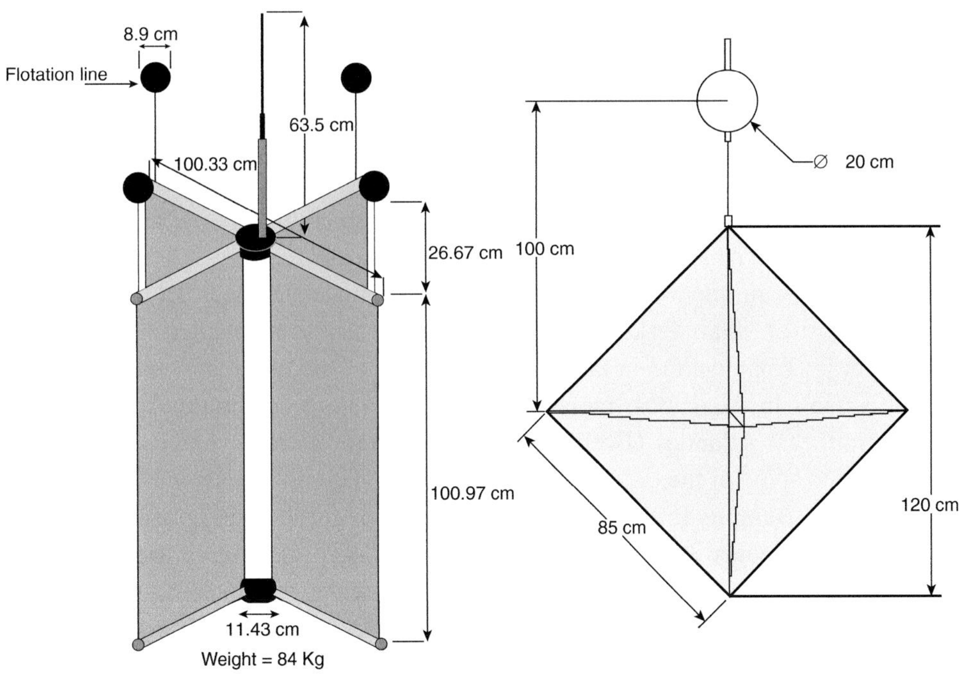

Figure 2.7. CODE and Tri-Star drifters. (Figures reproduced with permission). Left: From Poulain [359]. Right: A. Sybrandy, Pacific Gyre Inc, pers. communication.

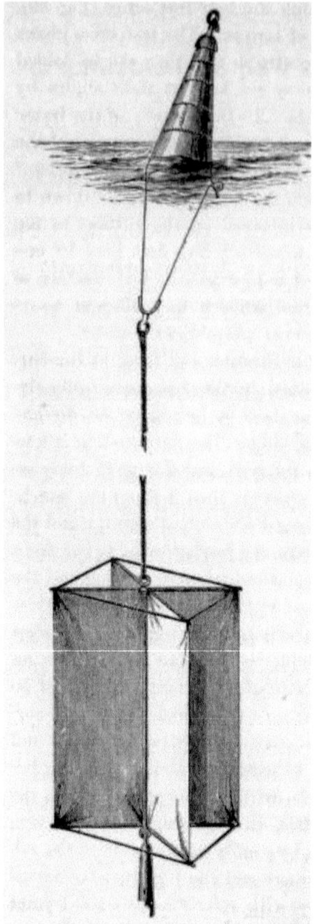

Figure 2.8. The "current drag" apparatus as deployed on the historic HMS *Challenger* expedition more than 150 years ago (Thomson 1878 [448].)

In addition to these most common surface and drogued drifters, there are a variety of other designs that have been used and may be still used for special purposes. These range from very simple low-tech units to highly sophisticated ones. Before the 20th-century, most of what we know about the surface currents was derived from ship drift records. This resulted in large errors because of the effect of wind and waves on the vessel. In the 1960s, more than 100,000 waterproof bottles were released in the Northeast Atlantic. Given the 10% return rate a large volume of data was collected and some of the currents we understand today in that area are derived from those results (Bumpus [59]). Many of the stories of these floating objects may be found in Ebbesmeyer and Scigliano [124], Hohn [199], and Burns [64]. There are also bottom drifters including simple units without any electronics that are designed to skim along the bottom (Woodhead and Lee [485]) as well as more sophisticated instruments that are tracked acoustically (Prater and Rossby [360]). As an alternative to the holey-socked drogues, parachute designs have also been deployed (McNally et al. 1983 [308]; Oeresland and Ulmestrand [339]) with some specifically designed for the nearshore surf zone (Johnson et al. [215]). In cases where one is interested in the very near surface of the ocean there are, for example, "FLAT" drifters such as those deployed in recent years in the Gulf of Mexico. There is a growing use of both visual and chemical tracer experiments (Ho et al. [197]) to look at small scale Lagrangian motion. These tracer experiments use image analysis on a variety of media. Milk, for example, was reported by Bezerra et al. [29] as a visual marker.

There are neutrally buoyant drifters (i.e., "floats") that can be ballasted for particular densities. A historical review of the early float development is given in Rossby [391]. Floats typically report their positions to an array of acoustic receivers (Rossby et al. [390]) or occasionally surface to report to a satellite. Some can alternate between profiling and Lagrangian modes (D'Asaro [98]). Even more sophisticated, there are now experiments with "autonomous underwater explorers" (Jaffe and Shurgers [208]). These are a collection of free-floating, self-ballasting floats that are untethered and can be tracked in the water column in 3-D over ranges of 3–5 km (Figure 2.9). There is a range of underwater vehicles (Davis et al. [105]) referred to as "gliders" that have a variety of propulsion mechanisms and are therefore not considered "Lagrangian" floats.

Figure 2.9. Autonomous underwater explorers. Prototypes of these free-floating drogue devices have been built and tested off the coast of Scripps Institute of Oceanography, (Jaffe and Shurgers [208]). They show promise in being an acoustically trackable swarm of Lagrangian-like particles in the ocean. (Image courtesy of J. Jaffee, C. Shurgers, pers. comm.) For a color version of this figure, please see the color plate section.

2.2.2 Communication

Before to the explosion in GPS technology within the last few decades, drifter positioning relied on various communication systems (Richardson [377]). Beginning with some radio transmitters such as that used by the Loran-C and Nimbus networks in the 1990's and 1980's, the ARGOS satellite system collected the majority of drifter data and is still a major contributor today. As additional satellite systems such as Irridium, OrbComm, Inmarsat, and GlobalStar have arrived in orbit, there are now a variety of alternatives for delivering GPS positions and auxiliary data. Because they vary in lag time, cost, and bandwidth, there is no one system that provides the best service. The sampling frequency has been consistently reduced over the decades to hourly averages or less for oceangoing drifters but some systems are now capable of minute-by-minute fixes in the coastal ocean within cell phone ranges.

2.2.3 Quality Control

Because each communication system and drifter design has its own set of issues, the raw data typically undergo a series of editing steps before the tracks are submitted to the master databases. Investigators typically remove bad data using multiple criteria, looking for outliers of position, velocity, and acceleration. It is often necessary to examine both the forward and backward differences and remove spikes greater than some number of standard deviations. This effort was particularly important before the advent of GPS-quality fixes but is still routinely needed for all tracks. Graphical user interfaces are used by various investigators to help clean the data given a variety of home-grown software packages in, for example, MATLAB or PYTHON. The most advanced processing routines provide the original data with quality control flags indicating those fixes that failed the "good data" criteria.

Data processing is particularly difficult for shallow water drifters, when there are often cases of the drifter running aground or getting hung-up on fixed gear (Lumpkin et al. [271]). These data require particular attention to maintain accurate flow statistics in a particular estuary or bay.

There is the issue of drifter "slip" due to the influence of wind and waves on the drifter. It is particularly important in the case of the "surface drifter" mentioned earlier where there may be errors on the order of a few centimeters per second in the resultant flow especially in times of high winds. In the case of the drogued drifter, the slip may be on the order of 0.5–4 cm/s for <10 m/s wind (Niiler et al. [329] Geyer [170]). Though it has never been measured, this slip may be much higher for strong winds. Each drifter design has specific water-following capabilities, so care is needed in devising new configurations that may not be consistent with others. As discussed later, this slip factor is easily modeled when comparing, for example, drifter vs. model-generated tracks.

Most of the major drifter programs submit their raw data to the Global Telecommunications System (GTS) so it is made available to many users. Archive centers have been set up, for example, at NOAA's Atlantic Oceanographic and Meteorological Lab (AOML) to process, store, and serve the data. Canada also processes most of the data at the Integrated Science Data Management (formerly MEDS) as reported in Keeley et al. [225]. While these two centers handle the bulk of the deep ocean drifters, there are other datasets from local areas such as that compiled at http://www.nefsc.noaa.gov/drifter for the drifters deployed off northeast coast of the United States and http://www.icess.ucsb.edu/drifter/deployments/ deployed off the northwest coast of the United States. In all cases, there is a balancing act between providing data as quickly as possible and maintaining proper quality control.

Real-time drifter tracks are gradually making their way on to the US Integrated Ocean Observing System "asset maps." They are plotted along with their numerical counterparts. On some of these websites, we are beginning to see the forecast drifter tracks overlaid with real drifter tracks, and, within the next few years, should see HF-radar-derived tracks routinely overlaid as well. These various Lagrangian displays will improve the general visualization and comparison of the flow fields beyond the standard vector and streamline views. To generate accurate model and HF-Radar-derived tracks however, requires substantial investments in computer coding in the form of "particle tracking," the subject that follows.

2.3 Particle Tracking

2.3.1 Basic Lagrangian Model

Our baseline for simulation is a massless particle that is passive physically and biochemically inert. Such a particle moves with the resolved (gridded) flow, with random walk added and parameterized by the local turbulence.

A Coastal Ocean Model is responsible for providing the resolved motion, including residual or low-frequency (seasonal) motion; tidal, wind-forced, and far-field boundary-driven motion; riverine inputs; and marine outfall effects. For the unresolved or subgrid motion, a statistical description is necessary, typically in terms of gridded eddy viscosity or diffusivity coefficients. Specifics are developed in Chapters 4 (random walk) and 6 (turbulence closure). Navigating within gridded fields, and interpolating them, is developed in Chapter 7.

Particle motion will generally require the integration of a Langevin equation,

$$\frac{d\mathbf{X}}{dt} = \mathbf{v}(\mathbf{X},t) + \mathcal{R}(\mathbf{X},t) \tag{2.1}$$

with $\mathbf{X}$ the location of a particle, $\mathbf{v}$ the Eulerian fluid velocity, and $\mathcal{R}$ the random walk contribution. This simple description may need some modifications. If, for example, a constant-depth track is desired, then the gridded vertical motion would be overriden (a common case). Some departures from passivity may be called for if the hydrodynamics of the drifter are to be incorporated. Typically this would require accounting for the forces acting on the drifter. With **V** the drifter velocity, M its mass,

$$\begin{aligned} M\frac{d\mathbf{V}}{dt} &= \mathbf{f}(\mathbf{V}-\mathbf{v})+\mathbf{g}(\mathbf{W})+\mathbf{B}+\mathcal{R} \\ \frac{d\mathbf{X}}{dt} &= \mathbf{V} \end{aligned} \tag{2.2}$$

Here **f** represents hydraulic drag and lift; **g** represents a potential "sail effect" (**W** being the surface wind); **B** represents buoyancy; and stochastic influences from all of these are bundled into $\mathcal{R}$. (**V**–**v**) is the slip velocity. Needless to say, this level of representation becomes complicated quickly.

Tidal motion is important in itself where tidal displacements are critical, for example, in the nearshore, and in generating important tide–topography interactions. In addition, tide is energetic and can dominate the generation of turbulence. It is common to expect numerical tide predictions of high quality; but of course where tidal motions are large, a high-accuracy prediction may still leave significant error relative to the smaller, aperiodic motions. If such a tide model is unavailable, the twin effects on velocity and turbulence must still be estimated.

A similar situation exists relative to surface motion. In some applications (oil spills, search-and-rescue), the surface component is quite critical. There can be significant contributions from unresolved motions, including Stokes drift, Langmuir circulation, and local atmospheric variations not available for modeling. Effects of these will typically be unresolved in available circulation models or data streams; yet they are known to influence surface trajectories, sometimes significantly. Vertical mixing requires surface boundary conditions on, for example mixing length; and lack of resolution can be serious for the turbulent variables. As in the tide case: if these surface effects are not present in a modeled circulation field, their effects still need to be accounted for.

This is the idealized "passive drifter" case. It is a common denominator in all particle modeling; specific applications (Chapters 9, 10, 11) illustrate the addition of features to this baseline. There are well-established methods to integrate Equation 2.1 or 2.2; some implementations are described in the accompanying electronic archive (http://www.nefsc.noaa.gov/drifter/particles.html). Chapter 3 provides background on random numbers and their generation.

In the following we discuss some of the practical simulation issues.

2.3.2 Practical Issues

Particle tracking models can range from simply following a fluid through Eulerian fields to some sophisticated routines that emulate random dispersion, vertical behavior, coherent structures, and a variety of other forces. There are a number of efforts to code various scenarios; as this is a rapidly evolving field, we maintain an electronic archive of references at http://www.nefsc.noaa.gov/drifter/particles.html and accessible through the publisher's website. Here we simply list the issues that come into play.

The first step in any particle tracking code is reading the flow field and the grid parameters. This step is becoming less arduous in recent years as modelers converge on common data formats for their output. There are often some issues with the method one interpolates around the grids, which obviously differs depending on whether the underlying grid is finite element (triangular) or finite difference. As discussed in Chapter 5, there are several different ways to handle a particle that has encountered a boundary and these vary depending on the type of boundary (surface, bottom, beach, etc.) The typical solution is a "reflective" condition whereby the particle bounces back into the domain but, as discussed later, there are several others. There is also the issue with what time steps to use (Dimou and Adams [117]). You want to ensure that a particle does not traverse an entire element within one time step but, at the same time, to make the time step smaller than the Lagrangian time scale (Brickman et al. [53]) The "Runge-Kutta" method of interpolating over time is the accepted method but there are variations that can be debated depending on the application. As noted in Brickman et al. [53], for example, higher order methods may not have a great advantage in runs where turbulence is parameterized. Simple, first-order forward differencing schemes like the "Euler" method may sometimes suffice but test are needed to verify these results. In some cases, the time steps can be adjusted in real time depending of the local flow fields and grid configurations (Knight et al. [232]).

One of the primary concerns in particle tracking is how to introduce the subgrid scale motions that are not resolved by the model. As discussed in later chapters, there are different "random walks" that can be introduced to simulate the natural dispersion. The usual strategy is to use the model's estimation of vertical and/or horizontal diffusivity to scale the random component. In all particle tracking experiments, it is necessary to test the robustness of the methods and consider the consequences of various assumptions. In areas of the ocean with strong stratification, it is known that numerical artifacts can cause artificial aggregation of particles, but methods are being developed to combat these problems (North et al. [335]). As shown in Griffa [185], for example, stochastic particle tracking is more than a simple advective–diffusion problem. In some cases, these small-scale processes can be parameterized as deterministic Lagrangian coherent structures (Haza et al. [192]). As addressed later in the chapter, there are a variety of studies underway to address these features. As our flow fields are becoming more realistic, there is some development in dynamic system theory where barriers to transport (Olascoaga [345]; Shadden et al. [408]; Coulliette et al. [86], can be identified in terms of hyperbolic points and extrema of finite size (FSLE) and time (FTLE). In some cases, a Lagrangian spin parameter can be introduced to help particles follow coherent eddy structures (Veneziani et al. [463];

Sawford [402]; Reynolds [374]). It is generally agreed that it is impossible to model a single drifter's track through a turbulent field but attempts can be made to reproduce the behavior in a statistical sense (Pasquero et al. [351]).

There is also the question of how many particles are needed to provide a statistically adequate representation of the flow field. As described in Graham and Moyeed's [176] application of an eddy interaction model, this can be resolved by conducting multiple runs, varying in the number of particles as well as repetition in each case, and following their technique in determining confidence intervals. Application of these ideas in the case of oceanic flow, however, is not always straightforward. By comparison with meteorology, for example, we are often concerned with time scales greater than the Lagrangian correlation time and therefore need alternative methods to define statistical significance (Brickman et al. [53]). Regardless of the methods we use, these particle tracking methods are still considered "experimental" as opposed to "theoretical." It is generally good practice to report the sensitivity of results to parameters such as the number of particles (Simons et al. [417]).

Because no model resolves all the processes, there are ways to add effects with particle tracking runs. If the ocean model was not forced by a good localize wind field, for example, the effects of this can be added (Paldor et al. [348]). If the drifters being simulated have a known "slip," this can be coded as well and is often a function of wind ranges (Edwards et al. [128]).

An infinite variety of "vertical behaviors" can be added to a particle's trajectory depending on, for example, the age of the particle, the time of day, the temperature of the water, and any number of modeled parameters. Practical oceanic scales allow little departure from passive horizontal motion. However, vertical motion is another matter, and there are many considerations that require closure. Options include fixed-depth; passive vertical motion; and depth regulation by hypothesis.

Bioregulation of depth is fundamental. This is the intersection of physical mixing and autonomous depth-regulation, behavior that develops as living "particles" mature and adapt to their environment (see Section 11.9). Most particle tracking routines address these issues and some include aspects of an animal's early life history such as spawning, growth, mortality, and recruitment (Lett et al. [248]). Chemical weathering of particles is similar. Buoyancy is the analog of "behavior " here; depth regulation occurs as buoyancy evolves. Finally, an important class of problems involves vertical behavior that overcomes physical influences at the expense of energy. Generally, timescales of motion and biochemical phenomena are superimposed on spatial structures; the Lagrangian particle moves through these and their joint representation is critical. Seasoned judgment about the limiting cases will allow very useful simplification.

Closing the vertical motion is quite critical in sheared flows, and can greatly affect horizontal trajectories. In such situations, the question "What would a passive trajectory look like?" is often a valuable starting point. The passive cases are the foundation for more complex phenomena – sedimentation, oil spills, IBMs – developed in subsequent chapters. In summary, there are variations in how a particle can be advected through a numerical field, so particle tracking exercises tend to demonstrate various scenarios and sensitivities to the assumptions involved.

To determine the source of particles arriving at one location, particle tracks can be run backwards (Batchelder [18]). In some cases they are subsequently run forward to determine the robustness of the results, given a prescribed randomness. This two-way tracking is described in Isobe et al. [207], for example, applied to the East China Sea. Many of these routines can be applied to high-frequency radar–derived velocities as well as numerical model output. This was done by Lipphardt et al. [253] in Monterey Bay, for example, where they describe their analysis of synoptic Lagrangian maps representing a particle's origin, fate, and residence times.

There has also been a trend in the recent literature to define the connectivity of different areas along the coast (Li et al. [251]). Concepts of probability density functions (Mitarai et al. [317]) and dispersal kernels (Edwards et al. [127]) have been introduced. Many of these studies include sensitivity analysis (Simons et al. [417]) on the effects of the number of particles, the particle release sites, initial depths, the time of year, the time of tracking, and other factors. More on "connectivity" is presented in Section 11.12.

Finally we conclude the discussion of particle tracking by admitting that there can be as many errors associated with tracking as there are with the underlying velocity fields. As noted by many of the meteorologists who paved the way before us, as in Stohl [440], for example, there are both truncation and interpolation errors. There may be undetected coding errors as well as debates on invalid assumptions as in Wolanski [484] and subsequent responses. As we gradually come together as oceanographers to share data and model output, we will also need to make a concerted effort to share code. With the explosion of both commercial and open source programming languages in recent years, there is a need to develop user communities that share and compare algorithms in a standardized way.

2.4 Model Validation with Drifters

Comparison of drifter data with the equivalent "modeled drifters" is a natural priority. Metrics of quality center on the model–data misfit – How small can the misfit be? What is zero misfit (perfect fit) statistically, given numerous imperfections? What prevents a misfit from getting smaller? What information can be deduced from a finite misfit under practical field conditions?

Conventional Eulerian data products have been the subject of long-term scrutiny, resulting in a sophisticated approach to misfit evaluation. Examples include the placement of in situ instruments to avoid extreme near-field effects; the use of volume-averaging to separate resolved motion from the subgrid in model fields; the coupling

Table 2.1. *RMS Displacement (σ) for horizontal diffusion alone, versus days from launch with $A_h = 1\ m^2/s$*

T (days)	RMS Displacement (km)	Apparent Mean Speed (km/d)
1	0.59	0.59
2	0.83	0.42
3	1.02	0.34
4	1.18	0.29

Note: The outcomes represent one standard deviation; dividing by the elapsed time is the corresponding "apparent mean speed."

between them; subgrid parameterization via turbulence closure; and the temporal- and spatial-averaging appropriate to both data and model. This development supports a statistical approach to data assimilation – using imperfect observations to supplement incomplete boundary or initial conditions to make statistically valid field estimates.

For Lagrangian data, comparable developments are still relatively young and the examination of misfit and data assimilation procedures remains an active front of investigation (Griffa et al. [186]; Taillandier et al. [444]).

Each drifter can be conceived as one random walk in a field with deterministic mean[3] motion plus a stochastic component. This is a *sample of one* from the ensemble of possible trajectories.

What can be expected? For typical coastal situations with tidal-time resolution we have horizontal eddy viscosity $A_h \simeq 1\ \mathrm{m}^2/\mathrm{s}$. For a group of drifters launched together, diffusion alone would widen the variance in their location linearly with time; $\sim 68\%$ would remain within one standard deviation of the origin, $\sqrt{4A_h t}$ in 2-D; 95% would stay within 2σ (Section 4.3). Example outcomes are evaluated in Table 2.1.

This places a limit on precision for a single drifter: *misfit with a model cannot be expected to be smaller*. Advanced grid resolution will reduce this stochastic part by resolving and simulating some of it, leaving less variability to the subgrid statistical closure.[4] The eddy viscosity A_h becomes less as a result; and $\sigma \propto \sqrt{A_h}$.

Of course this random walk would be superposed on other (tidal and subtidal) motions. Suppose tidal motion has velocity amplitude 1 m/s. For a rotary M2 tide this would result in a circular displacement amplitude of 10 km at roughly half-day frequency. Subtidal motion can be expected to be of order $0.1\ \mathrm{m/s} = 8.6$ km/d; in many ecological applications, this is the dominant concern. It is apparent that it is partly masked by the other motions.

[3] Ensemble mean.

[4] With mesh resolution comes the additional burden of refining the forcing, notably the atmospheric forcing.

2.4.1 Field Experience

In the GLOBEC experiments, Lynch et al. [277] suggested that the *rate of separation* between real and numerical drifters be taken as a metric of misfit. This time series of separation distance can be expected to display

- a periodic component at tidal frequencies indicating a tidal misfit;
- a linear-in-time trend indicating a subtidal misfit;
- an episodic component in the wind-band; and
- a square root growth indicating the irreducible diffusive part.

The tidal portions are hopefully recognizable as harmonic with known periods $\mathcal{O}(0.5d)$. The wind-band portion can be expected to be correlated with local observations. The diffusive cloud will ultimately be overtaken by the linear part – good news for subtidal skill assessment. However, this requires the passage of time, the aging of any forecast, and the inclusion of critical points in the computed circulation (hydrodynamics, forcing, and topography), all of which place upper limits on the forecast horizon. In [277] it is suggested that misfit growth rates of order 3–4 km/d could be achieved with data-assimilative forecasts made at sea in real time.

Field experiments show real and numerical drifter separation versus time. An example is presented in Figure 2.10 (Aretxabaleta et al. [8]). Note that there is a clear tidal signal, small in a relative sense. The mean separation rate is 2.35 km/d and the separation rate appears to blend a linear and square-root behavior with time. As suggested previously the former indicates a misfit in subtidal velocity, while the latter is indicative of diffusion. Tracking of the center of mass of dye released deep (5 m

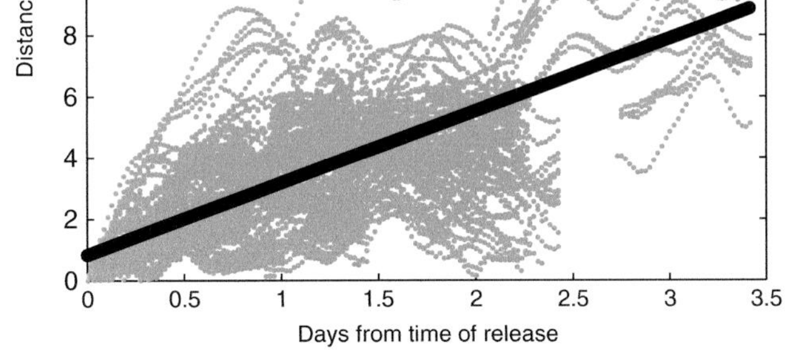

Figure 2.10. Drifter misfit growth over Georges Bank, from Aretxabaleta et al. [8]. (Reproduced with permission from Continental Shelf Research.)

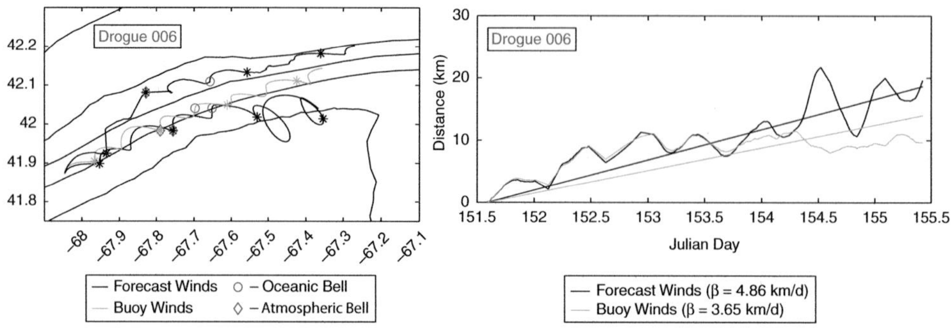

Figure 2.11. Northern Flank, Georges Bank, (From Proehl et al. [366]). (Reproduced with permission.) For a color version of this figure, please see the color plate section.

above bottom) gave smaller separation rates of 0.8 km/d, consistent with a smaller near-bottom velocity.

Figure 2.11 shows a tidally energetic case where significant contributors to the subtidal motion include nonlinear tidal rectification and local wind and stratification. The location is that of a bottom-trapped frontal system overlain by significant stratification. Several sources of misfit are illustrated; the tidal misfit is clearly visible, and two different wind products (hindcast and forecast) are shown and make a significant difference.

In the ECOHAB and MERHAB programs, comparable separations rates were observed. In the energetic Fundy region (Figure 2.12), the mean was higher, 3.72 km/d, and the same observations pertain to tidal, linear, and square-root features. In the coastal Gulf of Maine (Figure 2.13), the mean was 1.78 km/d and the observations are in fact more dominated by square-root growth, suggesting especially a small misfit with the subtidal flow. For completeness, we include the raw drifter trajectories from this experiment in Figure 2.13.

All of the field studies cited in the preceding used data assimilative methods for Eulerian data, notably the Acoustic Doppler Current Profile (ADCP) measurements available at sea in real time. Drifter tracks were used to evaluate the quality of the results. The modeling system is sketched in Figure 2.14. The data assimilation has been an essential ingredient, and the drifters critical to the evaluation of skill.

Assimilation of drifter tracks themselves is less developed. Notable is the work of Taillandier et al. in the coastal Adriatic Sea, where separation rates of roughly 3.5 km/d were achieved in 2-day simulations following assimilation of coastal drifters [444].

The separation distance versus time has been a useful metric of misfit in all of these studies. It is relevant for a few days into a coastal ocean forecast. It summarizes the model trajectories achieved and puts them in context of what is possible. Early GLOBEC experience [277] suggests rate 3.4 km/d was reachable with then-available

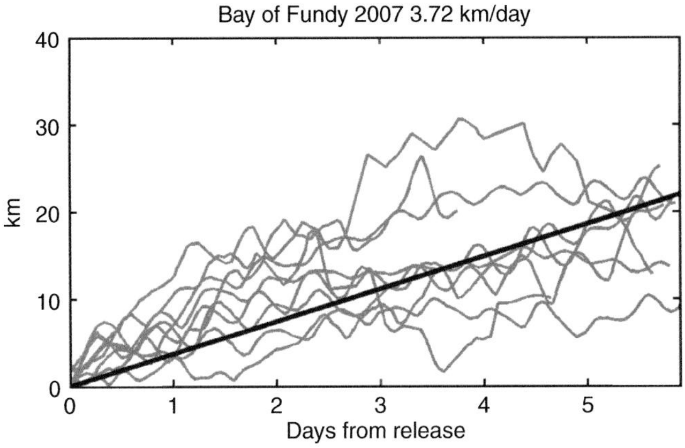

Figure 2.12. Drifter misfit growth in the Bay of Fundy. From Aretxabaleta et al. [11] (Reproduced with permission from the *Journal of Geophysical Research*.)

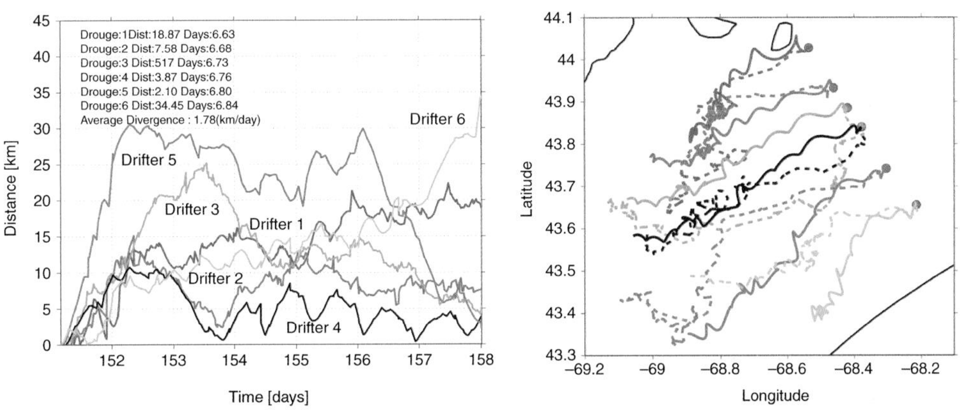

Figure 2.13. Drifter misfit growth (left) and trajectories (right) in the Maine coastal current. The solid lines are modeled; the dash lines, observed. From He et al. [193] (Reproduced permission from the *Journal of Geophysical Research*.) For a color version of this figure, please see the color plate section.

models. Subsequent results have shown steady improvement, and further progress can be expected in simulation models per se, resolution, data availability, and data assimilation procedures. In addition, the basic model of the drifter motion provides a frontier of possibilities. For example, Edwards et al. [128] found a halving of drifter misfit (to 1.4 km/d) by incorporating drifter hydraulics and estimates of unresolved local atmospheric forcing.

As documented by Huntley et al. [203], the misfit is obviously dependent on the launch locations and especially on the proximity to fronts where separation distances/day can be far greater. Barron et al. [17] examined the misfit from 30 different regions around the globe and found typical misfits often more than 100 km within a

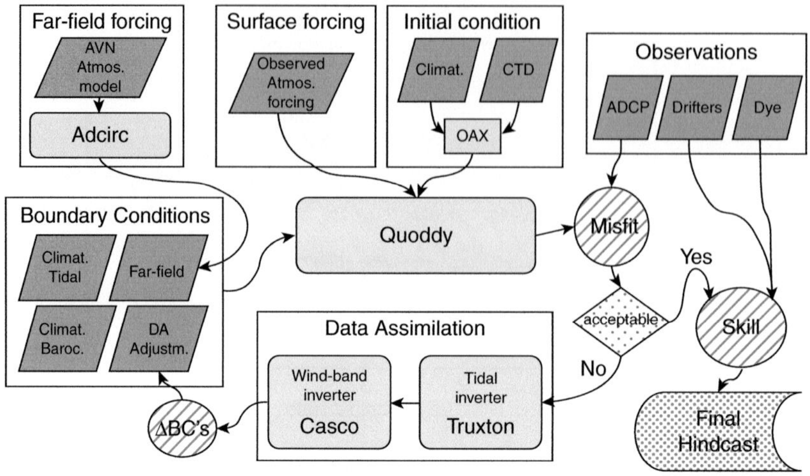

Figure 2.14. Shipboard modeling system. Aretxabaleta et al. [8] (Reproduced with permission.)

week's time. This separation distance metric is also used when comparing drifters with radar-derived tracks where values as low as 1 km/12 h (Molcard et al. [318]) and 1km/10h (Rypina et al. [395]) are reported especially in the case of very high frequency radar fields with 30–50 Mhz signals.

One may look at other metrics such as *separation after a specified time*, *time-in-a-circle* (Huntley et al. [203]), or *percentage of distance traveled* (Stohl [440]). It is noted in many papers (McClean et al. [304]) that a model's *Eulerian* predictive skill can be significantly different from its *Lagrangian* skill. Toner et al. [451] introduce a "type II" error that compares not the position of the tracks but the shape of the track. One might also compare the *Lagrangian time scales* of model versus observations. In field studies comparing clustered drifters or dye patches (Riddle [382]) to model simulations, one might report the patch dimensions versus time. Much of this work on metrics is actively under development especially due to the recent effort in the Gulf of Mexico.

As noted in the introduction to this chapter, there are various ways to examine the robustness of model flow fields using binned velocity data. This is considered an Eulerian comparison. The simplest is to compare seasonal or annual average flow fields by binning the information (Lacasce [236]) into gridded areas and examine the difference in statistics (mean, standard deviation, min, max, etc.) as well as the residence times in these areas. There are a variety of ways this can be done and one needs to be careful not to introduce aliasing due to tidal motions. Manning et al. [299], for example, truncated tracks to multiples of the M2 tidal period in any one grid. Huntley et al. [203] discuss differences in the metric results with and without tide. Depending on the quantity of tracks, the grid size must be chosen appropriately to

Figure 2.15. Aboard R/V Oceanus: preparation for drifter launches during the real-time data-assimilative modeling reported by He et al. [193]. Pictured are Ruoying He and Jim Manning (Reproduced with permission.) For a color version of this figure, please see the color plate section.

generate robust statistics. This bin-averaging was done in Li et al. [251], for example, comparing drifter data to ROMS model runs in the Gulf of Maine region (Figure 2.3). Another well-known issue associated with drifter-derived velocities is the drifter's tendencies to converge toward frontal regions and to flow toward regions of higher kinetic energy (Richardson [377]), which may create some artificial discrepancies in the model-data comparisons. Bin-averaged velocities can also be off if the data are highly inhomogeneous. Drifters will disperse and create a spurious high-to-low density mean flow in these fields (Davis [104]).

Another approach is to examine the connectivity of model and observed flows. One can compare the source and sink of particles as inferred by drifter versus modeled tracks. More on "connectivity" is presented in Section 11.12.

2.5 Drifter Applications

Drifters are deployed in a variety of studies usually to address a particular question. In the fall of 2013, for example, a group of marine biologists were concerned about certain species of sea turtle residing in Cape Cod Bay. These animals normally head south when the water starts getting cold but, because of a relatively warm October that year, there was concern that much of the population would be "cold stunned" as soon as the weather suddenly cooled. In these cases, the biologists want to know the general direction of surface currents to predict if and when the sickened animals come ashore. They addressed this question with satellite-tracked drifters and began to investigate the output of local ocean models.

There are a multitude of studies similar to this that address a particular question. In some cases, the question is purely scientific (i.e., addresses a particular hypothesis) and in other cases there is a real practical or economical reason to understand the flow fields. Because no amount of drifters could possibly resolve all the complexity of our coastal flows, it is now generally agreed, as stated earlier, that the primary reason to deploy drifters is to help validate the numerical simulations. In some cases drifter data are used together with model output in a "velocity reconstruction" (i.e. melding of the two) as in Chang et al. [75]. In many studies drifter data are used in combination with other data sources such as altimetry (Lumpkin and Garzoli [269]). Satellite imagery is now used to estimate "feature displacement velocities" (Yang et al. [493]) as another potential supplement to Lagrangian observations.

Regardless of whether numerical models are involved, drifter applications include aquaculture assessment, design of offshore structures (wind and wave power generators), larval transport, advection of pollutant materials including plastics (Law et al. [240]), tidal kinetic energy potential, power plant effluent dispersal, search-and-rescue, and much more. As we move closer to the regulation of ocean space, it is important to define these areas in terms of their connectivity, residence times, and the variability over multiple scales of time and space. At some point in the future, we envision routine deployments of drifters as part of our coastal ocean observing systems similar to the release of weather balloons for meteorological applications. These data can potentially be assimilated into our coastal ocean forecast models on a routine basis (see Section 2.5.1).

Another important application, as we discuss in more detail in Chapter 10, is oil spill tracking. Given the 2010 spill in the Gulf of Mexico, considerable funding and effort have been devoted to Lagrangian processes in that area. The Consortium for Advanced Research on Transport of Hydrocarbon in the Environment released more than 300 drifters in the summer of 2012. The Grand Lagrangian Deployment (GLAD) was the largest dispersion experiment of its kind, and several publications are in preparation at the time of this writing. As the cost of drifter technology is minimized, other large-scale clusters have been conducted and planned (Huntley et al. [203]; Rypina et al. [395]).

In recent years, there has been a call to deploy drifters purely as an educational tool where students are involved with all aspects of the design, building, assembly, deployment, and tracking of drifters. They learn basic concepts of physical oceanography in a hands-on activity that includes downloading, plotting, and analyzing the position and time data. Lesson plans for all levels are being developed from simple ideas of latitude and longitude to complex calculations of dispersion (Hochstaedter and Sullivan [198]; Manley [298]; Manning [300]). Successful examples include NOAA's Adopt-a-Drifter Program (http://www.adp.noaa.gov/) and Student Drifters (http://studentdrifters.org).

Clusters of drifters are now often deployed to estimate dispersion in addition to transport. In these cases, estimates of eddy diffusivity and other statistics (Schroeder et al. [406]; Manning and Churchill, 2006 [301]) are derived and often compared to the original ideas of Okubo [344]. The drifter-derived dispersion is often compared to very high frequency radar observations (Molcard et al. [318]; Rypina et al. [395]) in addition to model output.

Finally, it is hoped that drifter track archives can be maintained and well documented for years to come to make interdecadal comparisons of mean currents. As in Fratantoni [160] and Lumpkin and Garzoli [269], for example, we hope to assess climatic changes in both the large-scale and coastal flows.

2.5.1 Drifter Assimilation

Assimilation of drifter tracks is less developed than for more conventional Eulerian data. Some early approaches were to treat short drifter displacements as velocity measurements. Essentially the drifter becomes a moving current meter but the Lagrangian connectedness would be lost without further methodological development.

In recent years, some studies have shown that drifter data can now be assimilated into a model (Molcard et al. [319]; Ozgokmen et al. [347]; Taillandier et al. [443], [444], [442]) in real time to improve forecast results. There are multiple methods under development that include variational approaches and Kalman filtering as well as twin experiments that evaluate the process; but the general strategy is to minimize the distance between observed and modeled drifter tracks at each assimilation time step. Though most require these time steps to be smaller than the Lagrangian timescale, that restriction may be less using other methods (Salman et al. [397]). Some of these studies show improved model performance with directed drifter launch strategies (Molcard et al. [319]; Haza et al. [192]). So, just as in meteorology one assimilates the path of a weather balloon, ocean modelers are now incorporating drifter data in their simulations (Toner et al. [451]; Kuznetsov et al. [235]). Much of this work is focused in the Mediterranean, where an operational forecast system has been in place for more than a decade. While the model improvements are reportedly minor ($\sim$ 15%) given both surface drifters and deeper layer floats contributing data (Nilsson et al. [330]), the potential for more improvement with more Lagrangian data is clear and methods can be applied elsewhere. In the coastal ocean context, it is significant that Taillandier et al. [444] achieved separation rates of roughly 3.5 km/d over 2-day simulations by assimilating coastal drifters, a large reduction of the misfit prior to drifter assimilation ($\sim$ 15 km/d).

3

Probability and Statistics – A Primer

3.1 Basics – Random Numbers

A random number is generally unknowable in advance; but the properties of all possible random numbers in a certain class *are* knowable. These properties characterize the ensemble of all numbers that *might* occur – their type (integer, real, complex, ...), range, most common values, etc. – formally, the *distribution* of all possibilities across the relevant sample space. An uninformative distribution is one in which anything is possible, and all values are equally credible – this is the rare case, and in practical situations we expect greater specificity about the distribution.

There are some general tools; some common distributions that occur in analysis and have been found useful in science; and some standard ways to generate "random" numbers, or more precisely, pseudo-random numbers, whose distribution is predictable.[1]

We will reserve the symbol P for the probability of an event: $P[\ ]$ is the probability of [ ] occuring. By definition, P is a positive real number and lies between 0 (never) and 1 (certain).

3.1.1 Continuous Distributions: f and F

Let x be one sampling from a distribution $\mathcal{D}$ of random numbers:

$$x \sim \mathcal{D} \tag{3.1}$$

The distribution $\mathcal{D}$ is characterized by the functions $F(x)$ and $f(x)$ with the range of x covering all possible values of x – its sample space.

[1] We cannot make deterministic machines be truly random, so we deal with approximations to randomness that improve with computing power.

$f(x)$ is the probability density function (PDF). f is real and ≥ 0. The probability of x occurring on the infinitesimal interval dx is $dP = f(x)\,dx$. The probability that X occurs in the finite interval $[x_1, x_2]$ is $P = \int_{x_1}^{x_2} dP = \int_{x_1}^{x_2} f(x)\,dx$. For small Δx, $P \approx f(x)\Delta x$.

$F(x)$ is the cumulative distribution function (CDF), the integral of f from below:

$$F(x) = \int_{-\infty}^{x} f(u)\,du = \int_{-\infty}^{x} dP \tag{3.2}$$

$P(x \leq x_1) = F(x_1)$. F is monotonic with limits $F(-\infty) = 0$; $F(\infty) = 1$; this places a normalization requirement on the PDF $f(x)$. The probability of x occurring on the interval $[x_1, x_2]$ is $P = F(x_2) - F(x_1)$. The inverse of F is called the quantile function.

Examples 3.1–3.3 illustrate the functions $f(x)$ and $F(x)$.

Example 3.1 Uniform $f(x) = \frac{1}{\Delta x}$ on the interval $[-3, 2]$, and vanishes outside that interval. $F(x)$ is linear on the same interval and constant outside it. (For $x \leq -3$, $F = 0$; for $x \geq 2$, $F = 1$.) See Figure 3.1.

Example 3.2 $f(x) = \alpha x$ on interval $[0, 2]$, and vanishes outside that interval. $F(x) = \alpha\frac{x^2}{2}$ on the same interval. For $x \leq 0$, $F = 0$; for $x \geq 2$, $F(x)$ is constant. To normalize, the factor $\alpha = \frac{1}{2}$. See Figure 3.2.

Example 3.3 $f(x) = \alpha x[1 - \frac{x}{K}]$ on the interval $[0, K]$; vanishes outside. Then $F(x) = \alpha[\frac{x^2}{2} - \frac{x^3}{3K}]$ on that same interval. Normalizing to $F(K) = 1$ gives $\alpha = \frac{6}{K^2}$. This is illustrated in Figure 3.3. The independent variable is scaled by K: $\xi = \frac{x}{K}$. Hence,

$$f(x) = \frac{6}{K}[\xi][1 - \xi] \tag{3.3}$$

$$F(x) = \xi^2[3 - 2\xi] \tag{3.4}$$

and the range is $0 \leq \xi \leq 1$. See Figure 3.3.

3.1.2 Properties: Survival, Hazard Rate

There is a large class of situations where $t \geq 0$ represents a future time at which an event first happens. In these cases, $F(t)$ would represent the cumulative probability that something has happened before t and $f(t)dt$, the probability that it will happen in the coming dt. Examples might include the failure of an instrument or mortality in a certain population. In these cases the definition of the survival function S is natural,

$$S(x) = 1 - F(x) \tag{3.5}$$

This is the probability of surviving to or beyond x.

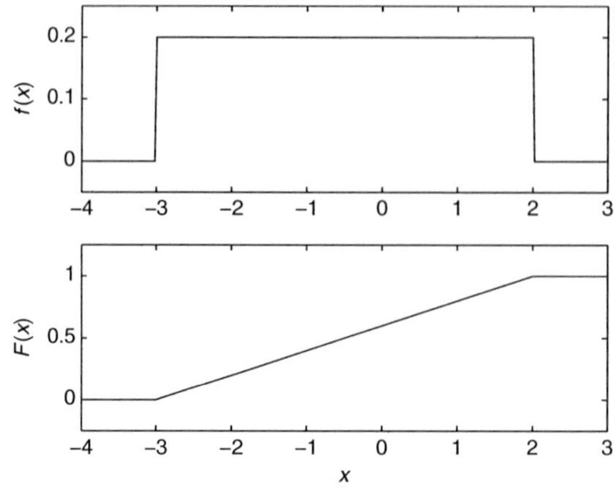

Figure 3.1. Plot of f and F for Example 3.1 with uniform PDF.

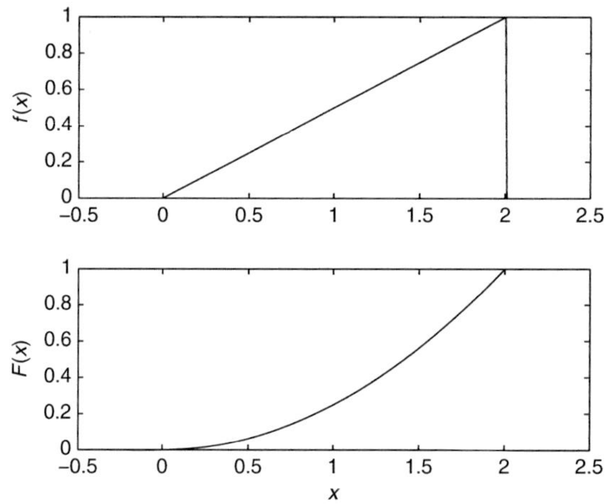

Figure 3.2. Plot of f and F for Example 3.2 with linear PDF.

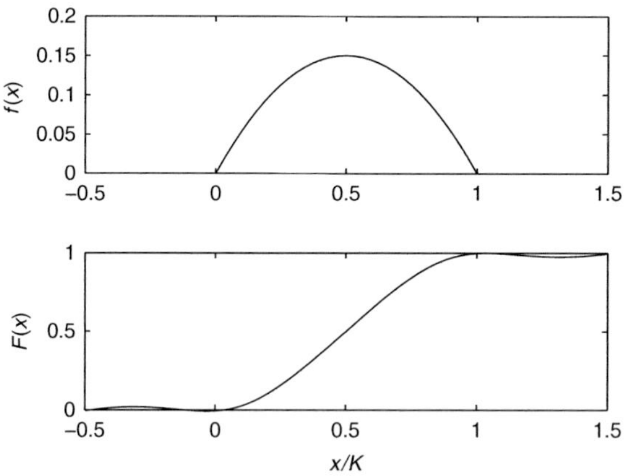

Figure 3.3. Plot of f and F for Example 3.3, with logistic PDF and $K = 10$. The independent variable is dimensionless, x/K.

The hazard dH over the increment dt is the change in S, normalized by S:

$$dH \equiv hdt = \frac{-dS(t)}{S(t)} \tag{3.6}$$

$$= \frac{dF}{S(t)} \tag{3.7}$$

so h is the hazard rate:

$$h(t) = \frac{f(t)}{S(t)} = \frac{f(t)}{1 - F(t)} \tag{3.8}$$

$h(t)$ has many names, depending on context: hazard rate, failure rate, replacement rate, mortality rate, stage completion rate, etc. These have broad application in individual-based modeling (IBM).

3.1.3 Properties: Mean, Variance, Moments

Let a random number x have mean $\overline{x}$ – the average of all possible occurrences. Conceptually, we can separate all occurrences of x into mean and variability,

$$x = \overline{x} + \widehat{x} \tag{3.9}$$

It is the variable part $\widehat{x}$ that is random; the mean is presumed constant and knowable. By definition, the mean variability is zero:

$$\overline{x} = \overline{\overline{x} + \widehat{x}} = \overline{x} + \overline{\widehat{x}} \tag{3.10}$$

and hence the latter term must vanish:

$$\overline{\widehat{x}} = 0 \tag{3.11}$$

In terms of the probability distribution function, the mean is the integral[2]

$$\overline{x} = \int_{-\infty}^{\infty} xf(x)\, dx \equiv E(x) \tag{3.12}$$

and this is termed the expected value or expectation of x – all possible occurrences weighted by their relative frequency f.

The expectation terminolgy is more precise; it indicates averaging over all *possible* values of x – as distinguished from those that actually occur in any particular context.

[2] Ordinarily, an average would be normalized; in this case by $\int f(x)\, dx$. For a properly defined distribution function f, this factor is unity.

The expectation is also referred to as the ensemble mean – emphasizing this distinction of the possible from the actual in any particular context (e.g., the actual average of a finite set of data). In this terminology, $E(x) = \overline{x}$; $E(\widehat{x}) = 0$; and $x - E(x) = \widehat{x}$.

The expectation generalizes – for any function $g(x)$,

$$E(g) = \int_{-\infty}^{\infty} g(x) f(x)\, dx \equiv \overline{g} \tag{3.13}$$

This is $\overline{g}$, the the ensemble mean value of $g(x)$.

The moments of a distribution function $f(x)$ are defined as the expected values of $\widehat{x}$ and its powers. For example, the nth moment is M_n:

$$M_n(x) = \int_{-\infty}^{\infty} (x - \overline{x})^n f(x)\, dx = E[(x - \overline{x})^n] \tag{3.14}$$

These are moments about the mean. The odd moments vanish for symmetric distributions.

The variance of x is the second moment, $n = 2$:

$$Var(x) = E[(x - \overline{x})^2] = M_2(x) \tag{3.15}$$

The variance is related to the expected value of x^2. If the mean is zero, they are identical. In general, since $x^2 = \overline{x}^2 + 2\overline{x}\widehat{x} + \widehat{x}^2$,

$$E[x^2] = \overline{x}^2 + Var(x) \tag{3.16}$$

It is entirely conventional to indicate by σ_x the RMS size of $\widehat{x}$; hence, $\sigma_x^2 \equiv Var(x)$.

A valid PDF always uniquely specifies its moments. It is problematic to go the other way, however – moments alone don't always determine a unique PDF.

Some standard metrics are phrased in terms of moments. The coefficient of variation (CV) normalizes variance by mean:

$$CV(x) = \left[\frac{E\left[(x - \overline{x})^2\right]}{\overline{x}^2} \right]^{\frac{1}{2}} = \frac{\sigma}{\overline{x}} \tag{3.17}$$

In other words, the CV is the ratio of the RMS size of $\widetilde{x}$, to the mean $\overline{x}$. This metric simply normalizes variability; if the mean is zero, it is meaningless. Its inverse is often termed the signal-to-noise ratio.

Higher moments are usefully normalized by σ. For example, the third moment is a metric of asymmetry; the skew coefficient is its normalization

$$Sk(x) = \frac{E[(x-\overline{x})^3]}{\sigma^3} \tag{3.18}$$

Similarly, the normalized fourth moment is the kurtosis

$$Kur(x) = \frac{E[(x-\overline{x})^4]}{\sigma^4} \tag{3.19}$$

The kurtosis is a metric of the pointedness or steepness of $f(x)$.

To illustrate: the familiar Gaussian distribution is symmetric; all its odd moments vanish. Its even moments have the property

$$M_p = a_p \cdot \sigma^p \tag{3.20}$$

with $a_2 = 1$ and $a_k = (k-1)a_{k-2}$ for even $k > 2$. The Gaussian distribution has $Sk = 0$, and $Kur = 3$, a standard of reference for other distributions. The *excess kurtosis* is relative to the Gaussian

$$Kur_e \equiv Kur - 3 \tag{3.21}$$

3.1.4 Properties: Median, Mode, Quartile

The *mean* of a distribution is the probabalistic expectation as covered previously. Two other metrics are important:

- The *median* of a distribution is that value below which half of all numbers occur – where $F(x) = 0.5$
- The *mode* of a distribution is that value that is most likely to occur – where $f(x)$ peaks.

All three metrics have distinctive, valid uses. For a symmetric, unimodal distribution, all three are the same. For more realistic cases, these differ in natural ways. For example, for the linear distribution in Example 3.2 the mode $= 2$, the median $= \sqrt{2}$, and the mean $= 4/3$.

Quartiles are a generalization of the median: the first quartile is the range $\infty < x \leq x_1$ that contains the lowest 25% of possible x values: $F(x_1) = 0.25$. The second quartile, $x_1 < x \leq x_2$, contains the next-largest 25%: $F(x_2) = 0.5$; and so forth. The median is the same as x_2.

For symmetric distributions, the mean and median are the same. More generally,

$$|\mu - med| \leq \sigma \tag{3.22}$$

Problem 3.1 For the uniform distribution on the interval $[0, 10]$:

1. What is the coefficient of variation?
2. What are the second, third, and fourth moments?
3. What is the skew coefficient?
4. What is the kurtosis?

Problem 3.2 For Example 3.2 (linear PDF):

1. Confirm the answers in the text: what is the mean, median, and mode?
2. What is the variance?
3. What is the CV?
4. What are the second, third, and fourth moments?
5. What is the skew coefficient?
6. What is the kurtosis?

3.1.5 Properties: Other Means

The mean as used earlier is the common one, more precisely the arithmetic mean. Other means are useful, including the harmonic and geometric means. To illustrate, we imagine a sample of N values x_i.

- The arithmetic mean $\frac{1}{N}\sum_{1}^{N} x_i$
- The geometric mean $\left[\prod_{1}^{N} x_i\right]^{\frac{1}{N}}$
- The harmonic mean $\left[\frac{1}{N}\sum_{1}^{N}\frac{1}{x_i}\right]^{-1}$

The continuous forms of these means are analogous. Gotelli and Ellison [175] provide a useful discussion of these means in an ecological context.

3.1.6 Bounding Theorems

Among the most useful inequalities are those that provide limits on deviates when only their first and/or second moments are known:

- **Markov's inequality:** For a non-negative deviate X, with mean μ and any $a > 0$

$$P[X \geq a] \leq \frac{\mu}{a} \tag{3.23}$$

- **Chebychev's inequality:** For a deviate X with mean μ and variance σ^2:

$$P\left[(X-\mu)^2 \geq b^2\right] \leq \frac{\sigma^2}{b^2} \tag{3.24}$$

for any real b.

In a different category are *asymptotic limits on distributions* when parameters approach extremes. The best known example of this is the Central Limit Theorem (Section 3.2.4). Similarly well known are various properties of the one-parameter Poisson distribution and its relation to the Gaussian and to the binomial distributions (Section 3.2.2).

3.1.7 Discrete Distributions: P_i and F_j

In cases where the random number in question is limited to a discrete set of integer values, we have an integer i drawn from a distribution $\mathcal{D}$:

$$i \sim \mathcal{D} \tag{3.25}$$

In describing the distribution $\mathcal{D}$ for integers, the continuous function $f(x)$ is replaced with the finite probability P_i – the probability that i occurs. There may be an infinity of possible values i; applications are common where i must be non-negative. P_i is analogous to $f(x)\Delta x$ for continuous variables. The set of P_i are sometimes referred to as the "probability mass function" – by analogy with $f(x)$ being the "probability density function." P_i is real, non-negative, ≤ 1.

The cumulative distribution function F need not be redefined here – it retains its sense of the accumulated finite probability of small occurrences, F increasing monotonically from 0 to 1. The sequence F_j is then the probability of an outcome *not larger than j*.[3] For the discrete case, then, we have the deviate i characterized by two sets of finite probabilities

$$P_i = P[i] \tag{3.26}$$

$$F_j = P[i \leq j] = \sum_{i \leq j} P_i \tag{3.27}$$

F_j is real, non-negative. Normalization requires $\sum_i P_i = 1$, guaranteeing that F increases monotonically from 0 to 1.

The simplest distribution in this category is the binary variable. There are only two outcomes, 0 and 1, and a single independent probability P_1. The rest derives from the normalization. (See Problems 3.3, 3.4.)

[3] In the discrete world, it is important to be clear about the finite possibility that $i = j$.

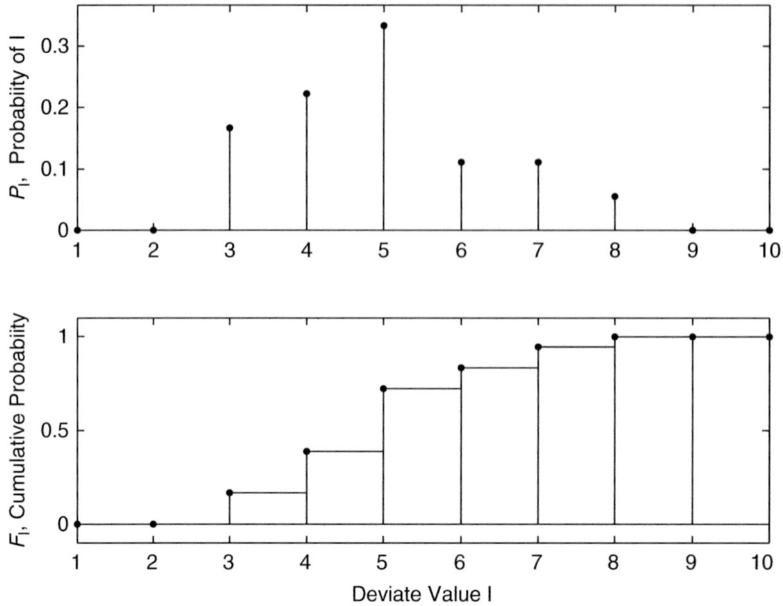

Figure 3.4. Plot of P_i and F_i for an example discrete distribution. The *unnormalized* probabilities are $[P_3 \ldots P_8] = [3\ \ 4\ \ 6\ \ 2\ \ 2\ \ 1]$. (See Problem 3.5.)

Figure 3.4 illustrates the more general discrete case for an arbitrary distribution.

The expectation and the various moments may be easily defined for the integer case, by analogy; summation over discrete outcomes replaces integration. For example, the mean is

$$E[i] = \sum_{-\infty}^{\infty} iP_i \equiv \bar{i} \tag{3.28}$$

with summation over all possible outcomes. (Recall the normalization requirement $\sum P_i = 1$.) The expectation of i is its ensemble mean $\bar{i}$; etc. The variance is similarly defined

$$Var(i) = E[(i-\bar{i})^2] = \sum_{-\infty}^{\infty} (i-\bar{i})^2 P_i \equiv \sigma_i^2 \tag{3.29}$$

Problem 3.3 For the binary variable $i = (0,1)$ with probability 0.3 of $i = 1$: what is f_i? $F(j)$? Plot them.

Problem 3.4 Consider a binary experiment with a 45% chance that $i = 1$; otherwise, $i = 0$. The experiment is repeated four times in sequence. The outcome of the four repeats is summed: $k = \sum i$.

1. What is the sample space for k?

2. What is P_k? F_k? Plot them.
3. What are the mean, variance, skew, and kurtosis of k?

Problem 3.5 For the example in Figure 3.4:

1. What is the mean, median, and mode?
2. What is the variance?
3. What is the CV?
4. What are the second, third, and fourth moments?
5. What is the skew coefficient?
6. What is the kurtosis?

3.2 Some Common Distributions

There is a good deal of literature defining distributions and their properties. Several common ones are summarized below – these are tractable, useful in applications, and can be generated on computers. Good discussions are found at Scheaffer and Young [403] and Ross [389] for theory; Gotelli and Ellison [175] for ecological applications; and Press et al. [362] for generation algorithms. Forbes et al. [157] provide an excellent summary of many distributions and their properties.

3.2.1 Continuous Distributions

Uniform Distribution: $\mathcal{U}$ The uniform distribution is a basic workhorse in working with real random numbers.

$$x \sim \mathcal{U} \tag{3.30}$$

The sample space is the interval $[0,1]$. On that interval, the PDF is uniform:

$$f(x) = 1 \qquad F(x) = x \tag{3.31}$$

Outside the interval, $f(x) = 0$ and $F(x)$ is constant (0 or 1). The moments:

$$\overline{x} = \frac{1}{2} \qquad \sigma_x^2 = \frac{1}{12} \tag{3.32}$$

This is easily generalized to the interval between a and b:

$$x \sim \mathcal{U}[a,b] \tag{3.33}$$

With $L \equiv b - a$

$$f(x) = \frac{1}{L} \qquad F(x) = \frac{x-a}{L} \tag{3.34}$$

for $a \leq x \leq b$; outside this interval, $f = 0$ and F is constant. The moments:

$$\bar{x} = \frac{a+b}{2} \qquad \sigma_x^2 = \frac{L^2}{12} \tag{3.35}$$

The uniform distribution's extreme values are at $x = \bar{x} \pm \sqrt{3}\sigma_x$. As suggested previously, the no-parameter distribution $\mathcal{U}$ indicates the default, $\mathcal{U}[0,1]$.

Gaussian Distribution: $\mathcal{G}$ Beyond $\mathcal{U}$, the Gaussian is the most commonly used distribution,

$$x \sim \mathcal{G}(\mu, \sigma^2) \tag{3.36}$$

$$f(x) = \frac{1}{\sqrt{2\pi\sigma^2}} e^{-\left[(x-\mu)^2/2\sigma^2\right]} \qquad F(x) = \frac{1}{2}\left[1 + \text{erf}\left(\frac{x-\mu}{\sigma\sqrt{2}}\right)\right] \tag{3.37}$$

where erf() is the error function. Properties:

$$\bar{x} = \mu \qquad \sigma_x^2 = \sigma^2 \tag{3.38}$$

The "normal" distribution is the Gaussian with ideal properties

$$\mathcal{N} = \mathcal{G}(0,1) \tag{3.39}$$

Although we tend to avoid this nomenclature, $\mathcal{N}$ is rightly singled out for normalcy. Any Gaussian can be manufactured from this starting point, equivalently stated:

$$\mathcal{G}(\mu, \sigma^2) = \mu + \sigma\mathcal{N} = \mu + \sigma\mathcal{G}(0,1) \tag{3.40}$$

Problem 3.6 What are the properties of the error function at the three values $x = -\infty$, $x = 0$, and $x = \infty$? Relate this to the definition of F. Does that make sense?

Problem 3.7 Compute and plot $f(x)$ for the Gaussian, Equation 3.37, for $\mu = 2$ and $\sigma = 3$.

Problem 3.8 Confirm the relation for F in Equation 3.37 analytically, by integrating the expression for f.

Problem 3.9 Numerically integrate the Gaussian F, starting from Problem 3.7. What properties do you expect? Do you obtain them? At what value of x is $F(x) = 0.5$?

Problem 3.10 Compute and the geometric distribution of $P = 0.2$ What is the continuous version? Compute and plot it. Comment on the comparison.

Problem 3.11 Refine Problem 3.10 with $P = 0.1, 0.05$ and 0.01. What do you expect? Do the combined results agree with theory?

Logistic: This distribution is similar to the Gaussian, but with higher kurtosis (stronger tails). The CDF is the logistic function. There are two parameters, μ and s^2. The distribution is symmetric about its mean μ ; s^2 is proportional to the variance.

$$x \sim \mathcal{L}(\mu, s^2) \tag{3.41}$$

$$f(x) = \frac{e^{-(x-\mu)/s}}{s\left[1+e^{-(x-\mu)/s}\right]^2} \qquad F(x) = \frac{1}{1+e^{-(x-\mu)/s}} \tag{3.42}$$

Properties:

$$\bar{x} = \mu \qquad \sigma_x^2 = (\pi s)^2/3 \tag{3.43}$$

Suppose we define the simple ideal logistic[4]:

$$y \sim \mathcal{L}(0,1) \tag{3.44}$$

y will have properties

$$\bar{y} = 0 \qquad \sigma_y^2 = \pi^2/3 \tag{3.45}$$

Then the transformation

$$x = \mu_x + \left[\frac{\sqrt{3}}{\pi}\sigma_x\right] y \tag{3.46}$$

produces the logistic distribution for x with the desired mean and variance.

The Hubbert distribution is a widely used example of the logistic distribution, common in reservoir engineering.

Problem 3.12 Compute and plot $f(x)$ and $F(x)$ for the logistic distribution. Compare it with the equivalent Gaussian distribution with the same mean and variance; plot their difference. What is the proper normalized metric of difference?

[4] Recall the normal distribution as an ideal from of the Gaussian.

Lognormal Distribution: $\mathcal{L}_{\mathcal{G}}$ This distribution is a logarithmic transform of the Gaussian distribution; it produces non-negative values. Let x be a Gaussian deviate with parameters (μ_x, σ_x).

$$x \sim \mathcal{G}(\mu_x, \sigma_x^2) \tag{3.47}$$

Then the lognormal deviate y is

$$y = e^x \tag{3.48}$$

"the natural logarithm of y is Gaussian." Hence the shorthand for this distribution:

$$y \sim \mathcal{L}_{\mathcal{G}}(\mu_x, \sigma_x^2) \tag{3.49}$$

The distribution, with $Q \equiv \frac{[\ln y - \mu_x]}{\sqrt{2\sigma_x^2}}$:

$$f(y) = \frac{1}{y\sqrt{2\pi\sigma_x^2}} e^{-Q^2} \qquad F(y) = \frac{1}{2}\left[1 + \mathrm{erf}(Q)\right] \tag{3.50}$$

The moments:

$$\bar{y} = e^{\left(\mu_x + \frac{1}{2}\sigma_x^2\right)} \qquad \sigma_y^2 = \phi^2 \left[e^{(2\mu_x + \sigma_x^2)}\right] \tag{3.51}$$

with $\phi^2 \equiv \left[e^{(\sigma_x^2)} - 1\right]$. These may be inverted for the parameters of x:

$$\mu_x = \ln\left(\frac{\bar{y}^2}{\sqrt{\sigma_y^2 + \bar{y}^2}}\right) \qquad \sigma_x^2 = \ln\left(\frac{\sigma_y^2}{\bar{y}^2} + 1\right) \tag{3.52}$$

This distribution will be skewed: y is unbounded on the positive side, but never negative. The skewness metric is the normalized third moment

$$Sk(y) \equiv \frac{E(y - \bar{y})^3}{\sigma_y^3} = 3\phi + \phi^3 \tag{3.53}$$

y will have minimum value 0. A final wrinkle is to introduce a minimum value τ for the shifted variable z

$$z = y + \tau \tag{3.54}$$

LogLogistic Distribution: $\mathcal{L}_L$ This is another logarithmic distribution with non-negative deviates. Its shape is similar but the tails are stronger. There are two non-negative parameters, a scale parameter α and a shape parameter β.

$$x \sim \mathcal{L}_L(\alpha, \beta) \tag{3.55}$$

$$F(x) = \frac{1}{1+\left(\frac{x}{\alpha}\right)^{-\beta}} \qquad f(x) = \left(\frac{\beta}{\alpha}\right)\left(\frac{x}{\alpha}\right)^{\beta-1} \frac{1}{\left[1+\left(\frac{x}{\alpha}\right)^{\beta}\right]^2} \tag{3.56}$$

$$\bar{x} = \frac{\alpha\pi}{\beta \sin\left(\frac{\pi}{\beta}\right)} \qquad \sigma_x^2 = \alpha^2 \left[\frac{2\pi}{\beta \sin\left(\frac{2\pi}{\beta}\right)} - \left(\frac{\pi}{\beta \sin\left(\frac{\pi}{\beta}\right)}\right)^2\right] \tag{3.57}$$

(The expression for the mean requires $\beta > 1$; that for variance, $\beta > 2$.)

Half-normal Distribution $\mathcal{H}_\mathcal{G}$ This is a simple perturbation of the normal, by rejecting negative outcomes.

$$x \sim \mathcal{G}(0, \sigma_x^2) \qquad y = |x| \tag{3.58}$$

The parameters are the Gaussian variance σ_x^2, with zero mean. The distribution f is the same as the Gaussian, but doubled for positive y and zero otherwise. F is similarly adjusted. The half-normal distribution will have nonzero mean and a reduced variance:

$$\bar{y} = \sigma_x \sqrt{\frac{2}{\pi}} \qquad \sigma_y^2 = \sigma_x^2 \left(1 - \frac{2}{\pi}\right) \tag{3.59}$$

The generalization for nonzero $\bar{x}$ is the folded normal distribution.

Exponential (waiting time, interarrival time) Distribution: $\mathcal{W}$ For discrete events occurring as a Poisson process (see later) with rate $\lambda \equiv \alpha t$: let t be the waiting time between one event and the next. t can have any positive real value.

$$t \sim \mathcal{W}(\alpha) \tag{3.60}$$

$$f(t) = \alpha e^{-\alpha t} \qquad F(t) = 1 - e^{-\alpha t} \tag{3.61}$$

$$\bar{t} = \frac{1}{\alpha} \qquad \sigma_t^2 = \frac{1}{\alpha^2} \tag{3.62}$$

The parameter $\frac{1}{\alpha}$ is the "mean waiting time" $\bar{t}$. The dimensionless time is αt, a type of scaling characteristic of this class of distributions. This distribution is skewed toward low values, with median = $0.693\bar{t}$; the probability that $t < \bar{t}$ is 0.632.

The probabilities of t occurring before or after t_1 are

$$P[t \leq t_1] = 1 - e^{-\alpha t_1} \qquad P[t > t_1] = e^{-\alpha t_1} \tag{3.63}$$

The probability of an event occurring in the interval (t_1, t_2) is

$$P[t_1 < t \leq t_2] = e^{-\alpha t_1}(1 - e^{-\alpha t_2}) \tag{3.64}$$

The small Δt linearization gives the probability of t occurring in the interval Δt:

$$P[t_1 < t < t_1 + \Delta t] \simeq e^{-\alpha t_1} \alpha \Delta t \tag{3.65}$$

This is exact in the small Δt limit.

In particular, if $t_1 = 0$ is "now," the probability of an event happening in the coming Δt is

$$P[0 < x < \Delta t] \simeq \alpha \Delta t \tag{3.66}$$

This is the famous "no memory" property of this distribution – given that nothing has happened before now, something will happen in the coming Δt with probability $\alpha \Delta t$ – *irrespective of history*. The mean waiting time until the next Poisson event is always the same; the future retains no memory of the past. This is a very useful distribution in queueing theory, where t is called the "interarrival" time (Hillier and Lieberman, [196]).

The exponential or waiting time distribution is the small-Δt limit of the discrete geometric distribution (see later).

In an oceanographic context, this and similar distributions could be used to represent spawning events, stage duration, or occurrence of unusual events.

Gamma Distribution: G_m This is a two-parameter distribution of positive real numbers. It generalizes the waiting time or exponential distribution. While the exponential $f(x)$ declines monotonically from $x = 0$, the gamma (assuming $\beta > 1$) builds up from $f(0) = 0$, then peaks before declining – making very small x unlikely.

$$x \sim G_m(\beta, \alpha) \tag{3.67}$$

The two parameters (β, α) are both > 0. Commonly, β is the *shape*; α is the *rate*, corresponding to its equivalent in the exponential distribution. The sample space is

$x \geq 0$. The distribution:

$$f(x) = \frac{\alpha e^{-\alpha x}(\alpha x)^{\beta-1}}{\Gamma(\beta)} \qquad F(x) = \frac{\gamma(\beta, \alpha x)}{\Gamma(\beta)} \tag{3.68}$$

Γ is the gamma function, γ the (lower) incomplete gamma function:

$$\Gamma(\beta) = \int_0^\infty e^{-u} u^{\beta-1} du \qquad \gamma(\beta, \alpha x) = \int_0^{\alpha x} u^{\beta-1} e^{-u} du \tag{3.69}$$

As in the exponential distribution, the independent variable x is always scaled by the rate parameter α; all occurrences of x appear as the dimensionless variable αx. As $\alpha x \to \infty$, γ saturates at $\Gamma(\beta)$. For integral values i, $\Gamma(i) = (i-1)!$, the familiar factorial function. The moments:

$$\overline{x} = \frac{\beta}{\alpha} \qquad \sigma_x^2 = \frac{\beta}{\alpha^2} \tag{3.70}$$

The peak of f occurs at $\alpha x = \beta - 1$ when $\beta > 1$.

The gamma distribution generalizes several other distributions:

- When the parameter $\beta = 1$ this is the *exponential* distribution.
- When the parameter β is a positive integer, this is the *Erlang* distribution.
- When the parameters $\beta = \frac{k}{2}$, $\alpha = \frac{1}{2}$, this is the *chi squared* distribution, also the *Pearson type III* distribution.

Erlang Distribution: $\mathcal{E}_r$ This is a continuous distribution, generalizing the exponential distribution. The deviate range is all positive real numbers. There are two parameters: "shape" k, a non-negative integer; and "rate" α, a non-negative real number.

$$x \sim \mathcal{E}_r(k, \alpha) \tag{3.71}$$

$$f(x) = \alpha \left[\frac{(\alpha x)^{k-1} e^{-\alpha x}}{(k-1)!} \right] \qquad F(x) = 1 - e^{-\alpha x} \sum_{m=0}^{k-1} \frac{(\alpha x)^m}{m!} \tag{3.72}$$

This is a subset of the gamma distribution when the shape parameter β is restricted to positive integers k. Again, the rate parameter α always scales x, which occurs only in the dimensionless form αx. The moments:

$$\mu_x = \frac{k}{\alpha} \qquad \sigma_x^2 = \frac{k}{\alpha^2} \tag{3.73}$$

When $k = 1$ this is the exponential distribution. Figure 3.5 illustrates the effect on $f(x)$ when k changes from 1 to 2. The exponential ($k = 1$) case peaks at the origin and

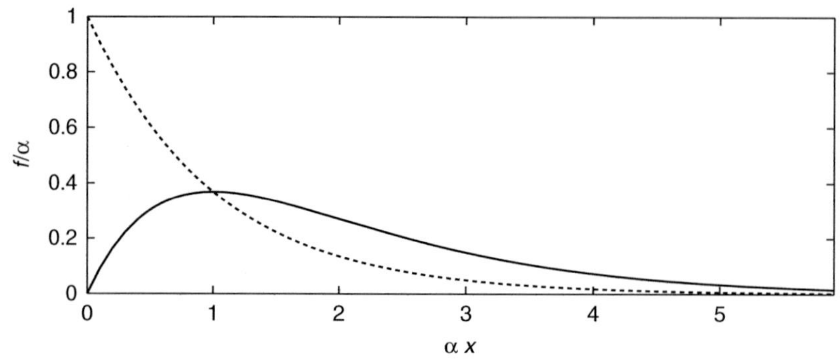

Figure 3.5. Plot of $f(\alpha x)$ for the gamma distribution for $\beta = 2$ (solid line). For reference, the dashed line is the exponential distribution, $\beta = 1$. (These distributions are the same as the Erlang distribution, $\mathcal{E}_r(\beta,\alpha)$ since β is an integer here.)

declines monotonically; the case $k = 2$ initially rises and peaks at $\alpha x = 1$, declining thereafter. The peaking phenomenon is characteristic of $k > 1$; the peak occurs later for higher k: $\alpha x = k - 1$.

We can go further in examining the hazard rate $h = \frac{f(x)}{1-F(x)}$ for this family of distributions. For $k = 1$, it is constant, $h_1 = \alpha$. This is the well-known "loss of memory" property of the exponential distribution. For $k = 2$ we have

$$h_2(x) = \alpha \left[\frac{\alpha x}{1 + \alpha x} \right] \tag{3.74}$$

For this case the hazard rate starts at 0 and rises to the asymptote α; the rise is 50% complete the "half-saturation" level $\alpha x^* = 1$. When $k = 3$, the half-saturation level is $\alpha x^* = 1 + \sqrt{3}$. Differences are pronounced when αx is small, maybe $\mathcal{O}(10)$ for low k; and certainly for αx below half-saturation.

More generally,

$$h_k(x) = \alpha \left[\frac{\frac{(\alpha x)^{k-1}}{(k-1)!}}{\sum_{m=0}^{k-1} \frac{(\alpha x)^m}{m!}} \right] \tag{3.75}$$

This is illustrated in Figure 3.6. All begin at zero and approach the same asymptote; the larger is k, the slower is the rise. For all $k \geq 2$ there is a sigmoid shape, that is, slow but accelerating growth in h near the origin.

Oceanographic applications of the hazard rate could include mortality or the failure of a drifter.

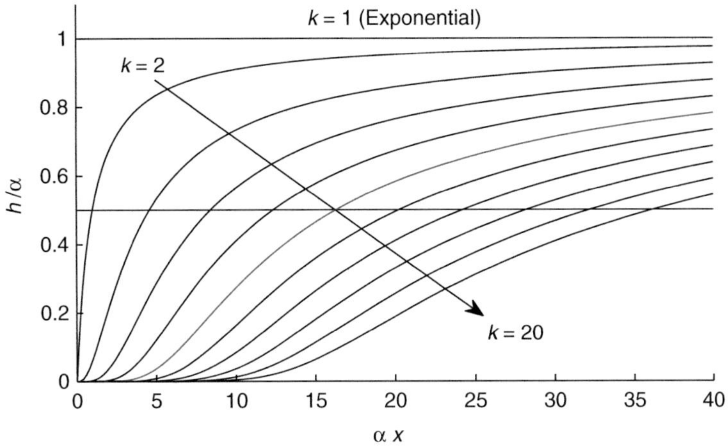

Figure 3.6. Normalized hazard rate h/α for the Erlang distribution, for $k = 2$ through 20 in increments of 2. For reference, the exponential distribution ($k = 1$, constant h) is shown. The additional reference line is the half-saturation level, $h = \frac{\alpha}{2}$.

Problem 3.13 Analytically confirm the mean and variance for the gamma distribution.

Problem 3.14 For the gamma distribution: does f peak when $\beta \leq 1$?

Problem 3.15 For the gamma distribution: what is the maximum value of f? What is the probability that this value of x is exceeded?

Problem 3.16 Numerically: evaluate and plot f, F and their mean and variance for the exponential and gamma distributions. Use $\alpha = 3$ and $\beta = [0.2, 1, 3, 5, 10]$. Show that they are the same when $\beta = 1$. What is the effect of varying β?

Problem 3.17 For $\mathcal{E}_r(k, \alpha)$: verify the results for $h(x)$ when $k = 1$ through 3 as given in the text.

Problem 3.18 For $k = 4, 8, 12$: obtain and plot the Erlang F(x) and h(x). Discuss the effect of k.

Problem 3.19 For the Erlang distribution: locate analytically the peak of f (value of αx and the values of f and F there); and plot same.

Problem 3.20 For the Erlang distribution: evaluate and and plot the half-saturation level versus k for the hazard rate. Compare with mean and mode of this distribution.

Problem 3.21 For the Erlang distribution: what is the survival function $1 - F(\alpha x)$? Find the quartiles of survival for $k = 2$, 4, and 6. Does this make sense?

Student's t Distribution: $\mathcal{S}_t$ This distribution governs samples from a Gaussian distribution when the mean and variance are unknown. It is invoked in estimating the mean and developing confidence intervals for estimates of the mean. Student's t is symmetric.

There is a single parameter, ν, the number of degrees of freedom. In context, $\nu > 2$. For high ν this distribution is insensitive to the parameter ν, and approaches the Gaussian.

$$t \sim \mathcal{S}_t(\nu) \tag{3.76}$$

$$f(t) = \frac{\Gamma(\frac{\nu+1}{2})}{\Gamma(\frac{\nu}{2})\sqrt{\nu\pi}} \left(1 + \frac{t^2}{\nu}\right)^{-\left(\frac{\nu+1}{2}\right)} \tag{3.77}$$

Γ is the gamma function. The moments ($\nu > 2$):

$$\bar{t} = 0 \qquad \sigma_t^2 = \frac{\nu}{\nu - 2} \tag{3.78}$$

This distribution is widely used in generating confidence intervals for sample-based estimates.

Chi-Squared Distribution: χ^2 This is a special case of the gamma distribution with $(\beta, \alpha) = (\frac{k}{2}, \frac{1}{2})$. There is a single positive integer parameter k, the "degrees of freedom." It is invoked in estimating variance and confidence intervals for same. $\chi^2(k)$ governs the distribution of the sum of k squares of independent Gaussian samples.

$$x \sim \mathcal{G}(0,1) \rightarrow y = \sum_{i=1}^{k} x_i^2 \tag{3.79}$$

The shorthand: "y is a χ^2 deviate with k degrees of freedom":

$$y \sim \chi^2(k) \tag{3.80}$$

χ^2 is asymmetric, always positive. The sample space is $y \geq 0$. The distribution:

$$f(y) = \frac{e^{-\frac{y}{2}} y^{\frac{k}{2}-1}}{2^{\frac{k}{2}} \Gamma(\frac{k}{2})} \qquad F(y) = \frac{\gamma(\frac{k}{2}, \frac{y}{2})}{\Gamma(\frac{k}{2})} \tag{3.81}$$

The moments:

$$\bar{y} = k \qquad \sigma_y^2 = 2k \tag{3.82}$$

The χ^2 distribution is the same as the *Pearson type III* distribution.

Gumbel Distribution: $\mathcal{G}_u$ The Gumbel is used in describing extreme values (max or min).

$$x \sim \mathcal{G}_u(m, \beta) \tag{3.83}$$

$$f = \frac{1}{\beta} e^{-(z+e^{-z})} \qquad F = e^{-e^{-z}} \tag{3.84}$$

with the composite variable $z = \frac{x-m}{\beta}$. Properties:

$$\mu = m + \gamma\beta \qquad \sigma = \frac{\pi\beta}{\sqrt{6}} \tag{3.85}$$

The constant $\gamma \simeq 0.577$. There are two parameters: m is the mode and β is proportional to σ. The standard Gumbel is $(m, \beta) = (0, 1)$. The Gumbel distribution is also called the log-Weibull distribtion.

The Gumbel is a particular case of the GEV (generalized extreme value, aka the Fisher–Tippett) distribution. The GEV unifies the description of three similar extreme-value distributions: Gumbel, Frechet, and Weibull.

Problem 3.22 Prove that the mode of the Gumbel distribution $= m$

Problem 3.23 Find the median of the Gumbel distribution.

Problem 3.24 For the Gumbel distribution, the mean is always $= m + a\sigma$, where a is a constant. What is a? Sketch and interpret.

Weibull Distribution: $\mathcal{W}_e$ The Weibull distribution is another extreme value distribution. Among other things, it has been used extensively in describing particle size distributions, in which context it is known as the Rosin–Rammler distribution. It is one of the particular cases covered by the GEV (generalized extreme value) distribution. It is defined for variates $x > 0$.

$$x \sim \mathcal{W}_e(\lambda, k) \tag{3.86}$$

$$f = \frac{k}{\lambda}\left(\frac{x}{\lambda}\right)^{k-1} e^{-\left(\frac{x}{\lambda}\right)^k} \qquad F = 1 - \exp\left[-\left(\frac{x}{\lambda}\right)^k\right] \tag{3.87}$$

There are two parameters, both non-negative. $\lambda > 0$ scales x; $k > 0$ is a shape parameter. Properties:

$$\mu = \lambda\Gamma\left(1 + \frac{1}{k}\right) \qquad med = \lambda\,[\ln(2)]^{\frac{1}{k}} \tag{3.88}$$

$$\sigma = \lambda \left[\Gamma\left(1+\frac{2}{k}\right) - \left[\Gamma\left(1+\frac{1}{k}\right)\right]^2 \right]^{\frac{1}{2}} \tag{3.89}$$

Γ is the gamma function.

The Rosin–Rammler distribution is an expression of this, applied often to particle sizes. If x_{50} is the median, then the CDF can be reexpressed:

$$F = 1 - \exp\left[\ln(0.5) \left(\frac{x}{x_{50}} \right)^k \right] \tag{3.90}$$

Problem 3.25 The Rosin–Rammler distribution is given in terns of the two parameters (x_{50}, k). You need to generate this distribuiton, but have only a Weibull generator. What are the equivalent Weibull parameters (λ, k) in terms the Rosin–Rammler parameters?

3.2.2 *Discrete Distributions*

Here the sample space is restricted to integers only.

Discrete Uniform Distribution $\mathcal{U}_D$ As the name implies, this is the discrete version of $\mathcal{U}$: all integers $0 \le i \le n$ are equally probable.

$$i \sim \mathcal{U}_D(n) \tag{3.91}$$

$$P[i=j] = \frac{1}{n+1} \qquad 0 \le \text{integer}\ j \le n \tag{3.92}$$

$$\overline{i} = \frac{n}{2} \tag{3.93}$$

$$\sigma_i^2 = \frac{n(n+2)}{12} \tag{3.94}$$

$$\overline{i^2} = \frac{2n(2n+1)}{12} \tag{3.95}$$

Notice that the outcome $i=0$ is equally possible. The nomenclature is variable on this point. A generalization is $\mathcal{U}_D(n_1, n_2)$ where the range is from n_1 to n_2, inclusively.

Problem 3.26 For the uniform discrete distribution $i \sim \mathcal{U}_D(n_1, n_2)$: find $\overline{i}$, $\overline{i^2}$ and σ_i^2. Show that these reduce to $\mathcal{U}_D(n_2)$ when $n_1 = 0$.

Binary (Bernoulli) Distribution: $\mathcal{B}e$ This is the most elementary discrete distribution. i is binary, 0 or 1. The probability that $i = 1$ is p; the probability that $i = 0$ is $1 - p$. There is a single intuitive parameter, $0 \le p \le 1$.

The binary distribution is often referred to as a "Bernoulli process" or "Bernoulli trial," with i the Bernoulli variable. The formal description is

$$i \sim \mathcal{B}e(p) \tag{3.96}$$

$$\overline{i} = p \qquad \sigma_i^2 = p(1-p) \tag{3.97}$$

The sample space is binary, $i = 0$ or 1; p is the probability that $i = 1$.

Binomial Distribution: $\mathcal{B}$ For n uncorrelated binary events: p is the probability of success for each event independently; $(1-p)$ is the probability of failure; i is the total number of successes.

$$i \sim \mathcal{B}(n,p) \tag{3.98}$$

$$\overline{i} = np \qquad \sigma_i^2 = np(1-p) \tag{3.99}$$

$$P[i = j] = \binom{n}{j} p^j (1-p)^{n-j} \qquad 0 \leq \text{integer}\, j \leq n \tag{3.100}$$

where the binomial coefficient $\binom{n}{j} = \frac{n!}{j!\,(n-j)!}$.

This distribution pertains to a sequence of n identical binary events. The outcome i is the total number of successes, *in any order*. Equivalently, the binomial distribution represents repeated Bernoulli trials. Naturally, $n = 1$ is the same as the binary (Bernoulli) distribution, $\mathcal{B}(1,p) \equiv \mathcal{B}e(p)$

For large n and small p: the binomial distribution approximates the Poisson distribution with the single parameter $\lambda \equiv np$:

$$\mathcal{B}(n,p) \approx \mathcal{P}ois(\lambda = np) \tag{3.101}$$

provided $p < 0.1$ and $np < 10$. Also for large n, the binomial approaches the Gaussian [157]:

$$\mathcal{B}(n,p) \approx \mathcal{G}(np, np(1-p)) \tag{3.102}$$

when $np(1-p) \geq 25$. Further, this Gaussian approximation holds if

$$np(1-p) > 5 \quad \textit{and} \quad 0.1 \leq p \leq 0.9 \tag{3.103}$$

or if

$$\min\left[np, n(1-p)\right] \geq 10 \tag{3.104}$$

Suitable adjustments are needed – Gaussian deviates are real; binomial deviates are integers.

Categorical Distribution: $\mathcal{C}$ This is an extension of the Bernouilli distribuiton, with k different mutually exclusive outcomes or "categories," each with probability p_j. Exactly one must occur: $\sum p_j = 1$. There is a vector of binary outcomes; only one is nonzero.

$$\mathbf{i} \sim \mathcal{C}(\mathbf{p}) \tag{3.105}$$

$$\overline{\mathbf{i}} = \mathbf{p} \qquad [Cov_{\mathbf{i}}] = \mathbf{p} - \mathbf{p}\mathbf{p}^T \tag{3.106}$$

The negative covariance reflects the equality constraint on the p_i.

There are other ways to describe this distribution, more natural in certain applications. This one is the vector form of the Bernouilli; it is consistent with their generalizaton, the multinomial. The idea is quite simple: drop a particle into one of k boxes, each having a probability of being the recipient.

Multinomial Distribution: $\mathcal{M}$ This is a generalization of the binomial distribution. There are n independent events ("trials," steps, experiments) as in the binomial case; *each event has exactly one of k outcomes*. (The binomial case is $k = 2$, either success or failure.) Each outcome occurs in any one event with probability p_k; all outcomes are exclusive. There are k integer deviates $i_j =$ the total number of times that outcome j occurs after all the n trials are done. Each i_j has the integer range $[0, 1, \dots\ n]$.

There are the obvious constraints – each trial must produce one positive outcome, and the outcomes are mutually exclusive:

$$\sum_{j=1}^{k} i_j = n \qquad \sum_{j=1}^{k} p_j = 1 \tag{3.107}$$

The moments

$$\mathbf{i} \sim \mathcal{M}(n, \mathbf{p}) \tag{3.108}$$

$$\overline{\mathbf{i}} = n\mathbf{p} \qquad [Cov_{\mathbf{i}}] = n\left(\mathbf{p} - \mathbf{p}\mathbf{p}^T\right) \tag{3.109}$$

There is negative covariance among the outcomes. They all must sum to n, and in each trial they are in competition with each other.

The multinomial generalizes the categorical to n trials and the binomial to k outcomes.

Problem 3.27 Confirm that the multinomial distribution replicates the binomial when $k = 2$.

Problem 3.28 Confirm that the multimomial distribution replicates the categoical when $n = 1$.

Poisson Distribution: $\mathcal{P}ois$ The occurrence of discrete events is at the "Poisson rate" λ over a certain interval; j is the number of events that actually occur in any such interval.

$$i \sim \mathcal{P}ois(\lambda) \tag{3.110}$$

$$P[i=j] = e^{-\lambda}\frac{\lambda^j}{j!} \qquad \text{integer } j \geq 0 \tag{3.111}$$

$$\overline{i} = \lambda \qquad \sigma_i^2 = \lambda \tag{3.112}$$

Some useful features of the Poisson distribution include

- *Additive property.* The sum of Poisson deviates, at different rates, is a Poisson deviate at the combined rate:

$$n_1 \sim Pois(\lambda_1) \qquad n_2 \sim Pois(\lambda_2) \tag{3.113}$$

$$n_1 + n_2 \sim Pois(\lambda_1 + \lambda_2) \tag{3.114}$$

- *Gaussian limit.* In the limit of large λ, the Poisson is approximately Gaussian, $\mathcal{P}ois(\lambda) \rightarrow \mathcal{G}(\lambda, \lambda)$. Suitable accommodations are needed, as Poisson deviates are integers while Gaussian deviates are real.
- *Binomial Limit.* The Poisson distribution approaches the binomial when the source is very large but the rate is finite:

$$\mathcal{B}(n,p) \approx \mathcal{P}ois(\lambda = np) \tag{3.115}$$

provided $p < 0.1$ and $np < 10$.
- *Inhomogeneous poisson distribution.* The additive property generalizes to the case of continuously varying intensity $\nu(\mathbf{x})$. For a finite region $\mathcal{R}$, the combined Poisson rate is $\lambda_{\mathcal{R}} = \int_{\mathcal{R}} \nu \, d\mathbf{x}$.

Geometric Distribution: $\mathcal{W}_g$ This is the discrete analog of the waiting time distribution. For a single individual: the per-step probability of a binary event is p; $(1-p)$ is the probability of the event not occurring. j is the number of consecutive steps at which the event *first occurs.*

$$i \sim \mathcal{W}_g(p) \tag{3.116}$$

$$P[i=j] = (1-p)^{j-1}p \qquad \text{integer } j \geq 0 \tag{3.117}$$

$$\overline{i} = \frac{1}{p} \tag{3.118}$$

$$\sigma_i^2 = \frac{1-p}{p^2} \tag{3.119}$$

Table 3.1. *Generalizations of the Binary Distribution*

	One Outcome per Trial i,p	Many Exclusive Outcomes per Trial $\mathbf{i}$, $\mathbf{p}$
One trial; $i=[0,1]$	$\mathcal{B}e(p)$ Bernouilli (binary)	$\mathcal{C}(\mathbf{p})$ Categorical
n Repeated trials; $i=[0,1,2,\ldots n]$	$\mathcal{B}(n,p)$ Binomial	$\mathcal{M}(n,\mathbf{p})$ Multinomial
Example	Coin toss	Die toss

Note: p is the probability of success *per trial*. i is the total number of successes over all trials. An unbiased coin would have $p=0.5$; a loaded die would have unequal p_i.

The probability of the event *not occurring* in j consecutive steps is

$$P(i>j)=(1-p)^j \tag{3.120}$$

and its complement is the probability that the first event *will occur* before or during j steps:

$$P(i\leq j)=1-(1-p)^j \tag{3.121}$$

For $p=\alpha\Delta t$, α constant: the limiting form of the geometric distribution as $\Delta t\to 0$ is the exponential distribution for $t=i\Delta t$.

Summary Many of the discrete distributions are simple generalizations of the basic Bernouilli (binary) distribution. There is room for confusion among them; but the ideas are simpler than their nomenclature. Table 3.1 displays four major discrete distributions in a simple 2×2 framework. Some gambling examples help illustrate the contexts.

Problem 3.29 Show that the binomial distribution, in the limit of large n, small p, becomes the Poisson distribution by comparing their expressions for $P(i)$. What is their difference?

Hint: You may want to make use of the limit

$$\lim_{n\to\infty}\left[1+\frac{z}{n}\right]^n \Rightarrow e^z \tag{3.122}$$

Problem 3.30 Prove that Equations 3.117, 3.120, and 3.121 for the geometric distribution approach their counterpart Equations 3.63 and 3.65 for the exponential distribution, in the limit of small Δt.

Problem 3.31 Find the cumulative distribution functions F_j for the binomial, Poisson, and geometric distributions.

Problem 3.32 It is asserted that the Poisson distribution is appoximately Gaussian for large λ. Confirm this by generating and plotting f and $P(i)$. Is this true? if so, how big does λ need to be?

Problem 3.33 Relate your finding in Problem 3.32 to the binomial distribution – is there a Gaussian appoximation for the binomial distribution? What are the conditions in terms of parameters?

3.2.3 Importance of $\mathcal{G}$, $\mathcal{U}$, $\mathcal{B}$, $\mathcal{P}ois$

It is clear that the Gaussian distribution is very well utilized and a standard of reference for these various distributions. Several distributions approach it asymptotically. It occurs often in problem formulation, whether intended a priori or as a finding. It and the uniform distribution play central roles in generating other distributions. And the central limit theorem (see later) is of fundamental importance in the treatment of data and in formulating nested processes. Its implementation in computer systems is typically at the hardware or firmware level, with libraries of routines that generate deviates in higher-level languages.

In the discrete category, the binomial distribution is fundamental, as is its generalization to the multinomial distribution. It is at the heart of describing aggregate outcomes when individual outcomes are binary. In addition, it is often employed in the generation of particles from a finite source. The Poisson distribution is its limiting form when the source is infinite for practical purposes, but the rate of supply is finite. This is an oft-used distribution for events of unknown origin.

These distributions occur in many contexts and form a baseline of analysis and formulation; and a computational system must have routines for generating deviates from these distributions.

3.2.4 The Central Limit Theorem

Suppose x is a random variable, with mean $\overline{x}$ and variance σ_x^2; but further details about its distribution are unknown. Let y be the average value of n independent samples of x.

$$y = \frac{1}{n}\sum_{i=1}^{n} x_i \tag{3.123}$$

It is easy to discover that

$$\overline{y} = \overline{x} \qquad \sigma_y^2 = \frac{1}{n}\sigma_x^2 \tag{3.124}$$

The averaging preserves the mean and tightens the variance. By the *central limit theorem*, the distribution of y approaches a Gaussian, asymptotically for large n:

$$\lim_{n \to large} y \sim \mathcal{G}\left(\bar{x}, \frac{\sigma_x^2}{n}\right) \tag{3.125}$$

for a wide range of non-Gaussian distributions of x. This remarkable result is the basis of many findings about estimating the parameters of x – under reasonable conditions, one can appeal to the Gaussian PDF to interpret sample means and variances. In the context of the random walk developed later, the total displacement is a sum of per-step random walks; so the accumulated displacement will approach a Gaussian distribution as time grows.

3.3 Generating Random Numbers

A *deviate* is one realization from a given distribution. Here we summarize some information about generating deviates on computers.

3.3.1 General Methods

Direct Simulation This is the most direct and potentially limiting method. If the desired PDF is built up of simpler probabalistic processes, simulate them and take the outcome as the deviate. An example: simulate n Bernoulli trials to obtain one binomial deviate. Needless to say this is laborious for large n, but remains a fallback option when there is a suitable hypothesis that relates a very simple distribution to a more complex one.

Inverse Transformation Method For continuous variables with known distributions, it is possible to generate deviates by (a) generating the uniform deviate and (b) transforming it. Specifically, if the desired deviate x has known $F(x)$, the the procedure is

$$y \sim \mathcal{U} \tag{3.126}$$

$$x = F^{-1}(y) \tag{3.127}$$

where F^{-1} is the inverse of F. (F^{-1} is known as the quantile function.)

Figure 3.7 shows a hypothetical distribution $y = F(x)$ that is desired. The idea of inversion is to generate a uniform deviate along the vertical axis, $y = F(x)$, and to find the corresponding value of x, working "in reverse." If y is sampled uniformly, then the *fraction* of deviates occurring on the small interval Δy will simply be $\Delta y = \Delta F$. Each has its corresponding x deviate, so ΔF is the fraction of x deviates occurring

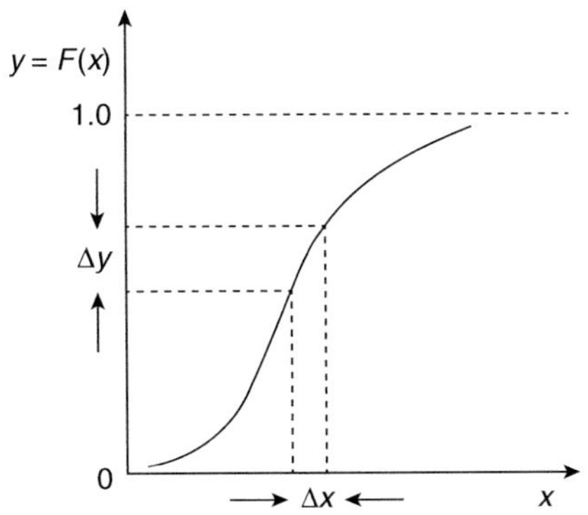

Figure 3.7. Inverse transformation method for general distribution $F(x)$.

in Δx. So,

$$f(x)\Delta x = \Delta F \tag{3.128}$$

$$f(x) = \frac{\Delta F}{\Delta x} \to \frac{dF}{dx} \tag{3.129}$$

and this completes the demonstration.

The waiting time deviate covered above is an example. Since the desired deviate x has distribution

$$F(x) = 1 - e^{-\alpha x} \tag{3.130}$$

we have in terms of the $F(x) = y \sim \mathcal{U}$:

$$e^{-\alpha x} = 1 - y \tag{3.131}$$

$$x = -\frac{1}{\alpha} \ln(1 - y) \tag{3.132}$$

An additional example is the linear distribution, Examples 3.4 and 3.5.

Example 3.4 Consider the linear distribution on the sample space $0 \le y \le b$. The distribution is

$$f(y) = 2\frac{y}{b^2} \qquad F(y) = \frac{y^2}{b^2} \tag{3.133}$$

Show that the inverse transformation method gives the following:

$$x \sim \mathcal{U} \qquad y = b\sqrt{x} \tag{3.134}$$

Example 3.5 Numerically generate some ensembles of the deviate y as in Example 3.4; confirm that the desired distribution is generated.

Rejection Method More generally, if F is not invertible: the rejection method requires only that f itself be evaluated, in addition to a simpler deviate that can be generated with known density f_1. Basically, generate deviates y from f_1, and then reject some of them so that f_1 gets trimmed back to the desired f.

Specifically:

1. Begin by selecting an arbitrary scaling factor κ such that $\kappa f_1 \geq f$ everywhere.
2. Sample y from f_1, and independently sample the uniform distribution for x:

$$y \sim \mathcal{F}_1 \qquad x \sim \mathcal{U} \tag{3.135}$$

3. Accept or reject y based on the acceptance criterion

$$x \leq \frac{1}{\kappa} \frac{f(y)}{f_1(y)} \tag{3.136}$$

   If true, accept the deviate y. Otherwise, reject it and try again. This will give the distribution $f(y), F(y)$.

Figure 3.8 illustrates this approach for the simple case where the uniform distribution is used for f_1. The method operates by generating a single candidate y, then reaching an accept/reject decision; and repeating until enough deviates are accepted. (See Figure 8.12 for a specific example.)

There will be two deviates per trial, a fraction of which are rejected. It is clear that this method will lead to many rejections unless the trial distribution f_1 is close to the desired f (Figure 3.9).

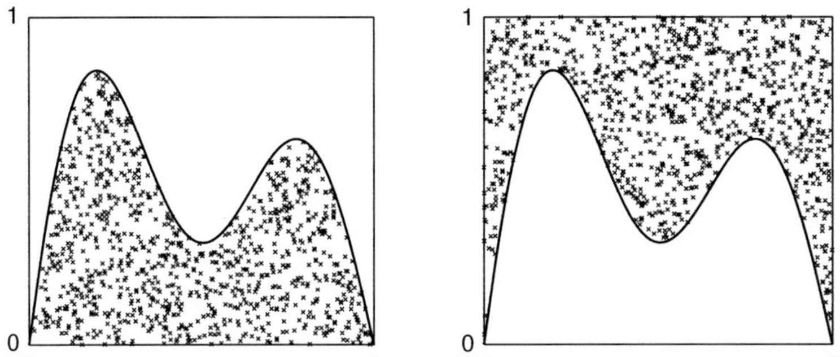

Figure 3.8. Example of the rejection method using the uniform deviate to generate a two-peaked distribution in 1D (the solid line). Each deviate requires two uniform deviates: a trial value (horizontal axis); and a rejection criterion based on the desired distribution (vertical axis). Left: Keepers: Right: Rejects.

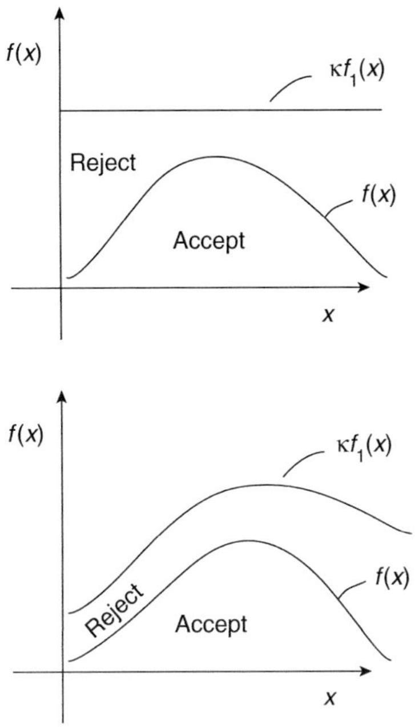

Figure 3.9. Rejection method. Top: The uniform distribution is the comparison f_1. Bottom: More generally with $f_1 \geq f$.

3.3.2 Some Specific Deviates

Uniform Deviate The most common random number generator produces the *uniform deviate* on the interval $[0, 1]$. In Matlab and Fortran the function is x=rand() with no arguments; there are further options. These are pseudo-random generators. Language- and/or machine-specific procedures are typically needed to get these generators started (the seeding problem) to help avoid the associated risks. In some cases the seeding is handled automatically, invisible to the user, by a machine clock call or other unpredictable metric. A good and practical discussion of algorithms for generating the uniform deviate can be found in [362]. The procedures described there can be implemented in portable software, thereby avoiding reliance on built-in deviates. They are recommended if routine scientific use is to be made of random number generators.

The mean of the uniform deviate $\mathcal{U}$ is by inspection 0.5; its variance is $1/12$.

$$x \sim \mathcal{U} \tag{3.137}$$

$$\bar{x} = \frac{1}{2} \qquad \sigma_x^2 = \frac{1}{12} \tag{3.138}$$

We can manufacture other uniform distributions for $a \le y \le b$ with the simple transformation

$$y = a + Lx \tag{3.139}$$

with the $L = b - a$ the interval length. This is the same as

$$y \sim \mathcal{U}(a,b) \tag{3.140}$$

with the transformed moments

$$\bar{y} = \frac{a+b}{2} \qquad \sigma_y^2 = \frac{L^2}{12} \tag{3.141}$$

Binary (Bernoulli) Deviate This is the basis of many operations with integer distributions. It is the counterpart of the uniform deviate for continuous distributions. $0 \le p \le 1$ is the probability of outcome 1; $(1-p)$, that for outcome 0.

It is reasonable to expect this to be available as a high-level call on all machines. If not, it is easily generated from the uniform deviate, $x \sim \mathcal{U}$. If $x \le p$, then $i = 1$. Otherwise, $i = 0$.

The Bernoulli deviate is a simple form of the more general generator for integers, uniformly distributed. In some implementations this is "randint()", by analogy with the uniform deviate "rand()".

Gaussian Deviate If a Gaussian deviate $\mathcal{G}$ is wanted, it can be generated from the uniform distribution $\mathcal{U}$ with the *Box–Müller transformation*. First, generate two independent uniform deviates

$$x_1 \sim \mathcal{U}(0,1) \tag{3.142}$$

$$x_2 \sim \mathcal{U}(0,1) \tag{3.143}$$

Then apply the deterministic transformations

$$y_1 = \sqrt{-2\ln(x_1)}\cos(2\pi x_2) \tag{3.144}$$

$$y_2 = \sqrt{-2\ln(x_1)}\sin(2\pi x_2) \tag{3.145}$$

Two values (y_1, y_1) will be independent Gaussian deviates with zero mean, unit variance:

$$y_i \sim \mathcal{G}(0,1) \tag{3.146}$$

A Gaussian distribution with mean μ, variance σ^2 is gotten by shifting and scaling:

$$z_i = \mu + \sigma y_i \tag{3.147}$$

This achieves

$$z_i \sim \mathcal{G}(\mu, \sigma^2) \tag{3.148}$$

The Gaussian deviate is popular enough to be supported in high-level languages.

Waiting Time – the Exponential Deviate Other important deviates can be generated by mapping simpler distributions. For example, for the exponential or waiting time deviate,

$$t \sim \mathcal{W}(\alpha) \tag{3.149}$$

one begins with the uniform deviate x and maps it:

$$x \sim \mathcal{U}(0,1) \tag{3.150}$$

$$t = \frac{-\ln(1-x)}{\alpha} \tag{3.151}$$

This achieves the desired deviate.

Lognormal Deviate The lognormal distribution is the logarithm of the Gaussian distribution.

$$y \sim \mathcal{L}G(\mu_x, \sigma_x^2) \tag{3.152}$$

This may be acieved by obtaining a Gaussian deviate, and mapping it:

$$x \sim \mathcal{G}(\mu_x, \sigma_x^2) \tag{3.153}$$

$$y = e^x \tag{3.154}$$

y will be lognormally distributed on the sample space $0 \leq y < \infty$. Adding an absolute minimum τ is not uncommon,

$$z = y + \tau \tag{3.155}$$

Problem 3.34 Generate some ensembles of *waiting time* deviate. Confirm that individual deviates have the distribution and moments given in the text. What size ensemble is needed to get 5% accuracy for the mean? For the variance?

Problem 3.35 Sample the waiting time ensembles generated above (Problem 3.34) for sample size $n = 5, 10, 20, 50, 100, 300, 1000$. Obtain in each case the sample average and the sample variance. Plot the results versus n. Explain the results.

Problem 3.36 For the sampled results in Problem 3.35 plot the distribution of means obtained, with $n = 5$ and $n = 100$. Explain.

Problem 3.37 Repeat problems (3.34, 3.35, 3.36) using *lognormal* deviates. What do you find? Explain.

3.4 Sampling; Finite N

It's a premise of simulation: that correctly distributed inputs produce valid distributions of outputs. This is a truism for big (effectively infinitely big) ensembles; barring unusual system behavior, it may be relied upon as a limit.

But on real machines, we have finite N. How big does N need to be? And, how far off are we?

On input: We conceive idealized distributions and sample them for inputs. The literature on random number generation and on sampling known processes provides guidance here.

On output: Many sources of variation are blended in nonlinear/inhomogeneous ways – ultimately too complex for analysis; that's the very reason for resorting to simulation! Here we are on shakier ground. Theoretically, results scale toward truth as N gets larger; and simulation metrics can quantify the effect of changing N. But we don't know either where we would ultimately end up, nor how far we are from it. A slow convergence rate does not mean that convergence has occurred!

This problem – of not knowing what we are asymptotically approaching – occurs in many forms in computation. For a familiar example, it could be compared to the resolution question in PDE simulation – finite resolution is necessary to get any approximation; how accurate is it? Here we're worsening the problem by adding the fact that all outcomes are stochastic and only partly sampled.

The "problem of N" is exacerbated by the inevitable dispersion that we seek to quantify, diluting any finite ensemble and exporting some of it across boundaries. Finite ensembles, initially adequate, can be spread over important features that are relatively small, effectively diluting N. And, mortality and reproduction introduce endogenous changes in N that are not predictable.

Simulation is experiment, and it is useful to adopt an experimentalist's perspective – we are sampling the output of a complex process. There are some critical differences, though. All inputs are statistically controllable and repeatable; and all processes and outcomes are completely observable, without error.

It is critical to adjust hypothesis testing to reflect these practical limits. Choose governing scientific questions carefully, and choose simulations that can address them with acceptable precision. A principle of parsimony is important here – solve the simplest problems that can shed light, first. A good scientific method is very careful about this. Overall, creativity, humility, and courage are necessary ingredients of success, as in all scientific inquiry.

3.4.1 Sample Statistics

We generate finite sets of deviates. On input, the samples are generated from distributions with *known* properties. On output, the samples are the result of a simulation; their distributions have *unknown* properties and the purpose is to estimate them. In either case, the governing question is classic: How well do the generated samples represent the distribution whence they came? This is the subject of Problems 3.34, 3.35, 3.36, and 3.37.

Suppose we have an input x with true mean and variance (μ,σ^2), and from this distribution, a set of N independent deviates x_i is generated and used. The properties of these numbers will be the *sample* mean and *sample* variance,

$$m = \frac{1}{N}\sum_{i=1}^{N} x_i \tag{3.156}$$

$$s^2 = \frac{1}{N-1}\sum_{i=1}^{N} (x_i - m)^2 \tag{3.157}$$

These are measured moments of one such set of N deviates. Every set of N deviates will have its own (m, s^2). Given N, what are the distributions of all possible values of the statistics m and s^2? In what sense is $m \simeq \mu$? $s^2 \simeq \sigma^2$?

The sample variance s^2 as defined in Equation 3.157 turns out to be an unbiased estimator of σ^2 – the denominator $N-1$ accommodating the loss of one degree of freedom due to the use of the estimated mean m. Another metric is commonly studied, the mean squared deviate,

$$S^2 = \frac{1}{N}\sum_{i=1}^{N} (x_i - m)^2 \tag{3.158}$$

There are various terminologies for these two metrics, s^2 and S^2, but good agreement on the distinction, 3.157 and 3.158. The distinction is important when N is not large. Needless to say, $S^2 = (\frac{N-1}{N})s^2$, an $\mathcal{O}(\frac{1}{N})$ bias. S^2 will on average be on the small side of σ^2, while s^2 will be, on average, correct.

Example 3.6 In passing, we note that the alternate formula for s^2

$$s^2 = \frac{1}{N-1}\left[\sum_{i=1}^{N} (x_i^2) - Nm^2\right] \tag{3.159}$$

This is an intuitive version of the exact relation

$$\sigma^2 = E[(x-\mu)^2] = E[(x^2)] - \mu^2 \tag{3.160}$$

and can be useful in certain practical situations.

Problem 3.38 Plot the ratio S/s versus N.

Problem 3.39 Confirm that the relation for the sample variance given in Equation 3.159 is exact for all $N \geq 2$.

3.4.2 Sample Mean

Any sample of N deviates x_i has an intuitive sample mean m as defined Equation 3.156. What is the distribution of m?

The expected m (the average over all possible sets of N deviates) is $E[m] = \mu$. This is good news! And, the variance of m is σ^2/N. The quantity $\frac{\sigma}{\sqrt{N}}$ is the *standard error* (*SE*) of the sample mean; and it is the basis of constructing a *confidence interval* for m – the interval that should contain most sample means.

$$m - \mu \approx \pm\zeta \cdot \frac{\sigma}{\sqrt{N}} \tag{3.161}$$

This confidence interval is a *metric of precision*; the parameter ζ (the "half-width") will be dependent on (a) the actual distribution of m (its PDF); and (b) the level of confidence we wish to assert. ζ increases with increasing confidence – as confidence that $(\mu - m)$ lies in an interval increases, the wider the interval must be.

The more precise statement concerns the probability that m lies within a given interval:

$$P\left[\mu - \zeta_\alpha \cdot \frac{\sigma}{\sqrt{N}} \leq m \leq \mu + \zeta_\alpha \cdot \frac{\sigma}{\sqrt{N}}\right] = 1 - \alpha \tag{3.162}$$

where $1-\alpha$ is the confidence level. The fraction α of all deviates m will lie outside this interval, in the tails of the m distribution; the balance, the fraction $1-\alpha$ of all sample means m, lies within the interval. Naturally, very tight intervals have smaller confidence levels. This is additional good news – the effect of enlarging N is to either tighten the interval or increase confidence – qualitatively, an intuitive result; the $\frac{1}{\sqrt{N}}$ factor quantifies it.

If generating Gaussian deviates, or for large samples from more general distributions (by the Central Limit Theorem): the *distribution* of m is Gaussian; $m \sim \mathcal{G}(\mu, \frac{\sigma^2}{N})$.

Table 3.2. *For m: Confidence Intervals about the True Mean*

Interval Half-Width ζ_α	Gaussian Confidence Level $1-\alpha$
1.0	0.6827
1.5	0.8664
2.0	0.9545
2.5	0.9876
3.0	0.9973

Note: N is the sample size; $(1-\alpha)$ is the confidence level – the probabilitiy that m will be in the interval $\mu \pm \zeta_\alpha \frac{\sigma}{\sqrt{N}}$. These are properties of the $\mathcal{G}(0,1)$ distribution.

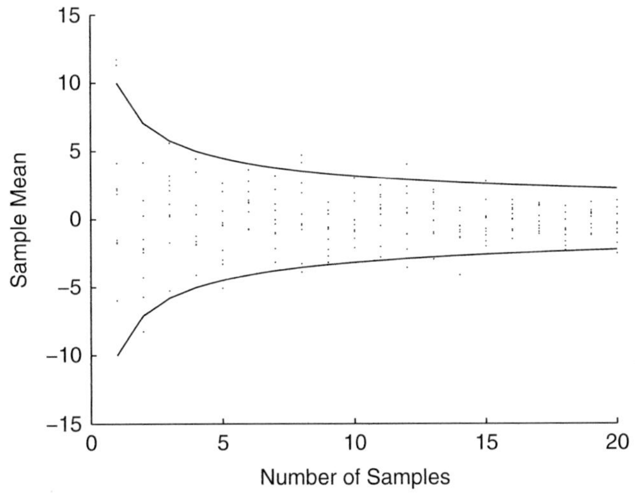

Figure 3.10. Computed sample means versus N, from $\mathcal{G}(0,\sigma^2)$; $\sigma = 5$. The solid lines enclose the confidence interval $\pm 2\frac{\sigma}{\sqrt{N}}$. Of the 200 samples shown, 13 fall outside the confidence interval.

The confidence levels are easily lifted from the Gaussian CDF, resulting in Table 3.2. The rule of thumb: the interval $\mu \pm 1$ SE contains the sample mean m, 68% of the time; ± 2 SE, 95% of the time.

Figure 3.10 shows some experimental confirmation. Samples of size N were taken from $\mathcal{G}(0,\sigma^2)$; the sample mean is shown along with the $\pm 2\frac{\sigma}{\sqrt{N}}$ envelope (roughly the 95% confidence interval). For each value of N there are 10 repeated samplings. Of the total of 200 samplings, this particular experiment yielded 13 means outside its interval. The convergence of the intervals is dominated by the $\frac{1}{\sqrt{N}}$ factor, which slows down at high N.

Sample Mean When σ Is Unknown: This is the situation when simulation output is being sampled; or where inputs are in fact outputs from another simulation, or other finite sample – perhaps physical data. Here we have *only* the sample statistics m and s. It remains common to construct confidence intervals using the sample statistics only, for example,

$$m - \mu \approx \pm \zeta_\alpha \frac{s}{\sqrt{N}} \tag{3.163}$$

and appropriate the results in Table 3.2 – effectively, using the sample standard error $\frac{s}{\sqrt{N}}$ as an approximation for the true value $\frac{\sigma}{\sqrt{N}}$ in Equation 3.161. This is a common approximation; ζ_α is independent of N.

More precisely: the blend of m and s obeys the Student's t distribution,

$$\frac{m - \mu}{s/\sqrt{N}} \sim \mathcal{S}_t(N-1) \tag{3.164}$$

and the appropriate statement is

$$P\left[m - \zeta_{\alpha,N} \cdot \frac{s}{\sqrt{N}} \leq \mu \leq m + \zeta_{\alpha,N} \cdot \frac{s}{\sqrt{N}}\right] = 1 - \alpha \tag{3.165}$$

Now the interval quantity $\zeta_{\alpha,N}$ depends on the distribution of $\mathcal{S}_t$, which generally *is* sensitive to N. Values of are ζ given in Table 3.3. For large N, $\mathcal{S}_t$ is asymptotically Gaussian and ζ loses its dependence on N. The Gaussian case of Table 3.2 is replicated in Table 3.3 for the extreme case.

Since any finite ensemble will have its own m and s, the sense of this confidence interval is: "of the possible intervals so generated, the fraction $(1-\alpha)$ will contain the true mean."

For reasonable sampling: the critical value ζ is not very sensitive to N, but very sensitive to α. So for a given α, the factor $\frac{1}{\sqrt{N}}$ dominates. There is a widely used rule-of-thumb, "The 95% confidence interval is roughly $\pm 2\frac{s}{\sqrt{N}}$, that is, ± 2 standard errors." This is good for $N \approx 25$ and greater.

Figure 3.11 shows the sample mean and sample standard deviation versus N, assuming the true σ is unknown. The confidence intervals are $\pm$ 2 standard errors $\left(\pm 2\frac{s}{\sqrt{N}}\right)$. The samples are taken from $\mathcal{G}(0,\sigma^2)$ with $\sigma = 5$. Of the 59 samples taken, 2 do not include the true mean.

Problem 3.40 Replicate Figures 3.10 but with $x \sim \mathcal{U}(a,b)$. (Start by converting the uniform deviate to have the same mean and variance.) What is the effect of changing distributions?

Problem 3.41 Same as Problem 3.40 but for Figure 3.11.

Table 3.3. *For μ: Half-width of Confidence Intervals about the Sample Mean*

Interval Half-Width	Confidence Level $1-\alpha$							
$\zeta_{\alpha,N}$	$N \to \infty$	$N=100$	$N=50$	$N=25$	$N=15$	$N=10$	$N=5$	$N=3$
1.0	0.6824	0.6803	0.6778	0.6727	0.6657	0.6566	0.6261	0.5774
1.5	0.8661	0.8632	0.8600	0.8533	0.8442	0.8321	0.7920	0.7276
2.0	0.9542	0.9518	0.9489	0.9431	0.9347	0.9234	0.8839	0.8165
2.5	0.9874	0.9859	0.9842	0.9803	0.9745	0.9661	0.9332	0.8704
3.0	0.9972	0.9966	0.9958	0.9938	0.9904	0.9850	0.9601	0.9045

Note: N is the number of data; $(1-\alpha)$ is the confidence level. For each value of N, there are five intervals in units of $\frac{s}{\sqrt{N}}$. These are properties of the Student's t distribution.

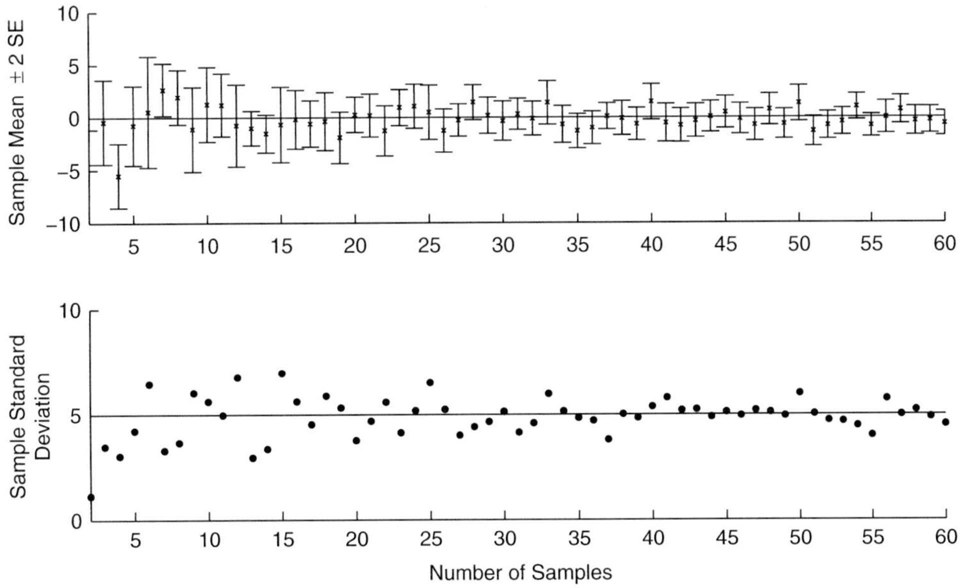

Figure 3.11. Sample means and standard deviations, versus N, from $\mathcal{G}(0,\sigma^2)$. The error bars indicate the confidence interval ± 2 SE; $\sigma = 5$ is used to generate the samples, but is not used in the sample statistics. The standard error is based on s: $\mathrm{SE} = \frac{s}{\sqrt{N}}$. The horizontal lines show true mean and σ. Two of the 59 confidence intervals fail to include the true mean.

3.4.3 Sample Variance

For a general distribution with mean μ, variance σ^2, and central moments[5] M_k: The sample variance is an unbiased estimator of the true variance

$$E[s^2] = \sigma^2 \tag{3.166}$$

and the variance of the sample variance is (Kenney and Keeping [226]; Weisstein [473])

$$Var(s^2) = \frac{1}{N}\left[M_4 - \frac{N-3}{N-1}M_2M_2\right] = \frac{\sigma^4}{N}\left[Kur_e + \frac{2N}{N-1}\right] \tag{3.167}$$

(*Kur* is the kurtosis, a metric of the pointedness or steepness of a distribution; Equation 3.19. Kur_e is the excess kurtosis relative to the Gaussian, $Kur - 3$.)

Problem 3.42 For the sample variance: What is its variance in terms of σ, N, and *Kur*?

Problem 3.43 For the sample variance: What is the coefficient of variation, in terms of σ, N, and *Kur*?

Problem 3.44 For the statistic S^2: What is the mean, variance, and standard error?

Gaussian Samples. In the special case $x_i \sim \mathcal{G}(\mu, \sigma^2)$, the moments $M_2 = \sigma^2$, $M_4 = 3\sigma^4$, and $Kur = 3$ (no excess kurtosis, $Kur_e = 0$). The mean and variance of the sample variance:

$$E(s^2) = \sigma^2 \qquad Var(s^2) = \left[\frac{2}{N-1}\right]\sigma^4 \tag{3.168}$$

The coefficient of variation and standard error:

$$CV_{s^2} = \sqrt{\frac{2}{N-1}} \qquad SE_{s^2} = \sigma^2\sqrt{\frac{2}{N(N-1)}} \tag{3.169}$$

The higher moments of s^2:

$$M_3(s^2) = 8\frac{(N-1)}{N^3}\sigma^6\left[\frac{N}{N-1}\right]^3 \tag{3.170}$$

$$M_4(s^2) = 12\frac{(N-1)(N+3)}{N^4}\sigma^8\left[\frac{N}{N-1}\right]^4 \tag{3.171}$$

5 $M_k = E[(x-\mu)^k]$; $M_2 = \sigma^2$; $Kur = M_4/M_2^2 = M_4/\sigma^4$. $Kur_e = Kur - 3$ is the excess kurtosis relative to the Gaussian value $Kur = 3$

The distribution for the ratio $\frac{s^2}{\sigma^2}$:

$$(N-1)\frac{s^2}{\sigma^2} \sim \chi^2(N-1) \tag{3.172}$$

Accordingly, the two-sided confidence interval is the integral of the χ^2 PDF:

$$P\left[\frac{\chi^2(N-1)|_{\frac{\alpha}{2}}}{N-1} \leq \frac{s^2}{\sigma^2} \leq \frac{\chi^2(N-1)|_{1-\frac{\alpha}{2}}}{N-1}\right] = 1-\alpha \tag{3.173}$$

These intervals narrow more slowly with N than their counterparts for the mean. The half-width for s remains above 10% beyond $N = 150$ (Table 3.4). In Figure 3.12 we generate and plot samples from a Gaussian, generally confirming the intervals in Equation 3.173.

Confidence Intervals. The same interval and sense of confidence can be restated as an interval for estimating σ^2. Given the sample statistic s^2, with probability $(1-\alpha)$, σ^2 is on the interval

$$s^2\left[\frac{N-1}{\chi^2(N-1)|_{1-\frac{\alpha}{2}}}\right] \leq \sigma^2 \leq s^2\left[\frac{N-1}{\chi^2(N-1)|_{\frac{\alpha}{2}}}\right] \tag{3.174}$$

The 95% confidence intervals ($0.025 \rightarrow 0.975$) are calculated in Table 3.5. A set of realizations is illustrated in Figure 3.13.

Problem 3.45 For the statistic S^2 based on sampling the Gaussian:

1. What is the distribution?
2. What is the mean, variance, and standard error?
3. What are the moments?

Non-Gaussian Samples. The results above assume sampling from a Gaussian distribution. More generally, If the population sampled is not Gaussian, we have from Equation 3.167:

$$E[s^2] = \sigma^2 \tag{3.175}$$

$$Var[s^2] = \sigma^4\left[\frac{2}{N-1} + \frac{Kur_e}{N}\right] \tag{3.176}$$

So we still have the $\sqrt{\frac{1}{N}}$ factor narrowing of the sample variance's coefficient of variation, as in the Gaussian case (Equation 3.169). Further development requires specification of the sampled distribution.

Table 3.4. *Two-Sided Intervals for s^2/σ^2, Equation 3.173 based on the χ^2 Distribution*

N	Confidence 95% $\le \frac{s^2}{\sigma^2} \le$	
10	0.3000	2.1136
20	0.4688	1.7291
30	0.5533	1.5766
40	0.6065	1.4903
50	0.6440	1.4331
70	0.6946	1.3602
100	0.7410	1.2972
150	0.7859	1.2395
250	0.8321	1.1831
500	0.8798	1.1278

Note: These intervals are asymmetric; the mean is 1 (unbiased). N is the sample size.

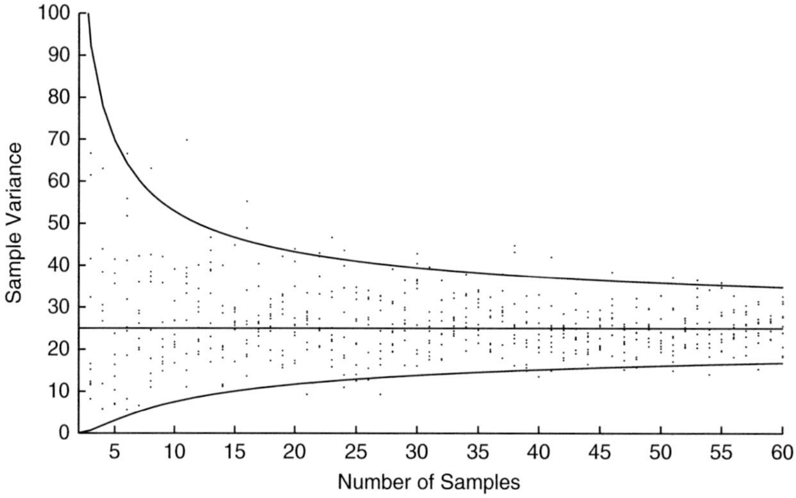

Figure 3.12. Sample variances versus N, from $\mathcal{G}(0,5^2)$. There are 590 realizations, 10 for each value of N. The three solid lines are the true mean σ^2, and the confidence intervals according the the χ^2 distribution, Equation 3.173. Thirty-three of the sample variances fell outside the 95% interval.

A reasonable, approximate interval for the occurrence of s^2 would be based on the statistic $\frac{s^2}{\sigma^2}$:

$$E\left[\frac{s^2}{\sigma^2}\right] = 1 \tag{3.177}$$

$$Var\left[\frac{s^2}{\sigma^2}\right] = \left[\frac{2}{N-1} + \frac{Kur_e}{N}\right] \equiv SE^2 \tag{3.178}$$

Table 3.5. *Two-Sided Confidence Intervals for $\frac{\sigma^2}{s^2}$, Equation 3.174, Based on the χ^2 Distribution*

N	Confidence 95%	
	$\leq \frac{\sigma^2}{s^2} \leq$	
10	0.4731	3.3329
20	0.5783	2.1333
30	0.6343	1.8072
40	0.6710	1.6487
50	0.6978	1.5528
70	0.7352	1.4398
100	0.7709	1.3495
150	0.8068	1.2724
250	0.8452	1.2018
500	0.8867	1.1367

Note: These intervals are asymmetric; the mean is $\frac{\sigma^2}{s^2} = 1$. (N is the sample size.)

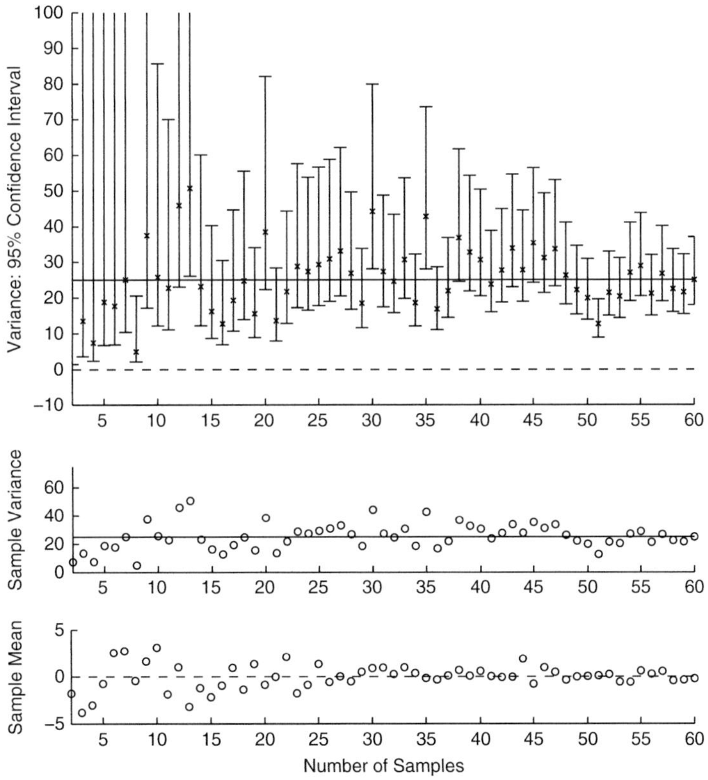

Figure 3.13. Confidence intervals for estimating variance (σ^2), based on samples of $\mathcal{G}(0,5^2)$, sample size N. Shown is the interval based on the χ^2 distribution as in Equation 3.174. Very small N does poorly, with very wide intervals. Overall, of the 58 intervals realized, 5 failed to capture the true variance.

Applying the "2 SE" approximation we get an approximate interval

$$1 - 2\ \mathrm{SE} \leq \frac{s^2}{\sigma^2} \leq 1 + 2\ \mathrm{SE} \tag{3.179}$$

The half-width 2 SE is a metric of the precision possible for individual sample variances. If the assumptions hold, we can expect roughly 95% of sample variances to fall within this interval. N needs to be reasonably large (> 9 for $Kur_e = 0$), or else this approximation fails (interval includes negative s^2).

This approximation is explored in Problem 3.46 *for Gaussian samples*. The correspondence with the more precise χ^2 version (Equation 3.173, reported in Table 3.4 and Figure 3.12) is surprisingly good when N is not too small.

Problem 3.46 Using the ± 2 SE as in Equation 3.179, construct (using *Gaussian* samples) the analog of the χ^2 intervals reported in Table 3.4 and Figure 3.12. Compare.

Problem 3.47 Same as Problem 3.46, but take the samples from the *lognormal* distribution. Construct the equivalents of Table 3.4 and Figure 3.12. What do you expect? Why? Compare your results with those for Problem 3.46. Explain.

Problem 3.48 Same as Problem 3.47 but start with the *uniform* distribution. Compare results with (1) the text and (2) your own lognormal results from Problem 3.47.

Confidence Intervals. Inverting Equation 3.179 gives a confidence interval based on the ± 2 SE rule of thumb:

$$\frac{s^2}{1 + 2\sqrt{\frac{2}{N-1} + \frac{Kur_e}{N}}} \leq \sigma^2 \leq \frac{s^2}{1 - 2\sqrt{\frac{2}{N-1} + \frac{Kur_e}{N}}} \tag{3.180}$$

This approximate interval is reported in Table 3.6 for $\frac{\sigma^2}{s^2}$ versus N, for $Kur_e = 0$. Experiments are plotted in Figure 3.14.

The very small N cases have confidence intervals that include negative numbers, clearly impossible. And, because the intervals given are proportional to the sample variance, there is the weakness that a small variance appears to be especially precise – a good case of being wrong and being very certain of it.

By comparison, the interval estimates based on χ^2, for *Gaussian* samples (Figure 3.13, Table 3.5, Equation 3.174) are asymmetric and the lower limit is always positive; it performed better in these experiments. (Problems 3.47 and 3.48 above explore the use of χ^2 intervals when the samples are not Gaussian.) There remains the problem that when s is small, so is the confidence interval.

Table 3.6. *Two-Sided Confidence Intervals about the Mean for $\frac{\sigma^2}{s^2}$, Based on the "±2 SE" Approximation 3.180*

N	Confidence, ±2 SE	
	$\leq \frac{\sigma^2}{s^2} \leq$	
10	0.5147	17.4853
20	0.6065	2.8481
30	0.6556	2.1063
40	0.6883	1.8279
50	0.7122	1.6780
70	0.7460	1.5163
100	0.7787	1.3972
150	0.8119	1.3016
250	0.8480	1.2184
500	0.8876	1.1450

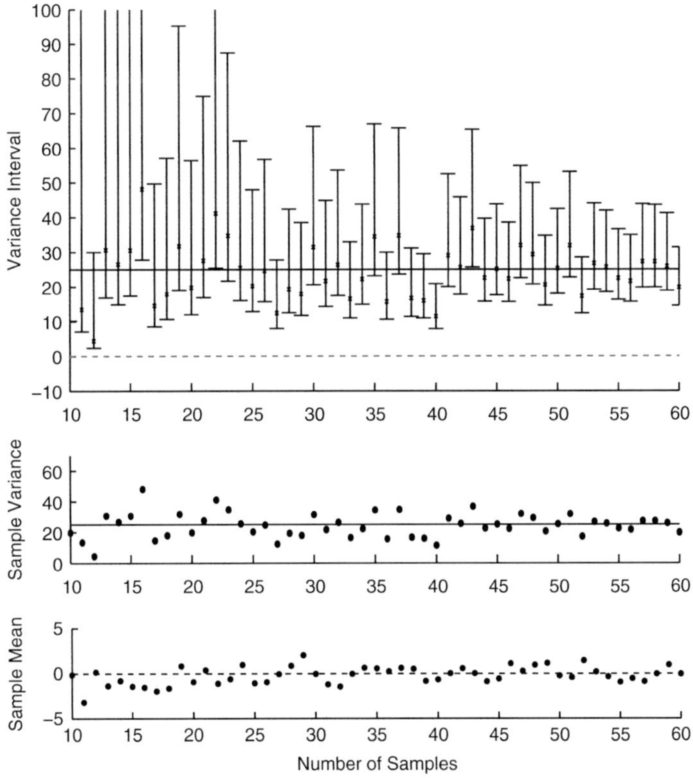

Figure 3.14. Approximate confidence intervals for estimating variance, based on samples of $\mathcal{G}(0,5^2)$, sample size N. Shown is the interval based on the "±2 SE" distribution as in Equation 3.180. Of the 51 intervals shown, 4 failed to capture the true mean σ^2. If we count the bogus negative intervals below $N = 10$, there were 8 extra failures; so a total of 12 failures out of 59. Selected data are reported in Table 3.6.

Problem 3.49 For non-Gaussian samples, confirm the approximate interval in Equation 3.180. How large must N be to keep the confidence limits always positive?

Problem 3.50 Same as Problem 3.49, but plot the minimum N versus *Kur*.

Problem 3.51 For $Kur = 5$: redo Figure 3.14 and interpret.

3.4.4 Recap

By analogy with physical experiments: "How many particles should one simulate?" is analogous to the question "How big need the sample be?" when devising experiments.

For both mean and variance, we have unbiased estimators that make common sense. For the mean, it is the simple sample average. For the variance, it is the sample variance s^2, with the $N-1$ factor in the denominator, a well-known adjustment to eliminate bias at small N.

The confidence interval is a *metric of precision.*

- On *Input*, one always knows what distribution is being sampled. The only question is, How do the sample properties differ from the true properties?
- On *Output*, one never knows what the distribution is. The *simulation model* is the generator; the job is to characterize its output in terms of mean and variability.
- Reliance on the central limit theorem is widespread; the established Gaussian sampling results provide a reasonable approximation in the absence of more information.

The interval ± 2 SE recurs in many analyses as a roughly 95% metric of precision. This is one of van Belle's [457] rules of thumb. For both mean and variance, the SE varies as $\frac{1}{\sqrt{N}}$. Precision improves as N increases.

The rate of interval reduction is slow for large N, placing practical limits on precision. Again, van Belle's [457] rule of thumb: make $N \geq 12$ as a practical compromise between large N and low precision.

For the sample mean, the known distribution is symmetric, Student's t. For the sample variance, the χ^2 distribution governs when the samples are from the Gaussian. More generally, we have the moments of sample variance, and can construct intervals based on the ± 2 SE rule.

3.5 Covariance

So far we've examined the ideal case of independent occurrences. This is the default. Unless there is a mechanism (hypothesized or accepted or evident in observation or theory), a justification for departing from independence is lacking.

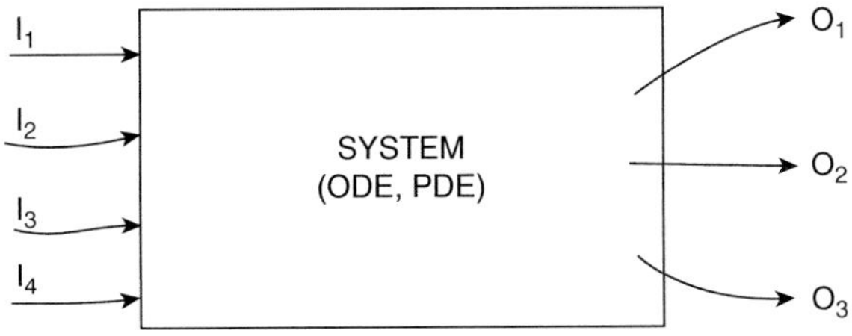

Figure 3.15. System with independent inputs. Covariance among outputs is created by the system processing.

There is a wide literature for reducing inputs to the smallest independent set possible. This is fundamentally an abstract, mathematical definition of variables; they may or may not correspond to natural quantities. This minimal set may be mapped to a more meaningful set, with the same number of different variables that may acquire covariance in the process. Or, it may be enlarged for convenience – maybe by adding some meaningful natural quantities, or those conveniently measurable. The interdependence of these "mapped" or "extra" input variables is assumed known. Mechanistic relations among variables – the "laws of motion" – create a correlation among outputs (Figure 3.15.) In the present context, it is the job of simulation to work these out.

Below we first establish some basic nomenclature, then discuss two kinds of statistical covariance:

- Temporal (Lagrangian if identified with a moving particle)
- Spatial (Eulerian; individual particles move through spatially covarying fields)

3.5.1 Definitions

The relevant definitions are generalizations of the variance definition provided earlier. The covariance of two random items x_1, x_2 is the *expected value of their demeaned product*:

$$Cov(x_1,x_2) = E\left[(x_1 - \mu_{x_1})\,(x_2 - \mu_{x_2})^*\right] \tag{3.181}$$

$$= E\left[\widehat{x_1}\,\widehat{x_2}^*\right] \tag{3.182}$$

When $x_2 = x_1$ this reduces to the definition of $Var(x)$.

Some care is useful at the start with respect to the form of the product.

- The asterisk ($*$) implies the complex conjugate.
- For complex quantities, the complex conjugate of x_2 is used. That is the sense of the comma notation on the left side.

- For convenience below, we drop the hatted notation, effectively assuming that x_i has already been demeaned.

More generally, for the vector $\mathbf{x}$ with m random components x_i, we have a square $(m \times m)$ matrix of scalar quantities

$$[Cov(\mathbf{x})] = \left[Cov(x_i, x_j)\right] \tag{3.183}$$

On the diagonal, we recover the variance, always real:

$$Cov(x_i, x_i) = Var(x_i) = \sigma_{x_i}^2 \tag{3.184}$$

The off-diagonals terms are the interactions. If they all vanish, then the x_i are completely independent. They are generally complex, with the Hermitian property

$$Cov(x_j, x_i) = Cov^*(x_i, x_j) \tag{3.185}$$

$[Cov(\mathbf{x})]$ is symmetric for real $\mathbf{x}$. The sum of diagonals of $[Cov(\mathbf{x})]$ (its Trace) is the sum of variances of all individual x_i, always real:

$$Tr\left([Cov(\mathbf{x})]\right) = \sum_{i=1}^{m} \sigma_{x_i}^2 \tag{3.186}$$

The transpose of $[Cov(\mathbf{x})]$ is its complex conjugate.

$$[Cov(\mathbf{x})]^T = [Cov(\mathbf{x})]^* \tag{3.187}$$

Generally, any deterministic linear transformation of $\mathbf{x}$ into $\mathbf{y}$ creates a covariance "sandwich":

$$\{\mathbf{y}\} = [A]\{\mathbf{x}\} \tag{3.188}$$

$$\left[Cov(\mathbf{y})\right] = E\left([A]\{\mathbf{x}\}\{\mathbf{x}\}^{*T}[A]^{*T}\right) \tag{3.189}$$

$$= [A][Cov(\mathbf{x})][A]^{*T} \tag{3.190}$$

The map $[A]$ need not be square, as when a set of x_i are mapped into a larger set of y_j. If $\mathbf{x}$ are all independent, then $[Cov(\mathbf{x})]$ is diagonal with entries $\sigma_{x_i}^2$.

Problem 3.52 There are five random variables, x_i. The set $\mathbf{x}$ is observed six times and the observations are as follows:

$$\mathbf{x} = \{5, -2, \ 7, \ 4, \ 2\} \tag{3.191}$$

$$\mathbf{x} = \{6, -2, \ 8, \ 4, \ 0\} \tag{3.192}$$

$$\mathbf{x} = \{6, -1, \ 7, \ 5, \ 2\} \tag{3.193}$$

$$\mathbf{x} = \{4, -2, \ 5, \ 4, \ 1\} \tag{3.194}$$

$$\mathbf{x} = \{5, -1, \ 6, \ 5, \ 2\} \tag{3.195}$$

$$\mathbf{x} = \{5, -3, \ 7, \ 5, \ 0\} \tag{3.196}$$

Using this data, estimate $[Cov(\mathbf{x})]\,1$.

Problem 3.53 Same as Problem 3.52 but for the three random variables y_i. The six sets of observations are:

$$\mathbf{y} = \{2, \ 1, \ 1\} \tag{3.197}$$
$$\mathbf{y} = \{5, -2, \ 7\} \tag{3.198}$$
$$\mathbf{y} = \{8, \ 4, \ 0\} \tag{3.199}$$
$$\mathbf{y} = \{6, -1, \ 2\} \tag{3.200}$$
$$\mathbf{y} = \{4, -3, \ 1\} \tag{3.201}$$
$$\mathbf{y} = \{5, \ 1, \ 6\} \tag{3.202}$$

Estimate $\left[Cov(\mathbf{y})\right]$.

Problem 3.54 Let $\mathbf{x}$ and $\mathbf{y}$ in Problems 3.52 and 3.53 be completely independent of each other. And let $\mathbf{z}$ be the concatenation of $\{\mathbf{x} \mid \mathbf{y}\}$. What is $[Cov(\mathbf{z})]$?

Problem 3.55 Let $\mathbf{y}$ be as defined in Problem 3.53. Let $\mathbf{z}$ be its transformation below. What is $[Cov(\mathbf{z})]$?

$$z_1 = y_1 + y_2 \tag{3.203}$$
$$z_2 = y_1 + 2y_2 \tag{3.204}$$
$$z_3 = y_1 - y_2 \tag{3.205}$$
$$z_4 = y_2 - y_3 \tag{3.206}$$
$$z_5 = -y_3 \tag{3.207}$$

3.5.2 Correlation and Autocorrelation

The correlation among x_i and x_j is their covariance, normalized by their standard deviations. If all x_i have uniform σ_x, and similarly for y,

$$Cor(x_i,x_j) = \frac{Cov(x_i,x_j)}{\sigma_{x_i}\sigma_{x_j}} \tag{3.208}$$

This is often referred to as the *correlation coefficient*; it is dimensionless, but generally complex unless $i = j$. The $m \times m$ *correlation matrix* is the normalized covariance matrix. The main diagonals are all $= 1$, real. The off-diagonals are less than 1 in magnitude, unless one variable is a replicate of another.

Problem 3.56 For Problems 3.52, 3.53, 3.54: What is the correlation matrix in each case?

When the components of $\mathbf{x}$ share a special identity (such as the ordered members of a single timeseries, or the values of a single Eulerian field), we have *autocorrelation* (self-correlation).

3.5.3 Autocorrelated Time Series

Suppose the variables x_i represent a sequence of observations of the same time-varying quantity, separated by uniform Δt. Then $\mathbf{x}$ is a discrete time series with potential for autocorrelation. Embedded in the notion of "time series" is a direction for the sequence – there is a "forward" direction, and "initial conditions" that propagate forward in time, setting the future into play.

For example, consider the simple first-order difference equation

$$x_i = \rho\, x_{i-1} + \epsilon_i \tag{3.209}$$

The random number x_i blends the memory of its predecessor, plus a new independent random number ϵ_i ("noise"). The scalar ρ is the memory term, assumed constant. ϵ_i is statistically constant with mean zero, variance σ^2, and no correlation to anything.

It is straightforward to run this series forward from x_0:

$$x_1 = \rho x_0 + \epsilon_1 \tag{3.210}$$

$$x_2 = \rho^2 x_0 + \rho\epsilon_1 + \epsilon_2 \tag{3.211}$$

$$x_3 = \rho^3 x_0 + \rho^2\epsilon_1 + \rho\epsilon_2 + \epsilon_3 \tag{3.212}$$

$$\vdots \tag{3.213}$$

$$x_n = \rho^n x_0 + \sum_{i=1}^{n} \rho^{n-i}\epsilon_i \tag{3.214}$$

x retains its initial condition plus all the intervening ϵ inputs as a weighted sum, accumulating noise as time proceeds. For $|\rho| < 1$ the initial value and the older inputs are eventually forgotten and the series converges, statistically. This is the "mature" time series, free from initial conditions and early noise, retaining only the more recent noise.

Figure 3.16 (Left) shows four random sequences generated with $\epsilon \sim \mathcal{U}[-.5,.5]$ and no memory. As expected, no pattern is visible except the properties of $\mathcal{U}$ itself. Figure 3.16 (right) is the same but memory is added, $\rho = 0.98$. There are three points to note:

- The sequences are smoother, and a time scale of smoothing is apparent.
- Each sequence is noticeably different – for example, when x becomes negative, it stays so for a while.

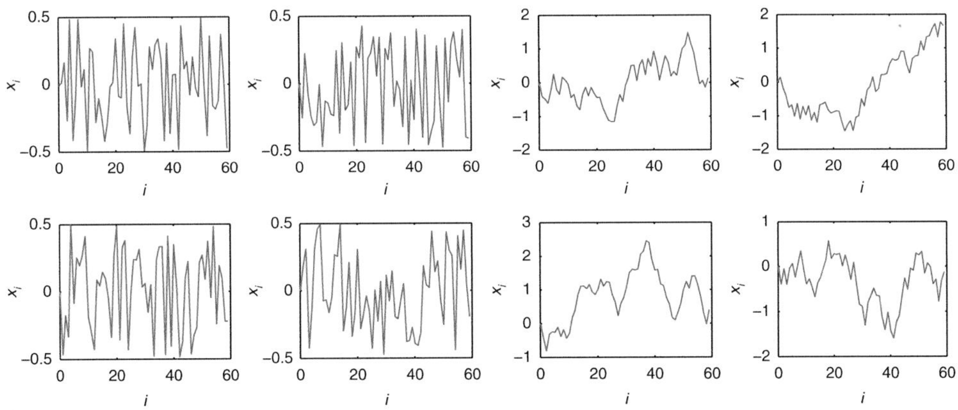

Figure 3.16. Random series generated as in Equation 3.209 with $\epsilon \sim \mathcal{U}[-0.5, 0.5]$ and $x_0 = 0$. Left: $\rho = 0$ (no memory). Right: $\rho = 0.98$.

- All sequences start at $x = 0$ and there is an initial transient during which this is retained.
- The overall size of x is larger once the initial transient is over.

These features are easily gotten. First, the expectation of Equation 3.209 reveals the mean x remains zero if so initialized. Second, multiplying Equation 3.209 by x_{i-1} we have

$$x_i x_{i-1} = \rho\, x_{i-1} x_{i-1} + \epsilon_i x_{i-1} \tag{3.215}$$

and taking the expectation,

$$Cov(x_i, x_{i-1}) = \rho \sigma_x^2 \tag{3.216}$$

Repeating this procedure with Equation 3.214 yields the generality

$$Cov(x_n, x_{n-k}) = \rho^k \sigma_x^2 \tag{3.217}$$

For the fully mature trajectory, we have $\rho^n \to 0$; squaring Equation 3.214 and taking expectation, with all ϵ independent,

$$\sigma_x^2 = \sigma_\epsilon^2 \sum_{i=0}^{\infty} (\rho^i)^2 = \sigma_\epsilon^2 \left[\frac{1}{1-\rho^2} \right] \tag{3.218}$$

Hence the input noise is amplified by the memory as suggested in Figure 3.16. For that case ($\rho = 0.98$) we find $\sigma_x \simeq 5\sigma_\epsilon$, consistent with the figure.

In terms of memory lapse: initial conditions decay in importance as ρ^k; the half-life H (number of time steps) is $\rho^H = 1/2$.

$$H = \frac{\ln(0.5)}{\ln(\rho)} \tag{3.219}$$

For the case in Figure 3.16 ($\rho = 0.98$), $H \simeq 34$, giving some scaling for the time axis. The same persistence pertains to any ϵ once introduced.

Needless to say, the parameters ρ and σ_ϵ must depend on the time step Δt. ρ is a decay *rate*; and ϵ reflects the *accumulation* of noise over Δt. We can expect as Δt decreases, that $\rho \to 1$ and ϵ gets smaller.

We explore this matter in a later chapter on random walk models (Chapter 4); for now, we note that the simple model we use here, for tiny Δt, converges to the ordinary differential equation (ODE)

$$\left(\frac{d}{dt} + \frac{1}{T} \right) x = \mathcal{R} \tag{3.220}$$

Simple Euler differencing leads to the discrete Equation 3.209, with memory term $\rho = \left(1 - \frac{\Delta t}{T}\right)$; and ϵ_i representing the noise increment $\int_{t_{i-1}}^{t_i} \mathcal{R}\, dt$. The latter is handled later by modeling the noise $\mathcal{R}$ as a Wiener process, with increment $\epsilon = |\mathcal{R}|\sqrt{\Delta t}$.

If this equation governs in the limit, then T is a physical parameter and the dimensionless quantity $\Delta t/T$ is the relevant discretization factor. When this metric is large, the noise dynamics cannot be resolved.

Problem 3.57 Redo the equivalent of Figure 3.16 with $\epsilon \sim \mathcal{G}(2,5)$. Are the results consistent with the text analysis? How do they differ from the text case where $\mathcal{U}$ was used?

3.5.3.1 Separation-Based Covariance and Correlogram

The preceding example illustrates a special case of interest: when autocorrelation depends solely on the separation between x_i and x_j. Specifically, let the separation "distance" be $|i-j|$.

$$Cov(x_i, x_j) = \rho^{|i-j|}\sigma_x^2 \tag{3.221}$$

The absolute values of i and j play no role, only the "distance" $r = |i-j|$ separating them. There are various ways to express this; for example, for this case,

$$Cor_{x,x}(r) = \rho^r \tag{3.222}$$

The plot of $Cor_{x,x}$ versus r is the *correlogram*, illustrated in Figure 3.17.

The covariance matrix for this distance-based formulation is

$$[Cov_x] = \sigma_x^2 \begin{bmatrix} 1 & \rho & \rho^2 & \rho^3 & \rho^4 & \cdots & \cdots & \cdots & \cdots & \cdots \\ \rho & 1 & \rho & \rho^2 & \rho^3 & \rho^4 & \cdots & \cdots & \cdots & \cdots \\ \rho^2 & \rho & 1 & \rho & \rho^2 & \rho^3 & \rho^4 & \cdots & \cdots & \cdots \\ \rho^3 & \rho^2 & \rho & 1 & \rho & \rho^2 & \rho^3 & \rho^4 & \cdots & \cdots \\ \rho^4 & \rho^3 & \rho^2 & \rho & 1 & \rho & \rho^2 & \rho^3 & \rho^4 & \cdots \\ \vdots & \vdots & \vdots & \vdots & \vdots & \vdots & \vdots & \vdots & \vdots & \ddots \end{bmatrix} \tag{3.223}$$

The correlogram appears in each row or column of this symmetric matrix, beginning on the diagonal.

Several forms of this distance-based correlation have been found useful in oceanographic contexts (Lynch [279]). Typical is the use of a geometric or exponential decay with distance r, coupled to a polynomial or other special function in r. The purely geometric decay ρ^r used in the preceding is perhaps the simplest, essentially an exponential in the small-Δt limit. Another form is

$$Cor_{x,x} = (1 + \alpha r)\, e^{-\alpha r} \tag{3.224}$$

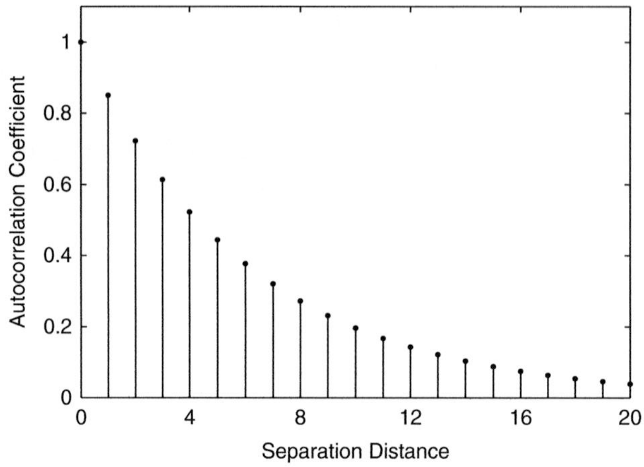

Figure 3.17. Correlogram example for $\rho = 0.85$.

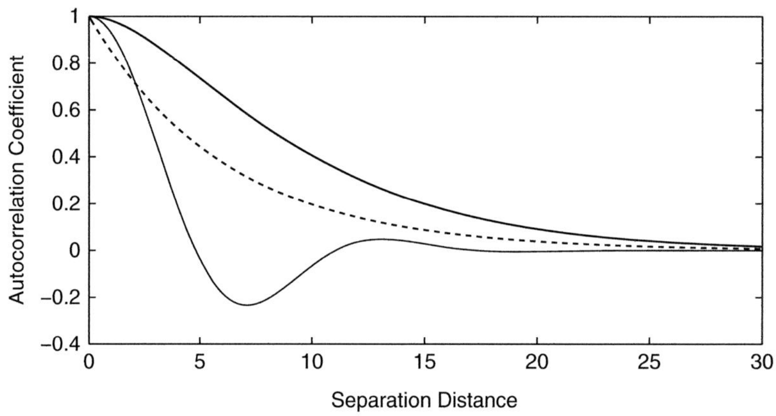

Figure 3.18. Correlogram examples for three distance-based forms: $(1+\alpha r)e^{-\alpha r}$ (top); ρ^r (middle, dashed line); and $J_0(\beta r)e^{-(\gamma r)^2}$ (bottom). Parameters: $\rho = 0.85$; $\alpha = 0.2$; $\beta = 0.5$; $\gamma = 0.1$.

which makes the correlogram smooth at the origin. Yet another form employing the Bessel function is:

$$Cor_{x,x} = J_0(\beta r)e^{-(\gamma r)^2} \tag{3.225}$$

which has an oscillatory pattern. r itself is typically a dimensionless measure of distance, and the normalizing length may be made locally variable in these formulations. Quite a bit of detail is possible in the choice of functional form and its parameterization. Examples of correlograms are shown in Figure 3.18.

These various forms represent different assumptions about the processes generating the random variable x. They may be reached on physical grounds, in terms of difference or differential equations, or in terms of observation. In every case the choice is substantive, a statement about the nature of randomness present.

These forms represent *unbounded* solutions. Their occurrence in timeseries analysis is natural (but not necessary). Appropriate initial conditions are necessary to get started, and the memory parameters inject a transient after which the series is "mature," independent of its starting point.

It is important to avoid a terminological trap here – "distance" or "separation" makes natural sense in many contexts, but generally it is "pseudo-distance" that is the defining idea here, whether in time, space, or both. The pseudo-distance includes anisotropic metrics of separation; it converts multidimensional situations into one-dimensional ones, a huge and very practical simplification.

3.5.3.2 Correlogram and Impulse Response

For linear, time-independent systems it is useful to describe the the **impulse response** series $[h_0, h_1, h_2, h_3, \dots]$, where h_i is the system response i time steps following a unit input.[6] If the forcing series is ϵ_i, then by superposition we have the composite response as the convolution of all prior inputs:

$$x_j = \epsilon_j h_0 + \epsilon_{j-1} h_1 + \epsilon_{j-2} h_2 + \epsilon_{j-3} h_3 + \epsilon_{j-4} h_4 + \epsilon_{j-5} h_5 + \dots \tag{3.226}$$

$$= \sum_{i=0} \epsilon_{j-i} h_i \tag{3.227}$$

If the ϵ are all independent with uniform properties, mean μ_ϵ, variance σ_ϵ^2, then the properties of x are

$$\mu_x = \mu_\epsilon \sum_{i=0}^{\infty} h_i \tag{3.228}$$

$$\sigma_x^2 = \sigma_\epsilon^2 \sum_{i=0}^{\infty} h_i h_i \tag{3.229}$$

$$Cov(x_j x_{j-k}) = \sigma_\epsilon^2 \sum_{i=k}^{\infty} h_i h_{i-k} \tag{3.230}$$

For lag k, Cov_x is the input variance times the sum of lagged products of the impulse response.

[6] By definition, $h_i = 0$ for $i < 0$.

3.5.4 Autocorrelated Eulerian Fields

The analysis of timeseries has an obvious forward bias in its analysis and generation, and we think in terms of initializing these series or their governing differential equations. In general, however, correlation notions are not limited to timeseries. In many cases, we are concerned with autocorrelation built into a Eulerian field – for example, temperature or wind stress, where we may find correlation in spatial patterns. The same notion of distance-based covariance is useful here as well, and very popular. We will use $\mathbf{x}$ to indicate spatial position, $u(\mathbf{x})$ to indicate an Eulerian field. Distance r will be the (positive) Euclidian distance scaled by the parameter L. There is no preferred direction.

The relationship between a governing equation, its probabilistic forcing, and the covariance it generates is quite general. Two of the correlograms described previously have the corresponding pairs

$$\left(\frac{d}{dx}+\frac{1}{L}\right)u=\mathcal{R} \quad \Rightarrow \quad Cor_{u,u}(r)=e^{-r} \tag{3.231}$$

$$\left(\frac{d^2}{dx^2}-\frac{1}{L^2}\right)u=\mathcal{R} \quad \Rightarrow \quad Cor_{u,u}=(1+r)\,e^{-r} \tag{3.232}$$

and the latter may be generalized to 2D:

$$\left(\nabla^2-\frac{1}{L^2}\right)u=\mathcal{R} \quad \Rightarrow \quad Cor_{u,u}=rK_1(r) \tag{3.233}$$

with K_1 the modified Bessel function of the second kind.[7] In all cases the distance r is the (positive) Euclidian distance separating points, scaled by L.

In dealing with spatial patterns, these solutions represent *unbounded* space; it is natural to exert boundary conditions on finite domains. These add additional detail, as do inhomogeneities in coefficients, etc. Typically one relies on numerical approximations for these continuous fields; for example, the matrix discretization of Equation 3.233 would be

$$[A]\{\mathbf{u}\}=\{\epsilon\} \tag{3.234}$$

with $[A]$ including the boundary conditions proper to the PDE, the inhomogeneities, and so forth, and ϵ a valid representation of the input noise. Under these conditions, $[A]$ is invertible and

$$[Cov(\mathbf{u})]=[A]^{-1}\,[Cov(\epsilon)]\,\left[A^*\right]^{-T} \tag{3.235}$$

[7] For large r, the asymptotic limit is $rK_1(r)\rightarrow\sqrt{\frac{\pi r}{2}}\left(1+\frac{3}{8r}\right)e^{-r}$.

$Cov(\mathbf{u})$ is a sandwich, with the PDE representation A^{-1} enveloping or packaging the noise driving it. As in the 1-D case, the rows are each a correlogram centered on the diagonal. Each correlogram is a 2-D map of correlation, centered on the diagonal of the covariance matrix; each row is the correlation of column j with row i.

When $[A]$ is a discretized PDE, it would be natural to distinguish different *independent* sources of forcing variability ϵ_i. For example, expanding the right side of Equation 3.234 as the sum of independent sources of variability in interior forcing ϵ_0, Dirichlet boundary forcing ϵ_1, and Neumann boundary forcing ϵ_2, we get three covariance components:

$$[Cov(\mathbf{u})] = [A]^{-1}[Cov(\epsilon_0)]\left[A^*\right]^{-T} + [A]^{-1}[Cov(\epsilon_1)]\left[A^*\right]^{-T} + [A]^{-1}[Cov(\epsilon_2)]\left[A^*\right]^{-T} \tag{3.236}$$

and so forth. Each independent source of variability is packaged by the system $[A]$ to produce components of the output covariance.

Example 3.7 Let a 1-D system have covariance-generating ODE as in Equation 3.232, $\left(\frac{d^2}{dx^2} - \frac{1}{L^2}\right) u = \mathcal{R}$. For a finite difference discretization with uniformly spaced

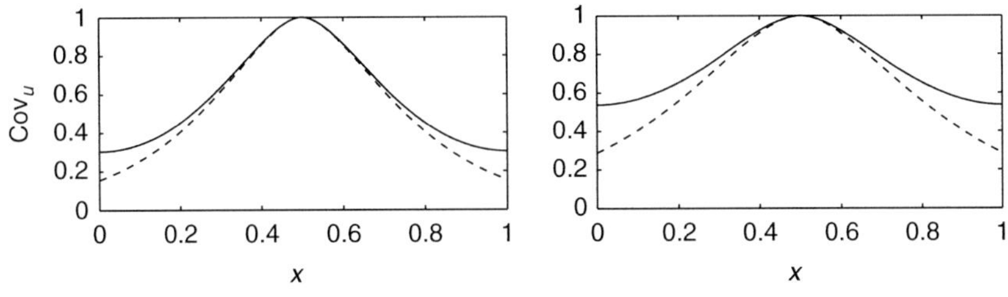

Figure 3.19. Correlograms computed as in Equation 3.235 for the discretized ODE equation 3.232 (Example 3.7). Unit length with 200 equal Δx. BC's: $\frac{du}{dn} = 0$ at both ends. Corellograms shown for node at the center, $x = 0.5$, along with the analytic solution for the unbounded case (dashed line). The panels are for $L = [0.05, 0.10, 0.15, 0.20]$. For small L, the unbounded solution is recovered. For the longer L, the analytic corelogram begins to be nonzero at the boundaries; and the bounded ODE-based correlogram begins to be different.

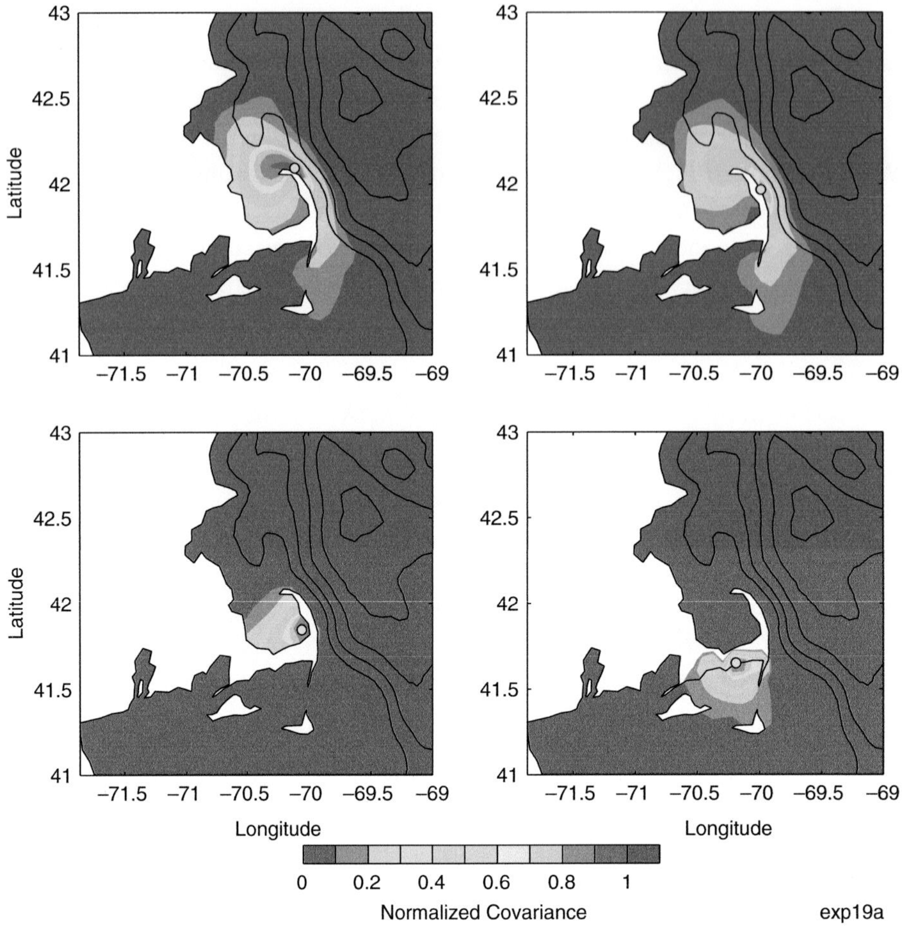

Figure 3.20. Correlogram examples for the PDE-based covariance with landward boundary conditions. Four different correlograms are shown, each centered as indicated. From Lynch and McGillicuddy, [285]. (Reproduced with permission.) For a color version of this figure, please see the color plate section.

nodes i, we get

$$u_{i-1} - \left(2 + \frac{\Delta x^2}{L^2}\right) u_i + u_{i+1} = \epsilon_i \tag{3.237}$$

and this would be row i of A. Make up this matrix for nodes $1 \to N$, subject to homogeneous Neumann boundary conditions at both ends (nodes 1 and N); and invert it. From Equation 3.235 above, with homogeneous independent noise $\epsilon_i \sim \mathcal{G}(0, \sigma_\epsilon^2)$, we get

$$[Cov(\mathbf{u})] = \sigma_\epsilon^2 [A]^{-1} \left[A^*\right]^{-T} \tag{3.238}$$

Each row of this result is a correlogram. A few are plotted in Figure 3.19, along with the unbounded result. They agree nicely unless the covariance extends to the finite boundaries (longer L), where the Neumann conditions enhance the covariance over the unbounded version.

These ideas greatly extend the potential for covariance to be used in problem formulation, such as objective analysis (Bretherton et al. 1976 [51]) and in the generation of random numbers. Lynch [279] presents this idea in a general PDE context; oceanographic applications are shown by McGillicuddy and Lynch [285]. Example 2-D correlograms appear in Figure 3.20. These were computed with the advective-diffusive-reactive equation for $[A]$, and no-flux boundary conditions on land – the latter dominate near shore.

Additional points about the covariance matrix:

- The famous Gauss-Markov Theorem is pervasive in estimation of partly or imperfectly observed fields. The covariances of field variability and of observation error play central roles. Optimal estimation, objective analysis and optimal interpolation all rely on this theorem for the sense of "optimality" or "statistical objectivity" (Wunsch [489]).
- A related item is the variogram for autocorrelated fields, $E[(x-y)^2]$ instead of $E[xy]$. The overlap is evident. The variogram is used extensively in geostatistical studies (field estimation by "kriging") while covariance is used extensively in forecasting.
- In least squares studies, the proper quadratic form to be minimized is $Q = \{\mathbf{u}\}^T [Cov(\mathbf{u})]^{-1} \{\mathbf{u}\}$. Q is the metric of the size of $\mathbf{u}$.

3.5.5 Generating Covariance

For time series autocorrelation, difference equations (discretized ODEs) in time are common. One postulates a process, driven by its own history and an independent series of random numbers. Typically, this results in "distance-based" correlation, that is, the separation in time of two correlates is the only variable that matters. This holds once the timeseries has achieved independence of its initial conditions. There are analytic versions of this autocorrelation.

For example, an autocorrelated sequence x_k can be constructed with the generator

$$x_k = \rho x_{k-1} + \left(\sigma_x \sqrt{1-\rho^2}\right) \mathcal{R}_k \tag{3.239}$$

- σ_x^2 is the unconditional variance of the series x_k.
- $\rho = E(X_{k+1}X_k)/\sigma^2$ is the autocorrelation coefficient of the series x_k.
- $\mathcal{R} \sim \mathcal{G}(0,1)$.

The resulting autocorrelation is $Cor_x(k) = \rho^k$, with k the time interval between two correlates.

There will be an initial transient – prominent when starting from rest with no history. To overcome the latter, let the initial x be random with the long-run variance σ_x^2, that is, $x_0 \sim \mathcal{G}(0, \sigma_x^2)$, and proceed as earlier. Effectively, this embeds in x_0 a fictitious history for $k < 0$.

For Eulerian fields, a similar distance-based correlation is popular and analytic forms are available. These assume an unbounded domain. In many practical situations, covariance near a boundary cannot be ignored, and the importance of the boundary conditions becomes critical. Under these conditions, the discretized PDE and its boundary conditions lead to a well-conditioned matrix representation of the system.

For example: let **u** be a linear transformation of ϵ:

$$\{\mathbf{u}\} = [L]\{\epsilon\} \tag{3.240}$$

$$[Cov(\mathbf{u})] = [L][Cov(\epsilon)][L]^{*T} \tag{3.241}$$

To *generate* a random set **u**: start by generating independent deviates ϵ.

- If we know L, then Equation 3.240 is simple to implement. The covariance from Equation 3.241 will result. When the generating PDE is known in discrete form $[A]$ as in Example 3.7, $[L] = [A]^{-1}$.
- If instead, we know what covariance $Cov(\mathbf{u})$ is desired, then $[L]$ needs to be:

$$\sigma_\epsilon^2 [L][L]^{T*} = [Cov(\mathbf{u})] \tag{3.242}$$

$$[L] = \frac{1}{\sigma_\epsilon}[Cov(\mathbf{u})]^{1/2} \tag{3.243}$$

The symmetry of the covariance matrices makes the square root meaningful here. Two proven methods for the matrix square root are the Cholesky decomposition and the singular value decomposition. Given $Cov(\mathbf{u})$, we find its "square root"; and multiply it by an uncorrelated set of deviates ϵ as in Equation 3.240. The result **u** will have the desired covariance.

In either case, the generation of independent $\epsilon \sim \mathcal{G}(0, \sigma_\epsilon^2)$ is straightforward.

Example 3.8 For the case in Example 3.7, we know $[L]$ since we know $[A]^{-1}$. Some **u** generated with this procedure appear in Figures 3.21 and 3.22.

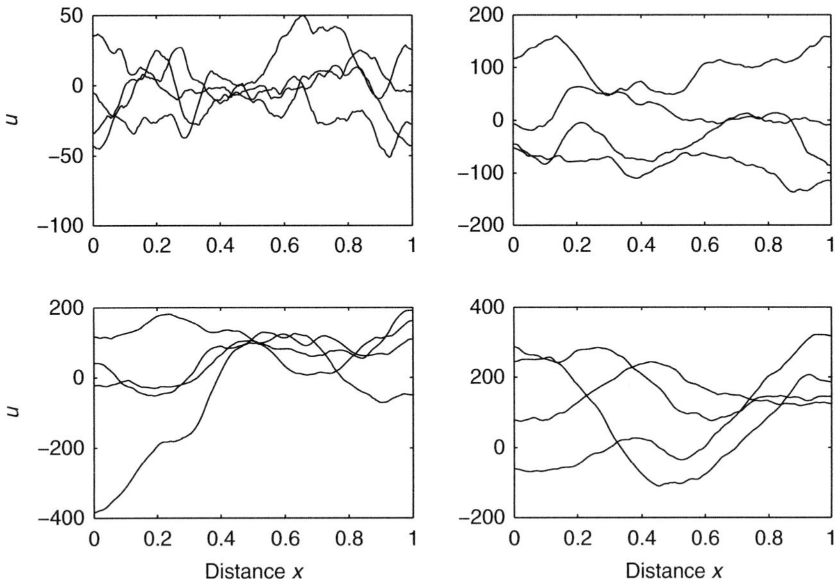

Figure 3.21. Random numbers u generated with ODE-based covariance from Example 3.8. Each pane shows four realizations of $u(x)$. The governing equation is the 1-D Helmholtz equation; $[A]$ is its discrete form (Equation 3.237) with 200 Δx separating Neumann boundaries. The four different panes use $L = [0.05, 0.10, 0.15, 0.20]$. In every case, the nodal $\epsilon \sim \mathcal{G}(0,1)$. Note: as L increases, the size of u and its smoothness both increase.

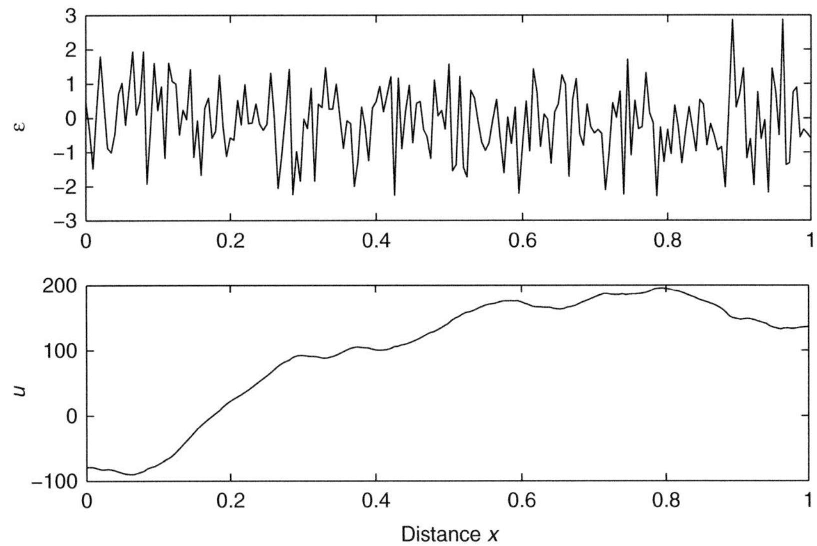

Figure 3.22. As in Figure 3.21, showing nodal ϵ (top) and u (bottom) for one case of $L = 0.15$. There are $200\Delta x$.

Problem 3.58 A 1-D string of equally spaced nodes has analytic distance-based covariance $Cor_{u,u} = (1+r)e^{-r}$, where r is the separation distance (number of Δx) scaled by L. Boundary conditions are unknown, so this is the unbounded covariance. Using this analytic function:

1. Make the covariance matrix for a string of length 100 (i.e., 100 nodes). Ignore the unknown boundary conditions.
2. Generate and plot 10 strings of random numbers $\mathbf{x}$ for arbitrary L, driven by $\epsilon \sim \mathcal{G}(0,1)$.
3. Do you results make sense? What is the effect of L?
4. The text (Example 3.8) used the bounded PDE; here you are using the unbounded analytic solution, in a bounded context. Expect artifacts near the boundaries. Do you see this?

3.5.6 Summary – Covariance

In population studies, abundance and distribution are commonly used to characterize survey data. Abundance corresponds roughly with the spatial mean, while distribution corresponds roughly with the spatial variance.

Covariance among variables is typically avoided on input if possible, but is a necessary feature on output. The intervening system is the covariance creator.

- Covariance is a generalization of variance.
- Correlation is normalized covariance.
- Autocorrelation is self-correlation, for example, among members of the same timeseries or samples of the same Eulerian field.
- A common model of autocorrelation depends on scaled separation distance ("pseudo-distance") between correlates, not on absolute position per se.

Analytical models of autocorrelation arise in differential equations (DEs). Uncorrelated noise, processed, bears the autocorrelation trademark of the DE process.

- Unbounded models are common, reflecting unbounded domains of the DE.
- Timeseries autocorrelation may be generally conceived as being generated by an ODE, with initial conditions.
- Eulerian field autocorrelation may be conceived as generated by a PDE, with boundary conditions.

General PDE models may be formulated numerically with stochastic forcing, boundary conditions, and inhomogeneities.

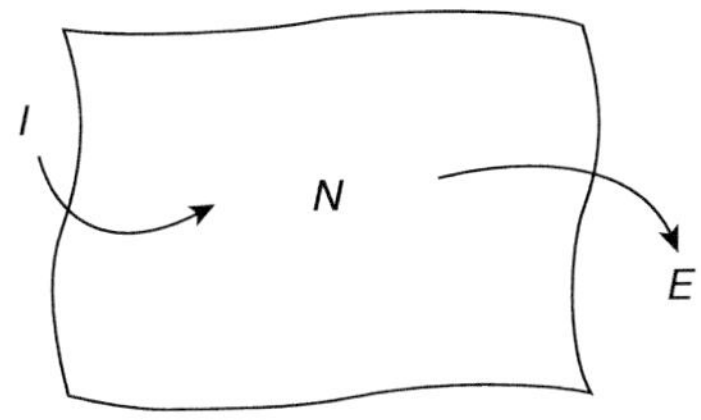

Figure 3.23. N particles in a box. For an individual in the box, the retention probability is P_r over the coming single time step. The escape probability is $P_e = 1 - P_r$.

- The system matrix transforms input to output.
- The output covariance is accordingly the input covariance surrounded by, or packaged by, the system matrix.

The covariance-creating processes are of substantive importance. One of the goals of simulation is to generate this covariance directly.

3.6 Particles in a Box

To close this chapter, we offer a detailed look at a very simple problem. Forms of this problem recur in the examples given in later chapters: drifters in a gyre, sediment particles on the bottom, stage duration for growing organisms, mortality. This exercise serves as a generic baseline for those situations and applies many of the statistical ideas introduced in this chapter.

Suppose we have the simple system illustrated in Figure 3.23. There are N particles in the box; I and E are the rates of import and export of particles. I is given; once in the box, each particle independently has probability P_r of *retention* in the box during the coming time step. $P_e = 1 - P_r$ is the probability of *export*. A retained particle has the same P_r for the next time step.

Nothing else is required. Simulation, one time step at a time, takes over and produces the time history of N and E for a given I. In this simple case it is possible to obtain analytic results over many time steps, in order to provide some baseline against which to view results of more complex simulations.[8]

[8] P_e in this context is a surrogate for the *hazard rate h*, accumulated over the coming time step: $P_e = h\Delta t = \Delta H$. The hazard in this case corresponds to particle export. Section 3.1.2 describes h in more generality; here, h is constant over time.
Needless to say, constant h is an extremely simple case. Especially when biotic outcomes are concerned, the hazard rate could well depend on many things including accumulated "residence-time-in-box" and all aspects of particle history. Example: if the hazard is mortality, one might make it relatively high for very young particles, and again for very old particles, and complete the picture with individual variation. These matters are quite substantive and require care in model representation; of course the particle simulation approach allows h to be sensitive to time, place, and individual particle state and history, readjusted during each time step for each particle.

The case is encountered in several contexts where a particle occupies a physical, biotic, or logical state with a binary, irreversible outcome – to exit or not. Examples: sediment particle settling or resuspension; IBM mortality or stage completion; boundary encounter or escape.

Often there is more than one destination on exit, E_1, E_2 and so forth with escape probabilities P_{e_i}, but the probability of retention remains as the unique alternative to all of these – not any of the exit options. $P_r = 1 - \sum P_{e_i}$. This is taken up in Section 3.6.6.

3.6.1 *Individual Residence Time*

From the view of an *individual particle in the box*: the probability of retention through the next k time steps is P_r^k (the repeated mutiplication of P_r, k times); the probability of export in the next following step is $P_r^k(1-P_r)$. The *geometric distribution* describes individual residence time T in the box:

$$\overline{T} = \frac{1}{1-P_r} \qquad \sigma_T^2 = \frac{P_r}{(1-P_r)^2} \tag{3.244}$$

T is the waiting-time-to-escape for an individual particle present in the box. The average is over all possible simulations. If all particles are alike, then T and its distribution would be a property of the system. If not, then it remains an individual property and system properties would require aggregation over all particles in it.

Overall system behavior requires aggregation over all particles. We will assume all particles are alike. What can be said about N and E, and what is their dependence on the properties of I?

3.6.2 *Aggregate Properties: Relaxation of Initial Condition*

Let N_0 be the number of such identical particles initially in a box, with no inputs. Each exits in a time step with probability P_e. What is the mean and variance of N and E over time?

(a) Let N_k be the number of particles retained through k time steps. The *binomial distribution* governs

$$N_k \sim \mathcal{B}(N_0, P_r^k) \tag{3.245}$$

The mean and standard deviation are

$$\overline{N_k} = N_0\, P_r^k \qquad \sigma_{N_k} = \sqrt{N_0\, P_r^k(1-P_r^k)} \tag{3.246}$$

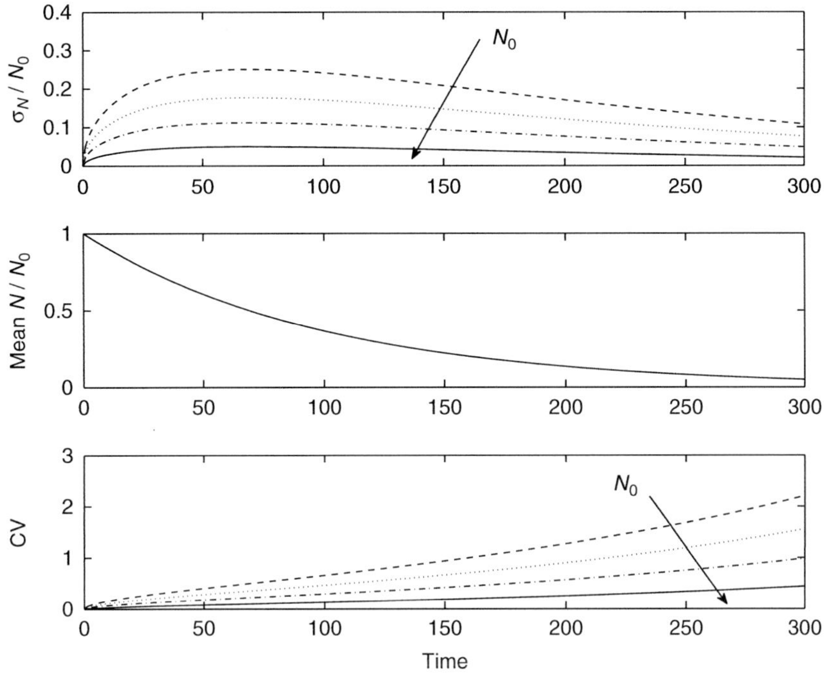

Figure 3.24. Plots of binomial retention in a single box with N_0 particles initially and no subsequent inputs; $P_r = 0.99$ and mean residence time $\overline{T} = 100$. The parameter $N_0 = [4, 8, 20, 100]$. Top: σ/N_0 grows initially, then decays slowly. Middle: Mean retention – a decreasing geometric series. Bottom: CV rises monotonically with time. Variability decreases as N_0 increases from 4 (top line) to 100 (lower line). In all cases, the peak variance occurs at $k = 69$ irrespective of N_0.

The mean decreases monotonically with time k. The variance is initially zero; it increases, peaks, and then goes back to zero as ultimately all particles exit. The peak σ occurs when $P_r^k = \frac{1}{2}$, with $\frac{\sigma}{N_0} = \sqrt{\frac{1}{4N_0}}$.

The coefficient of variation $\sigma/\overline{N}$ is

$$CV = \sqrt{\frac{(1 - P_r^k)}{N_0\, P_r^k}} \tag{3.247}$$

While N decreases monotonically with time, σ peaks and the CV increases monotonically.

Note the $\sqrt{N_0}$ factor in σ and the CV. This is the effect of the finite ensemble size N_0 – increasing N_0 decreases both the peak in relative variance, and the growth rate of CV.

Figure 3.24 plots these factors versus time. For small N_0, the CV becomes significant at large time.

Problem 3.59 Derive this relationship for the time of peak variance in terms of $\overline{T}$:

$$k = \frac{-\ln 2}{\ln\left(1 - \frac{1}{\overline{T}}\right)} \tag{3.248}$$

Problem 3.60 For the example in Figure 3.24, what value does the formula 3.248 predict? Is it realized?

Problem 3.61 For relaxation of an initially loaded box: what is the half-life of the initial resident population of particles? Compare with individual residence time as derived in Equation 3.244.

Problem 3.62 For relaxation of an initially loaded box: what is the initial mean export rate? If that rate were held constant, how long would it take to completely empty the box ? Compare with individual residence time as derived in (Equation 3.244).

3.6.3 Export Rate

Consider only a single time step with N known at the start. From the previous section, the *conditional* export rate $E|_N$ has mean and variability

$$E|_N = \overline{E|_N} + \widehat{E|_N} \tag{3.249}$$

The properties, assuming known N:

$$\overline{E|_N} = (1-P_r)N \qquad Var(E|_N) = NP_r(1-P_r) \tag{3.250}$$

Problem 3.63 Find an expression for the mean squared $E|_N$.

The complete export rate is

$$E = (1-P_r)N + \widehat{E|_N} \tag{3.251}$$

The long-run mean export is the mean over N of all these conditional means:

$$\overline{E} \equiv \overline{\overline{E|_N}} = (1-P_r)\overline{N} \tag{3.252}$$

and E is alternately expressed as

$$E = \overline{E} + \widehat{E|_N} + (1-P_r)\widehat{N} \tag{3.253}$$

$\overline{E}$ is the unconditional mean; the long-run balance requires it to balance mean input, $\overline{E} = \overline{I}$.

$\overline{E}$ is the unconditional variability. There are two separate contributions. The first is the conditional variability during a time step with N known, as expressed previously; the second isolates the effect of $\overline{N}$.[9]

3.6.4 Long-Run Balance

Next we activate the input rate I with a known distribution but independent of N and E. The mass balance for the system with superscript k indicating time level:

$$N^{k+1} = N^k - E^k + I^k \tag{3.254}$$

with the rates E^k and I^k pertaining to the single time step *following* time k.

Assume a stochastic steady state, where a meaningful $\overline{N}$ is present and stable. The long-run balance requires that the long-term means balance:

$$\overline{I} = \overline{E} = \overline{N}(1 - P_r) \tag{3.255}$$

Subtracting $\overline{N}$ from the mass balance gives

$$\widehat{N}^{k+1} = \widehat{N}^k - E^k + I^k \tag{3.256}$$

and with a little rearrangement we have

$$\widehat{N}^{k+1} = \widehat{N}^k P_r + \widehat{I}^k - \widehat{E|_{N^k}} \tag{3.257}$$

There are two sources of variability – the usual per-step variability present when N is known and that due to departures in N from its long-term mean. Both are independent with zero mean.

Squaring both sides, and isolationg the uncorrelated terms,[10] we have

$$(\widehat{N}^{k+1})^2 = (\widehat{N}^k)^2 P_r^2 + (\widehat{I}^k)^2 + (\widehat{E|_{N^k}})^2 + (\text{uncorrelated terms}) \tag{3.258}$$

The last term has variance conditional on N^k; the mean over all N will give

$$\overline{Var(\widehat{E|_{N^k}})} = \overline{N} P_r (1 - P_r) \tag{3.259}$$

The mean of Equation 3.258 is the result:

$$Var(N) = \left[\frac{1}{1 - P_r^2}\right] \left[Var(I) + P_r \overline{I}\right] \tag{3.260}$$

[9] From another view, the $\overline{N}$ term is accounting for the difference between unconditional and conditional means.

[10] For the cross term $\widehat{N^k}\widehat{E|_{N^k}}$ – the zero-mean property is clearest if one first takes the mean with N fixed; it vanishes. Its aggregation over all possible N must then vanish, regardless of the distribution of N.

The term $\overline{I}$ clearly reveals a dependence on overall particle abundance. The CV_N is the variability relative to mean:

$$CV_N^2 = \left[\frac{1-P_r}{1+P_r}\right]\left[CV_I^2 + \frac{P_r}{\overline{I}}\right] \tag{3.261}$$

Increases in $\overline{I}$ cause the relative variability in N to decrease. In the limit, high mean input drives CV_N to its minimum.

From Equations (3.252, 3.253) we have the **unconditional** E:

$$\overline{E} = (1-P_r)\overline{N} = \overline{I} \tag{3.262}$$

$$\widehat{E}^k = \widehat{N}^k(1-P_r) + \widehat{E|_{N^k}} \tag{3.263}$$

Squaring the latter and taking its mean as earlier[11] provides the unconditional variance:

$$Var(E) = Var(N)(1-P_r)^2 + \overline{N}P_r(1-P_r) \tag{3.264}$$

In terms of the input:

$$Var(E) = \left[\frac{1-P_r}{1+P_r}\right]\left[Var(I) + \frac{2P_r}{1-P_r}\overline{I}\right] \tag{3.265}$$

$$CV_E^2 = \left[\frac{1-P_r}{1+P_r}\right]\left[CV_I^2 + \frac{2P_r}{1-P_r}\frac{1}{\overline{I}}\right] \tag{3.266}$$

Again, the $1/\overline{I}$ term indicates increasing relative variability at low abundance. By comparison with CV_N, its relative effect is enhanced by the factor $2/(1-P_r)$. For example, for $P_r = 0.8$, this is a factor of 10. Essentially, $Var(N)$ already contains a sensitivity to the mean; $Var(E)$ adds more.

Problem 3.64 Express CV_E in terms of CV_N and $\overline{N}$.

3.6.5 Summary – Steady State

A general summary appears in Table 3.7. Two common examples follow to illustrate.

Example 3.9 *Uniform input:* Suppose $Var(I) = 0$:

$$Var(N) = \left[\frac{P_r}{1-P_r^2}\right]\overline{I} \qquad CV_N^2 = \left[\frac{1-P_r}{1+P_r}\right]\frac{P_r}{\overline{I}} \tag{3.267}$$

[11] The cross term has zero mean, as noted earlier.

Table 3.7. *Summary of Relations among Input, Export, and Number of retained Particles for a Box (Figure 3.23) under Stationary Conditions*

	Mean	Variance	CV^2
Input rate I (given)	$\bar{I}$	$Var(I)$	$Var(I)/(\bar{I})^2$
Storage N	$\frac{\bar{I}}{1-P_r}$	$\left[\frac{1}{1-P_r^2}\right]\left[Var(I)+P_r\bar{I}\right]$	$\left[\frac{1-P_r}{1+P_r}\right]\left[CV_I^2+\frac{P_r}{\bar{I}}\right]$
Export rate E	$\bar{I}$	$\left[\frac{1-P_r}{1+P_r}\right]\left[Var(I)+\frac{2P_r}{1-P_r}\bar{I}\right]$	$\left[\frac{1-P_r}{1+P_r}\right]\left[CV_I^2+\frac{2P_r}{1-P_r}\frac{1}{\bar{I}}\right]$

Note: The uniform per-step probability of retention is P_r for particles in the box; the statistical properties of I are given; N and E are the response.

With no variability on input, the process still generates variability – this is the pure effect of finite I. At high abundance, CV_N becomes negligible, recovering the limiting case – the concentration of many small particles, vanishing variability.

Example 3.10 *Poisson input:* Suppose $Var(I)=\bar{I}$:

$$Var(N)=\left[\frac{1+P_r}{1-P_r^2}\right]\bar{I} \qquad CV_N^2=\left[\frac{1-P_r}{1+P_r}\right]\left[\frac{1+P_r}{\bar{I}}\right] \tag{3.268}$$

Adding Poisson variability on input more that doubles the variance. The size effect remains unchanged.

Problem 3.65 (1) Express $Var(E)$ and CV_E in terms of the input properties for (a) constant input and (b) Poisson input. (2) Evaluate these for $\bar{I}=50$ and $P_r=0.85$.

3.6.6 Exit Paths

Suppose that a particle in the box can exit in two ways; the exit rate is split into E_1 and E_2, with per-step probabilities P_{e1} and P_{e2}. The previous analysis of E is unchanged, with

$$P_r=1-(P_{e1}+P_{e2}) \tag{3.269}$$

For convenience, let $p_i=P_{ei}/P_e$ be the relative probability of following route 1 on release, and likewise for p_2.[12] We have then

$$p_1+p_2=1 \tag{3.270}$$

[12] p_i is a conditional probability – given that a particle is released, which path will it take?

Given E, we again have a binomial distribution of particles to path i, with *conditional* mean $\overline{E_i|_E} = p_i E$. Averaging over all E we get the intuitive *unconditional* results:

$$\overline{E_i} = p_i\overline{E} \qquad \widehat{E_i} = p_i\widehat{E} \tag{3.271}$$

$$\overline{E_1} + \overline{E_2} = \overline{E} \qquad \widehat{E_1} + \widehat{E_2} = \widehat{E} \tag{3.272}$$

The variances:

$$Var(E_1) = p_1^2\, Var(E) \qquad Var(E_2) = p_2^2\, Var(E) \tag{3.273}$$

$$Var(E_1) + Var(E_2) + 2\overline{\widehat{E_1}\widehat{E_2}} = Var(E) \tag{3.274}$$

In accord with intuition, the unconditional standard deviation of E is simply apportioned among E_1 and E_2: $\sigma_{Ei} = p_i\sigma_E$.

Routine manipulation gives the cross-correlation term:

$$2\overline{\widehat{E_1}\widehat{E_2}} = \left[1 - (p_1^2 + p_2^2)\right] Var(E) \tag{3.275}$$

This is the unconditional correlation. It is positive; increasing E increases both E_1 and E_2, on average. By contrast, when E is known, E_1 and E_2 are exactly opposed, negative covariance. That effect is overcome by variation in E. In other words, a rising tide lifts all boats.

Finally, we find the CV in each case inherits the CV of the total:

$$CV_{E1}^2 = CV_{E2}^2 = CV_E^2 \tag{3.276}$$

$$\frac{\overline{\widehat{E_1}\widehat{E_2}}}{\overline{E_1}\,\overline{E_2}} = \left[\frac{1 - (p_1^2 + p_2^2)}{2p_1p_2}\right] CV_E^2 \tag{3.277}$$

Problem 3.66 A box has three exit paths with relative probabilities p_1, p_2, p_3. Follow the text analysis and generate the unconditional mean, variance, and CV of each, in terms of the total export E.

These results generalize to the case of M **exit pathways**. Given E, the multinomial distribution governs the sorting; the conditional mean is $\overline{E_i|_E} = p_i E$; for the unconditional export,

$$\overline{E_i} = p_i\overline{E} \qquad \widehat{E_i} = p_i\widehat{E} \tag{3.278}$$

$$\sum\overline{E_i} = \overline{E} \qquad \sum\widehat{E_i} = \widehat{E} \tag{3.279}$$

The variances:

$$Var(E_i) = p_i^2 Var(E) \qquad \sum Var(E_i) = Var(E) \sum p_i^2 \tag{3.280}$$

Individual variances do not account for the complete variance; some is apportioned among the covariances. Squaring Equation 3.279,

$$\sum_{i,j} \overline{\widehat{E_i}\widehat{E_j}} = Var(E) \tag{3.281}$$

Isolating the covariance terms, we get

$$\sum Var(E_i) + \sum_{i \neq j} \overline{\widehat{E_i}\widehat{E_j}} = Var(E) \tag{3.282}$$

The covariance terms are a net positive fraction, $\left[1 - \sum p_i^2\right]$, of the total variance.

3.6.7 Input Paths

For this system, the case of multiple independent inputs I_j aggregates readily:

$$I = \sum I_j \qquad \overline{I} = \sum \overline{I_j} \qquad \widehat{I} = \sum \widehat{I_j} \tag{3.283}$$

and the variances simply add,

$$Var(I) = \sum Var(I_j) \tag{3.284}$$

These are the relations among the moments; the complete distribution is not specified.[13]

3.6.8 Autocorrelation

From Equation 3.257: the product of both sides with $\widehat{N^k}$ gives, on average, the covariance

$$\overline{\widehat{N^{k+1}}\widehat{N^k}} = P_r \, Var(N) \tag{3.285}$$

This is a "lag-1 Markov process," with correlation P_r among adjacent-in-time N values. It is a metric of smoothness or persistence in the presence of variability. These

[13] If each input is itself a Poisson process, their sum is a Poisson process with single mean and variance $\overline{I}$. That assumption is common in analytical work; in the simulation context, such an assumption is not necessary when I itself is the output of some other process.

ideas are developed in more generality elsewhere; here we note the obvious and intuitive relation to the individual residence time, $\overline{T} = \frac{1}{1-P_r}$ as noted in Equation 3.244 – higher retention, more autocorrelation, longer residence time.

Problem 3.67 For the case of steady (sustained) input: derive the expression for the filling time – the time it would take for mean input to equal mean storage. Compare this aggregate, steady result with individual residence time as derived in Equation 3.244; and for the half-life of a group of particles (Problem 3.61).

3.6.9 Example – Branch Point

Imagine a coastal branch point where floating particles are attracted by a downwelling zone and trapped there at the surface (Figure 3.25). Particles on average spend 20 time steps there before exiting along two different paths, one of which is four times more probable than the other.

This system was modeled as a box with $P_r = 0.95$ and exit probabilities $p_1 = 0.8$, $p_2 = 0.2$, and a Poisson input with $\overline{I} = 30$.

A simulation was started empty and run to stochastic steady state. The results appear in Figure 3.26. The initial transient lasts about 50 steps; the steady N settles around 600 with significant smoothing of the input; the export occurs in the right proportion, and with less variability than the input, due to the smoothing effect of the retention.

Problem 3.68 For the simulation illustrated in Figure 3.26: compute the mean, variance, and CV using the formulas given, for all the quantities plotted. Does this simulation produce what you expect?

3.6.10 A Network of Boxes

Figure 3.27 illustrates several boxes interchanging particles. Each has its own external input and an export out of the network from which there is no return. Examples: a

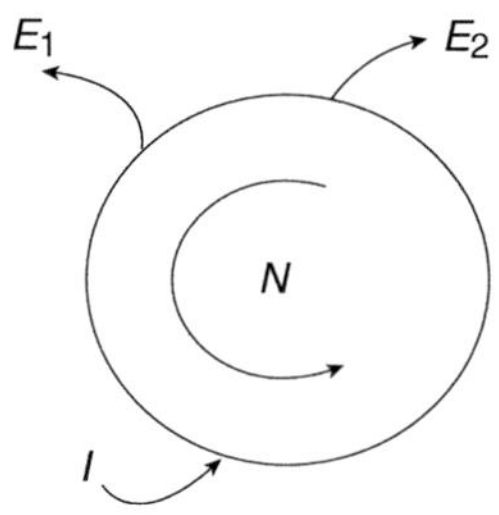

Figure 3.25. Definition sketch for the example: Particles trapped over a downwelling area with two exit paths.

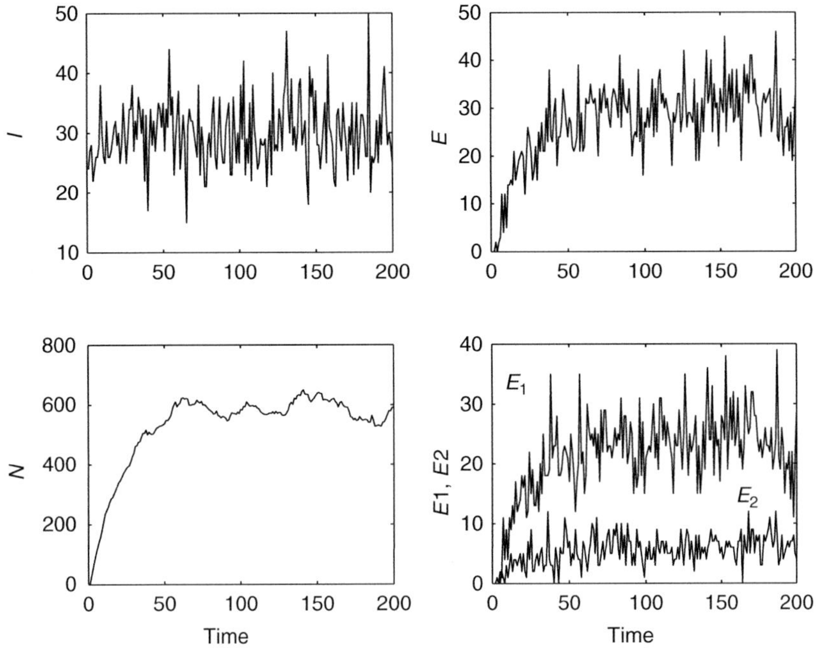

Figure 3.26. Simulation results for the downwelling example sketched in Figure 3.25. The two exit paths have relative probabilities in the proportion 4 : 1.

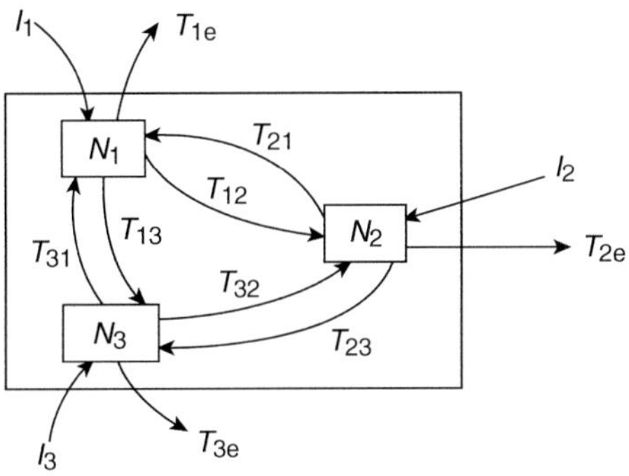

Figure 3.27. Three boxes in a network, exchanging particles.

drogue that fails, mortality of a living particle. For a single box, this is the system developed earlier; for multiple boxes, there are internal transfers of particles. These transfers are one-way, non-negative, pertaining to an individual particle already in a particular box. Transition probabilities for individual particles need to be generalized, reflecting origin and destination.

Nomenclature:

N_j^k is the number of particles in box j at the start of time step k.

I_j^k is the external input rate of particles to box j during step k.

T_{ij}^k is the transfer rate of particles from box i to box j during step k. These are non-negative, one-way transfers for particles already in box i. $T_{ij} \geq 0$; generally $T_{ij} \neq -T_{ij}$ when $j \neq i$. T_{ii} represents "self-retention," that is, retention of a particle in box i.

P_{ij} is the probability of a particle in box i to move to box j in the coming time step. These are conditional probabilities; $0 \leq P_{ij} \leq 1$ and $P_{ij} \neq P_{ji}$. More properties are listed below.

T_{ie} is the rate of particle transport from box i, *out* of the system, with no return.

About the P_{ij}:

- $\sum_j P_{ij} + P_{ie} = 1$ or equivalently, $\sum_j P_{ij} \leq 1$ and $P_{ie} = 1 - \sum_j P_{ij}$.
- These reduce to the one-box versions used previously, with the distinction that P_{ii} represents self-retention for box i while $\sum_j P_{ij}$ represents system-wide retention.
- Another version: $P_{ii} + \sum_{j \neq i} P_{ij} + P_{ie} = 1$. This separates self-retention from retention via transfer to a different box.
- Three matrices are used below; all depend exclusively on the P_{ij}:

$$[\mathbf{P}] = \left[P_{ij}\right]^* \qquad [\mathbf{Q}] = \left[P_{ij}(1 - P_{ij})\right]^* \qquad [\mathbf{R}] = \left[P_{ij}^2\right]^* \tag{3.286}$$

The * notation indicates transposition.[14]

There is no T_{ei} and that will allow its elimination. It is implied by the inequality form of the P constraints.

We will follow the path used for the one-box case, with suitable generalizations.

3.6.10.1 Transfer Rate

Again considering only a single time step with all N_i known at the start: the multinomial distribution governs all the T_{ij}. The properties are dependent on N in the originating box:

$$T_{ij} = \overline{T_{ij}} + \widehat{T|_{N_i}} \tag{3.287}$$

$$\overline{T_{ij}|_{N_i}} = N_i P_{ij} \qquad Var(T_{ij}|_{N_i}) = N_i P_{ij}(1 - P_{ij}) \tag{3.288}$$

These are conditional, given N_i. The long-run or unconditional mean transports are the average over all N_i:

$$\overline{T_{ij}} = \overline{\overline{T_{ij}|_{N_i}}} = \overline{N_i} P_{ij} \tag{3.289}$$

[14] Here we are using a (row, column) index for matrices and preserve the (from, to) convention for the P_{ij}.

and the transfer rate is

$$T_{ij} = \overline{T_{ij}} + \widehat{T_{ij}|_{N_i}} + \widehat{N}_i P_{ij} \tag{3.290}$$

This is the long-term, unconditional mean plus two perturbations: the first is the conditional contribution with N_i known ; the second isolates the effect of variable N_i. The latter diminishes in importance for very large N_i; it carries the important effect of finite N.

3.6.10.2 Steady State

The long-term mean import and export balance, for each box, is[15]

$$\overline{I_i} + \sum_j \overline{N_j} P_{ji} = \overline{N_i} \tag{3.291}$$

and the system as a whole is summarized by the matrix equation

$$[\mathbf{I} - \mathbf{P}]\left\{\overline{N}\right\} = \left\{\overline{I}\right\} \quad \Longleftrightarrow \quad \left\{\overline{N}\right\} = [\mathbf{I} - \mathbf{P}]^{-1}\left\{\overline{I}\right\} \tag{3.292}$$

[**I**] is the identity matrix, not related to the inputs I. In this manner, the case of a single box, Equation 3.255, is generalized.

Example 3.11 Consider a system of two boxes. On average, Box 1 transfers 20% of its contents to Box 2; it exports 7%. Box 2 transfers 10% to Box 1, and exports 5%. Input is to Box 1 only, at $I_1 = 100$. We have the probability matrix

$$[\mathbf{P}] = \begin{bmatrix} 0.73 & 0.10 \\ 0.20 & 0.85 \end{bmatrix} \tag{3.293}$$

The solution is: $\left\{\overline{N_1}, \overline{N_2}\right\} = \overline{I_1}\{7.317, 9.756\}$.

If the input rate were switched to I_2 only, the solution would be $\left\{\overline{N_1}, \overline{N_2}\right\} = \overline{I_2}\{4.878, 13.17\}$.

For an arbitrary blend of inputs, these solutions add.

Problem 3.69 For Example 3.11, evaluate the matrices [**Q**] and [**R**].

Problem 3.70 Redo Example 3.11 with the loss rate eliminated in Box 1. Specifically, find [**P**] and $\overline{N_i}$ in terms of I_i.

Problem 3.71 For Example 3.11: find the export rates.

15 The self-retention transport occurs as both inport and export in this representation; removing the term $\overline{N_i} P_{ii}$ from both sides is another way to express this balance.

Problem 3.72 Generate and solve the system of equations for the steady state among three boxes. Per time step: Box 1 transfers 10% of its particles in the ratio 3:1 to Boxes 2 and 3, on average; 1% are lost. For Box 2: 15% go to Box 3 only, with 5% lost. Box 3: 4% total transfer, split evenly between Boxes 1 and 2; no loss.

(This system is taken up again in Problem 3.73.)

For the time series N_i^k: since here we have self-retention in [**P**], we can express the balance during a time step as the accumulation of input + internal transfers from all boxes,

$$N_i^{k+1} = I_i^k + \sum_j T_{ji}^k|_{N_j^k} \tag{3.294}$$

This form accounts automatically for T_{ii}^k, the retained portion of N_i^k. Expanding, and canceling the steady-state balance, we have

$$\widehat{N_i^{k+1}} = \widehat{I_i^k} + \sum_j \widehat{T_{ji}|_{N_j^k}} + \sum_j \widehat{N}_j P_{ji} \tag{3.295}$$

Squaring both sides produces many uncorrelated terms that vanish on average. We are left with the equations for the unconditional variances:

$$Var(N_i) = Var(I_i) + \sum_j \overline{N_j} P_{ji}\left(1 - P_{ji}\right) + \sum_j Var(N_j) P_{ji}^2 \tag{3.296}$$

All the $Var(N_i)$ are linked implicitly; a single matrix equation governs.

$$[\mathbf{I} - \mathbf{R}]\{Var(N)\} = [\mathbf{Q}]\left\{\overline{N}\right\} + \{Var(I)\} \tag{3.297}$$

$\overline{I}$ may be substituted for $\overline{N}$ using the mean balance:

$$[\mathbf{I} - \mathbf{R}]\{Var(N)\} = [\mathbf{Q}][\mathbf{I} - \mathbf{P}]^{-1}\left\{\overline{I}\right\} + \{Var(I)\} \tag{3.298}$$

This relation reduces to 3.260 in the one-box case. The familiar effect of $\overline{N}$ (or $\overline{I}$) is retained; it diminishes in importance at high throughput.

Example 3.12 For the two-box system in Example 3.11, we have

$$\begin{bmatrix} 0.467 & -0.01 \\ -0.04 & 0.277 \end{bmatrix} \left\{ \begin{array}{c} Var(N_1) \\ Var(N_2) \end{array} \right\} = \left\{ \begin{array}{c} 2.32 \\ 2.41 \end{array} \right\} \overline{I_1} + \left\{ \begin{array}{c} Var(I_1) \\ 0 \end{array} \right\} \tag{3.299}$$

For $\overline{I_2} \neq 0$, there would be additions of $\overline{I_2}$ and $Var(I_2)$; the system is linear in them.

Example 3.13 A coastal segment is represented by four boxes: surface and benthos, upstream and downstream. The two surface boxes contain a partial recirculation cell.

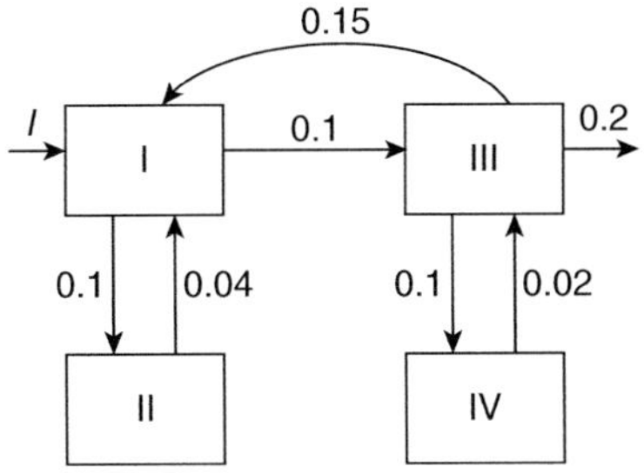

Figure 3.28. A simple system with horizontal fluxes of particles at the surface, and vertical transfers to/from the benthos. See Example 3.13.

The benthic boxes have no horizontal transfers. The P_{ij} are shown in Figure 3.28 representative of sediment particles.

$$[\mathbf{P}]^* = \begin{bmatrix} 0.8 & 0.1 & 0.1 & 0 \\ 0.04 & 0.96 & 0 & 0 \\ 0.15 & 0 & 0.55 & 0.1 \\ 0 & 0 & 0.02 & 0.98 \end{bmatrix} \tag{3.300}$$

For this system, $\overline{N} = [175, 437.5, 50, 250]$ and $Var(N) = [147, 434, 49.6, 249.9]$.

Problem 3.73 Evaluate $Var(N_i)$ for the three-box system described in Problem 3.72.

Problem 3.74 Solve the system in Example 3.12 and interpret the results.

Problem 3.75 Complete the system in Example 3.12 by adding the possibility of nonzero I_2.

Problem 3.76 For Example 3.13: add a bedload flux from box 2 to 4, with $P_{24} = 0.03$. What is the effect on mean and variance of the benthic boxes?

Problem 3.77 For Example 3.13: add $Var(I) = 20$; find the mean and variance of the N.

3.6.11 Closing Ideas – Particles in Boxes

In these simple situations we can develop properties of the system performance over time. Simulation one step at a time needs the probabilities of events for the coming time step only; the system performance would emerge, as predicted by these analyses.

In more realistic situations, each particle would carry its complete history. Simulation allows individual probabilities to be variable over particles, time, history, accumulated biotic state, and so forth. That information would accompany an individual particle and be sensitive to its environment.

The routing of particles here is given; in more comprehensive situations, the routing itself would be part of a coupled system. For example, a coastal ocean simulation model would provide the motion connectivity.

3.7 Closure

Metrics. There are two generic types of metrics: abundance and distribution.

In the abundance category, we have an excellent discussion in Gotelli and Ellison [175] of various metrics. Included are the geometric mean, arithmetic mean, bounds, moments, median, mode, and so forth – and discussion of when each is useful. A major concern is that a sampling area be large enough to support meaningful metrics of abundance. In ecological systems, it is common to have massive individual death ("the average larva is a dead one"); or entire areas that are not populated. In such systems, the problem of small numbers is unavoidable. With particle methods, it is always possible to aggregate individuals when they are superabundant and to disaggregate them when they are scarce. This is one way to reduce the loss of information to "small N" as time proceeds.

In the distribution arena, we are dealing with the general field of spatial statistics. There has been a very broad activity in this area. Silverman [416] and Wand and Jones [466] provide simple introductions; an advanced reference is Cressie [92]. Field estimation techniques include optimal estimation, objective analysis, statistical interpolation, data inversion, and kriging; these share an underlying unity despite diversity of vocabulary, and in fact there is a significant bridge to discretized PDEs and data assimilation. Some current sources include Wunsch [489]; Lynch [279].

Undersampling. Brickman and Smith [52] discussed two types of undersampling errors. Type U-I represents undersampling of the particle source region; type U-II represents undersampling the portion of the source region that eventually intersects the target region. These may coalesce, in a site- and problem-specific way, to the "size" of the target relative to the source. A large target, as in a retention problem, risks U-I undersampling and a relatively small N may suffice. On the other hand, a small target, as in a dispersion or loss problem, risks U-II undersampling and is likely to require a larger N. In both cases, these ideas depend greatly on the dynamics embedded in a particle: the likelihood of growth, death, reproduction, biochemical transformation, and the transit time from source to target. These ideas are inspired by practical concerns with the management of oceanic resources in a highly heterogeneous coastal sea, where resolution in space, time, and N all conspire and present serious challenges.

Experimental Protocol. A great deal of oceanic consideration is quite problem specific relative to outputs and outcomes. "What is the question?" is always the first issue; and second, "What questions can be addressed?", given machines, data, and time. Though there are some analytic and heuristic guidelines, ultimately one resorts to an experimental protocol – repeat the simulation, look for robust conclusions.

3.8 General Sources

- Early summaries of useful distributions appear in Abramowitz and Stegun [2]. Devroye [115] offers a comprehensive early review. Good current references include Ross [389] and Scheaffer and Young [403]. An excellent current summary of distributions is available at Forbes et al. 2010 [157].
- Cressie [92] gives a thorough review of statistical processes. Excellent depth in linear algebra is available in [114, 453].
- A good practical reference is Press et al. [362] covering algorithms for generating the uniform deviate and general computational practice. A practical set of contemporary rules of thumb has been collected by van Belle [457].
- For a useful text on ecological statistics, see Gotelli and Ellison [175]. Hilborn and Mangel provide a general discussion of modeling random processes in ecology [195]; Mangel offers a "Biologist's Toolbox" [296]. Okubo and Levin [343] provide a thorough examination of diffusion processes in living systems.
- Pioneering work by Fiering [146, 147] was done in Water Resources; a good contemporary reference in that field is Loucks et al. [259].
- Hillier and Lieberman [196] provide an excellent introduction to the use of random processes in queueing theory.

4

Dispersion by Random Walk

In this chapter we describe some simple random walk models. The size of the random walk is fundamentally related to the macroscale diffusion coefficient. This is assumed available as a gridded field, representing the microscale or subgrid motions. The models are classified according to the form of memory implied. These are AutoRegressive (AR) models; individual particles retain memory of their own history.

Here we develop some basic ideas about random walk through a known Eulerian environment. We assume that particle motion is a sum of resolved or "macroscale" motion (the statistical mean motion – resolved on a finite grid); and a stochastic component representing the unresolved subgrid processes (the microscale). For exposition, we assume the mean motion is zero; it can be added later, without loss of generality.

The classic literature relates the ensemble of random walks to a macroscale diffusion process; the size of the walk and the macroscale diffusion coefficient are fundamentally linked. Practical extensions are reported in Hunter et al. [202], Proehl [365], Visser [464], Riddle [382], Skellam [420], Ross and Sharples [388]; and used in the applications by Fischer et al. [153], Dimou and Adams [117], Proehl [366], to name a few. The current frontier of Lagrangian work in the coastal ocean is reported in the LAPCOD volume [186]; the summary therein [302] is particularly useful.

The literature on timeseries analysis is immediately relevant and deeper than the exposition here. Box et al. [49] is a classic work in this area. There are several more introductory texts, for example, Chatfield [79]. In particular, these include the important topics related to identifying random walk models from observed timeseries.

We begin with a simple introductory development. Next we go into the more general continuous processes and address a hierarchy of random walk models with memory.

At the macroscale level, turbulence closure provides the essential link from the macroscale diffusivity to the random walk details. In Chapter 6 the closure problem is described.

4.1 Introduction: Discrete Drunken Walk

Here we provide a brief introduction to the simplest of random walks. Readers familiar with these developments should skip this section.

Consider the 1-D drunken walk process: in any time interval, move 1, 0 , or -1 units with equal probability; do this repeatedly. All steps are completely independent. The mean step will be zero; the mean squared step σ^2 will be 2/3. This might be described as discrete hopping on a fixed "meter" or time increment. There will be n time steps accumulating in each trajectory.

Five separate simulated trajectories appear in Figure 4.1 (left). The horizontal axis is time (n, the number of time steps); the vertical axis is the total displacement versus time for a single trajectory. Each trajectory is an individual (Lagrangian) sum of n accumulated random steps. Each is clearly "random" at the per-step displacement level; each accumulates its own specific Lagrangian history. On average, the group of five gets nowhere. But they clearly disperse. All five are *statistically the same*, members of the ensemble of all possible trajectories.

The accumulated displacement X for any trajectory will have zero mean. Its square will be the *square of sums*. On average the product of different steps is independent and will vanish when averaged over the ensemble of possible trajectories. So we end up with a *sum of squares*. The result: the mean X^2 will be $n\sigma^2$ – that is, the mean X^2 will grow linearly with time. This is illustrated in the right panel of Figure 4.1.

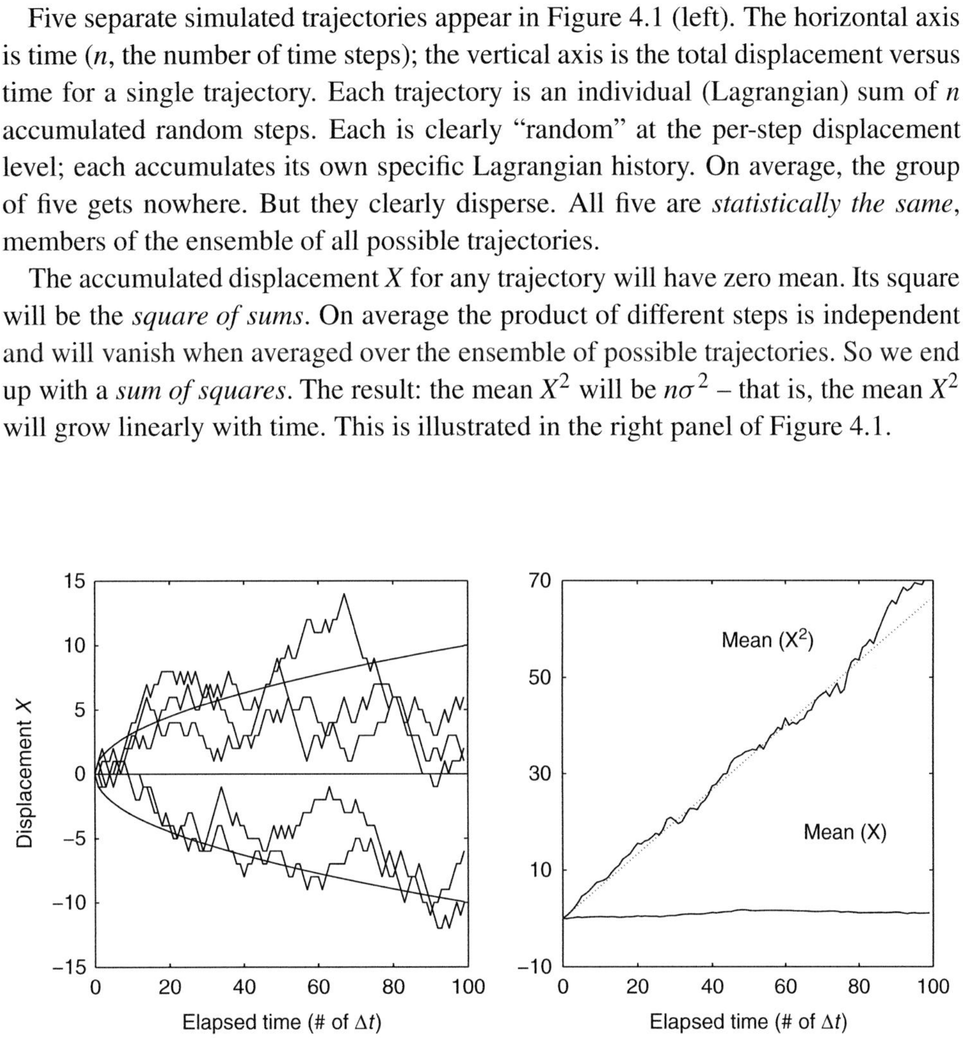

Figure 4.1. Simplest random walk. left: Five individual walks, verses time. The smooth analytic envelope $\sigma_X = \pm\sqrt{n}\,\sigma$ is plotted for reference; it represents $\pm$ one standard deviation of the ensemble mean displacement. Right: ensemble mean and variance of 100 individual walks, versus number of time steps. As on the left plot, the analytical result $\sigma_X^2 = n\sigma^2$ is plotted for reference (light dotted line).

Table 4.1. *Unnormalized Probability of Occurrence of Displacement Sampled at Different intervals*

−4	−3	−2	−1	0	1	2	3	4
				1				
			1	1	1			
		1	2	3	2	1		
	1	3	6	7	6	3	1	
1	4	10	16	19	16	10	4	1

Note: The horizontal dimension is the displacement; the vertical is the sampling interval. The number of possible displacements broadens as the sampling interval increases.

So far we have relied on a discrete probability distribution of walks, chosen for simplicity. There are many embellishments – variable waiting time; variable step size, and so forth.

Relative to the accumulations X: By the central limit theorem, "all large sums are normal." So the individual accumulations X *become* Gaussian as time progresses: $X \sim \sqrt{n\sigma^2} \cdot \mathcal{G}(0,1)$ The ensemble average properties (averaged across *all possible trajectories*) are Gaussian even at low n.

Subsampling. Suppose we observe every second step without altering the underlying trajectory. Over two sequential steps, the possible outcomes for a single particle are [2, 1, 0, 1, 2] with (unnormalized) probabilities [1, 2, 3, 2, 1]. The possible outcomes are broadening. Observing every third step, the outcomes broaden further to [1, 3, 6, 7, 6, 3, 1]; and so forth as in Table 4.1. As the sampling interval grows, the outcome distribution develops a central peak and tails. At the longer sampling intervals, the distribution is less abrupt, although the underlying process and trajectory is the same. The limiting distribution is the Gaussian.

The generalization beyond 1-D is intuitive. Three 2-D trajectories are shown in Figure 4.2; each was achieved with independent walks in both x and y. With two independent walks, we can expect their variances to add; if $R = \sqrt{X^2 + Y^2}$ is the distance of a particle from its origin, then

$$\overline{R^2} = \overline{X^2} + \overline{Y^2} \tag{4.1}$$

$$= \sigma_X^2 + \sigma_Y^2 \tag{4.2}$$

So far, everything is fundamentally discrete. The time step and the walk size are fixed, built into the underlying process. In what follows, we relax this assumption and consider the underlying processes to be continuous.

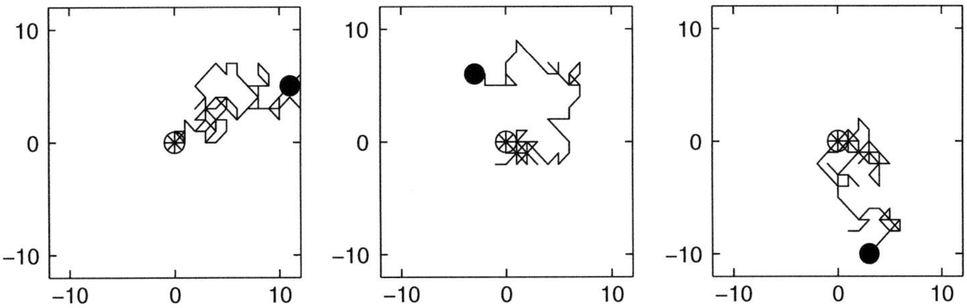

Figure 4.2. Three independent drunken walks in x and y. The crosshairs mark the start point (the origin) in each case. The solid dot marks the termination after 100 time steps.

Problem 4.1 Confirm the results given for mean and variance of the simple discrete walk.

Problem 4.2 Generate some walks as it Figure 4.1. Plot the ensemble statistics (mean and variance) versus ensemble size n. How big should an ensemble be?

Problem 4.3 Generate some walks as in Figure 4.1 and observe every third position. Plot the observed trajectories. Compare with theory.

4.2 Continuous Processes

4.2.1 Resolved and Subgrid Motion

Let $\mathbf{v}(\mathbf{x},t)$ be the fluid velocity, continuous-in-time and discretized on a mesh. There will be a spatially resolved part $\overline{\mathbf{v}}$, presumed known and integrable. The remainder will be unresolved, the subgrid motion $\widehat{\mathbf{v}}$. A stochastic model for it is required. Turbulence closure relations link the two parts.

Displacement will be the integral of both, and it is similarly partitioned:

$$\Delta\mathbf{X}_i = \int_t^{t+\Delta t} \mathbf{V}_\mathrm{i}\, dt = \int_t^{t+\Delta t} \overline{\mathbf{V}}_i\, dt + \int_t^{t+\Delta t} \widehat{\mathbf{V}}_i\, dt \tag{4.3}$$

$$= \Delta\overline{\mathbf{X}}_i + \Delta\widehat{\mathbf{X}}_i \tag{4.4}$$

These are Lagrangian integrals. The subscript indicates "particle i"; and $\mathbf{V}_i$ indicates the velocity sampled by a particle, $\mathbf{V}_i = \mathbf{v}(\mathbf{X}_i)$. Each particle trajectory will be the Lagrangian integration of the resolved field $\mathbf{v}$ plus one sampling of the subgrid $\widehat{\mathbf{v}}$. Displacement, like velocity, will have a deterministic or mean part, plus a stochastic part.

It is the per-step displacement $\Delta\widehat{\mathbf{X}}_i$ that we wish to describe in this chapter – its properties, how it is to be generated, and the underlying assumptions. Accordingly, we need a probabilistic model of $\Delta\widehat{\mathbf{X}}_i$, compatible with the basics of motion (Equations 4.3, 4.4). Several articles address this canonical problem, that of determining a proper *Lagrangian stochastic model* (LSM). Mariano and Ryan [302] p. 453ff, and Pasquero et al. [351] provide good summaries of recent progress, including that of Griffa [185], Rupolo et al. [393], Rodean [385], and Brickman and Smith [52].

In practical work, Δt must be finite, small but not zero, assumed to be converging toward zero. The immediate implication is that the parameters of $\Delta\widehat{\mathbf{X}}_i$ depend on the time step Δt. As this is a major numerical variable, but not a physical parameter, the dependence on Δt must be explicit, not hidden in other LSM parameters.

For simplicity we will drop the subscript for particle i; it is implied.

4.2.2 A Hierarchy

Below we discuss a **hierarchy of Lagrangian stochastic models:** ARO, AR1, AR2, and AR1-s. The AR part of this hierarchy refers to the common description "AutoCorrelated" random walk. These all have the discrete form[1]

$$\Delta\widehat{X}^{k+1} = \rho_1 \Delta\widehat{X}^{k} + \rho_2 \Delta\widehat{X}^{k-1} + \mathcal{R}^{k+1} \tag{4.5}$$

that is, the current random displacement is a blend of at most its two predecessors, plus a new disturbance $\mathcal{R}^{k+1}$. All have the independent random displacement $\mathcal{R}$ added in each time step. The hierarchy provides an increasing level of Lagrangian memory among the random disturbances. AR0 is the baseline random walk model, with no memory of previous displacements – ρ_1 and ρ_2 both zero. AR1 adds one level of additional memory, ρ_1, but keeps $\rho_2 = 0$. AR2 invokes both memory terms. Each of these has independent x- and y-components; that is implied throughout. AR1-s is a hybrid – one level of memory is represented, with a spinning tendency built-in.

Of course we are dealing with increments to position $\Delta\widehat{X}$, so the Lagrangian memory of all past displacements *is* accumulated as the current position $\widehat{X}$ in any trajectory, even for AR0:

$$\widehat{X}^{k} = \sum_{j=1}^{k} \Delta\widehat{X}^{j} \tag{4.6}$$

In the context of coastal ocean simulation, Δt is always finite but small relative to natural time scales; and attention is always paid to the limiting differential equations. The random walk processes as in Equation 4.5 are commonly conceived as discrete processes, but it is a mistake to ignore the dependence of parameters on Δt. Accordingly, we consider the limiting form of the underlying processes in order to develop this dependence explicitly, and require the convergence of random walk processes as

[1] Throughout, $\Delta\widehat{X}^{k+1} = \widehat{X}^{k+1} - \widehat{X}^{k}$ is the displacement during time interval $k+1$.

Δt becomes arbitrarily small. There is a key reliance on the form of noise operating in the continuum – throughout this is assumed to be *Wiener noise*, which has very special integral properties.

4.3 The AR0 Model – Uncorrelated Random Walk and Simple Diffusion

4.3.1 The Displacement Process

The simplest stochastic model is the direct specification of the statistics of $\Delta\widehat{X}$:

$$\Delta\widehat{X}^{k+1} = \int_t^{t+\Delta t} \widehat{V} dt = \mathcal{R}^{k+1} \tag{4.7}$$

The stochastic displacement from time t to $t+\Delta t$ is random, a realization of a *Wiener process*. This critical process accumulates, in $\mathcal{R}$, uncorrelated noise on the interval Δt.

The defining characteristics of the Wiener process: the discrete sequence $\mathcal{R}$ is Gaussian with zero mean, no autocorrelation, and variance $\sigma^2_{\mathcal{R}}$ proportional to the integration time Δt:

$$\mathcal{R}^{k+1} \sim S\sqrt{\Delta t}\cdot\mathcal{G}(0,1) \tag{4.8}$$

$$\sigma^2_{\mathcal{R}} \equiv E[\mathcal{R}^{k+1}\mathcal{R}^{k+1}] = S^2\Delta t \tag{4.9}$$

There is no autocorrelation within the sequence $\mathcal{R}^{k+1}$, $\mathcal{R}^{k}$, $\mathcal{R}^{k-1}$, ... S here is the RMS size of the noise being accumulated. It is *not* dimensionless: $\mathcal{R}$ has units of displacement, so S^2 has units of X^2/t (More discussion about Wiener processes occurs in Appendix C.)

It is a quick matter to discover that $\Delta\widehat{X}$ inherits the statistics of $\mathcal{R}$:

$$\sigma^2_{\Delta\widehat{X}} \equiv E[\Delta\widehat{X}^{k+1}\Delta\widehat{X}^{k+1}] = \sigma^2_{\mathcal{R}} \tag{4.10}$$

and the trajectory $\widehat{X}$ is the sum of these increments

$$\widehat{X}^L = \sum_{k=1}^{L} \Delta\widehat{X}^k \tag{4.11}$$

$\widehat{X}^L$ has the usual properties of a sum of independent random variables:

$$E(\widehat{X}^L\widehat{X}^L) = L\sigma^2_{\Delta\widehat{X}} = S^2 L\Delta t = S^2 t \tag{4.12}$$

that is, the variance of $\widehat{X}(t)$ grows linearly with t.[2]

[2] $t = L\Delta t$; S is the size of the Wiener process, Equation 4.9.

This is a classic finding. It is occurs in many early expositions that invoke random walk processes in, for example, populations (Okubo [343]), turbulence (Taylor [445]), or molecular motions of gases (Present [361]). The contexts are diverse, and the mathematical development similar.

The finite integration of random signals is developed extensively in the context of Wiener processes for study of dynamics and control. *The properties 4.8, 4.9 are crucial*; as Δt is refined, the RMS displacement scales with $\sqrt{\Delta t}$. The scale S is not dependent on Δt.

Use of this model is extremely simple: integrate the mean displacement of a particle $\Delta\overline{X}$ given the known mean field $\overline{V}$, using a conventional method; and add a random displacement $\Delta\widehat{X}$ generated with zero mean and $\sigma_{\Delta\widehat{X}} = \sqrt{S^2\Delta t}$. Repeat this for all particles and update their individual positions; then go on to the next time step. The ensemble properties given above will emerge. This random walk model has a single physical parameter S, plus the numerical parameter Δt.

4.3.2 Correspondence to Diffusion

It is well known that this procedure is equivalent to solving the diffusion equation,

$$\frac{\partial C}{\partial t} = D\frac{\partial^2 C}{\partial x^2} \tag{4.13}$$

when $\sigma_{\Delta\widehat{X}} = \sqrt{2D\Delta t}$, that is, the RMS size of the random displacement over Δt. In other words, $S^2 = 2D$ and the random displacement $\mathcal{R}$ has mean size $\sqrt{2D\Delta t}$:

$$\mathcal{R} \sim \sqrt{2D\Delta t}\cdot\mathcal{G}(0,1) \tag{4.14}$$

The mysterious size parameter S is actually a surrogate for the conventional macroscale diffusivity.[3] A great many contributors have demonstrated this equivalence. (For a recent study, see [365].)

This theory is well developed, both for continuous and discrete Lagrangian trajectories. Of course, in simulation we are concerned with discrete processes; but their convergence as $\Delta t \to 0$ is critical. And, the simulations will require stochastic parameters that exhibit dependence on Δt.

As an illustration, Figure 4.3 shows some individual trajectories simulated with $\Delta\widehat{X} \sim \sqrt{2D\Delta t}\cdot\mathcal{G}(0,1)$. The envelope shown is the analytic result for linear variance growth in particle positions, $\widehat{X} = \pm\sigma_{\widehat{X}}(t) = \pm\sqrt{2Dt}$. The per-step displacements have no autocorrelation; and although the displacements are very erratic, each contains the Lagrangian accumulation of all prior random displacements.

[3] Alternatively put: if S is known, then the random walk generates ensemble variance as in a diffusion process with $D = S^2/2$.

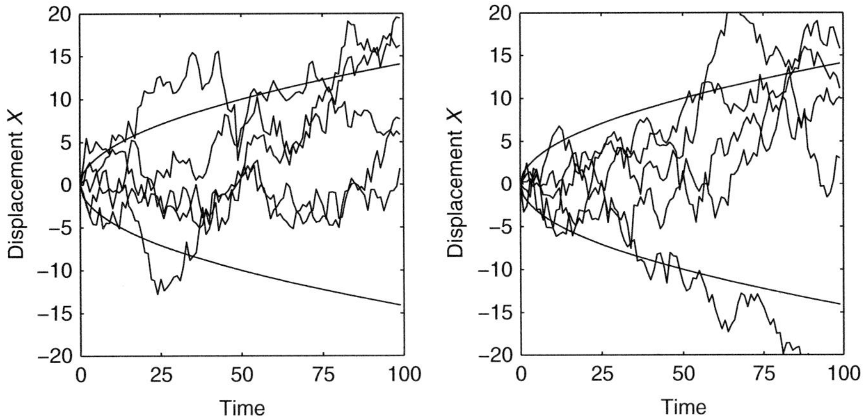

Figure 4.3. AR0: Individual 1-D particle trajectories computed with $\Delta\widehat{X} \sim \sqrt{2D\Delta t} \cdot \mathcal{G}(0,1)$. $D = 1$, $\Delta t = 1$. Each panel depicts five different random walks; each is statistically the same. The bounding line is $\pm$ one analytic standard deviation of all possible particle positions at that time ($\pm\sqrt{2Dt}$).

For clarity at this point, we recall that by definition the random forcing is a Wiener process:

$$\mathcal{R} = \sqrt{2D}\Delta\mathcal{W} \sim \sqrt{2D\Delta t} \cdot \mathcal{G}(0,1) \tag{4.15}$$

The 1-D diffusion correspondence that we will use is

$$\frac{d}{dt}\sigma^2_{\widehat{X}} = 2D \tag{4.16}$$

For independent isotropic walk in two dimensions,

$$\frac{d}{dt}\left(\sigma^2_{\widehat{X}} + \sigma^2_{\widehat{Y}}\right) = 4D \tag{4.17}$$

and so forth. Generally the more complex models described below show variance growth rates that are slower at the outset, if the process generating the motion is truly at rest to start. It is the asymptotic limit of this growth rate that connects to the diffusion coefficient.

Problem 4.4 Generate an ensemble of N simple random walks, using the Gaussian deviate. Show that your results are in accord with theory in terms of particle variance growth.

Problem 4.5 Show the dependence of your results on N.

Problem 4.6 Redo Problem 4.4 using the uniform deviate with the same mean (0) and variance. What differences do you observe?

Problem 4.7 Redo Problem 4.4 with spatially dependent diffusion coefficient $D(x) = 1 + |x|/L$, were L is a scaling length. Describe the outcome.

This is the AR0 model – no memory of past stochastic influences.

AR0

$$\Delta\widehat{X}^N = \mathcal{R}^N \tag{4.18}$$

$$\mathcal{R} \sim \sqrt{2D\Delta t} \cdot \mathcal{G}(0,1) \tag{4.19}$$

There is an obvious problem as Δt becomes small – it is not credible that random displacements are instantly forgotten or have no staying power, energy, or inertia. Below we deal with this in the AR1 and AR2 models. But first, we address some natural complications – inhomogeneity, anisotropy, shear, and convergence – that are common in the coastal ocean.

4.3.3 Multi-Dimensions

These same ideas are applicable to 2- and 3-dimensional situations, under the common assumption of *independent random walk in orthogonal directions.*

In 2-D, for an ensemble starting at the origin, the properties of the x and y distributions will be independently preserved. There will be no preference for any direction. The per-step displacement is the simple generalization for two independent scalars,

$$\Delta\widehat{X} \sim S_{\Delta\widehat{X}} \sqrt{\Delta t} \cdot \mathcal{G}(0,1) \tag{4.20}$$

$$\Delta\widehat{Y} \sim S_{\Delta\widehat{Y}} \sqrt{\Delta t} \cdot \mathcal{G}(0,1) \tag{4.21}$$

In terms of the distance R from the origin, we have for the ith particle

$$\widehat{R}_i^2 = \widehat{X}_i^2 + \widehat{Y}_i^2 \tag{4.22}$$

Because the increments $\Delta\widehat{X}_i$ add, we have

$$\widehat{X}_i = \sum_{k=1}^{N} \Delta\widehat{X}_i^k \tag{4.23}$$

$$\overline{\widehat{X}_i} = 0 \qquad \overline{\widehat{X}_i^2} = N\Delta t S_{\Delta\widehat{X}}^2 \tag{4.24}$$

and analogously for y. N is the number of time steps.

The mean $\widehat{R}^2$ will be $\sigma^2_{\widehat{X}} + \sigma^2_{\widehat{Y}} = t(S^2_{\Delta\widehat{X}} + S^2_{\Delta\widehat{Y}})$. The variances add; but both the mean $\widehat{R}$ and its RMS value will grow with time; for the latter

$$\sqrt{\overline{\widehat{R}^2}} = \sqrt{\sigma^2_{\widehat{X}} + \sigma^2_{\widehat{Y}}} = \sqrt{t(S^2_{\Delta\widehat{X}} + S^2_{\Delta\widehat{Y}})} \tag{4.25}$$

The RMS $\widehat{R}$ grows with $\sqrt{t}$. When $S_{\Delta\widehat{X}} = S_{\Delta\widehat{Y}}$ we have the special case of isotropic diffusion. If not, we have simple scaling of the two orthogonal axes – this is a very special case of anisotropy, which is treated in more generality later.

Problem 4.8 Generate an ensemble of 2-D Cartesian random walks under conditions analogous to Problem 4.1 above. Show that (1) the x- and y-displacements from the origin have zero mean and proper variance growth and (2) that the circumferential distributions are uniform.

4.3.4 Inhomogeneous Diffusion

In the case where D is not constant, there will be a pileup of trajectories walking away for regions of high D and falsely accumulating in regions of low D. One adds the pseudo-velocity $\frac{dD}{dx}$ (in multidimensional form, ∇D) to the deterministic velocity $\overline{V}$ – that is, add a pseudo-advection term that counteracts this artificial tendency.

$$\Delta X = \int_t^{t+\Delta t} \left(\overline{V} + \frac{dD}{dx} \right) dt + \Delta\widehat{X}\left(X^*\right) \tag{4.26}$$

$$X^* = X + \frac{\Delta t}{2} \left(\overline{V} + \frac{dD}{dx} \right) \tag{4.27}$$

The pseudo-advective term contributes directly; in addition, the random displacement is evaluated at the advected position X^*, approximately at the center of the time step.

This is very practical problem in real simulations. The idea has been used successfully in numerous instances: Skellam [420], Hunter et al. [202], Ross and Sharples [388], Proehl et al. [366], and many others discuss and apply it. The presentation by Visser [464] is particularly clear and simple.

Rodean [385] summarized the necessary requirement in terms of the *well-mixed condition*: an initially uniform distribution of particles should remain so over time. This can seriously constrain practical applications where there is insufficient resolution to make gradients of D computable. (Brickman and Smith [52].) North et al. [335] extend these findings and suggest a vertical interpolation scheme for diffusivity, using continuous functions suitable for differentiation.

This accommodation for variable diffusivity is critical; it is among the most important practical matters in coastal ocean simulation. If there were a list of key items, this would be joined by the $\sqrt{D\Delta t}$ sizing of random walk, the proper handling of autocorrelation, and the link between diffusivity and turbulence closure.

Problem 4.9 Revisit Problem 4.7 with this adjustment for $\frac{dD}{dx}$. Describe the results.

4.3.5 Anisotropic Diffusion

Directional diffusion would be characterized by the generalization of the gradient–flux relation

$$\begin{Bmatrix} F_x \\ F_y \\ F_z \end{Bmatrix} = \begin{bmatrix} D_{xx} & D_{xy} & D_{xz} \\ D_{xy} & D_{yy} & D_{yz} \\ D_{zx} & D_{zy} & D_{zz} \end{bmatrix} \begin{Bmatrix} G_x \\ G_y \\ G_z \end{Bmatrix} \tag{4.28}$$

The diffusive flux vector is $\{\mathbf{F}\}$; $\{\mathbf{G}\}$ is the gradient of concentration. $[\mathbf{D}]$ is the diffusion tensor, comprising the scalar directional diffusivities D_{ij}. In the standard representation, $[\mathbf{D}]$ is symmetric ($D_{xy} = D_{yx}$ etc.)

When the off-diagonals vanish (and the axes are orthogonal), we can treat this as three independent scalar random walks without further problems. These three directions are the *principal directions* of the process.

In the coastal ocean, the main anisotropy is in the vertical, where entirely different length and time scales govern the subgrid-scale turbulent mixing. The mixing process is typically divided into independent vertical (z) and horizontal (x, y) parts, with no interaction: $D_{zx} = D_{zy} = 0$:

$$[\mathbf{D}] = \begin{bmatrix} D_{xx} & D_{xy} & 0 \\ D_{yx} & D_{yy} & 0 \\ 0 & 0 & D_{zz} \end{bmatrix} \tag{4.29}$$

z is one of the principal directions. The vertical diffusion may be treated separately and independently of the horizontal. So we are left with the problem of 2-D horizontal dispersion:

$$\begin{Bmatrix} F_x \\ F_y \end{Bmatrix} = \begin{bmatrix} D_{xx} & D_{xy} \\ D_{xy} & D_{yy} \end{bmatrix} \begin{Bmatrix} G_x \\ G_y \end{Bmatrix} \tag{4.30}$$

and its conversion into proper horizontal random walk processes.[4] It is the off-diagonals D_{xy} that are our concern. They couple the x- and y-fluxes. If they vanish, we have independent walks in independent directions, each with its own directional diffusion coefficient but no interaction. Otherwise, they represent fluxes that must be accounted for.

[4] Of course, the remaining vertical flux $F_z = D_{zz}G_z$ may not be ignored; but it is not coupled to the horizontal transport.

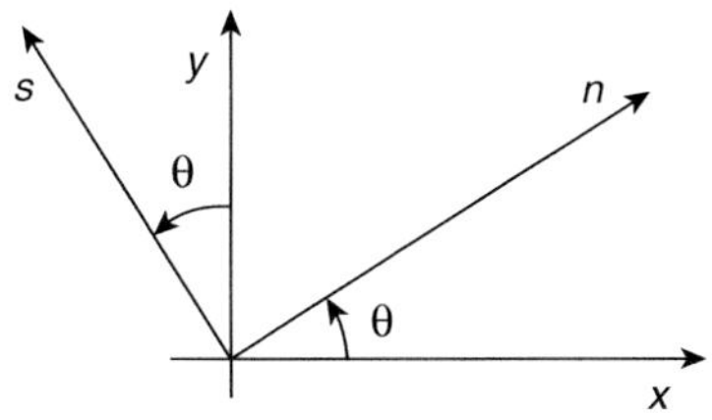

Figure 4.4. Coordinate rotation definition.

How to proceed: find the *principal directions* (n,s) of $[\mathbf{D}]$, along which diffusion occurs independently; produce independent random walks in these directions; and reexpress them in the original (x,y) coordinate system.

The rotation from one coordinate system to another is illustrated in Figure 4.4. Any vector $\{\mathbf{F}\}$ in the (x,y) system is equal to its expression in the (n,s) system:

$$\begin{Bmatrix} F_n \\ F_s \end{Bmatrix} = \begin{bmatrix} \underline{\mathbf{n}}\cdot\underline{\mathbf{x}} & \underline{\mathbf{n}}\cdot\underline{\mathbf{y}} \\ \underline{\mathbf{s}}\cdot\underline{\mathbf{x}} & \underline{\mathbf{s}}\cdot\underline{\mathbf{y}} \end{bmatrix} \begin{Bmatrix} F_x \\ F_y \end{Bmatrix} \tag{4.31}$$

where $\underline{\mathbf{x}}$ is the unit vector in the x direction, etc. The transformation matrix $[\mathbf{T}]$ is thus defined:

$$[\mathbf{T}] = \begin{bmatrix} \underline{\mathbf{n}}\cdot\underline{\mathbf{x}} & \underline{\mathbf{n}}\cdot\underline{\mathbf{y}} \\ \underline{\mathbf{s}}\cdot\underline{\mathbf{x}} & \underline{\mathbf{s}}\cdot\underline{\mathbf{y}} \end{bmatrix} = \begin{bmatrix} \cos\theta & \sin\theta \\ -\sin\theta & \cos\theta \end{bmatrix} \tag{4.32}$$

and the inverse transformation similarly defined, invoking $[\mathbf{T}]^{-1}$. By construction, $[\mathbf{T}]$ is nonsingular and orthogonal: $[\mathbf{T}]^{-1} = \left[\mathbf{T}^*\right]$, that is, its inverse equals its transpose. The rotations are

$$\{\mathbf{F}\}_{ns} = [\mathbf{T}]\{\mathbf{F}\}_{xy} \qquad \{\mathbf{F}\}_{xy} = \left[\mathbf{T}^*\right]\{\mathbf{F}\}_{ns} \tag{4.33}$$

The rotation matrix $[\mathbf{T}]$ has two orthogonal eigenvectors $\{\mathbf{W}\}_1, \{\mathbf{W}\}_2$ and their associated eigenvalues λ_1, λ_2, such that

$$[\mathbf{T}]\{\mathbf{W}_i\} = \lambda_i\{\mathbf{W}_i\} \tag{4.34}$$

When rotated into the (n,s) coordinate system, the gradient-flux law becomes

$$\{\mathbf{F}\}_{ns} = [\mathbf{T}]\left[\mathbf{D}_{xy}\right]\left[\mathbf{T}^*\right]\{\mathbf{G}\}_{ns} \tag{4.35}$$

$$= [\mathbf{D}_{ns}]\{\mathbf{G}\}_{ns} \tag{4.36}$$

with $[\mathbf{D}_{ns}]$ the diffusion matrix in the (n,s) coordinate system:

$$[\mathbf{D}_{ns}] = [\mathbf{T}]\left[\mathbf{D}_{xy}\right]\left[\mathbf{T}^*\right] \tag{4.37}$$

If in Figure 4.4, $\underline{\mathbf{n}} = \underline{\mathbf{w}}_1$ and $\underline{\mathbf{s}} = \underline{\mathbf{w}}_2$, then $[\mathbf{D}_{ns}]$ will be diagonal:

$$[\mathbf{D}_{ns}] = \begin{bmatrix} \lambda_1 & 0 \\ 0 & \lambda_2 \end{bmatrix} \tag{4.38}$$

We have found the principal directions of this diffusion system! The rows of $[\mathbf{T}]$ are the eigenvectors $\{\mathbf{W}\}^*$ of $\left[\mathbf{D}_{xy}\right]$; the transformation matrix is

$$[\mathbf{T}] = \begin{bmatrix} \underline{\mathbf{w}}_1 \cdot \underline{\mathbf{x}} & \underline{\mathbf{w}}_1 \cdot \underline{\mathbf{y}} \\ \underline{\mathbf{w}}_2 \cdot \underline{\mathbf{x}} & \underline{\mathbf{w}}_2 \cdot \underline{\mathbf{y}} \end{bmatrix} \tag{4.39}$$

and the diffusivity is diagonal: $[\mathbf{D}] = \left[diag(\lambda)\right]$.

To sum up the procedure: Given $\left[\mathbf{D}_{xy}\right]$,

1. Find the eigenvectors $\{\mathbf{W}\}_1, \{\mathbf{W}\}_2$ of $\left[\mathbf{D}_{xy}\right]$ and the associated eigenvalues λ_1, λ_2. These are the principal directions.
2. The λ_i are the diffusivities in the principal directions: $\left[\mathbf{D}_{1,2}\right] = \left[diag(\lambda)\right]$.
3. The *rows* of $[\mathbf{T}]$ (the *columns* of $\{\mathbf{T}^*\}$) are the $\{\mathbf{W}\}$.
4. Make independent random walks in the two orthogonal directions, $\Delta_i \sim \sqrt{2\lambda_i \Delta t} \cdot \mathcal{G}(0,1)$.
5. Transform these walks into the (x, y) system:

$$\begin{Bmatrix} \Delta\widehat{X} \\ \Delta\widehat{Y} \end{Bmatrix} = \left[\mathbf{T}^*\right] \begin{Bmatrix} \Delta_1 \\ \Delta_2 \end{Bmatrix} \tag{4.40}$$

If one is given the principal directions in terms of the rotation angle θ and the principal diffusivities (λ_1, λ_2): form $[\mathbf{T}]$ according to Equation 4.32; then follow steps 4 and 5 above.

Problem 4.10 Suppose we have $D_{xx} = D_{yy} = 1$, giving the illusion of isotropy, but $D_{xy} = 0.8$. The eigenvalues in this case are $\lambda_1 = 1.8$, $\lambda_2 = 0.2$; and $\theta = \pi/4$. Because the ensemble lengths scale as $\sqrt{D}$, the ensemble will have aspect ratio $\sqrt{1.8/.2}$ – it will be elongated by a factor of 3, and rotated 45 degrees! Confirm the numbers for this case.

Problem 4.11 Simulate this case. Do the simulation results agree with theory?

4.3.6 Shear and Convergence

Lynch and Smith [289] described simple homogeneous random walk in sheared and/or convergent Eulerian flows. The effects of these features can distort the usual simple outcomes of random walk – the ensemble displacement patterns show the effect of randomly walking over important gradients in the Eulerian flow. This leads to various forms of distortion, including the elongation of ensembles by shear

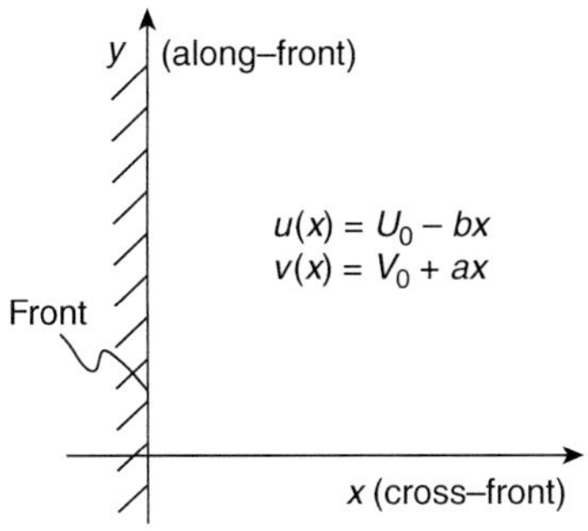

Figure 4.5. Simple frontal system with Shear and Convergence.

dispersion, and to limited variance growth in the convergent direction as random walk is "trapped" in that direction. Figure 4.5 illustrates this setup. Following [289] we pose a simple 2-D situation crudely representing a front with x the distance across the front, y the distance along it. For the Eulerian velocity (u, v) we have uniformity in the y direction, and a gradient in the x direction. There is a mean trajectory, embodying both shear and convergence effects. Superposed are independent AR0 walks in x and in y:

$$\Delta\widehat{X} = -b\Delta t\widehat{X} + \mathcal{R}_x \tag{4.41}$$

$$\Delta\widehat{Y} = a\Delta t\widehat{X} + \mathcal{R}_y \tag{4.42}$$

with independent and uncorrelated disturbances $\mathcal{R}$:

$$\mathcal{R}_x \sim \sqrt{2D_x\Delta t}\cdot\mathcal{G}(0,1) \tag{4.43}$$

$$\mathcal{R}_y \sim \sqrt{2D_y\Delta t}\cdot\mathcal{G}(0,1) \tag{4.44}$$

Because of the underlying gradients in the Eulerian velocity, covariance develops between $\widehat{X}$ and $\widehat{Y}$ – as in anisotropic diffusion – even when the random walk is isotropic ($D_x = D_y$).

$$Cov_{\widehat{X}\widehat{Y}} = 2D_x t \begin{bmatrix} \overline{\widehat{X}\widehat{X}} & \overline{\widehat{X}\widehat{Y}} \\ \overline{\widehat{Y}\widehat{X}} & \overline{\widehat{Y}\widehat{Y}} \end{bmatrix}$$

The main diagonals initially dominate, reflecting uncorrelated random walks. The off-diagonal terms $\overline{\widehat{X}\widehat{Y}}$ grow over time.

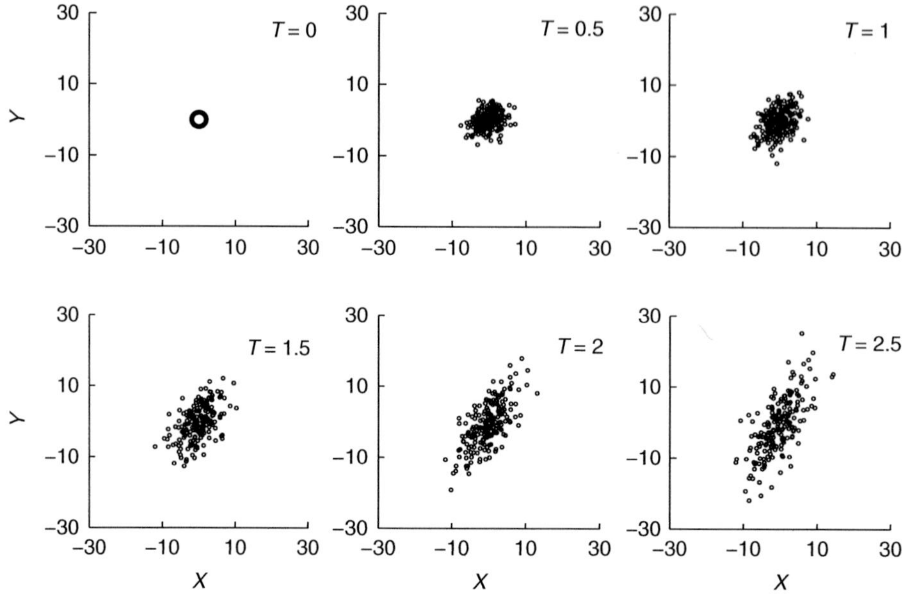

Figure 4.6. Shear-only case; simulation of ellipse growth versus T.

Two dimensionless times emerge, shear time $T = at$ and convergence time $\tau = bt$. With $\mathcal{D}$ the ratio of diffusivities $\frac{D_y}{D_x}$, and $\beta = \frac{b}{a} \equiv \tau/T$, the general result is

$$Cov_{\widehat{X}\widehat{Y}} = 2D_x t \begin{bmatrix} \left[\frac{1-e^{-2\tau}}{2\tau}\right] & \frac{T}{2}\left[\frac{1+e^{-2\tau}-2e^{-\tau}}{\tau^2}\right] \\ \frac{T}{2}\left[\frac{1+e^{-2\tau}-2e^{-\tau}}{\tau^2}\right] & T^2\left[\frac{\tau+\frac{1}{2}(1-e^{-2\tau})-2(1-e^{-\tau})}{\tau^3}\right]+\mathcal{D} \end{bmatrix}$$

Appendix F contains other equivalent expressions of this – limiting forms, Taylor series, and so on.

Figure 4.6 shows a simulation outcome for **shear only** and isotropic walk. The elongation in the along-front (y) direction is clear and intuitive. There is also an early-time tilting of the principal directions (45°); the progressive along-front elongation makes the tilt angle diminish as time advances.

When we have no shear ($a = T = 0$) but **only convergence**, the mean trajectory is

$$\overline{X} = \frac{U_o}{b}\left[1 - e^{-\tau}\right] \qquad \overline{Y} = V_o t \tag{4.45}$$

and the covariances are

$$\overline{\widehat{X}\widehat{X}} = \frac{D_x}{b}\left[1 - e^{-2\tau}\right] \qquad \overline{\widehat{Y}\widehat{Y}} = 2D_y t \tag{4.46}$$

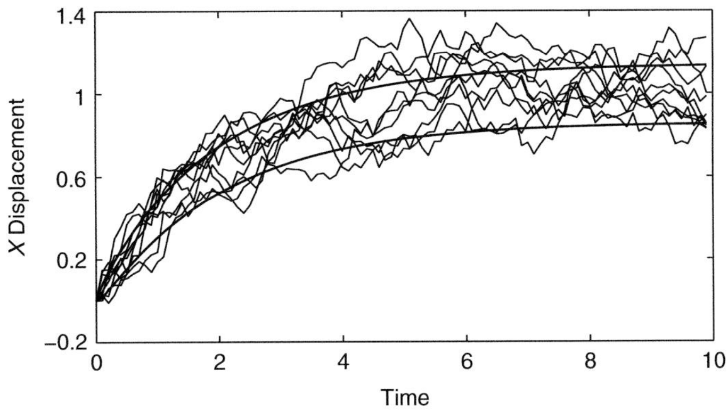

Figure 4.7. Convergence-only case. 10 X trajectories are plotted versus time; the envelope is $\overline{X}(t) \pm 1$ standard deviation, using the analytical formulas 4.45, 4.46. The trapping of trajectories due to convergence is plainly evident. Parameters: $[U_o, a, b, D, \Delta t] = [0.5, 0, 0.5, 0.01, 0.1]$.

The off-diagonal entries vanish. The mean position seeks the front, with constant along-front motion. σ_x^2 is trapped at $\frac{D_x}{b}$; and σ_y^2 grows linearly with time, as in the normal case. The decay to equilibrium goes as $e^{-\tau} = e^{-bt}$. (Form 2 in Appendix F is the most convenient for this case.)

Simulations are plotted in Figure 4.7, confirming the theory.

4.3.7 Metrics of Resolution

The selection of Δt needs scaling criteria. Several useful dimensionless metrics, familiar in numerical analysis of transport problems, are readily adapted here. Each needs to be kept below unity, or the ratio of small integers:

- Keep the random displacement on the order of the spatial resolution:

$$\sigma < \Delta x \Longrightarrow R = \frac{D\Delta t}{\Delta x^2} < \frac{1}{2} \tag{4.47}$$

  R is the familiar "Diffusion number," in widespread use in the analysis of parabolic differential equations.
- the Courant number measures advective wave speed as resolved numerically:

$$C = |\mathbf{V} + \nabla D| \frac{\Delta t}{\Delta x} \tag{4.48}$$

  Here the extra term ∇D is the pseudovelocity, a necessary element in simple random walk models to avoid artificial particle attraction to low-D regions.

- Peclet number: another metric of balance between advective and diffusive motion; in the present context the pseudovelocity is added:

$$P_e = \frac{|\mathbf{V} + \nabla D|\Delta x}{D} \tag{4.49}$$

This is usefully expressed as the sum of the conventional Peclet number, plus the relative change in D over a grid element:

$$P_e = \frac{|\mathbf{V}|\Delta x}{D} + \frac{\Delta D}{D} \tag{4.50}$$

This is a criterion of spatial resolution, whether advection and diffusion are resolved in balance on a grid. The relative change in D, on a per-element basis, is familiar in grid generation where a similar criterion occurs concerning bathymetric resolution and regions of steep gradients in depth require enhanced resolution.
- There are some dependencies among these. *Pe* is not independent of the other two:

$$P_e = \frac{C}{R} \tag{4.51}$$

The Courant number can be expressed as the sum of its conventional expression plus a term in the local change in D:

$$C = |\mathbf{V}|\frac{\Delta t}{\Delta x} + \frac{\Delta D}{D}R \tag{4.52}$$

Δx appears throughout these. In 2-D, a useful local metric is $\Delta x = \sqrt{A_e}$ for quadrilaterals; for triangles it would be $\sqrt{2A_e}$, A_e being the local (horizontal) element area.

4.3.8 Stepsize and the Need for Autocorrelation

This has been a brief exposition of simple random walk simulation. Many have explored this in theory and practice (e.g., Csanady [96], Fischer [153], Dimou and Adams [117], Proehl and Lynch [365]).

In practical simulation work, the discrete version of this process is absolutely necessary. The discrete parameter Δt is assumed to be small in these cases, arbitrarily small. But it is never zero! Of course, there are random walk processes where the stepsize Δt *is* a meaningful parameter in itself. But in marine simulation and especially fluid motions, one normally deals with models based on the continuum PDEs; in this case Δt is arbitrary and assumed to be as small as possible, not having a real significance in the processes represented. One needs to avoid endowing Δt

with increasing significance on its own; its presence in the random walk representation must be clearly visible, so that this numerical parameter can be adjusted as the numerics requires, without hazard to the formulation.

In particle simulation, each particle can be integrated separately, without sharing a common time step. That is a great advantage. Resolution can be adjusted as the need arises, and the full set of ODE integration methods may be used when appropriate. These include automatic error estimation and control.

Ensemble size N is ultimately limited by Δt: for a fixed-duration simulation, the work required for all trajectories scales as $\frac{N}{\Delta t}$. This matter is trivial for trial simulations, but serious and potentially limiting for practical work. This practical limitation puts ensemble size in competition with resolution of the motion.

More fundamentally, the simple AR0 model of random walk has random displacements or jumps; the mean size of the jump decreases as $\sqrt{\Delta t}$. The implied velocity in a step, $\frac{\Delta \widehat{X}}{\Delta t}$, *increases* as $\frac{1}{\sqrt{\Delta t}}$! In 2- and 3-D, velocity can change both speed and direction nearly instantaneously. These are inescapable simulation effects as $\Delta t \rightarrow 0$.

It is not credible that $\Delta \widehat{X}$ can change so abruptly or arbitrarily, and the problem is exacerbated when Δt needs to be small. Extensions of this formulation are needed; this brings us to the AR1 model.

4.4 The AR1 Model – Autocorrelated Velocity

The first extension is to have sequential $\Delta \widehat{X}$ correlated with their predecessors, for the same particle (not with other particles). If the uncorrelated AR0 provides a random *walk*, then from a motion point of view, AR1 is a random *flight*. The stochastic disturbance affects particle velocity $\widehat{V}$.

The canonical form will be discrete:

$$\widehat{V}^k = \rho \widehat{V}^{k-1} + \mathcal{R}_1^k \tag{4.53}$$

$$\Delta \widehat{X}^k = \Delta t \widehat{V}^k \tag{4.54}$$

$$\widehat{X}^k - \widehat{X}^0 = \Delta t \sum_{i=1}^{k} \widehat{V}^i \tag{4.55}$$

We will concentrate first on the velocity series $\widehat{V}$. Afterward we will develop its summation for particle position $\widehat{X}$.[5]

[5] The time convention is always: $\Delta \widehat{X}^k = \widehat{X}^k - \widehat{X}^{k-1}$.

This discrete form is the practical simulation approach. Its continuous counterpart is conveniently written as first-order ODEs

$$\frac{d\widehat{V}}{dt} + \frac{\widehat{V}}{T} = \mathcal{R}_v \tag{4.56}$$

$$\frac{d\widehat{X}}{dt} = \widehat{V} \tag{4.57}$$

$$\widehat{X}(t) - \widehat{X}(0) = \int_{\tau=0}^{t} \widehat{V} d\tau \tag{4.58}$$

or, as the equivalent second-order ODE

$$\frac{d^2\widehat{X}}{dt^2} + \frac{1}{T}\frac{d\widehat{X}}{dt} = \mathcal{R}_v \tag{4.59}$$

$\mathcal{R}_v$ is the fundamental stochastic input, imported through a first-order filter with time constant T. $\mathcal{R}_1$ is the integrated result; it drives discrete changes in velocity over Δt (Equation 4.53).[6]

These continuous forms, essential in many analytic contexts, are always discretized in simulation work. The continuum is the limit of all discretizations for tiny $\frac{\Delta t}{T}$. The links among Δt, ρ, T, $\mathcal{R}_v$, and $\mathcal{R}_1$ are essential in relating the theoretical continuum to the practical discrete form. These are developed below.[7]

4.4.1 AR1: Continuous Form and Its Discretization

Exact Integration. We have the first-order ODE

$$\frac{d\widehat{V}}{dt} + \frac{\widehat{V}}{T} = \mathcal{R}_v \tag{4.60}$$

with parameter T and initial condition $\widehat{V}_0$. The solution is an integration of the input:

$$\widehat{V}(t) = \widehat{V}_0 \widehat{V}_h(t) + \int_{\tau=0}^{\tau=t} \widehat{V}_h(t-\tau)\mathcal{R}_v(\tau)\, d\tau \tag{4.61}$$

with homogeneous solution $\widehat{V}_h$

$$\widehat{V}_h(t) = e^{-t/T} \tag{4.62}$$

[6] $\mathcal{R}_v$ need not be differentiable; but it must be integrable.

[7] We assume throughout that $T > 0$, else we have an unstable process.

The parameter T is prominent. It scales the response time for relaxing (forgetting) the starting point $\widehat{V}_0$; and it modulates the assimilation of the random input $\mathcal{R}_v$. Equivalently, T is the persistence time for a disturbance $\mathcal{R}_v$ once introduced. T is also commonly referred to as the velocity decorrelation time.

$\mathcal{R}_v dt$ is a *Wiener increment*, scaled by the intrinsic size S_v:

$$\mathcal{R}_v dt = S_v d\mathcal{W} \tag{4.63}$$

So we have

$$\widehat{V}(t) = \widehat{V}_0 \widehat{V}_h(t) + S_v \int_{\tau=0}^{\tau=t} \widehat{V}_h(t-\tau)\, d\mathcal{W} \tag{4.64}$$

$$= \widehat{V}_0 \widehat{V}_h(t) + S_v S_h \Delta\mathcal{W} \tag{4.65}$$

The *Wiener integral* $\Delta\mathcal{W}$ is Gaussian,

$$\Delta\mathcal{W} \sim \sqrt{\Delta t} \cdot \mathcal{G}(0,1) \tag{4.66}$$

S_h is the RMS size of the homogeneous response $V_h(t)$ on the interval $[0, \Delta t]$. It is close to unity for small $\frac{\Delta t}{T}$. From Equation 4.62,

$$S_h = \sqrt{\left[\frac{1-e^{-2\Delta t/T}}{2\Delta t/T}\right]} = \sqrt{1 - \frac{\Delta t}{T} + \cdots} \tag{4.67}$$

We note that S_v has units of $\Delta V/\sqrt{t}$. S_h is dimensionless.

Discretizations. Alternatively, we can simply integrate Equation 4.60 over the finite interval Δt:

$$\widehat{V}^{k+1} - \widehat{V}^k + \frac{1}{T}\int_{t_k}^{t_{k+1}} \widehat{V} dt = \int_{t_k}^{t_{k+1}} \mathcal{R}_v \, dt = S_v \Delta\mathcal{W}^{k+1} \tag{4.68}$$

Use of 4.68 requires numerical approximation of $\int \widehat{V} dt$, without splitting up the homogeneous and particular parts. Results for two common discretizations appear below, along with the exact result:

Euler:

$$\widehat{V}^{k+1} = \left(1 - \frac{\Delta t}{T}\right) \widehat{V}^k + S_v \Delta\mathcal{W}^{k+1} \tag{4.69}$$

Crank–Nicolson:

$$\widehat{V}^{k+1} = \left(\frac{1-\Delta t/2T}{1+\Delta t/2T}\right) \widehat{V}^k + \left(\frac{1}{1+\Delta t/2T}\right) S_v \Delta\mathcal{W}^{k+1} \tag{4.70}$$

Exact (Equation 4.65):

$$\widehat{V}^{k+1} = \widehat{V}^k e^{-\Delta t/T} + S_h S_v \Delta \mathcal{W}^{k+1} \tag{4.71}$$

The Wiener integral $\Delta\mathcal{W}$ is Gaussian as in 4.66. S_v is the (constant) size of the Wiener input, $\mathcal{R}_v dt = S_v d\mathcal{W}$. S_h is the RMS size of the homogeneous solution over the first Δt (Equation 4.67).

4.4.2 AR1: Discrete Canonical Form

All of these discrete representations share the same canonical form introduced above (Equation 4.53). There are two parts, the memory of the predecessor and a new random effect:

$$\widehat{V}^{k+1} = \rho \widehat{V}^k + \mathcal{R}_1^{k+1} \tag{4.72}$$

$$\mathcal{R}_1 \sim S_1 \sqrt{\Delta t} \cdot \mathcal{G}(0,1) \tag{4.73}$$

There are two parameters, ρ and S_1 (the size of $\mathcal{R}_1$), summarized in Table 4.2. Both depend on $\frac{\Delta t}{T}$, the resolution of the decay time in the governing ODE.

- ρ is the dimensionless autoregressive coefficient. For the simplest Euler representation,

$$\rho = 1 - \frac{\Delta t}{T} = e^{-\frac{\Delta t}{T}} - \frac{1}{2!}\left(\frac{\Delta t}{T}\right)^2 + \cdots \tag{4.74}$$

  clearly an $\mathcal{O}(\Delta t^2)$ approximation to the exact result. The Crank–Nicolson representation is in between these, $\mathcal{O}(\Delta t^3)$.
- $\mathcal{R}_1$ is a random input series, Wiener noise with size S_1 accumulated over one time step. S_1 itself is the product of the noise size Sv, intrinsic to the continuous Wiener process and bearing dimensions, and a dimensionless time-discretization factor (close to unity for small $\Delta t/T$). In the limit as $\Delta t \to 0$, $S_1 \to S_v$ for all cases.

AR1 Properties. By definition, the input $\mathcal{R}_1$ has zero mean and is not correlated with anything. Its variance: $\sigma^2_{\mathcal{R}_1} = S_1^2 \Delta t$.

It is straightforward to discover the moments of the output $\widehat{V}$. For the variance[8]

$$E(\widehat{V}^{k+1}\widehat{V}^{k+1}) = \rho^2 E(\widehat{V}^k \widehat{V}^k) + \sigma^2_{\mathcal{R}_1} \tag{4.75}$$

[8] We describe the *unconditional* variance of $\widehat{V}$ here, $E(\widehat{V}^{k+1}\widehat{V}^{k+1}) = E(\widehat{V}^k\widehat{V}^k) = \sigma^2_{\widehat{V}}$, independent of k. If $\widehat{V}^k$ were known, then the *conditional* variance of $\widehat{V}^{k+1}$ would be simply that of $\mathcal{R}_1$.

Table 4.2. *AR1 Parameters Depending on Discretization Method*

Method	ρ	S_1
Euler	$\left[1-\frac{\Delta t}{T}\right]$	S_v
C-N	$\left[\frac{1-\Delta t/2T}{1+\Delta t/2T}\right]$	$\left[\frac{1}{1+\Delta t/2T}\right] S_v$
Exact	$e^{-\Delta t/T}$	$\sqrt{\left[\frac{1-e^{-2\Delta t/T}}{2\Delta t/T}\right]} S_v$

Note: S_v is the intrinsic size of the Wiener noise.

and thus

$$\sigma^2_{\widehat{V}} = \frac{1}{\left[1-\rho^2\right]} \cdot \sigma^2_{\mathcal{R}_1} \tag{4.76}$$

The input variance is amplified by the memory. The lag-1 covariance is

$$E(\widehat{V}^{k+1}\widehat{V}^k) = \rho E(\widehat{V}^k\widehat{V}^k) + E(\mathcal{R}_1^{k+1}\widehat{V}^k) \tag{4.77}$$

$$= \rho\, \sigma^2_{\widehat{V}} \tag{4.78}$$

(The second term vanishes by definition.) The autocorrelation coefficient, $E(\widehat{V}^{k+1}\widehat{V}^k)/\sigma^2_{\widehat{V}}$, is simply ρ. The same process leads to the easy conclusion that the lag-n covariance is

$$E(\widehat{V}^{k+n}\widehat{V}^k) = \rho^n\, \sigma^2_{\widehat{V}} \tag{4.79}$$

AR1: Convolution of Input Noise. The general AR1 discrete form, Equation 4.72, is equivalent to the discrete convolution of all previous noise inputs $\mathcal{R}_1^i$:

$$\widehat{V}^N = \sum_{i=1}^{i=N} \mathcal{R}_1^i Z_{N-i} \tag{4.80}$$

where we assume the starting conditions of no motion, $\widehat{V}^0 = 0$; and the $\mathcal{R}_1^i$ series begins at $i = 1$. The Z_i are the discrete convolution coefficients (the backward-facing impulse response). Advancing the convolution one step,

$$\widehat{V}^{N+1} = \left[\sum_{i=1}^{i=N} \mathcal{R}_1^i Z_{N+1-i}\right] + \mathcal{R}_1^{N+1} Z_0 \tag{4.81}$$

Substitution into Equation 4.72 gives

$$\sum_{i=1}^{i=N} \mathcal{R}_1^i \left[Z_{N-i+1} - \rho Z_{N-i}\right] + \mathcal{R}_1^{N+1} [Z_0 - 1] = 0 \tag{4.82}$$

and we have the requirements for the Z_i:

$$Z_0 = 1 \tag{4.83}$$

$$Z_j = \rho Z_{j-1} = \rho^j \qquad (j \geq 1) \tag{4.84}$$

The general AR1 autoregression 4.72 amounts to the discrete convolution of all previous noise. Assuming $|\rho| < 1$, the influence of a disturbance $\mathcal{R}_1$ introduced at time j is felt in all future times, at geometrically decreasing strength.

AR1 Stability, Oscillations, Resolution The AR1 model is stable for $|\rho| < 1$. For the Euler expression 4.74, $\rho = 1 - \frac{\Delta t}{T}$ so we have stability when $\frac{\Delta t}{T} < 2$ (we assume T is non-negative). As soon as $\frac{\Delta t}{T}$ exceeds 1, however, we get negative ρ and there will be an unintended oscillation in $\Delta \widehat{X}$ from one time step to the next; but it remains stable until $\frac{\Delta t}{T}$ exceeds 2. Clearly in the range $1 < \frac{\Delta t}{T} < 2$, Δt is too big for physical realism. These considerations vanish when the continuous version $\rho = e^{-\Delta t/T}$ is used; in that case, the AR1 model is always stable, never oscillates; but it is still not resolved when $\Delta t/T$ is large.

Problem 4.12 Confirm the AR1 stability limits given in the text, by generating some sequences of $\widehat{V}$ using (1) the Euler form $\rho = \left[1 - \frac{\Delta t}{T}\right]$ and (2) the continuous form $\rho = e^{-\Delta t/T}$. Do your results agree with theory?

AR1 Covariance. The AR1 input $\mathcal{R}_1$ has variance $\sigma^2_{\mathcal{R}_1}$; it is not correlated with itself nor with any previous $\widehat{V}$,

$$E\left[\mathcal{R}_1^i \mathcal{R}_1^j\right] = \sigma^2_{\mathcal{R}_1} \delta(i-j) \tag{4.85}$$

$$E\left[\mathcal{R}_1^k \widehat{V}^k\right] = \sigma^2_{\mathcal{R}_1} \tag{4.86}$$

$$E\left[\mathcal{R}^k \widehat{V}^{k-i}\right] = 0 \qquad (i > 0) \tag{4.87}$$

Let C_i be the lagged covariance of $\widehat{V}$:

$$C_i \equiv E\left[\widehat{V}^k \widehat{V}^{k-i}\right] \tag{4.88}$$

with $C_i = C_{-i}$, and $C_0 \equiv \sigma^2_{\widehat{V}}$. The covariances for AR1 obey the recursion

$$C_0 = \left[\frac{1}{1-\rho^2}\right]\sigma^2_{\mathcal{R}_1} \tag{4.89}$$

$$C_n = \rho C_{n-1} = \rho^n C_0 \qquad (n \geq 1) \tag{4.90}$$

Assuming $0 < \rho < 1$, we have C_i declining geometrically with i.

4.4.3 AR1: Displacement

In terms of particle displacement: the velocity disturbance $\mathcal{R}_1$ is by definition the stochastic displacement $\Delta\widehat{X}$ divided by the interval Δt. As a matter of formulation (Equations 4.53–4.55), $\mathcal{R} = \mathcal{R}_1 \Delta t$; $\Delta\widehat{X} = \Delta t\widehat{V}$; and $\sigma_{\mathcal{R}} = \Delta t \cdot \sigma_{\mathcal{R}_1}$

$$\Delta\widehat{X}^k = \rho\Delta\widehat{X}^{k-1} + \mathcal{R}^k \tag{4.91}$$

$$\mathcal{R} \sim \Delta t S_1 \sqrt{\Delta t}\mathcal{G}(0,1) \tag{4.92}$$

The total displacement will be the *summation* of the per-step displacements $\Delta t \sum \widehat{V} = \sum \Delta\widehat{X}$.

$$\widehat{X}^k - \widehat{X}^0 = \mathcal{R}^1(1+\rho+\rho^2+\rho^3+\cdots+\rho^{k-2}+\rho^{k-1}) \tag{4.93}$$

$$+\mathcal{R}^2(1+\rho+\rho^2+\rho^3+\cdots+\rho^{k-2}) \tag{4.94}$$

$$+\cdots \tag{4.95}$$

$$+\mathcal{R}^{k-1}(1+\rho) \tag{4.96}$$

$$+\mathcal{R}^k \tag{4.97}$$

These partial sums amount to a convolution of all previous random disturbances,

$$\widehat{X}^k - \widehat{X}^0 = \sum_{i=1}^{k} \mathcal{R}^i B_{k-i} \tag{4.98}$$

with B_n the partial sum

$$B_n = \sum_{i=0}^{n} \rho^i = \frac{1-\rho^{n+1}}{1-\rho} \tag{4.99}$$

B_n increases monotonically with n, from unity toward its upper limit:

$$B_\infty = \frac{1}{1-\rho} \tag{4.100}$$

reflecting the retention in $\widehat{X}$ of the earliest disturbances. The recursion is

$$B_0 = 1 \tag{4.101}$$

$$B_n = B_{n-1} + \rho^n \tag{4.102}$$

$$= \rho B_{n-1} + 1 \tag{4.103}$$

$$= \left(1 - \rho^{n+1}\right) B_\infty \tag{4.104}$$

AR1 Variance Growth and Diffusivity

Shift the temporal axis so that $\widehat{X}^0 = 0$: an ensemble "launch" at $t = 0$. The variance of position is

$$\sigma^2_{\widehat{X}} = E\left(\widehat{X}^k \widehat{X}^k\right) = \sigma^2_{\mathcal{R}} \sum_{i=0}^{k-1} B_i^2 \tag{4.105}$$

The increment in variance growth in the kth step will be

$$\Delta\sigma^2_{\widehat{X}} = \sigma^2_{\mathcal{R}} B^2_{k-1} \tag{4.106}$$

Initially, there is no past to remember and the variance growth per step is simple, $B_0 = 1$ and $\Delta\sigma^2_{\widehat{X}} = \sigma^2_{\mathcal{R}}$. For a well-developed particle ($k \to \infty$), the per-step variance growth is

$$\Delta\sigma^2_{\widehat{X}} = \sigma^2_{\mathcal{R}} \left[\frac{1}{1-\rho}\right]^2 \tag{4.107}$$

This asymptote represents a significant increase relative to the no-memory case. For example, $\rho = 0.9$ leads to $\Delta\sigma^2_{\widehat{X}} = 100\sigma^2_{\mathcal{R}}$ in this asymptotic limit of mature memory. Disturbances introduced are remembered for a long time.

Relative to the continuum, the diffusion coefficient D is proportional to the growth rate of variance $\sigma^2_{\widehat{X}}$. In 1-D:

$$2D = \frac{\Delta\sigma^2_{\widehat{X}}}{\Delta t} = \frac{\sigma^2_{\mathcal{R}}}{\Delta t} \left[\frac{1}{1-\rho}\right]^2 \tag{4.108}$$

in the limit of mature memory. A little rearrangement gives

$$S_1 \Delta t = (1-\rho)\sqrt{2D} \tag{4.109}$$

So restating Equation 4.92, the basic recursion for displacement is

AR1

$$\Delta \widehat{X}^k = \rho \Delta \widehat{X}^{k-1} + \mathcal{R}^k \tag{4.110}$$

$$\mathcal{R} \sim (1-\rho)\sqrt{2D\Delta t} \cdot \mathcal{G}(0,1) \tag{4.111}$$

$$\rho = e^{-\Delta t/T} \simeq 1 - \frac{\Delta t}{T} \tag{4.112}$$

Both D and ρ are needed to parameterize the subgrid motion. D is routinely available in coastal ocean models. ρ is a surrogate for the velocity decorrelation timescale T, and one needs a way to close this. The turbulence decay rate is a candidate for memory in the subgrid, so one could estimate T beginning with that characteristic time τ. In the terminology of the level 2.5 closure model (Chapter 6):

$$\frac{1}{\tau} = \frac{1}{q^2}\frac{dq^2}{dt} \implies \tau = \frac{B_1 l}{2q} \tag{4.113}$$

AR1: $\mathcal{R}_1$ Half-Life

As noted previously, input disturbances $\mathcal{R}_1$ persist in influencing future V, at geometrically decreasing strength ρ^k. The half-life H is given by $\rho^H = 0.5$ and therefore

$$H = \frac{ln(0.5)}{ln(\rho)} \tag{4.114}$$

For ρ near unity, the Taylor series gives the approximation

$$H \simeq -\frac{ln(0.5)}{1-\rho} = \frac{0.693}{1-\rho} \tag{4.115}$$

Values appear in Table 4.3.

AR1 Simulations. AR1 simulations were made with a Gaussian random generator. Generally, large ensemble sizes (N=10,000) recovered the analytic variance growth results above and returned zero mean as expected.

Some representative individual AR1 trajectories are shown in Figure 4.8. For reference, the plots include 1 standard deviation of the ensemble position according to the analytic expression derived previously. The early part of the plots show the slower growth of variance while the memory is still accumulating. Each trajectory is statistically the same. Individuals disperse, increasingly with time; collectively, they get nowhere on average.

Table 4.3. *Input half-life H (number of Δt) versus $\rho = e^{-\frac{\Delta t}{T}}$*

ρ	$T/\Delta t$	Half-life H
0.1	.43	0.30
0.5	1.44	1.0
0.8	4.48	3.1
0.9	9.5	6.6
0.95	19.5	13.5
0.98	49.5	34.3
0.99	99.5	69.0

Note: T is the velocity decorrelation time; $\frac{T}{\Delta t}$ is its resolution.

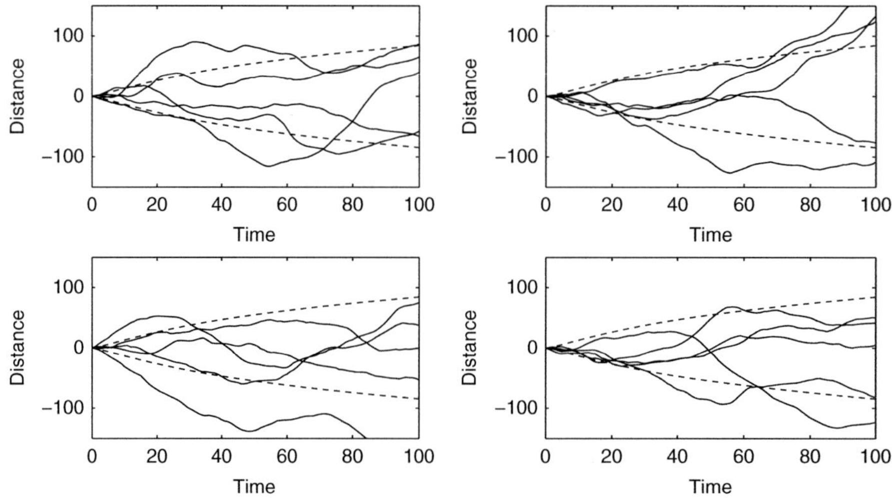

Figure 4.8. AR1 simulation results for $\rho = 0.9$; $\sigma_{\mathcal{R}} = 1$; $\Delta t = 1$; $B_\infty = 10.0$; $T = 9.5$. There are five 1-D trajectories in each frame. The envelope shown is $\pm$ one analytic standard deviation for $\widehat{X}$.

4.4.4 Some Summary Observations about the AR1 Model

Equations 4.53, 4.54, and 4.55 define the discrete AR1 model. Their implementation is quite straightforward. When the variance growth in $\overline{X}$ is linked to the eddy diffusivity D, the recursion is given in Equations 4.110, 4.111, and 4.112.

The timing in the random process is all relative to the velocity decorrelation time T. The process becomes a simple random walk when $\Delta t >> T$; random influences persist while the elapsed time $t < T$ and are forgotten when $t >> T$. The question, "Is there memory in the random flight?" is really, "How big is $\frac{\Delta t}{T}$?".

In continuous time, the autocorrelation function for $\widehat{V}(t)$ is $e^{-t/T}$. This was examined very early (e.g. Taylor, [445]) who called attention to the importance of T and the lower growth rate of variance in early time. Brickman and Smith [52], Pasquero et al. [351](pp. 122 ff) and others report typical decorrelation time $T=$ a few days in the ocean.

Disturbances, once introduced, are persistent. For the velocity disturbances, the half-life H (number of Δt) of a velocity disturbance is $H = \ln(0.5)/\ln(\rho)$.

It is plain to see that ρ tends toward unity when the time step is small compared to the natural time scale T. This is very important in simulation, where Δt is always a discretization parameter, "as small as possible." It is critical that ρ be properly dependent on Δt.

Displacement is the sum of random velocities – as in AR0, these accumulate and early and late displacements affect a complete trajectory equally.

Stability of the AR1 model requires $|\rho| < 1$. For the Euler representation, assuming T is not negative, this means $\Delta t/T < 2$. Clearly, violating stability would be tantamount to huge Δt relative to T, in which case the Euler discretization would seem foolish. In that case the continuous version $\rho = e^{-\Delta t/T}$ would restore stability and revert to AR0 gracefully at large Δt and/or small T. Other discrete approximations can retain unconditional stability (e.g., Crank–Nicolson), but still lose accuracy by embedding discretization error into the AR1 model, leading to Nyquist oscillations for large Δt.

Parameterization of the AR1 model links $\rho = e^{-\frac{\Delta t}{T}}$ to the turbulent velocity autocorrelation time. The rate of variance growth for mature series (many T from rest) is linked to the turbulent diffusion coefficient D, in setting the random step size for displacement: $\mathcal{R} \sim (1-\rho)\sqrt{2D\Delta t}\cdot\mathcal{G}(0,1)$ (Equation 4.111). We can expect both to be available from turbulence closure; at present the computation of D is routine and used in the calculation of other gridded fields. The selection of T requires some creativity.

Relative to Δx (gridsize); simulated motion will be partitioned between resolved Eulerian fields and subgrid motion that requires a stochastic model. Clearly, more motions are resolved on finer grids in (x,t); with resolution comes the switching of some motions from the subgrid into the resolved Eulerian gridded fields. For motions in tidally energetic shelves, resolution of intertidal motion is common (*e.g.* Xu et al. [491]), and random walk models for these systems are likely to have shorter, intertidal decorrelation time T, as the motion at tidal timescales is resolved.

Finally, AR1 introduces persistence to the randomness; but it does *not* create oscillatory behavior. Clearly, some processes at the subgrid scale could be properly modeled to include, for example, rotational or wave-like motion in $\widehat{V}$ and/or $\widehat{X}$. AR2 and AR1-s introduce the possibility of such behavior, being second-order representations of the random dynamic.

4.5 The AR2 Model – Autocorrelated Acceleration

In the previous section we argued that $\frac{\Delta\widehat{X}}{\Delta t} \equiv \widehat{V}$ must have autocorrelation at some time scale T, else large instantaneous jumps would be occurring without a physical basis when Δt is small. The random input to that process, $\mathcal{R}_v$, shows up as a random "force" in the pseudo-momentum equation 4.56. Along the same lines, we can argue that $\frac{\Delta\widehat{V}}{\Delta t}$ must be subject to similar constraints; or equivalently that the process generating the pseudo-force $\mathcal{R}_v$ be an autocorrelated process itself, with time scale T_a. Hence we get the following equation set for generating random motion:

$$\frac{d\widehat{A}}{dt}+\frac{\widehat{A}}{T_a}=\mathcal{R}_a \tag{4.116}$$

$$\frac{d\widehat{V}}{dt}+\frac{\widehat{V}}{T}=\widehat{A} \tag{4.117}$$

$$\widehat{X}(t)-\widehat{X}(0)=\int_{t'=0}^{t}\widehat{V}dt' \tag{4.118}$$

(Here A indicates acceleration; A is used elsewhere to indicate an eddy viscosity.) Equivalently, the second-order system can be consolidated:

$$\left(\frac{d}{dt}+\frac{1}{T_a}\right)\left(\frac{d}{dt}+\frac{1}{T}\right)\widehat{V}=\mathcal{R}_a \tag{4.119}$$

$$\widehat{X}(t)-\widehat{X}(0)=\int_{t'=0}^{t}\widehat{V}dt' \tag{4.120}$$

The randomness enters at $\mathcal{R}_a$ and is processed twice to get to velocity. Velocity is simply integrated to get to displacement. $\mathcal{R}_a$ is the only stochastic input; the rest is internal processing. We have moved the random input one dynamic level further from $\widehat{V}$. If ARO is random *jump* and AR1 random *flight*, this might be random *push*. $\widehat{V}$ will be a doubly autocorrelated AR2 process.

We continue the notational conventions used in AR1 and AR0, summarized in Table 4.4.

4.5.1 Discrete Canonical Forms

Practical implementation will be discrete, and there are many possibilities. There will be only two *canonical forms*, however. The primary object in both is the $\widehat{V}$ series; total displacement $\widehat{X}$ is the simple accumulation of $\widehat{V}$.

Table 4.4. *Notation Conventions Used in AR2, AR1, and AR0*

- $\mathcal{R}_a$ is the stochastic inputs to the *continuous* AR2 process. It plays the role that $\mathcal{R}_v$ did in AR1. Both $\mathcal{R}_a$ and $\mathcal{R}_v$ are the basic stochastic inputs for their respective continuous models.
- AR2:

  $\mathcal{R}_a$ is the noise forcing $\frac{dA}{dt}$; its size is S_a.
  $\mathcal{R}_a dt = S_a d\mathcal{W}$ is a Wiener increment.
  $\mathcal{R}_2$ will be its integration over Δt: $\mathcal{R}_2 \sim S_2\sqrt{\Delta t}\cdot\mathcal{G}(0,1)$.
  $S_2 \approx S_a$, depending on the discretization strategy.
  $\mathcal{R}_2$ forces the discrete series A.
- AR1:

  $\mathcal{R}_v$ is the noise forcing $\frac{dV}{dt}$; its size is S_v.
  $\mathcal{R}_v dt = S_v d\mathcal{W}$ is a Wiener increment.
  $\mathcal{R}_1$ will be its integration over Δt: $\mathcal{R}_1 \sim S_1\sqrt{\Delta t}\cdot\mathcal{G}(0,1)$.
  $S_1 \approx S_v$, depending on the discretization strategy.
  $\mathcal{R}_1$ forces the discrete series V.
- AR0:

  X is discrete from the outset.
  $\mathcal{R} \sim S\sqrt{\Delta t}\cdot\mathcal{G}(0,1)$ forces the discrete series ΔX.
- The variables $\mathcal{R}_v$ and $\mathcal{R}_a$ are continuous random functions, integrable but not differentiable.
- The variables $\mathcal{R}$, $\mathcal{R}_1$, and $\mathcal{R}_2$ are discrete series of scaled Gaussian deviates.
- All these random inputs are uncorrelated with themselves or anything else.
- In all cases, X simply accumulates V: $\Delta X = \Delta t V$; $X = \sum \Delta t V$.

Canonical Form 1. This form is based on the coupled first-order differential equations (4.116, 4.117):

$$\mathcal{R}_2^{k+1} \sim S_2\sqrt{\Delta t}\cdot\mathcal{G}(0,1) \tag{4.121}$$

$$\widehat{A}^{k+1} = \rho_a\widehat{A}^k + \mathcal{R}_2^{k+1} \tag{4.122}$$

$$\widehat{V}^{k+1} = \rho_v\widehat{V}^k + \widehat{A}^{k+1}\int_0^{\Delta t} V_h\, dt \tag{4.123}$$

$$= \rho_v\widehat{V}^k + \widehat{A}^{k+1}\Delta t\overline{V}_h \tag{4.124}$$

with $\overline{V}_h$ the mean of the homogeneous $\widehat{V}$ solution over Δt:

$$\overline{V}_h = \frac{1-e^{-\Delta t/T}}{\Delta t/T} = 1 - \frac{1}{2}\frac{\Delta t}{T} + \mathcal{O}\left(\frac{\Delta t}{T}\right)^2 \tag{4.125}$$

This is *canonical form 1* of the AR2 model. One random number $\mathcal{R}_2^{k+1}$ plus the current values of $\widehat{A}^k$ and $\widehat{V}^k$ are needed to move forward.

This form carries time series for both $\widehat{A}$ and $\widehat{V}$. It injects random forcing into the $\widehat{A}$ series, which in turn drives the $\widehat{V}$ series. It is evident that $\widehat{A}$ is an AR1 process by itself; the AR1 developments above are directly applicable.

Canonical Form 2. This form uses a single second-order equation for $\widehat{V}$. Only the $\widehat{V}$ series is computed; the random forcing is injected directly into it. $\widehat{A}$ is implied but not explicitly computed.

From form 1, we can eliminate $\widehat{A}^k$ in favor of $\widehat{V}^{k-1}$. Form the quantity

$$\widehat{V}^{k+1} - \rho_a \widehat{V}^k = \rho_v \widehat{V}^k + \widehat{A}^{k+1} \Delta t \overline{V}_h - \rho_a \left(\rho_v \widehat{V}^{k-1} + \widehat{A}^k \Delta t \overline{V}_h \right) \tag{4.126}$$

Since $\widehat{A}^{k+1} - \rho_a \widehat{A}^k = \mathcal{R}_2^{k+1}$, we get

$$\widehat{V}^{k+1} - \rho_a \widehat{V}^k = \rho_v \widehat{V}^k - \rho_a \rho_v \widehat{V}^{k-1} + \mathcal{R}_2^{k+1} \Delta t \overline{V}_h \tag{4.127}$$

So we arrive at the result:

$$\widehat{V}^{k+1} = (\rho_v + \rho_a) \widehat{V}^k - \rho_v \rho_a \widehat{V}^{k-1} + \mathcal{R}_2^{k+1} \Delta t \overline{V}_h \tag{4.128}$$

This is *canonical form 2* of the AR2 model:

$$\widehat{V}^{k+1} = \rho_1 \widehat{V}^k + \rho_2 \widehat{V}^{k-1} + \mathcal{R}_1^{k+1} \tag{4.129}$$

One random number plus $\widehat{V}^k$ and $\widehat{V}^{k-1}$ are required to advance one step. The equivalence between the two canonical forms is

$$\rho_a = e^{-\Delta t/T_a} \qquad \rho_v = e^{-\Delta t/T} \tag{4.130}$$

$$\rho_1 = \rho_v + \rho_a \qquad \rho_2 = -\rho_v \rho_a \tag{4.131}$$

$$\mathcal{R}_1 \sim S_1 \sqrt{\Delta t} \cdot \mathcal{G}(0,1) \qquad \mathcal{R}_2 \sim S_2 \sqrt{\Delta t} \cdot \mathcal{G}(0,1) \tag{4.132}$$

$$S_1 = \Delta t \overline{V}_h S_2 \tag{4.133}$$

Finite Difference Discretizations – AR2

Various differencing approximations lead from the continuum to the discretized canonical forms. The discrete canonical parameters ρ, S, etc. will reflect the approximations; they will depend on finite Δt, T, and T_a.

Scheme 1: $\widehat{A}$ and $\widehat{V}$ For the coupled first-order dynamics, 4.116–4.117, we have in difference form,

$$\widehat{A}^{k+1} = \left[1 - \frac{\Delta t}{T_a}\right] \widehat{A}^k + \mathcal{R}_2^{k+1} \tag{4.134}$$

$$\widehat{V}^{k+1} = \left[1 - \frac{\Delta t}{T}\right] \widehat{V}^k + \Delta t \widehat{A}^{k+1} \tag{4.135}$$

This is an instance of *canonical form 1*.

Scheme 2: Staggered $\widehat{A}$ and $\widehat{V}$ With an appropriate staggering of the $\widehat{A}$ and $\widehat{V}$ series, one can center the difference equations to obtain construct $\mathcal{O}(\Delta t^2)$ discretizations; and then construct time-marching solutions that leapfrog each other:

$$\widehat{A}^{k+1} = \left[\frac{1 - \frac{\Delta t}{2T_a}}{1 + \frac{\Delta t}{2T_a}}\right] \widehat{A}^k + \frac{1}{1 + \frac{\Delta t}{2T_a}} \mathcal{R}_2^{k+1} \tag{4.136}$$

$$\widehat{V}^{k+1} = \left[\frac{1 - \frac{\Delta t}{2T}}{1 + \frac{\Delta t}{2T}}\right] \widehat{V}^k + \frac{\Delta t}{1 + \frac{\Delta t}{2T}} \widehat{A}^{k+1} \tag{4.137}$$

This is a close cousin of the previous scheme, another instance of *canonical form 1*.

Scheme 3: $\widehat{V}$ Only, Explicit For the second-order dynamical equation, 4.119, simple explicit differencing gives

$$\frac{\widehat{V}^{k+1} - 2\widehat{V}^k + \widehat{V}^{k-1}}{\Delta t^2} + \left(\frac{1}{T} + \frac{1}{T_a}\right)\left(\frac{\widehat{V}^k - \widehat{V}^{k-1}}{\Delta t}\right) + \frac{1}{TT_a}\widehat{V}^{k-1} = \frac{1}{\Delta t}\int_k^{k+1} \mathcal{R}_a dt = \frac{1}{\Delta t} S_a \Delta \mathcal{W} \tag{4.138}$$

This leads to *canonical form 2* with

$$\rho_1 = 2 - \left(\frac{\Delta t}{T} + \frac{\Delta t}{T_a}\right) \tag{4.139}$$

$$\rho_2 = -1 + \left(\frac{\Delta t}{T} + \frac{\Delta t}{T_a}\right) - \frac{\Delta t}{T}\frac{\Delta t}{T_a} \tag{4.140}$$

$$\mathcal{R}_1^{k+1} = \Delta t S_a \Delta \mathcal{W} \tag{4.141}$$

$$\sim \Delta t S_a \sqrt{\Delta t} \cdot \mathcal{G}(0,1) \tag{4.142}$$

Scheme 4: $\widehat{V}$ only, Implicit Other differencing approximations are clearly possible [279]; for example, a centered, implicit scheme might be

$$\frac{\widehat{V}^{k+1} - 2\widehat{V}^k + \widehat{V}^{k-1}}{\Delta t^2} + \left(\frac{1}{T} + \frac{1}{T_a}\right)\frac{\widehat{V}^{k+1} - \widehat{V}^{k-1}}{2\Delta t} + \frac{1}{2TT_a}\left(\widehat{V}^{k+1} + \widehat{V}^{k-1}\right) = \frac{1}{\Delta t} S_a \Delta \mathcal{W} \tag{4.143}$$

This is Scheme 4, pursued in the exercise below. It is another instance of *canonical form 2*.

Provided the differencing is consistent with the ODE, these schemes are all dynamically equivalent, although they have different discretization errors. All will converge at small $\frac{\Delta t}{T}$ and $\frac{\Delta t}{T_a}$. Generally, centered differences can be expected to have higher convergence rates than uncentered ones.

Equivalence of Canonical Forms. In general, algorithmic differences distinguish the two different canonical forms of AR2; but mathematically, all reduce to *canonical form 2*,

$$\mathcal{R}_1 \sim S_1\sqrt{\Delta t}\cdot \mathcal{G}(0,1) \tag{4.144}$$

$$\widehat{V}^{k+1} = \rho_1 \widehat{V}^k + \rho_2 \widehat{V}^{k-1} + \mathcal{R}_1^{k+1} \tag{4.145}$$

where the series $\widehat{A}$ is implied but never computed. The coefficients ρ_1 and ρ_2 are the memory coefficients. The random forcing is $\mathcal{R}_1$. It scales with the size S_a, the size of the acceleration noise on input.

Accordingly, **we will concentrate on canonical form 2 from here on.**

Example 4.1 In Scheme 1, elimination of $\widehat{A}$ from the series produces the same result as Scheme 3. From Equations 4.134 to 4.135, we have

$$\widehat{A}^{k+1} - \rho_a \widehat{A}^k = \mathcal{R}_2^{k+1} \tag{4.146}$$

$$\widehat{V}^{k+1} - \rho_a \widehat{V}^k = \rho_v \left[\widehat{V}^k - \rho_a \widehat{V}^{k-1}\right] + \Delta t \left[\widehat{A}^{k+1} - \rho_a \widehat{A}^k\right] \tag{4.147}$$

$$= \rho_v \left[\widehat{V}^k - \rho_a \widehat{V}^{k-1}\right] + \Delta t \mathcal{R}_2^{k+1} \tag{4.148}$$

With $\rho_v = 1 - \frac{\Delta t}{T}$ and $\rho_a = 1 - \frac{\Delta t}{T_a}$, this reduces exactly to Scheme 3.

Problem 4.13 Verify that the explicit discretization 4.138 leads to the expressions of ρ_1 and ρ_2 in Equations 4.139 and 4.140

Problem 4.14 Replace the explicit FD treatment of Equation 4.138 with the centered implicit approximation 4.143 and generate the AR2 expressions for ρ_1 and ρ_2 for this case. Compare with the explicit ones, 4.139 and 4.140. This is **Scheme 4**.

4.5.2 AR2 Velocity

Homogeneous Solution and Characteristic Equation. The AR2 recursion (Equation 4.145) in homogeneous form is the *characteristic equation*:

$$\widehat{V}^{k+1} - \rho_1 \widehat{V}^k - \rho_2 \widehat{V}^{k-1} = 0 \tag{4.149}$$

There will be solutions with the property $\widehat{V}^k = \lambda \widehat{V}^{k-1}$, representing unforced relaxation within the recursion. λ must satisfy

$$\lambda^2 - \rho_1 \lambda - \rho_2 = 0 \tag{4.150}$$

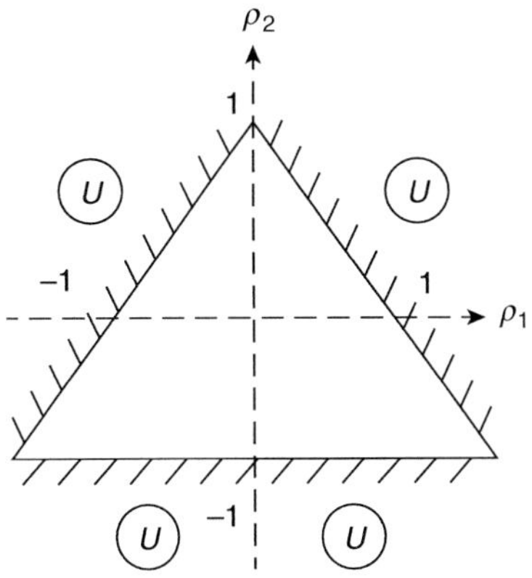

Figure 4.9. Stability range for AR2. Points (ρ_1, ρ_2) within the triangle are stable; U indicates an unstable region, where one of the inequalities in Equation 4.152 is violated.

There will be two roots:

$$\lambda = \frac{\rho_1}{2} \pm \sqrt{\frac{\rho_1}{2}^2 + \rho_2} \tag{4.151}$$

AR2 Stability. There will be stability limits on the AR2 model. The necessary and sufficient stability condition for this second-order process is $|\lambda| < 1$ [279]. The characteristic equation constrains λ so this requirement is[9]

$$-1 \leq \rho_2 \leq 1 - |\rho_1| \tag{4.152}$$

The stability range of ρ_1 and ρ_2 is displayed in Figure 4.9. Compared with AR1, there is a little more stability range for ρ_1, if ρ_2 is negative.

Example 4.2 For the explicit FD treatment, Scheme 3, the AR2 stability constraints become

$$0 \leq \frac{\Delta t}{T} \frac{\Delta t}{T_a} \tag{4.153}$$

$$\frac{\Delta t}{T} \frac{\Delta t}{T_a} \leq \left(\frac{\Delta t}{T} + \frac{\Delta t}{T_a} \right) \tag{4.154}$$

$$2\left(\frac{\Delta t}{T} + \frac{\Delta t}{T_a} \right) \leq 4 + \frac{\Delta t}{T} \frac{\Delta t}{T_a} \tag{4.155}$$

Problem 4.15 Confirm the AR2 stability limits given in Equation 4.152 by generating some numerical sequences of $\widehat{V}$. Do your experimental results agree with theory?

[9] We assume ρ_1 and ρ_2 are both real.

Problem 4.16 Translate the stability limits on (ρ_1, ρ_2), Equation 4.152, into their equivalent for:

(1) the explicit finite difference approximation 4.138 (The results are given in the text; verify them.)
(2) the centered implicit finite difference approximation 4.143

AR2 – Convolution of Input Noise. The general AR2 discrete form, Equation 4.145, is equivalent to the discrete convolution of all previous noise inputs $\mathcal{R}_1^i$:

$$\widehat{V}^N = \sum_{i=1}^{i=N} \mathcal{R}_1^i Z_{N-i} \tag{4.156}$$

where we assume the starting conditions of no motion, $\widehat{V}^{-1} = \widehat{V}^0 = 0$; and the $\mathcal{R}_1^i$ series begins at $i = 1$. Proceeding as in the AR1 demonstration, we have Equation 4.81 for $\widehat{V}^{N+1}$, and

$$\widehat{V}^{N+2} = \left[\sum_{i=1}^{i=N} \mathcal{R}_1^i Z_{N+2-i}\right] + \mathcal{R}_1^{N+1} Z_1 + \mathcal{R}_1^{N+2} Z_0 \tag{4.157}$$

Substitution into the basic relation, 4.145, gives

$$\sum_{i=1}^{i=N} \mathcal{R}_1^i \left[Z_{N-i+2} - \rho_1 Z_{N-i+1} - \rho_2 Z_{N-i}\right] + \mathcal{R}_1^{N+1} [Z_1 - \rho_1 Z_0] + \mathcal{R}_1^{N+2} [Z_0 - 1] = 0 \tag{4.158}$$

and we obtain the requirements for the Z_i:

$$Z_0 = 1 \tag{4.159}$$

$$Z_1 = \rho_1 \tag{4.160}$$

$$Z_j = \rho_1 Z_{j-1} + \rho_2 Z_{j-2} \qquad (j \geq 2) \tag{4.161}$$

The convolution 4.156 is equivalent to the general autoregressive form 4.145. The convolution coefficients Z_j satisfy the characteristic equation, 4.149.

AR2 Covariance. The input $\mathcal{R}_1$ has variance $\sigma^2_{\mathcal{R}_1}$; it is not correlated with itself or with any previous $\widehat{V}$. For the covariance of $\widehat{V}$ we use the C_i notation defined

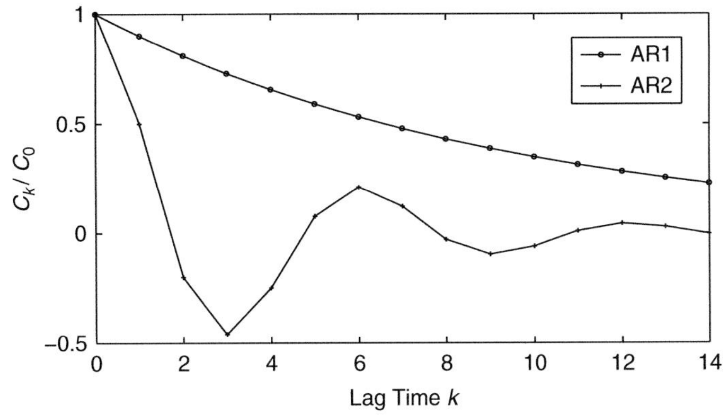

Figure 4.10. Covariance C_i versus lagtime i, AR1 versus AR2, both with $C_0 = 1$. The AR1 has $\rho = 0.9$; the AR2 has $(\rho_1, \rho_2) = (0.8, -0.6)$.

previously.[10] The recursion is

$$\sigma^2_{\widehat{V}} \equiv C_0 = \left[\frac{1}{1 - \rho_1^2\left(\frac{1+\rho_2}{1-\rho_2}\right) - \rho_2^2}\right] \sigma^2_{\mathcal{R}_1} \tag{4.162}$$

$$C_1 = \left[\frac{\rho_1}{1-\rho_2}\right] C_0 \tag{4.163}$$

$$C_n = \rho_1 C_{n-1} + \rho_2 C_{n-2} \qquad n \geq 2 \tag{4.164}$$

The entire series is linear in the driving variance $\sigma^2_{\mathcal{R}_1}$. These results reduce to AR1 ($\rho_2 = 0$) and AR0 ($\rho_2 = \rho_1 = 0$), as expected. Relative to C_0, there is a trend among the AR models. For AR0, the input variance in $\mathcal{R}_1$ is transmitted to $\widehat{V}$ without change. For AR1, the input variance is amplified by the added memory, and in AR2, it is further amplified.

Once again, we have recovered the characteristic equation, this time in the C_i. For a stable process, $|C_i|$ decreases with i.

Figure 4.10 illustrates the C_i series for AR1 and AR2.

AR2 Oscillations. The behavior of the C_i series is simple for AR1: $C_i = \rho^i$; it will decay monotonically when $0 < \rho < 1$. The AR2 series is more complicated; the example in Figure 4.10 oscillates.[11]

10 $C_i \equiv E\left[\widehat{V}^k \widehat{V}^{k-i}\right]$ with $C_i = C_{-i}$, and $C_0 \equiv \sigma^2_{\widehat{V}}$.

11 We are assuming that ρ_1 and ρ_2 are both real.

The series C_i satisfies the characteristic equation. As above (Equation 4.151) it has solutions of the form

$$C_n = \lambda C_{n-1} \tag{4.165}$$

with characteristic roots

$$\lambda = \frac{\rho_1}{2} \pm \sqrt{\left(\frac{\rho_1}{2}\right)^2 + \rho_2} \tag{4.166}$$

There are two important possibilities. When the roots are complex, they are conjugates and have the same size. When they are real, one is larger in size than the other.

1. **Complex Roots.** The roots will be complex conjugates when ρ_2 is sufficiently negative:

$$\rho_2 < -\left(\frac{\rho_1}{2}\right)^2 \tag{4.167}$$

   In that case, the per-step attenuation $|\lambda|$ and period M of the covariance series is

$$|\lambda|^2 = -\rho_2 \tag{4.168}$$

$$\tan(\theta) = \sqrt{-1 - 4\frac{\rho_2}{\rho_1^2}} \tag{4.169}$$

$$M = \frac{2\pi}{\theta} \tag{4.170}$$

$$|\lambda|^M = [-\rho_2]^{\frac{\pi}{\theta}} \tag{4.171}$$

   ($|\lambda|^M$ is the per-period attenuation rate.) Lines of constant ρ_2 are constant attenuation; lines of constant ρ_2/ρ_1^2 are constant period. In the limit of high resolution, $M \to \infty$, $\rho_2 \to -\left(\frac{\rho_1}{2}\right)^2$. At the other extreme, $M = 4$, we have $\rho_1 = 0$.
   For the parameters in Figure 4.10, this comes in at $M = 6.11$, $|\lambda| = 0.77$, $|\lambda|^M = 0.21$, consistent with the simulation results.
2. **Real Roots.** In the rest of the stable range, we have two real roots. The largest in size is

$$|\lambda| = \left|\frac{\rho_1}{2}\right| + \sqrt{\left|\frac{\rho_1}{2}\right|^2 + \rho_2} \tag{4.172}$$

   When $\rho_2 > 0$, one root is positive, representing monotone decay. The other root is negative, representing a red/black or Nyquist mode, reversing sign every Δt.

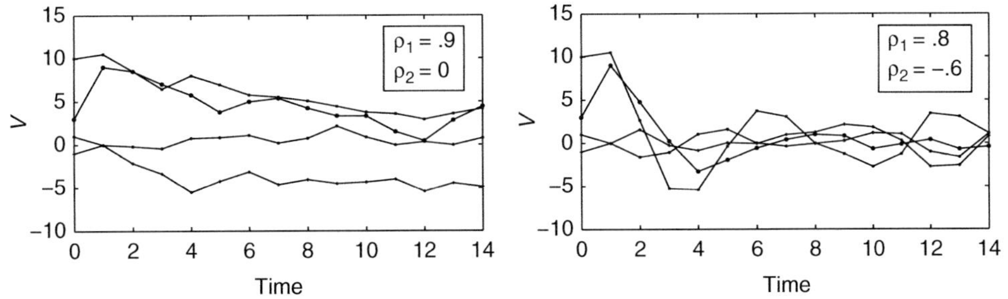

Figure 4.11. Simulated AR2 velocity series for various initializations and $\sigma = 1$. The values of ρ are as in Figure 4.10. The limiting AR1 cases (left) show basic decay in the series. The AR2 series (right) show oscillation at the period predicted by the analysis and Figure 4.10.

When $\rho_2 < 0$, we get two roots of the same sign. If ρ_1 is positive, both are monotone. If ρ_1 is negative, both roots are negative – these are both Nyquist modes, changing sign every Δt.

The characteristic equation governs homogeneous solutions for both C and $\widehat{V}$ – so both will exhibit oscillatory behavior under the conditions given. Simulated AR2 velocity series appear in Figure 4.11 and confirm these findings.

Problem 4.17 Confirm the conditions for non-oscillatory AR2 solutions given previously.

Problem 4.18 Translate the conditions for non-oscillatory AR2 solutions into constraints on $\frac{\Delta t}{T}$ and $\frac{\Delta t}{T_a}$ for the Euler discretization.

4.5.3 AR2 Displacement and Link to Diffusion

Convolution. Starting from rest, $\widehat{X}^{k\leq 0} = \widehat{V}^{k\leq 0} = 0$, let the forcing $\mathcal{R}_1$ begin at $\mathcal{R}_1^1$. As developed, we have $\widehat{V}$ as the convolution of all previous inputs $\mathcal{R}_1$

$$\widehat{V}^N = \sum_{i=1}^{i=N} \mathcal{R}_1^i Z_{N-i} \tag{4.173}$$

The per-step displacement is

$$\Delta\widehat{X}^k \equiv \widehat{X}^k - \widehat{X}^{k-1} = \Delta t \widehat{V}^k \tag{4.174}$$

and the total displacement is the sum of the velocities:

$$\widehat{X}^N = \Delta t \sum_{k=1}^{k=N} \widehat{V}^k = \Delta t \left[\mathcal{R}_1^1 \cdot B^{N-1} + \mathcal{R}_1^2 \cdot B^{N-2} + \mathcal{R}_1^3 \cdot B^{N-3} + \cdots \right] \quad (4.175)$$

$$= \Delta t \sum_{k=1}^{k=N} \mathcal{R}_1^k \cdot B^{N-k} \quad (4.176)$$

with B^k being the partial sum of the velocity convolution coefficients Z ("the sum of Z terminating at k") :

$$B^k = \sum_{i=0}^{i=k} Z_i \quad (4.177)$$

This is a simple extension of the nomenclature defined for the AR1 analysis. $\widehat{X}$ is the convolution of the partial convolutions of $\mathcal{R}_1$. Initially, $B^0 = 1$. B^k reaches a finite upper limit as $k \to \infty$. The early $\mathcal{R}_1$ ultimately mature in their cumulative influence; the later $\mathcal{R}_1$ are younger and their influence has accumulated over a shorter duration.

Variance Growth. With the $\mathcal{R}$ all independent, we have the variance of $\widehat{X}$:

$$\overline{\widehat{X}^N \widehat{X}^N} = \Delta t^2 \sigma_{\mathcal{R}_1}^2 \sum_{i=0}^{i=N-1} \left[B^i\right]^2 \quad (4.178)$$

This will grow as length N increases. The per-step variance growth will be

$$\overline{\widehat{X}^{N+1} \widehat{X}^{N+1}} - \overline{\widehat{X}^N \widehat{X}^N} = \Delta t^2 \sigma_{\mathcal{R}_1}^2 \left[B^N\right]^2 \quad (4.179)$$

For a stable AR2 process, the B will saturate at high N – they represent the cumulative memory of old events, and if old events are no longer actively affecting current velocity, their persistence in $\widehat{X}$ has reached a plateau and is no longer changing. So, the per-step change in the variance of position is constant.

Put another way, B^k is a partial sum of decreasing contributions (assuming a stable process); the sum reaches a finite asymptote B^∞, independent of k.[12]

[12] By way of comparison, the AR1 coefficient B_n is a partial sum of convolution coefficients ρ^i; it admits the closed form

$$B_n = \sum_{i=0}^{n} \rho^i = \frac{1 - \rho^{n+1}}{1 - \rho} \quad (4.180)$$

$$\Delta\sigma_{\widehat{X}}^{2\,N+1} = \Delta t^2 \sigma_{\mathcal{R}_1}^2 \left[B^\infty\right]^2 \tag{4.181}$$

To find B^∞: we have already the recursion among the Z, Equations 4.159 –4.161:

$$Z_0 = 1 \qquad\qquad Z_1 = \rho_1 \tag{4.182}$$

$$Z_j = \rho_1 Z_{j-1} + \rho_2 Z_{j-2} \tag{4.183}$$

Routine substitution gives the recursion among B:

$$B^0 = 1 \qquad\qquad B^1 = 1 + \rho_1 \tag{4.184}$$

$$B^N = 1 + \rho_1\, B^{N-1} + \rho_2\, B^{N-2} \tag{4.185}$$

For large N, B reaches the asymptote

$$B^\infty = \frac{1}{[1 - \rho_1 - \rho_2]} \tag{4.186}$$

When $\rho_2 = 0$ we recover the AR1 relation, Equation 4.100.

Link to Diffusion. As in the AR1 discussion, the growth rate of variance is by definition twice the macroscopic diffusion coefficient. From Equation 4.181, assembling the preceding discussion leads to

$$S = \Delta t S_1 = (1 - \rho_1 - \rho_2)\sqrt{2D} \tag{4.187}$$

$$\mathcal{R} \sim S\sqrt{\Delta t} \cdot \mathcal{G}(0,1) \tag{4.188}$$

So the basic AR2 recursion in terms of displacement is

AR2

$$\Delta\widehat{X}^N = \rho_1 \Delta\widehat{X}^{N-1} + \rho_2 \Delta\widehat{X}^{N-2} + \mathcal{R}^N \tag{4.189}$$

$$\mathcal{R} \sim (1 - \rho_1 - \rho_2)\sqrt{2D\Delta t} \cdot \mathcal{G}(0,1) \tag{4.190}$$

$$\rho_1 = e^{-\Delta t/T_a} + e^{-\Delta t/T} \tag{4.191}$$

$$\rho_2 = -e^{-\Delta t/T_a} e^{-\Delta t/T} \tag{4.192}$$

Deficit and Half-Life. Define the deficit D as the gap between B^j and its asymptote:

$$D^j = B^\infty - B^j \tag{4.193}$$

It is easy to confirm the recursion

$$D^j = \rho_1 D^{j-1} + \rho_2 D^{j-2} \tag{4.194}$$

Again we have recovered the characteristic equation! D will have solutions of the form

$$D^{j+M} = \lambda^M D^j \tag{4.195}$$

λ will be the roots studied above (Section 4.5.2, Equation 4.151):

$$\lambda = \frac{\rho_1}{2} \pm \sqrt{\left(\frac{\rho_1}{2}\right)^2 + \rho_2} \tag{4.196}$$

The mode corresponding to the largest $|\lambda|$ will persist longest; its per-step decay rate is $|\lambda|$. So, for large j, if we want the deficit cut in half, $|\lambda|^H = 1/2$, and the half-life will be

$$H = \frac{\log(.5)}{\log|\lambda|} \tag{4.197}$$

Asymptotically, this is the timing (number of steps) that scales the approach to "steady state" growth in $\widehat{X}$ variance. Specific analyses for λ are given in Section 4.5.2.

Example 4.3 When $\rho_2 < -\left(\frac{\rho_1}{2}\right)^2$, λ is complex and $|\lambda|^2 = -\rho_2$. The half-life is

$$H = \frac{\log(.5)}{0.5\log(-\rho_2)} \tag{4.198}$$

Problem 4.19 Find an expression for the AR2 half-life when λ is real.

Problem 4.20 Make a contour plot of H versus (ρ_1, ρ_2).

Summary of Recursions. The AR2 recursions are summarized in Appendix G. These are generalizations of those for AR1, which are recovered when $\rho_2 = 0$.

4.6 The AR1-s Model – Spinning Walk

Above the horizontal coordinates are treated independently. Several authors [46, 45, 373, 402] have developed a coupled or rotating representation of random walk wherein a *spin* term is introduced. A version of this type of model:

$$\widehat{U}^{k+1} = \rho\widehat{U}^k + \omega\widehat{V}^k + \mathcal{R}_{u_1} \tag{4.199}$$

$$\widehat{V}^{k+1} = -\omega\widehat{U}^k + \rho\widehat{V}^k + \mathcal{R}_{v_1} \tag{4.200}$$

where the $\mathcal{R}$ are independent; and the antisymmetric term ω represents spin. A positive ω corresponds to a velocity that decays and bends to the right (clockwise), with uncorrelated perturbations added in each step.

Equations (4.275, 4.276) are recognizable as an Euler discretization of

$$\frac{d\widehat{U}}{dt} = -\frac{\widehat{U}}{T} + \Omega\widehat{V} \tag{4.201}$$

$$\frac{d\widehat{V}}{dt} = -\frac{\widehat{V}}{T} - \Omega\widehat{U} \tag{4.202}$$

with parameters

$$\rho = 1 - \frac{\Delta t}{T} \tag{4.203}$$

$$\omega = \Omega\Delta t \tag{4.204}$$

The spin period T_s and the spin frequency Ω are surrogates for each other:

$$\Omega T_s \equiv 2\pi \tag{4.205}$$

$$\omega \equiv 2\pi\frac{\Delta t}{T_s} \tag{4.206}$$

4.6.1 AR1-s Complex Velocity

Let the complex velocity $\widehat{Z}$ be[13]

$$\widehat{Z}^k = \widehat{U}^k + \mathrm{j}\widehat{V}^k \tag{4.207}$$

with $\mathrm{j} = \sqrt{-1}$; and let $\mathcal{R}_{z_1}$ be the analogous complex blend of the two independent scalar forcings,

$$\mathcal{R}_{z_1}^{k+1} = \mathcal{R}_{u_1} + \mathrm{j}\omega\mathcal{R}_{v_1} \tag{4.208}$$

Then the two real equations can be combined

$$\widehat{Z}^{k+1} = r\widehat{Z}^k + \mathcal{R}_{z_1}^{k+1} \tag{4.209}$$

with complex r

$$r = \rho - \mathrm{j}\omega \tag{4.210}$$

This is the complex form of the AR1 model, Equation 4.72. We have the complex velocity remembering a complex blend of its predecessor, and being forced by a

[13] A very brief summary of some complex number properties appears in Appendix B. For a general review of complex numbers, see, for example, Hildebrand [140] or Boas [41].

complex blend of independent real and imaginary parts. The real part of each is the x-component; the imaginary part, the y-component. The term $j\omega$ achieves the spin, converting part of the previous $\widehat{U}$ into $\widehat{V}$, and vice versa.

The results of the AR1 section, adapted for complex algebra, are applicable – with the understanding that variances and covariances are defined in the conjugate sense. Specifically, let $\otimes$ indicate conjugate multiplication of two complex quantities:

$$A \otimes B = AB^* \tag{4.211}$$

with B^* the complex conjugate of B.[14] Using these, we have the definitions

$$C_n \equiv \overline{\widehat{Z}^{k+n} \otimes \widehat{Z}^k} \tag{4.212}$$

$$\sigma^2 \equiv C_0 \equiv \overline{\widehat{Z} \otimes \widehat{Z}} \tag{4.213}$$

These reduce to the simpler definitions when everything is real. C_n will generally be complex; σ^2 will always be real.

Some results:

$$\sigma^2_{\widehat{Z}} = \frac{\sigma^2_{\mathcal{R}_Z}}{1 - r \otimes r} \tag{4.214}$$

or in terms of the scalar components of velocity:

$$\sigma^2_{\widehat{U}} + \sigma^2_{\widehat{V}} = \frac{\sigma^2_{\mathcal{R}_{u_1}} + \sigma^2_{\mathcal{R}_{v_1}}}{1 - \rho^2 - \omega^2} \tag{4.215}$$

For the covariance,

$$C_n = r^n \sigma^2_{\widehat{Z}} \tag{4.216}$$

The covariances will generally be complex:

$$r^n = |r|^n e^{jn\theta} \tag{4.217}$$

$$|r| = \sqrt{\rho^2 + \omega^2} \tag{4.218}$$

$$\theta = \tan^{-1}\left(-\frac{\omega}{\rho}\right) \tag{4.219}$$

The stability condition:

$$\rho^2 + \omega^2 < 1 \tag{4.220}$$

[14] $A \otimes B =$ "A multiplied by the conjugate of B" There is an antisymmetry: $B \otimes A = (A \otimes B)^*$.

or equivalently,

$$\left(\frac{\Delta t}{T}\right)^2 + (\Omega \Delta t)^2 < 2\frac{\Delta t}{T} \tag{4.221}$$

The per-step spin will be θ and the number of Δt to complete one period is

$$N = -\frac{2\pi}{\theta} \tag{4.222}$$

The parameter ratio ω/ρ is $\mathcal{O}(\Delta t)$. For small Δt the Taylor series for $\tan^{-1}$ gives

$$\theta = -\frac{\omega}{\rho}\left[1 - \frac{1}{3}\left(\frac{\omega}{\rho}\right)^2 + \frac{1}{5}\left(\frac{\omega}{\rho}\right)^4 - \cdots\right] \tag{4.223}$$

$$N = \frac{2\pi}{\omega}\left[\frac{\left[1 - \frac{\Delta t}{T}\right]}{\left[1 - \frac{1}{3}\left(\frac{\omega}{\rho}\right)^2 + \frac{1}{5}\left(\frac{\omega}{\rho}\right)^4 - \cdots\right]}\right] \tag{4.224}$$

and

$$N\Delta t = \frac{2\pi}{\Omega}\left[1 + \mathcal{O}(\Delta t)\right] \tag{4.225}$$

This is a first-order approximation to the continuous-time identity

$$\Omega T_s = 2\pi \tag{4.226}$$

where T_s is the spin period defined at the outset (Equation 4.205).

Finally, the basic recursion (4.209) indicates that in real terms, the velocity vector will rotate the amount θ in each Δt. For positive ω, this amounts to a veering to the right for any trajectory – clockwise looping.

4.6.2 AR1-s Displacement

The AR1-s displacement development follows that in the AR1 section with its extension to complex r. The complex displacement χ is $\Delta t \sum Z$. It is the convolution of inputs; its variance is

$$\sigma_\chi^2 = \Delta t^2 \sigma_{\mathcal{R}Z}^2 \sum [B \otimes B] \tag{4.227}$$

$$= \sigma_{\mathcal{R}\Delta\chi}^2 \sum [B \otimes B] \tag{4.228}$$

(By definition $\mathcal{R}_{\Delta X} = \Delta t\ \mathcal{R}_Z$; i.e., $\Delta\widehat{X} = \Delta t\ \Delta\widehat{V}$). The convolution coefficients for position are unchanged in form; in particular, the asymptote is

$$B_\infty = \frac{1}{1-r} \tag{4.229}$$

and the per-step increment in position variance is, asymptotically,

$$\Delta\sigma_X^2 = \Delta t^2 \sigma_{\mathcal{R}_Z}^2 \left[\frac{1}{(1-r)\otimes(1-r)}\right] \tag{4.230}$$

$$= \sigma_{\mathcal{R}_{\Delta X}}^2 \left[\frac{1}{(1-\rho)^2+\omega^2}\right] \tag{4.231}$$

In scalar form, we have the 2-D results for AR1-S in the asymptotic limit.

$$\sigma_{\widehat{U}}^2 + \sigma_{\widehat{V}}^2 = \frac{\sigma_{\mathcal{R}_{\widehat{U}}}^2 + \sigma_{\mathcal{R}_{\widehat{V}}}^2}{1-\rho^2-\omega^2} \tag{4.232}$$

For the accumulated displacement, the per-step change in variance is

$$\Delta\sigma_{\widehat{X}}^2 + \Delta\sigma_{\widehat{Y}}^2 = \frac{\sigma_{\mathcal{R}_{\Delta\widehat{X}}}^2 + \sigma_{\mathcal{R}_{\Delta\widehat{Y}}}^2}{(1-\rho)^2+\omega^2} \tag{4.233}$$

The growth rate of variance defines the diffusivity D; and for the usual Wiener forcing, $\sigma^2 = S^2\Delta t$

$$4D\Delta t = \frac{\left(S_{\Delta\widehat{X}}^2 + S_{\Delta\widehat{Y}}^2\right)\Delta t}{(1-\rho)^2+\omega^2} \tag{4.234}$$

And so the forcing is

$$\left[(1-\rho)^2+\omega^2\right] 4D = S_{\Delta\widehat{X}}^2 + S_{\Delta\widehat{Y}}^2 \tag{4.235}$$

For independent, isotropic forcing $\mathcal{R}$, we have the AR1-s in operational form:

AR1-s

$$\Delta\widehat{X}^{k+1} = \rho\Delta\widehat{X}^k + \omega\Delta\widehat{Y}^k + \mathcal{R}^{k+1} \tag{4.236}$$

$$\Delta\widehat{Y}^{k+1} = -\omega\Delta\widehat{X}^k + \rho\Delta\widehat{Y}^k + \mathcal{R}^{k+1} \tag{4.237}$$

$$\mathcal{R} \sim \sqrt{\left[(1-\rho)^2+\omega^2\right]}\sqrt{2D\Delta t}\cdot\mathcal{G}(0,1) \tag{4.238}$$

The AR1 results are recovered when $\omega = 0$ (Equations 4.110, 4.111).

4.6.3 AR1-s Results

Veneziani et al. [462] successfully applied this model to drifter data from the Gulf Stream recirculation region, where significant oscillatory patterns are observed. Berloff and McWilliams [25] examined similar data and were successful with independent random AR2 models in x and y without spin. It remains an open question as to the relative merits of either – the AR1 with spin, or the AR2 without.

Both AR1-s and AR2 require the same amount of work – two random numbers $\mathcal{R}_x, \mathcal{R}_y$ per time step for the two random walks $\widehat{U}, \widehat{V}$; and four parameters $\rho, \omega, \sigma_{\mathcal{R}_x}, \sigma_{\mathcal{R}_y}$. In the AR1-s case there is a built-in tendency to veer in the same direction each time step, thereby achieving the spin. The AR2 model has no preferred direction, it is random.

Figure 4.12 displays a few samples of this "spinning walk." AR1-s parameters: $\left[\sigma_{\mathcal{R}_x}, \sigma_{\mathcal{R}_y}, \Delta t, T, T_s\right] = \left[1.0\sqrt{\Delta t},\ 1.0\sqrt{\Delta t},\ 0.1,\ 1,\ 2\right]$. For these particular parameters, we have the speed variance $= 2.19$, and the growth in displacement variance $= 1.84$. Figure 4.13 shows trajectories with the same parameters, except that the rotation is zero. The effect is clear: the spinning walk stays nearer the origin. Removing the spin reduces the speed variance to 1.05; and increases the displacement variance growth rate to 20. The removal of spin is responsible for roughly halving the speed

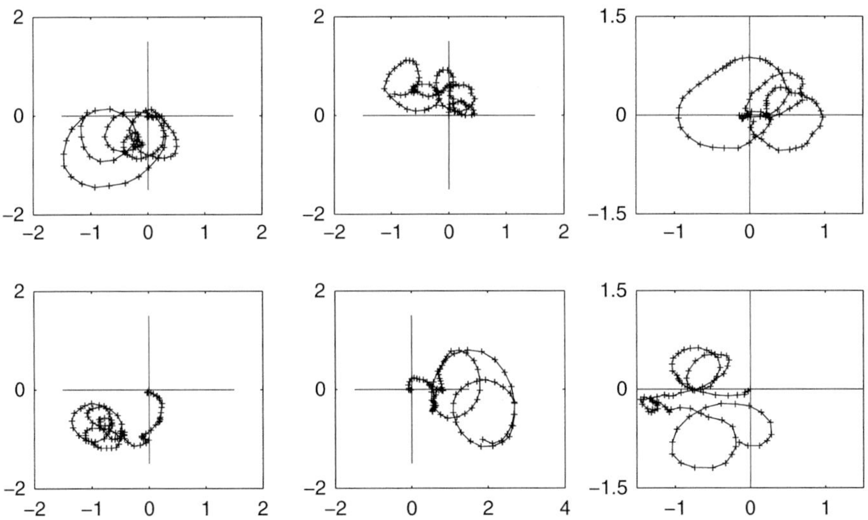

Figure 4.12. (X, Y) displacement for six random walks with the spinning AR1-s model. Parameters: $[\Delta t,\ T,\ T_s] = [0.1,\ 1,\ 2]$; $\sigma_{\mathcal{R}_x} = \sigma_{\mathcal{R}_y} = \sqrt{\Delta t}$. All trajectories were started at the origin with no memory; duration was $100\Delta t$. $\rho = 1 - \frac{\Delta t}{T}$; $\omega = \frac{\Delta t}{T_s}$. The crosshairs mark the origin ± 1.5

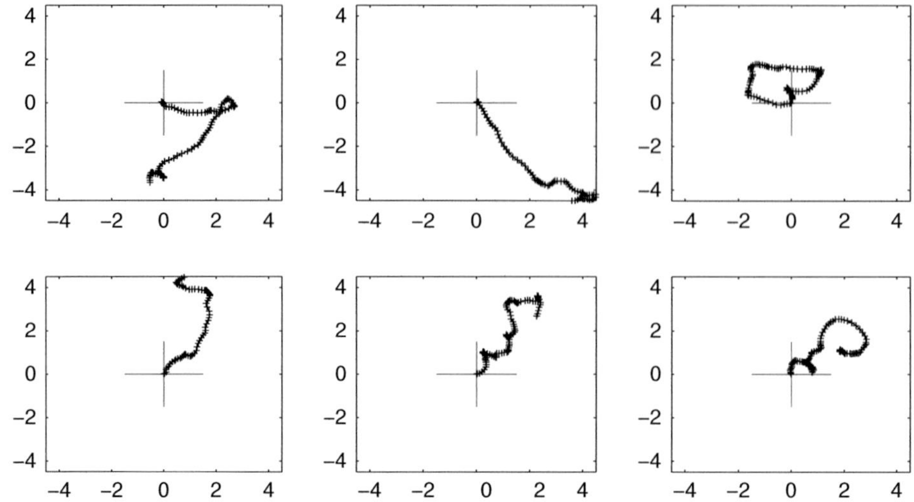

Figure 4.13. As in Figure 4.12 but with no rotation. $\rho = 1 - \frac{\Delta t}{T}$; $\omega = 0$.

variance; and for allowing the displacement variance growth to increase by roughly a factor of 10.

Spin in this case retains the trajectories nearer the origin, by roughly $\sqrt{10}$, evident in 4.12 and 4.13; and increasing their speed.

4.6.4 Vorticity Sources

A bridge is implied between vorticity generation and its role in subgrid motion.

Suppose that vorticity is generated locally, resulting in the Eulerian field $\overline{\omega}(\mathbf{x},t)$. Particles move within this field with *individual* ω_i driven by $\overline{\omega}$:

$$\omega_i^{k+1} = \overline{\omega} + \rho_\omega(\omega_i^k - \overline{\omega}) + \mathcal{R}_\omega \tag{4.239}$$

This is a separate AR1 process governing the individual parameter ω_i. It has Lagrangian memory and seeks equilibrium with local sources $\overline{\omega}$, nonzero near sources of subgrid vorticity. Particles that encounter a source area, acquire spin which then decays when they move away.

The composite model: Individual particle velocity $(U,V)_i$ governed by an AR1-s process with $r = \rho - \mathrm{j}\omega_i$; and individual particle spin parameter ω_i governed by a separate AR1 process, driven locally.

For example, each panel of Figure 4.14 shows three particles undergoing AR1-s random walk with mean x-motion and initially zero spin. When a particle enters the box indicated, spin initiates; on exit, it decays back to zero.

The vorticity generation region effectively "traps" particles. Figure 4.15 shows a similar setup with a longer box. Nine trajectories are overlain in each panel, with

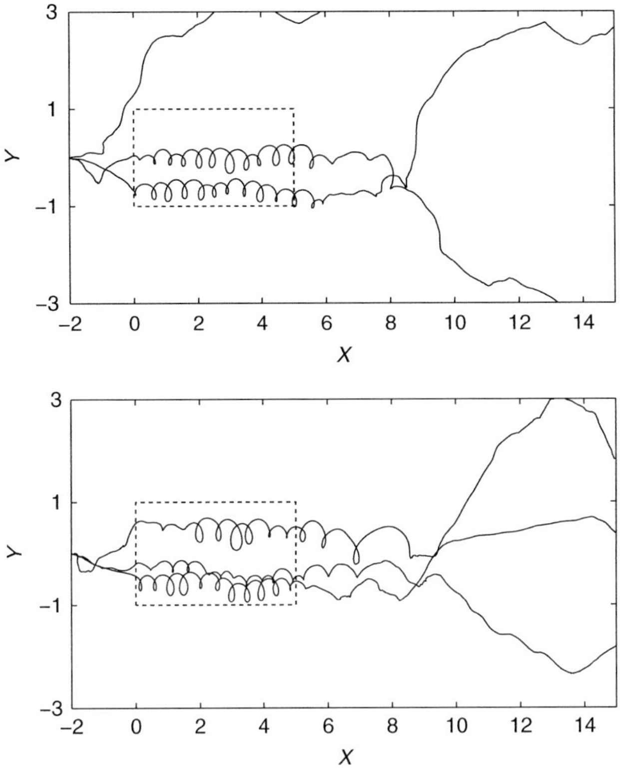

Figure 4.14. AR1-s particles with mean speed $U = 1$ encounter a vorticity source in the indicated box. Initially, individual spin ω is zero. On encounter with the vorticity box, spin increases toward equilibrium $\overline{\omega} = \frac{2\pi\,\Delta t}{T_s}$, with time scale $T_\omega = 2$. On exit, it decays back to zero. Of the six trajectories shown, one misses the vorticity source; the others initiate spinning. The spin decays downstream of the source, being visually absent for $X > 9$ or so. Parameters: spin period $T_s = 0.5.$; AR1 velocity time scale $T = 1$.

$\overline{\omega} = 0$ everywhere except in the box in the lower panel. The spin dynamics are given by Equation 4.239 with constant spin noise in all cases. When the source $\overline{\omega}$ is nonzero, variance growth is constrained. The growth and decay of ω on entrance/exit from the box is evident.

Many substantive closure issues arise in parameterizing such calculations; these results are illustrative only.

Problem 4.21 Implement the 2-D AR1 model with spin as in Equations 4.275–4.276. Plot the cumulative positions for 10 drifters on the $x - y$ plane. What is the effect of spin?

Problem 4.22 Perform an experiment where the spin is reversed. Do the trajectories agree with your intuition? Why or why not?

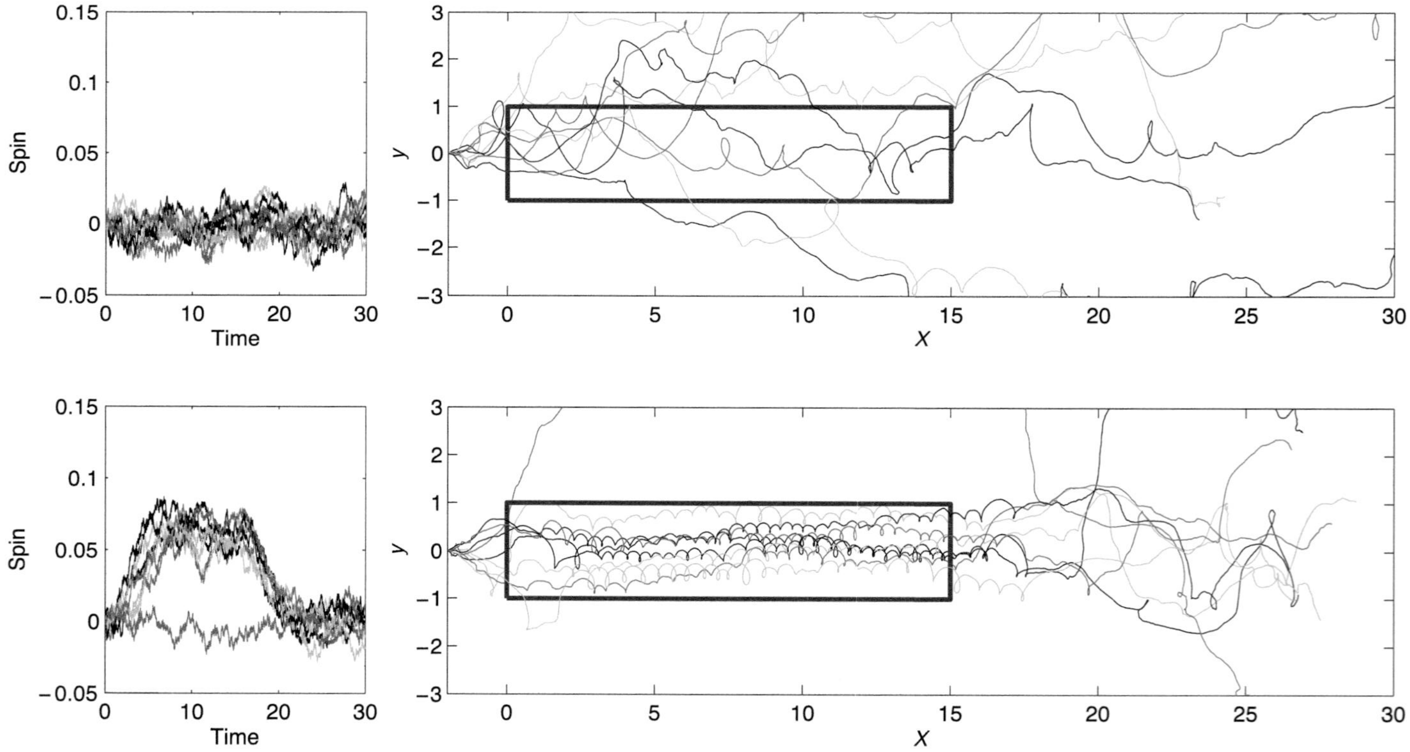

Figure 4.15. As in Figure 4.14, but with full AR1 dynamics for ω_i. σ_ω is everywhere constant. Top: No vorticity generation in the box. Bottom: Constant $\overline{\omega}$ in the box, as before. Particles that walk into the boxed region, are trapped there relative to those that evade it or those in the upper panel. For a color version of this figure, please see the color plate section.

Problem 4.23 Construct an AR2 model without spin, and compare the output with the AR1 model with spin. Are there qualitative differences?

Problem 4.24 Derive the expressions for the AR1-s displacement variance. Use the results to explain the trajectories shown in Figures 4.12 and 4.13.

Problem 4.25 In Figure 4.13 the displacement variance has become larger than that observed in Figure 4.12 due to the lack of rotation. If instead, the displacement variance is to remain the same, what value of ρ should be used? (Zero rotation is to be retained). Confirm this by experiment.

Problem 4.26 Continuing Problem 4.25: under this change, what is the comparison of speed variance observed between the spinning and non-spinning cases? Does this make sense? Is it consistent with theory?

Problem 4.27 Redo Figure 4.13, using no spin, but making the variance of displacement equal to the spinning version, Figure 4.12 by making proper adjustments to the diffusion coefficient. Do your observations agree with intuition and with theory? What is the effect on speed variance?

4.7 Summary – Four Random Walk Models

It is the job of exposition to make things simple. We seek here to summarize *how to do it in simulation*. Specifically, (a) What random numbers are input? (b) What is the time-stepping process? c) What ensemble properties emerge? The growth rate of displacement variance is to be matched to conventional diffusion coefficients, by sizing the random inputs properly at the start.

Always, there are the per-step and cumulative identities:

$$\Delta\widehat{X}^k = \widehat{X}^k - \widehat{X}^{k-1} \tag{4.240}$$

$$= \Delta t \widehat{V}^k \tag{4.241}$$

$$\widehat{X}^k = \sum_1^k \Delta\widehat{X}^i = \Delta t \sum_1^k \widehat{V}^i \tag{4.242}$$

$$\sigma_{\Delta\widehat{X}} = \Delta t \; \sigma_{\widehat{V}} \tag{4.243}$$

and the timing conventions assume all particles at rest at the origin at $t = 0$:

$$\widehat{V}^0 = \widehat{X}^0 = 0 \tag{4.244}$$

$$\sigma_{\widehat{V}} = \sigma_{\widehat{X}} = 0 \tag{4.245}$$

For quantities related to random numbers: the subscript 1 indicates velocity (first derivative of position) – for example, $\mathcal{R}_1$ has units of speed and drives the velocity equation. No subscript, *e.g.* $\mathcal{R}$, indicates displacement. Both $\mathcal{R}$ and $\mathcal{R}_1$ are scaled Wiener increments:

$$\mathcal{R} \sim S\sqrt{\Delta t} \cdot \mathcal{G}(0,1) \tag{4.246}$$

$$\mathcal{R}_1 \sim S_1\sqrt{\Delta t} \cdot \mathcal{G}(0,1) \tag{4.247}$$

Because displacement and velocity are always locked together (Equation 4.241), they differ by the simple factor Δt.

$$\mathcal{R} = \Delta t\, \mathcal{R}_1 \tag{4.248}$$

$$S = \Delta t\, S_1 \tag{4.249}$$

$$\sigma_{\mathcal{R}} = \Delta t\, \sigma_{\mathcal{R}_1} \tag{4.250}$$

In one of the forms of AR2, an analogous subscript 2 indicates acceleration, "the second derivative of position."

In making the connection to diffusion, we assume a mature trajectory and tie the asymptotic limit of variance growth to D. The basic relations below are for 1-D. In 2-D, the appropriate changes for orthogonal, independent diffusion are:

- $\sigma^2_{\mathcal{R}}$ is replaced by $(\sigma^2_{\mathcal{R}_x} + \sigma^2_{\mathcal{R}_y})$.
- $2D$ becomes $(2D_x + 2D_y)$.

4.7.1 AR0 Model

This is the simplest random walk model – where particle displacements are not correlated with their predecessors. There is no memory, or equivalently, the time step is long relative to the correlation time T for subgrid motion. No accounting for A or V is needed.

AR0

$$\Delta\widehat{X}^k = \mathcal{R}^k \tag{4.251}$$

$$\mathcal{R} \sim \sqrt{2D\Delta t} \cdot \mathcal{G}(0,1) \tag{4.252}$$

This Lagrangian formulation is equivalent to the Eulerian diffusion equation, diffusivity D.

Per-step variance is constant:

$$\sigma^2_{\Delta\widehat{X}} = \sigma^2_{\mathcal{R}} \tag{4.253}$$

Accumulated variance is linear in t:

$$\Delta\sigma^2_{\widehat{X}} = \sigma^2_{\mathcal{R}} \tag{4.254}$$

4.7.2 AR1 Model

It is natural to add memory of past disturbances to the walk. This is achieved by accounting for particle velocity V and letting *it* receive stochastic disturbances, with relaxation time T:

$$\widehat{V}^{k+1} = \rho\widehat{V}^k + \mathcal{R}_1^{k+1} \tag{4.255}$$

$$\mathcal{R}_1 \sim S_1\sqrt{\Delta t}\cdot\mathcal{G}(0,1) \tag{4.256}$$

The parameters: $\rho = e^{-\frac{\Delta t}{T}}$; S_1 is linked to diffusivity D.

In terms of displacement:

AR1

$$\Delta\widehat{X}^k = \rho\Delta\widehat{X}^{k-1} + \mathcal{R}^k \tag{4.257}$$

$$\mathcal{R} \sim (1-\rho)\sqrt{2D\Delta t}\cdot\mathcal{G}(0,1) \tag{4.258}$$

$$\rho = e^{-\Delta t/T} \simeq 1 - \frac{\Delta t}{T} \tag{4.259}$$

The V series has autocorrelation ρ

$$C_0 \equiv \sigma^2_{\widehat{V}} \qquad C_n = \rho^n C_0 \tag{4.260}$$

and the velocity variance reflects the accumulated memory:

$$\sigma^2_{\widehat{V}} = \sigma^2_{\mathcal{R}_1}\left[\frac{1}{1-\rho^2}\right] \tag{4.261}$$

Stability demands $|\rho| < 1$.

AR1 reduces to AR0 when $\frac{\Delta t}{T}$ is large. Effectively, the process forgets its history. At the opposite extreme, for small $\frac{\Delta t}{T}$ the memory becomes very important.

4.7.3 AR2 Model

The AR2 model has random acceleration input instead of the AR1 velocity input. A second relaxation time T_a is needed for acceleration, in addition to the relaxation time T for velocity, already used in AR1. There are two equivalent canonical forms:

Form 1 consists of two first-order difference equations, retaining acceleration A explicitly:

$$\widehat{A}^k = \rho_a \widehat{A}^{k-1} + \mathcal{R}_2^k \tag{4.262}$$

$$\widehat{V}^k = \rho_v \widehat{V}^{k-1} + \widehat{A}^k \Delta t \overline{V}_h \tag{4.263}$$

$$\mathcal{R}_2 \sim S_2 \sqrt{\Delta t} \cdot \mathcal{G}(0,1) \tag{4.264}$$

there are two decorrelation times, T_a and T; $\rho_a = e^{-\Delta t/T_a}$ and $\rho_v = e^{-\Delta t/T}$. **Form 2** eliminates A, resulting in a single second-order difference equation:

$$\widehat{V}^k = \rho_1 \widehat{V}^{k-1} + \rho_2 \widehat{V}^{k-2} + \mathcal{R}_1^k \tag{4.265}$$

$$\mathcal{R}_1 \sim S_1 \sqrt{\Delta t} \cdot \mathcal{G}(0,1) \tag{4.266}$$

S_1 is linked to diffusivity. The memory parameters: $\rho_1 = \rho_a + \rho_v$; $\rho_2 = -\rho_a \rho_v$.

In terms of displacement:

AR2

$$\Delta \widehat{X}^k = \rho_1 \Delta \widehat{X}^{k-1} + \rho_2 \Delta \widehat{X}^{k-2} + \mathcal{R}^k \tag{4.267}$$

$$\mathcal{R} \sim (1 - \rho_1 - \rho_2) \sqrt{2D\Delta t} \cdot \mathcal{G}(0,1) \tag{4.268}$$

$$\rho_1 = e^{-\Delta t/T_a} + e^{-\Delta t/T} \tag{4.269}$$

$$\rho_2 = -e^{-\Delta t/T_a} e^{-\Delta t/T} \tag{4.270}$$

Stability requires

$$-1 \leq \rho_2 \leq 1 - |\rho_1| \tag{4.271}$$

For the covariance of $\widehat{V}$, the recursion is

$$\sigma_{\widehat{V}}^2 \equiv C_0 = \left[\frac{1}{1 - \rho_1^2 \left(\frac{1+\rho_2}{1-\rho_2} \right) - \rho_2^2} \right] \sigma_{\mathcal{R}_1}^2 \tag{4.272}$$

$$C_1 = \left[\frac{\rho_1}{1 - \rho_2} \right] C_0 \tag{4.273}$$

$$C_n = \rho_1 C_{n-1} + \rho_2 C_{n-2} \qquad n \geq 2 \tag{4.274}$$

AR1 and AR0 are recovered from AR2 by setting ρ_2 and/or ρ_1 to zero as appropriate.

4.7.4 AR1-s Model

In the AR1-s model, velocity evolves with only the memory of its predecessor; but in addition to ρ there is a second parameter ω. When positive, it causes trajectories to veer right during each time step.

$$\widehat{U}^k = \rho\widehat{U}^{k-1} + \omega\widehat{V}^{k-1} + \mathcal{R}^k_{u_1} \tag{4.275}$$

$$\widehat{V}^k = -\omega\widehat{U}^{k-1} + \rho\widehat{V}^{k-1} + \mathcal{R}^k_{v_1} \tag{4.276}$$

where the $\mathcal{R}$ are independent. AR1-s is the complex generalization of the scalar AR1 model; with complex memory parameter $r = \rho - \mathrm{j}\omega$.

In terms of the displacements:

AR1-s

$$\Delta\widehat{X}^k = \rho\Delta\widehat{X}^{k-1} + \omega\Delta\widehat{Y}^{k-1} + \mathcal{R}^k \tag{4.277}$$

$$\Delta\widehat{Y}^k = -\omega\Delta\widehat{X}^{k-1} + \rho\Delta\widehat{Y}^{k-1} + \mathcal{R}^k \tag{4.278}$$

$$\mathcal{R} \sim \sqrt{\left[(1-\rho)^2 + \omega^2\right]}\sqrt{2D\Delta t}\cdot\mathcal{G}(0,1) \tag{4.279}$$

The standard AR1 results are recovered when $\omega = 0$.

4.8 Concluding Remarks – Random Motion

We began this exposition with the common example of drunken walks. In extending it, we have used the Newtonian motion analogy – force, acceleration, velocity, displacement, and added memory to the process. We concluded with an example of spinning walk. These are all easy to simulate, and raise important questions about the subgrid motions they are representing.

Enormous advances in random motion have occurred in the fluid mechanical arena. Early work (Taylor [445], Richardson [376]) connected turbulent fluid motion to random walk processes, and highlighted autocorrelated processes. Atmospheric turbulence and random walk processes are nicely summarized by Rodean [385]. Oceanographic applications are summarized by Griffa et al. [185]. Recent oceanographic advances are assembled by Griffa et al. [186]; the contributions by Pasquero et al. [351] and Mariano and Ryan [302] are noteworthy. Bennett [24] has made important advances in contributions to Lagrangian fluid mechanics generally.

In the ecological context, Okubo and Levin [343] discuss a very wide range of behavioral processes within turbulent environments. Specific biotic processes need to be represented in these cases. Murray [322, 323] discusses a wide range of these processes. Werner et al. [474] discuss practical oceanographic applications in terms of individual-based models – the union of random motion in both physical and biological space.

These processes are easy to implement! Experiment is encouraged. Among the key elements:

- The random walk size scales with $\sqrt{\Delta t}$ as a consequence of the process being driven by Wiener noise.
- The random walk size is proportional to $\sqrt{D\Delta t}$ to match the macroscopic diffusion outcome.
- The adjustment for variable diffusivity requires a pseudo-velocity ∇D to be applied to the random walk.
- For small Δt the need for memory becomes important.
- Adding memory decreases the random input per step, needed to achieve the same variance growth rate
- The random walk parameters describe subgrid turbulence. D is generally computed in coastal ocean models and used; it may be considered known. The memory parameters ρ require care in selection on the part of the analyst.
- Spin may be added to the random walk to represent small, unresolved features, and so forth, that are known but not present in gridded model products

Criteria for valid random motion models are presented in Thomson [449] and in Rodean [385]. The ultimate conclusion is that satisfaction of the *well-mixed condition* (WMC) is required:

> *If the particles of a tracer are initially well-mixed (in both position and velocity), they will remain so.*

Brickman and Smith [52] showed that for contemporary (2002) shelf circulation models, the WMC can be demonstrated only if resolution is sufficiently high. Tests using *tidally averaged* diffusivities on the New England Shelf indicated that available resolution is insufficient to resolve their gradients. It is suggested that these gradients be obtained by imposing an analytical structure on the turbulent quantities, which would render them continuous and differentiable. This is suggested by Ross and Sharples [388], who also discuss the implementation of boundary conditions. North et al. [335] describe the water column profile scheme, in which a spline fit in the vertical for diffusivity is superior to linear interpolation where stratification is strong.

Generally, turbulence closure models will describe the statistics of the subgrid motion in terms of the resolved motion. These are reviewed in Chapter 6.

5

Boundary Conditions, Boundary Layers, Sources

An oceanic particle collides with a boundary. On both theoretical and practical grounds, this is inevitable. What happens? And how is it to be represented in particle simulation?

Particle – boundary interactions occur at the subgrid scale, in an unresolved boundary layer. Motion into, out of, and within such a layer needs appropriate closure. One needs to supplement even the simplest random walk models with proper boundary relations. Needless to say, transport and fate of biochemically active particles make further demands. The rules governing such encounters – Boundary Conditions or BC's – are quite substantive, and their identification is often part of the basic purpose of a modeling study.

The general problem of transport processes in turbulent boundary layers is vast. We offer a general framework from a particle point of view; at the very least, the vocabulary it supports needs to be clear and comprehensive. Some examples are used to illustrate its use.

Included in this discussion are horizontal boundaries including beaches, marshes, river inlets, and open-sea boundaries; and vertical boundaries at the top (air–sea interface) and bottom (sediment interface) of the water column. All these surfaces themselves are potentially in motion – flooding, surface waves, sediment motion.

5.1 Boundary Layers: Continuum and Discretization

Fluid boundary layers are collapsed to lines in 2-D models; surfaces in 3-D. The governing PDE formulations recognize generic types of conditions governing heat, mass, momentum fluxes across them. This is a widely recognized form of aggregation of boundary phenomena into relations among resolved quantities – those expressed in PDE fields. There are four classic boundary types in this category:

- Type I: the value of an oceanic variable u is given. These are commonly referred to as "Dirichlet" boundaries.

$$u = f(\mathbf{X}, t) \tag{5.1}$$

- Type II: the value of an oceanic gradient, normal to the boundary, is given. These are commonly referred to as "Neumann" boundaries.

$$\frac{\partial u}{\partial n} = g(\mathbf{X}, t) \tag{5.2}$$

  Sometimes the Neumann classification implies the homogeneous form, $\frac{\partial u}{\partial n} = 0$.
- Type III: this is a mixed condition:

$$\frac{\partial u}{\partial n} + a(\mathbf{X}, t) = b(\mathbf{X}, t) \tag{5.3}$$

  These have many names, depending on the phenomenon described. Type III is a generalization of Type I and II and recovers those for extreme values of the coefficients a and b.
- A fourth "radiation" boundary is similarly mixed:

$$\frac{\partial u}{\partial n} + a(\mathbf{X}, t)\frac{\partial u}{\partial t} = b(\mathbf{X}, t)u + c(\mathbf{X}, t) \tag{5.4}$$

  This Type IV boundary arises in cases where oceanic wave phenomena are allowed to exit an open boundary, unimpeded.

The derivative *normal* to the boundary is very important in these conditions and arises naturally for the most common oceanic phenomena. These would be described by PDEs with gradient and divergence operators, and flux-type boundary integrals $\oint \nabla U \cdot \underline{\mathbf{n}}\, ds$. When the basic phenomenon involving circulation of vector fields $\mathbf{U}$, the derivative boundary terms are often required in the direction *tangential* to the boundary, and this would accompany PDEs with $\nabla \times \mathbf{U}$ terms and circulation boundary integrals $\oint \mathbf{U} \times \underline{\mathbf{n}}\, ds$.

In all cases the coefficients (f, g, a, b, c) are assumed known, imposed on the solution they are bounding. There are many generalizations for vector fields and nonlinear phenomena. Presentations of PDEs are always accompanied by the blend of boundary conditions necessary and sufficient to guarantee a unique solution; and the constraints that must be satisfied by the coefficients.

Physically, these are idealizations of transport phenomena occurring across a boundary layer, parameterized in terms of the resolved oceanic variables and their derivatives. Heat, mass, momentum, and energy are among the classic variables that would have continuous representation in PDEs bounded by these conditions.

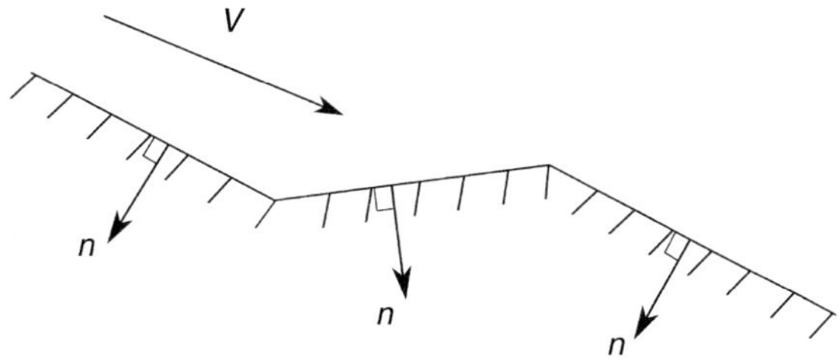

Figure 5.1. Collision between a particle and a discretized boundary.

Needless to say, the coefficients represent the aggregation of unresolved phenomena occurring in an infinitesimally thin boundary layer.

5.2 Discretized Boundaries

Practical model boundaries are not smooth. Typical is a set of connected line segments with abrupt change in slope. Particles immersed in continuous velocity fields will necessarily risk collision with kinked model boundaries. At that point, the particle has entered a boundary layer. A disposition of its future is necessary (Figure 5.1).

The simplest characterization of the boundary is in terms of line segments (surfaces in 3-D) that support local ($\underline{\mathbf{n}}, \underline{\mathbf{s}}$) definition.[1] These are not globally straight, but kinked in some way that allows their definition in terms of small pieces ("boundary elements"), each with its own local coordinate system. Ultimately one needs to address the question of discontinuous boundary coordinate systems and what to do about it – that is a fundamental issue in the discretization of PDEs. Here, it is enough to observe that discrete simulations will always have particles colliding with a boundary segment; and the location of such a collision is computable by considering all possible boundary segments in a particle's neighborhood. If all segments are properly classified, then the disposition of a colliding particle is straightforward.

5.2.1 Particle Motion and Change

Particle navigation on non-square grids is covered in Sections 7.1.8.3 and 7.3. One can always detect the encounter of a particle with a boundary element – time and location. The use of boundary elements to locate collisions is a natural use of the necessary data structures (see Section 7.4.2). There are three generic considerations.

First, one separates consideration of normal (n) and tangential (s) *motions*. A convention is needed; we use that in Figure 5.2. The normal is directed outward. This

[1] Here and elsewhere, (n,s) is a local, orthogonal coordinate system. ($\underline{\mathbf{n}}, \underline{\mathbf{s}}$) are the unit vectors; $\underline{\mathbf{n}}$ is directed *out* of the ocean.

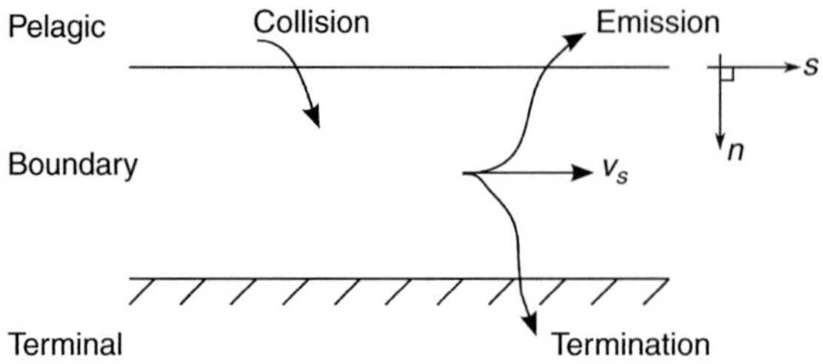

Figure 5.2. Boundary states: In the ocean (pelagic); in a boundary layer, idealized as a line in a discretized model (boundary) ; and a formerly active particle now removed from the ensemble (terminal).

is a standard calculus convention and used in typical PDE discretizations. This local geometry is a property of a boundary element itself, not of a particle.

Second, distinguish between particle *states*: is a particle *pelagic* (in the ocean) or on the *boundary* (or in a boundary layer)? This two-state classification needs a complement, that of the particle that has been eliminated from further action – its dynamic is over, its history remains a matter of record; it joins an archive of *terminal* particles. (As a simulation outcome, it would be a mistake to discard a particle and its history at the point of termination.)

Third, relative to boundary exchange, separate incoming particles from outgoing ones – collisions from emissions. This incoming/outgoing terminology is in reference to the boundary layer. A pelagic particle may collide; a boundary particle may be emitted or terminated.

5.2.2 States and Transitions

A particle in the boundary state has the three discrete options: emission into the pelagic state; terminate; or continue (stay in the boundary). These transitions all have probabilities. In addition, a boundary particle needs to have its tangential motion specified.

State: It is useful to identify a particle as either pelagic or boundary. The former may collide with a boundary; the latter may emit or terminate. Boundary elements contain particles bearing individual identities and histories.

Transitions: Incoming: If an oceanic particle encounters a boundary: it may either be dealt with immediately or be arrested at the point of encounter, awaiting further orders.

The termination option is a one-way ticket to the archive of spent particles; otherwise, a particle remains active and its normal and tangential disposition needs specification.

For a boundary encounter: a porous boundary admits some but not all incoming particles to be terminated. An incoming individual will either terminate with probability P_t, or be retained with probability $(1-P_t)$. In the latter case, continued presence on a boundary is subject to per-step transition probabilities. Such particles constitute potential sources in the future.

5.2.3 Boundary Types

Below is a list are some types of boundaries. These are best viewed as a logical catalog of boundary element types, illustrated in Figure 5.3.

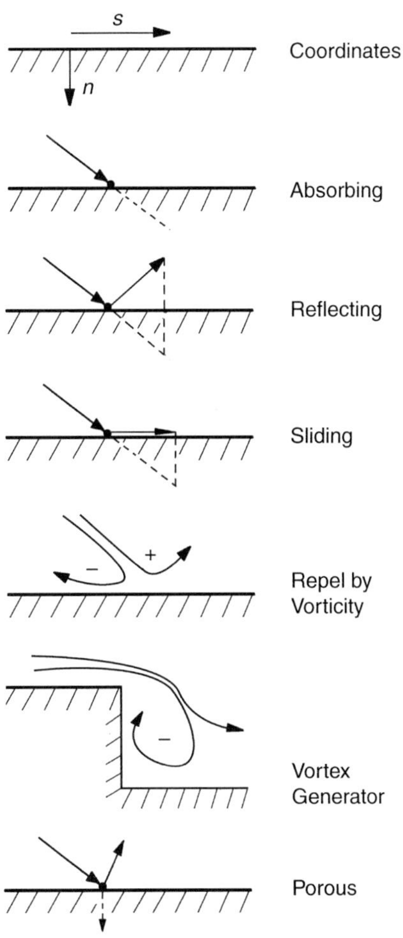

Figure 5.3. Boundary types.

Absorbing: Incoming particles disappear forever, with perfect certainty. Examples: death; perfect porosity; complete immobilization with no hope of re-emission. "The wall of death."

Arresting: Incoming particles are completely immobilized with zero porosity; remain active and may be emitted in the future. Example: sinking cysts reaching bottom.

Arresting/sliding: Arrested particles may move tangentially as in sediment particles in a bottom boundary layer awaiting vertical activation by bottom stress. Example: bedload particles rolling or bumping along bottom.

Slip: Like the sliding case – no escape; but tangential motion at some fraction of the free-stream velocity.

Porous: One form of the arresting case with instant disposition: re-emit with probability P_e; or terminate with probability $1-P_e$.

Reflecting: The opposite of the absorbing case: incoming particles "bounce" off a boundary with normal components reversed and tangential components unchanged. (This is the extreme of the arresting case with immediate, certain reemission.) This would be specular reflection; diffuse reflection would need a direction of emission added.

Stranding: An Arresting boundary where particles are trapped on a boundary, awaiting hydraulic conditions for re-activation. Examples: flooding/dewatering and shoreline motion with finite-depth drifters or sediment particles; shoaling boundaries with fixed-depth drifters.

Downwelling A particle encounters a horizontal boundary, but cannot follow the vertical motion. Example: offshore wind, passive particles.

Repelling: This nonporous boundary would repel particles while they are still pelagic, preventing contact. One approach: bias the random walk against normal walk into a boundary. Another: AR1-s with mean vorticity generation in the boundary layer, vanishing at long distance from the wall. The latter involves boundary-generated vorticity injected locally into the subgrid, causing particles to execute helical trajectories adjacent to a boundary.

Vortex generator: As in Figure 5.3, the generation of intermediate-size vortices (large, predictable, but still unresolved) at abrupt changes in boundary curvature. The flow direction biases the unresolved vorticity generated.

5.2.4 Basic Needs for Boundary Particles

Everywhere along an unresolved boundary one must specify:

- The local normal and tangential coordinate system
- P_t the probability of terminating a boundary particle
- P_e the probability of emitting (activating) a boundary particle

- If emitting, the emission distance normal to the boundary
- V_s the tangential velocity of a boundary particle

(The probability of a boundary particle remaining on the boundary is $[1 - P_t - P_e]$.) In addition to the local geometry, there are four parameters: two state transition probabilities, the emission distance, and one tangential velocity (a 2-D vector in 3-D models).

5.2.5 Boundary Sources

There are boundary and internal sources of particles. Focusing on boundary sources, there are three options for boundary sources:

- Let particles occur first as pelagic particles, which collide with a boundary; then emit them according to P_e. These particles would retain the full memory of their pre-boundary pelagic history in addition to their history while resident in the boundary layer.
- Place them initially in the boundary layer; then emit them according to P_e. Particle memory would be limited to initial placement location and time of activation, plus their boundary history.
- Make new particles truly exogenous – inject them as sources across boundaries into either the boundary or directly into the pelagic state, according to given source relations.

In the first two cases, the inventory of boundary particles evolves with the system; the "source" of emissions from the boundary layer is internal to the system. In the third case, a spatially variable binomial source rate could be deployed along a boundary. When the abundance of potential source particles is effectively unlimited, this could be represented as a spatially variable Poisson source (Section 3.2.2). A blend of these ideas is natural. (More at Section 5.6.)

5.3 The Law of the Wall

The law of the wall is among the most robust relations in turbulent fluid mechanics. It is supported by both observation and analysis in many contexts. It governs the inner regions of turbulent boundary-layer flow near an impermeable, rough boundary. General overviews occur in any elementary fluid mechanics text (e.g., Fay [139], Sabersky et al. [396]; Csanady [94]). Recent articles attest to its enduring value, practical utilization, and contemporary needs for improvement. (Perlin [355]; Bradshaw [50]).

Turbulence closure models will always incorporate this "law" at solid boundaries, forming the basis of a "slip" boundary condition along unresolved solid surfaces that is nearly universally used.

The essential points:

- The "wall" is a rough surface, plane on average, with roughness size y_r.
- Near the wall there is a layer of constant shear stress τ_o between the arrested flow and the "free stream" flow.
- The tangential velocity profile in this layer is logarithmic with distance from the wall.
- The mixing length l is linear in distance from the wall.
- The shear stress is quadratic in local velocity U.

With y the distance from the wall (Figure 5.4):

$$U = \frac{U^*}{\kappa} \ln \frac{y}{y_r} \tag{5.5}$$

$$U^* = \sqrt{\frac{\tau_o}{\rho}} \tag{5.6}$$

$$l = \kappa y \tag{5.7}$$

$$\frac{\tau_o}{\rho} = c_d U|U| \tag{5.8}$$

$$c_d = \left[\frac{1}{\kappa} \ln \frac{y}{y_r}\right]^{-2} \tag{5.9}$$

$$A_y = \kappa U^* y \tag{5.10}$$

U is the tangential velocity, averaged over turbulent fluctuations. U^* is the friction velocity. l is the turbulent mixing length. τ_0 is the shear stress exchanged between fluid and the boundary. The quadratic drag coefficient c_d varies with distance from the wall. $\kappa = 0.41$ is von Karman's constant. A_y is the eddy viscosity normal to the wall.

These are included the discussion of turbulence boundary conditions (Section 6.3.1.2) and used in connection with the sediment application, Chapter 9. Here we use the coordinate y as the distance normal to the computational boundary; y_w is the unresolved gap (Figure 5.4).

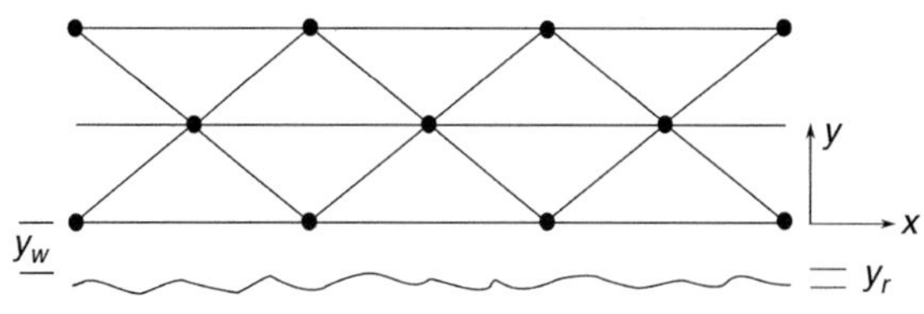

Figure 5.4. Horizontal computational mesh adjacent to an impermeable wall; the distance y_w is unresolved.

Example 5.1 Extensive experience in the Gulf of Maine supports parameterizing the bottom boundary layer with $(y_w,\ y_r) = (1.0,\ 0.035)\ m$ with 1 m vertical resolution and bottom quadratic drag coefficient $c_d = 0.005$ [275].

5.4 A Repellant Boundary Layer

In practical ocean models, the boundary layer ("log layer") is never resolved. The flow slips along a surface separated by y_w from the wall; the small gap transmits the shear stress unchanged, τ_0; there is no normal velocity (Figure 5.4).

The gridded velocity represents the ensemble mean; both normal and tangential parts have turbulent fluctuations and individual fluid particles move into/out of the log layer with zero mean.

Eddy diffusivity D is used to determine random walk in the grid, normal to the boundary.[2] There is an implied strong gradient in D from zero at the true wall (where all motion is arrested), to the finite gridded value D_b at the computational boundary. As discussed in Section 4.3.4, random walk needs to recognize this gradient.

5.4.1 Boundary Particles

A particle already within the log layer will see the pseudo-velocity $\frac{\partial D}{\partial n} \simeq \frac{D_b}{y_w}$ directed away from the wall. Hence its displacement in a time step will be

$$\Delta y = \Delta t \frac{D_b}{y_w} + \mathcal{R}_y \tag{5.11}$$

Assume the particle is in the center of the log layer. $\mathcal{R}_y$ will be based on $\frac{D_b}{2}$, with zero mean and $\sigma = \sqrt{D_b \Delta t}$. If the displacement fails to exceed the threshold

$$\Delta t \frac{D_b}{y_w} + \mathcal{R}_y = \frac{y_w}{2} \tag{5.12}$$

then it remains on the boundary. Otherwise, it is emitted with probability P_e:

$$P_e = P\left[\mathcal{R}_y \geq \frac{y_w}{2} - \Delta t \frac{D_b}{y_w}\right] \tag{5.13}$$

$$= P\left[\frac{\mathcal{R}_y}{\sigma} \geq \frac{y_w}{2\sigma} - \frac{\sigma}{y_w}\right] \tag{5.14}$$

This depends on the distribution of $\mathcal{R}_y$, the time step, the gridded D, and the boundary parameter y_w. Of these, only y_w is new here, a property of the boundary. The other quantities are already internal to any simulation. For real systems, this provides a

[2] Note that here, D is the eddy diffusivity; in other chapters, it denoted by A_n.

one-parameter BC format even when the 'wall' is only an idealization. In practice, a single generation of $\mathcal{R}_y$ and a binary comparison (Equation 5.13 or 5.14) is needed. (The normalization is useful, although it masks the independence of the parameter Δt.)

5.4.2 *The Gaussian Case*

Suppose $\mathcal{R}_y \sim \sigma_w \mathcal{G}(0,1)$. Then

$$P_e = P\left[\mathcal{G}(0,1) \geq \frac{1}{2}\frac{y_w}{\sigma_w} - \frac{\sigma_w}{y_w}\right] \tag{5.15}$$

This is plotted in Figure 5.5. The breakpoint ($P_e = 0.5$) occurs at $\frac{y_w}{\sigma_w} = \sqrt{2}$. This is the point where pseudo-advection alone ($\frac{\partial D}{\partial y}$) would move a particle out of the boundary layer. Adding diffusion to this case makes $0.5 < P_e < 1$. The range $0 < P_e < .5$ is the other case: advection alone is not capable of activating a particle (Figure 5.6). These features remain true for any unbiased distribution $\mathcal{R}_y$; the shape reflects the distribution details.

If conditions persist over many Δt, the *waiting time* in the boundary layer will obey the geometric distribution with mean $\frac{\Delta t}{P_e}$. (This is the finite version of the exponential distribution.)

Intuitively, large Δt favors advection over diffusion, as the latter displacements scale with $\sqrt{D\Delta t}$ while the former scale linearly with $D\Delta t$. Hence increasing this computational parameter moves P_e to the right side of the curve in Figure 5.5.

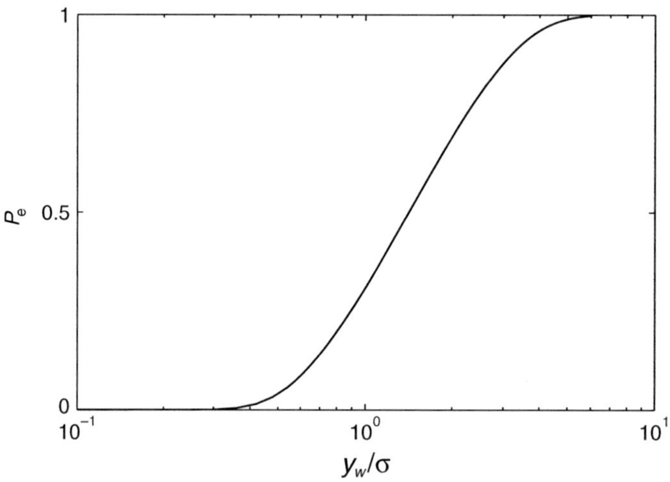

Figure 5.5. Probability of emission of a boundary particle, P_e, versus $\frac{y_w}{\sigma_w}$; Equation 5.14, for the Gaussian case.

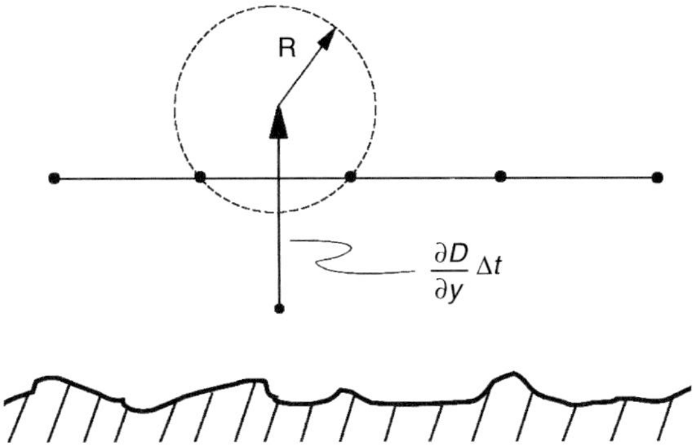

Figure 5.6. The balance of pseudo-advection and diffusion reflected in P_a, Equation 5.13. Note that here, D is the eddy diffusivity normal to the boundary (indicated A_n in other chapters.) The situation illustrated leads to $P_e > 0.5$.

5.4.3 Pelagic Particles

A particle in the resolved fluid will experience increases in gridded D as it approaches such a boundary, providing a natural repellant. This of course assumes that the proper boundary condition on $\frac{\partial D}{\partial y}$ is enforced, spanning the boundary. This amounts to a boundary condition connecting gridded D to its normal gradient:

$$\frac{\partial D}{\partial y} = \frac{D}{y_w} \tag{5.16}$$

Advocates of numerical analysis will recognize Equation 5.16 as the type III BC on the numerical differentiation of the gridded field D; or equivalently, the necessary BC for projecting gridded D onto smooth ∇D. A standard example would be the Galerkin projection.

Overall, a single parameter y_w is needed everywhere on a wall boundary to produce an effective wall that either repels or absorbs fluid particles from the interior, and either emits or retains particles in the boundary layer.

5.4.4 The Steady State

Consider a plane uniform boundary adjacent to a semi-infinite fluid (Figure 5.7). There are N_b boundary particles per unit of boundary surface; the density of pelagic particles is ρ per unit volume, uniform. A probabilistic steady state will occur when the number of pelagic particles incoming, balances the number of boundary particles outgoing in a single time step. The latter will average $\overline{N_b P_e}$ per time step. For a pelagic particle at distance y from the boundary, the probability of a collision P_c is the probability of displacement to or beyond the boundary

$$P_c(y) = 1 - F(y) \tag{5.17}$$

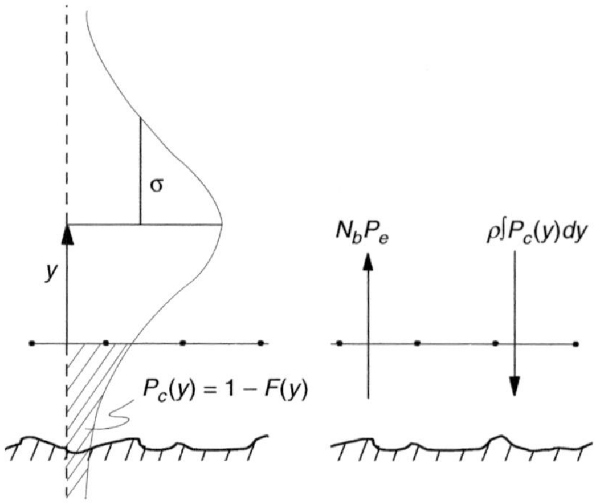

Figure 5.7. Definition sketch for the steady-state balance of probabilities at a semi-infinite plane wall. Left: A single particle at distance y from the wall. Right: Steady-state balance of pelagic collisions with boundary activations.

The balance is the summation of all pelagic particles,

$$P_e\overline{N}_b = \rho \int_{y=0}^{\infty} P_c(y)\, dy = \rho \int_{y=0}^{\infty} (1 - F(y))\, dy \tag{5.18}$$

For the infinite integral, the integrand falls off with increasing y, so we can expect a finite limit. For a Gaussian random walk, size σ_p, the integral is $0.4\sigma_p$, so we have the mean boundary density $\overline{N}_b$:

$$\overline{N}_b = 0.4\rho\frac{\sigma_p}{P_e} \tag{5.19}$$

This simple result confirms intuition: $\overline{N}_b$ scales directly with the pelagic density ρ and σ_p; and inversely with the boundary layer ejection probability P_e.

Changing the distribution of the pelagic random walk makes detailed changes in this finding, but not substantive ones. For example, if the random walk is a uniform distribution, half-width W, the balance is

$$\overline{N}_b = \rho\frac{W}{4P_e} = \rho\frac{\sqrt{3}}{4}\frac{\sigma_p}{P_e} \tag{5.20}$$

($\sigma_p^2 = \frac{W^2}{3}$ in this case.)

Problem 5.1 For a boundary layer 1 m thick, roughness height 0.01 m: what is c_d?

Problem 5.2 For the above conditions, if the bottom velocity is 1 m/s, what is the vertical diffusivity? Use level 2.5 closure.

Problem 5.3 Prove Equation 5.19.

Problem 5.4 Prove Equation 5.20.

Problem 5.5 Simulate vertical random walk in a 1-D vertical system 100 m deep, with Gaussian random walk, uniform $D = 200$ cm^2/s. At top and bottom use walls 1 m from true boundaries; and $P_a = 0.4$ at bottom, 0.95 at top.

1. What is P_a? P_c?
2. Does the system come to a steady state?
3. If so, is it compatible with the analysis in Section 5.4? Explain.

This analysis is developed more fully for the bottom boundary condition, in the context of sediment particles (Chapter 9).

5.5 Examples

To illustrate the application of boundary conditions, we offer some simple examples.

5.5.1 Oiling the Coast

A 2-D channel has length $L = 10$ km (x), width $W = 0.5$ km (y). Steady flow is along-channel, with parabolic velocity distribution $u(y)$. Hydraulic slip conditions are in force at both sides, $-du/dn = ku$.

$$u(y) = u_0[1 - (y/y*)^2] \tag{5.21}$$

$$y*^2 = [W/k + W^2/4] \tag{5.22}$$

Along the centerline $y = 0$ and $u_0 = 2.0$ m/s. The wall condition is $k = 2$, effectively no-slip. Horizontal eddy viscosity is constant and isotropic, $A_h = 1.5$ m^2/s.

Upstream, particles of oil are introduced, each with oil mass M_k. They are advected with an AR0 random walk added.

If a particle encounters a top or bottom boundary:

- Its normal and tangential motion is arrested.
- It begins oiling the coast, that is, shedding oil onto the boundary, at the rate $dM_k/dt = -\alpha M_k$; $\alpha = 0.005$ per second.
- The boundary oil is left immobile; it builds up there.
- This continues until (a) $M_k \to 0$; or (b) the particle with its remaining oil is ejected back into the flow.
- Each boundary particle has ejection probability $P_e = 1/25$ (high) or $1/500$ (low).
- On ejection, the intrusion distance is $\sqrt{2D\Delta t}$.
- Particles reinjected into the flow may encounter a boundary again.

The downstream boundary allows all particles to flow freely out, never to return. From a particle simulation viewpoint, this is an absorbing boundary. For particle k, the point of exit is $(x,y,t)_k$; it marks the release of the remaining oil M_k to the downstream environment.

One implementation uses two kinds of particles: pelagic and boundary. Oil is introduced upstream as pelagic particles. When a pelagic particle hits a boundary, a new "boundary particle" is initiated with $M = 0$. It will never move; a zero-sum transfer begins between the two. The pelagic particle will eject back into the flow with probability P_e.

All oil will be conserved. It will occur in two forms: in boundary particles immobilized at the sides of the channel or in pelagic particles exiting downstream.

Results: Figure 5.8 shows 10 trajectories for the low P_e case. Five of the particles encountered the boundary; 4 remained there long enough to transfer all their oil without return to the channel flow. One was reinjected with some oil left; it subsequently hit the boundary again, and deposited its remaining oil. Overall, half of all the oil ended up on the boundary.

If P_e is increased to $1/25$, then boundary stays are shorter. Less oil is transferred before reinjection; and more particles have multiple boundary encounters. Figure 5.9 shows this. Figure 5.10 shows M_k versus time for several individual particles in this simulation. The horizontal (constant M_k) sections are when a particle is pelagic –

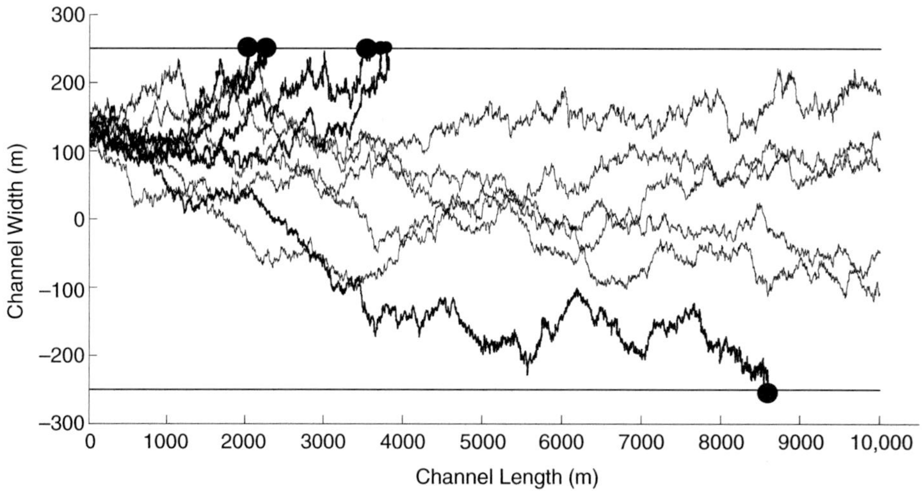

Figure 5.8. *Oil example:* Particle trajectories for low $P_e = 1/500$. Ten particles are released. The heavy boundary dots are proportional to the amount of oil deposited. The heavy lines mark the five particles that hit the boundary; one of the five was reemitted with some oil remaining; it hit the boundary a second time and deposited its remaining oil. Overall, 50% of the initial oil was deposited along the shoreline.

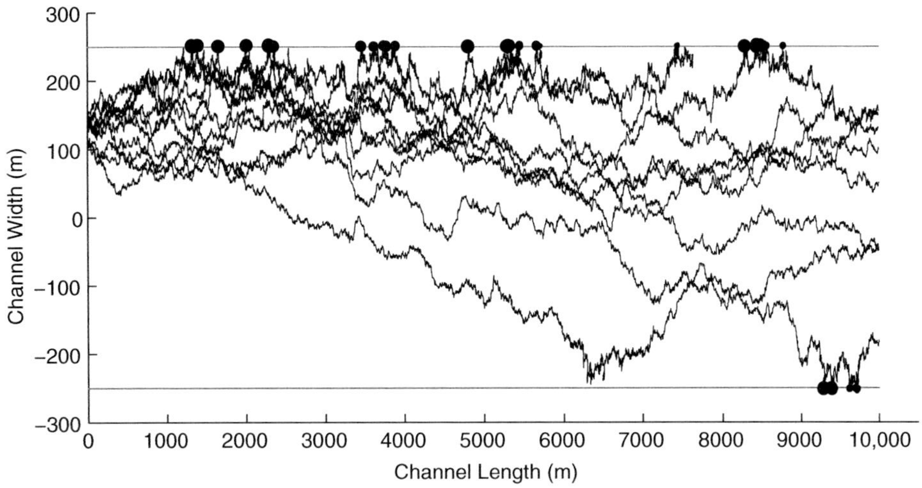

Figure 5.9. *Oil example:* Particle trajectories for high $P_e = 1/25$. There are many boundary encounters, but no particles are completely depleted of oil, owing to the low residence time at the boundary. The dots mark the boundary deposits, proportional to the amount left. Overall, 24.5% of the total oil is deposited on the boundary; all particles exit downstream with some oil remaining.

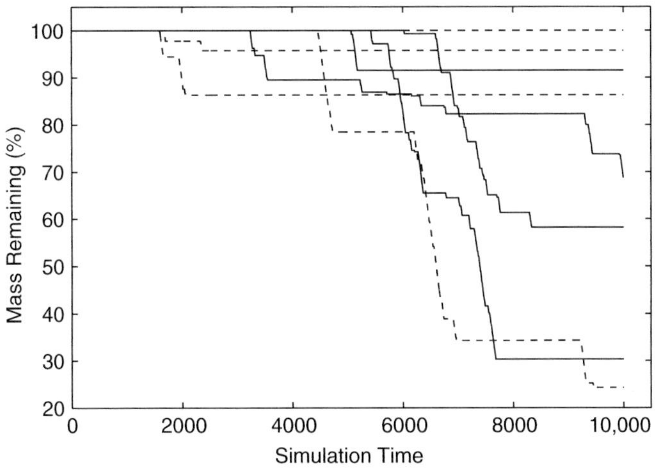

Figure 5.10. *Oil example:* The oil mass remaining with a particle, versus time, for several individual particles. The horizontal sections are when a particle is pelagic – no loss. M_k decreases during boundary encounters. This is the high ejection rate case, $P_e = 1/25$.

no loss. M_k declines during boundary encounters. All particles exit to the right-hand (downstream) boundary with some retained oil.

This is the “base case,” with high P_e. The problems below explore perturbations to it.

Problem 5.6 For the oil example, what is u at the boundary as a function of k? Plot it and explain.

Problem 5.7 Continuation of Problem 5.6. How does the boundary oil accumulation rate depend on k? (1) What do you expect and why? (2) Simulate the problem with various values of k and plot the results. Do your results make sense?

Problem 5.8 For the oil example: with no-slip, near-boundary particles slow down, spending more time near the boundary. Change k so that the boundary u is larger, $u_0/2$. What do you expect? Simulate this and explain.

Problem 5.9 For the oil example with no slip at the boundary: suppose that A_h is proportional to u_0^2. Relative to the base case, suppose u_0 is doubled. What will be the effect on (1) the total amount of oil exiting downstream? (2) the distribution of the boundary oil?

Problem 5.10 For the oil example, base case. If the oil is injected upstream at the centerline, what will happen? Specifically, how much oil will be left on the boundary, and where? Do this by simulation.

Problem 5.11 For the oil example, base case. Simulate the base case using AR1 for the particle trajectories. How does the AR1 memory parameter ρ affect the total amount of oil that exits downstream?

Problem 5.12 For the oil example, base case. Simulate the base case but add an evaporative loss to all non-boundary particles, $dM_k/dt = -k_e M_k$. What is the effect of k_e? Specifically, what value of k_e would result in cutting the amount of oil left on the boundaries by 50%?

5.5.2 Bioaccumulation

Here we concentrate on the boundary segment alone, representing the coastal ocean as a source of particles.

A shellfish bed is located along a boundary segment. These take in ambient water and filter it, at a constant rate. There is no possibility of disaggregating these organisms.

Offshore, a toxin-producing planktonic organism is present in large numbers, each with M_k of toxin on board. These are represented as individual particles. They collide with the shellfish boundary at the Poisson rate r_c particles per day. While on the boundary, individual particles have probability $P_{in}\Delta t$ of being ingested and probability $P_e\Delta t$ of being emitted back into the coastal ocean.

Ingested toxin is accumulated in the shellfish tissue, M_s, and slowly metabolized into benign products at the rate k_m:

$$\Delta M_s = \sum_{k_{new}} M_k \left[1 - \frac{M_s}{K}\right] - k_m \Delta t M_s \tag{5.23}$$

The summation includes only the newly ingested particles. The accumulation rate is limited by the total already accumulated; when M_s reaches K the shellfish are saturated.

As long as the supply is active, the boundary will build up and maintain an inventory of particles N. The steady state can be found by balancing the rates:

$$N = \frac{r_c}{P_{in} + P_e} \tag{5.24}$$

M_s will approach its asymptote K.

Simulation results appear in Figure 5.11. Parameters: $(M_k, K) = (1, 50)$ mg; $(r_c, P_{in}, P_e) = (50, 2, .5)$ particles per day; and $k_m = 2\%$ per day. N reaches its asymptote 20 in a day or two; $M_s \rightarrow 50$ more slowly, taking about 4 days. At 7 days the supply is shut off (r_c is abruptly set to 0). N responds quickly; the accumulation M_s persists; a slow decay can be seen, the effect of k_m.

Problem 5.13 In Equation 3.1, there is a term proportional to Δt. What is its effect on the steady state? Is it an algorithmic effect? If so, devise a time-stepping scheme that avoids this anomaly.

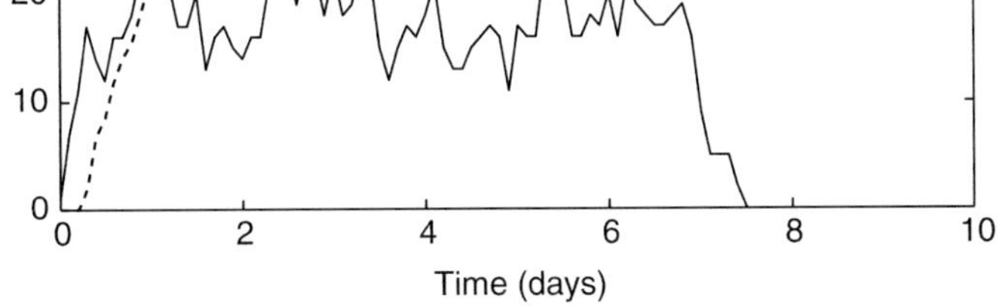

Figure 5.11. Bioaccumulation example. The Poisson source is started instantly at $t = 0$. The system builds up to a stochastic steady state over about 5–7 days. At 7 days the source is removed; the boundary particles exit, leaving behind the accumulated M_b, decaying with a characteristic time 50 days.

Problem 5.14 Bioaccumulation example: make the input a poisson rate.

Problem 5.15 Bioaccumulation example: allow the particles to grow at the rate g per day following boundary contact. (1) What is the effect of g on the steady state? Simulate this system for $g = 10\%$ per day; do your results confirm your analysis?

Problem 5.16 Bioaccumulation example: Reexpress the problem in Eulerian form, by introducing M_b as the total toxin present on the boundary and doing away with the individual particles.

5.5.3 Wetland Harvesting

Suppose a wetland is present in a boundary segment. Its water volume is V. Its grass grows at the bulk logistic rate

$$\frac{dH}{dt} = gH\left[1 - \frac{H}{H*}\right] \tag{5.25}$$

H is the average height of the grass; g is the growth rate; H^* is the carrying capacity, that is, the maximum height the grass can achieve. As $H \rightarrow H^*$, net growth slows and ultimately stops. The amount of nutrient taken up by the grass, per unit of H grown, is a constant, α.

Growth is nutrient limited: $g = g_0\left[1 - e^{-C/C^*}\right]$, where C is the aqueous nutrient concentration in the wetland and C^* is the exponential saturation concentration, a constant. g is linear in C at low levels; at high concentration it saturates at g_o as the nutrient ceases to limit.[3]

The adjacent ocean is a nutrient source, in the form of particles incident on the wetland boundary, at Poisson rate r_c. Each incident particle has volume V_k and nutrient concentration C_o, the fixed oceanic concentration. Wetland particles have probability of emission P_e; when emitted, they will have the same particle volume V_k but the wetland concentration C.

Let M_n be the total dissolved nutrient in the wetland, concentration $C = M_n/V$. The wetland nutrient and water balances in a time step are

$$\Delta M_n = -\alpha\Delta H + \sum_{\text{in}} V_k C_0 - \sum_{\text{out}} V_k C_w \tag{5.26}$$

$$\Delta V = \sum_{\text{in}} V_k - \sum_{\text{out}} V_k \tag{5.27}$$

[3] The Taylor series approximation for low C is $\frac{g}{g_o} = \frac{C}{C^*}$.

with the summations over the water particles in transition during Δt. In addition,

$$\Delta H = g_0 H \left[1 - e^{-C_w/C^*}\right]\left[1 - \frac{H}{H^*}\right]\Delta t \tag{5.28}$$

$$C = \frac{M_n}{V} \tag{5.29}$$

The growing grass incorporates nutrient at the rate $\alpha \Delta H$; when growth reaches H^*, the nutrient concentrations will equilibrate ($C \to C_o$) and there will be no more net nutrient input or net grass growth.

Let V_o be the temporal mean of V. There is a mass balance requirement if V is to be statistically steady:

$$P_e = r_c \frac{V_k}{V_o} \tag{5.30}$$

Parameters: $H^* = 100$; $(g_o, r_c) = (0.15, 30)$ per day; $(C^*, C_o) = (5, 20) \times 10^{-6}$ mg/l; $\alpha = .02$; $(V, V_k) = (10^5, 10)m^3$.

Simulation is shown in Figure 5.12. Initially, $C = C_o$ and H is low. H grows over the first month, then levels out near H^*. In the process, C is drawn down as it is incorporated into the growing grass; after 40 days it begins to recover slowly, as the ocean exchange with the wetland is the only continuing process. The nutrient recovers eventually to C_o.

If one wanted to keep C low, it would be necessary to have growth continuing. Figure 5.13 shows the case where the grass is harvested and removed from the system when it reaches $C_*/2$. There are three cycles shown. C declined continually; during the third cycle, the nutrient limitation slows the growth, so the third harvest is delayed.

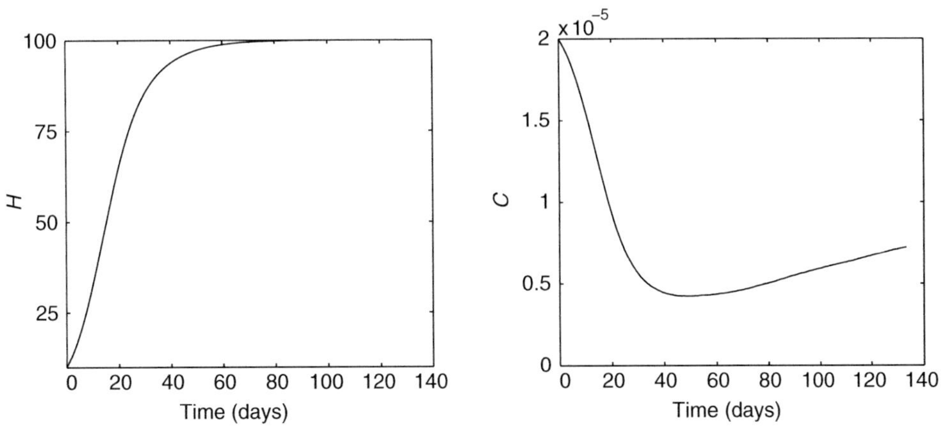

Figure 5.12. Wetland example, base case.

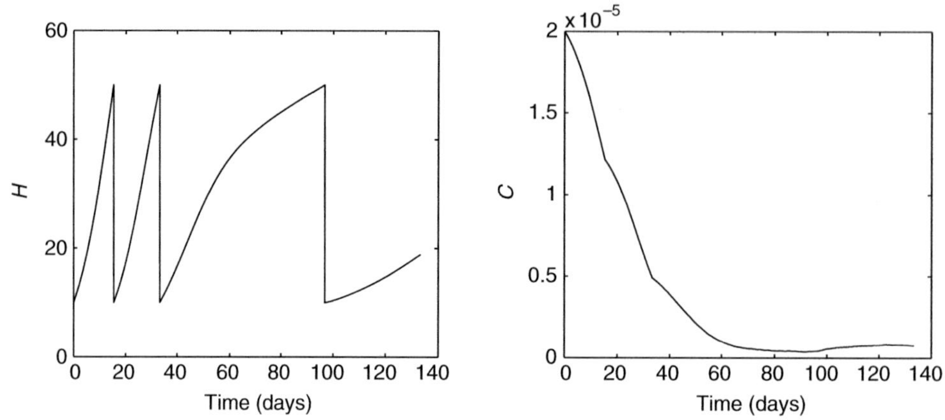

Figure 5.13. Wetland example, with harvesting when H reaches $H^*/2$.

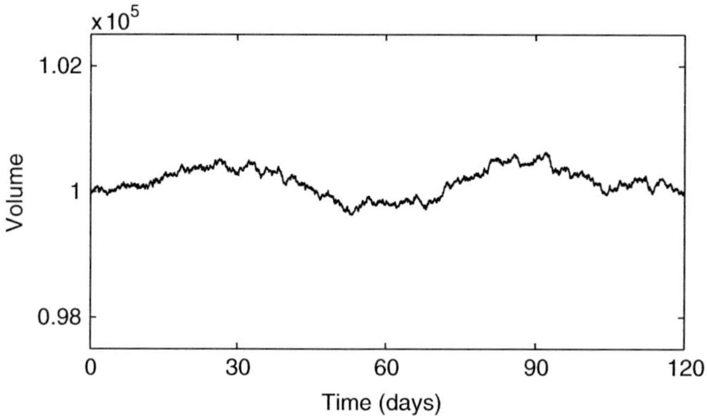

Figure 5.14. Wetland example. Volume of water in the wetland. The variability is entirely due to the random exchange with the ocean.

Continuing this cycle would keep C low and result in extracting nutrient from the ocean, incorporated in the harvest.

Figure 5.14 shows the time history of V. It was started at V_o. There is intermittent inflow and outflow, balancing each other on average.

Problem 5.17 Wetland example: Let T be exchange time for the grassland, that is, the time it takes for input to equal V_o. What is the exchange rate, in terms of percent of V_o per day?

Problem 5.18 Wetland example: Suppose the particles of water being exchanged were of different size: V_k^i and V_k^e. Change the formulation of this problem. What changes are required? What relation will guarantee a statistical mass balance if the average wetland volume is V_o?

Problem 5.19 Wetland example: Change the formulation as follows: the incoming particles of water are simultaneously exchanged with the identical amount or water exiting.

Problem 5.20 Wetland example: Continuation of Problem 5.19: Change the formulation so the incoming particles vary in quantity and frequency.

Problem 5.21 Wetland example: This formulation employs two rate limiters, in logistic and exponential saturation form. Consult Appendix D for some alternatives; discuss at least two and their relative strengths and flexibilities.

5.6 Beyond the Boundary – Exogenous Sources

Outside the boundary layer there are potential sources of particles, truly exogenous to any simulation. Examples include particles entering from rivers, across oceanic boundaries on the shelf, and particles buried in the benthos and below the benthic boundary layer where active exchange between pelagic and boundary particles is routinely occurring. These external sources need a description that expresses physical hypotheses, and is sensitive to discretization parameters Δx, Δt. We will concentrate on the *source rate* $\nu(\mathbf{x},t)$ of these new or external particles – number of particles per unit of time, per unit of boundary surface area (surface length in 2-D), being *injected* into a particular boundary element.

Consider a finite supply of source particles n per unit of boundary length, each having probability p of being injected during a given Δt. The number of injected particles N_{in} per unit of boundary area is governed by the binomial distribution:

$$N_{in} \sim \mathcal{B}(n,p) \tag{5.31}$$

This is a discrete distribution with positive integer outcomes, mean np and variance $np(1-p)$, for number of particles injected over the step Δt. For small p and/or n, zeros and ones will be common; but larger N_{in} do occur. In the limit of large n and small p, this distribution is close to the Poisson distribution

$$N_{in} \sim \mathcal{P}ois(\lambda) \tag{5.32}$$

with the Poisson rate $\lambda = np$. Both mean and variance of the Poisson deviate are the same:

$$\overline{N}_{in} = \lambda \tag{5.33}$$

$$\sigma^2_{N_{in}} = \lambda \tag{5.34}$$

This is quite a practical distribution – one parameter, no accounting for an assumed large number of potential injectees, equally likely. In addition to the convenience in formulation, there is a clear operational advantage to not having to account separately for a large number of potential (but unlikely) source particles simply to get the much smaller sum of injectees.

The Poisson distribution has useful properties, notably the additive property that implies that λ scales with boundary element area A. The consequence is that one can define a Poisson *intensity* ν per unit area on the continuous boundary, such that can λ is its aggregate over a boundary element

$$\lambda = \int \nu \, da = \overline{\nu} A \tag{5.35}$$

The integral and average here is over a boundary element. An analogous aggregation process leads λ to be proportional to Δt.

So the Poisson process can be used to inject source particles at known intensity into a boundary layer as desired. What is needed is the Poisson intensity function along the boundary; it is readily aggregated to finite boundary elements and time step.[4]

The idealized steady state defined in Figure 5.7 is readily modified to account for N_{in} being injected "from nowhere" into the boundary layer. The result for Gaussian pelagic particles is

$$\overline{N_b} = \frac{0.4\rho\sigma_p + \lambda}{P_e} \tag{5.36}$$

In accord with intuition, the addition of Poisson sources adds to the expected density of boundary particles.

Problem 5.22 Derive Equation 5.36.

Problem 5.23 Plot the Gaussian result (Equation 5.36) for various values of λ. Specifically, plot $P_e\overline{N_b}$ versus $\rho\sigma_p$, for the parameter $\frac{\lambda}{\rho\sigma_p} = [0,\ 0.25,\ 0.5,\ 1.0,\ 1.5,\ 2.0,\ 5.0]$. Are the results in accord with intuition? Why or why not?

Problem 5.24 Equation 5.36 is for Gaussian pelagic particles. What is the comparable result when the uniform deviate governs the pelagic random walk?

5.7 Summary Comments

A simulation includes pelagic and boundary particles. The former can collide with boundaries following oceanic transport; at that point they become boundary particles. Boundary particles accumulate on the boundary, potentially moving along it, and either reenter the pelagic state according to boundary processes, or terminate there.

[4] It is an easy generalization to conceive of Poisson sources on the interior of a domain, that is, source particles injected directly into the pelagic system.

The probabilities of these transitions depend on boundary processes and the ambient ocean conditions. On emission, a boundary particle changes state and pelagic processes resume. On termination, a particle should be removed to an archive, without losing its accumulated history.

High resolution near an impermeable boundary will create a repulsive tendency for pelagic particles via the ∇D pseudo-velocity term necessary in particle motion. For coarser resolution, this gradient can be internalized in boundary element parameterization.

Boundary processes are dependent on the complete local environment – tide, light, heat, salt, nutrients, biota. Some of these are oceanic variables; some are associated with the boundary per se; some reflect the populations resident there. Biochemically active particles will respond to all these things over time. The boundary environment itself can be affected by contact with particles, illustrated in the bioaccumulation and harvesting examples. There are important opportunities for creative linking of particle methods to population and pollution models of this kind. (The latter are developed in several sources, e.g., Lynch [280].)

6
Turbulence Closure

The parameters for random displacement models – $[D, \rho_1, \rho_2, T, T_a, \Omega]$ – need to be available as Eulerian fields if a particle is to undergo Lagrangian motion. It is the job of turbulence closure to separate the motions into a mean part $\overline{V}$ and a perturbation $\widehat{V}$; and to describe their interaction. That is the subject of this chapter.

The simplest representation of random displacement is in terms of a diffusion coefficient, the product of characteristic turbulent velocity and length scale (or alternately, the product of a turbulent kinetic energy and a characteristic time). The simplest characterization of memory is the decay time associated with turbulent energy.

We can expect to have at least three parameters available: a turbulent energy, a length scale, and a decay time. The diffusion coefficient is the product of the first two, and *all contemporary shelf models use these diffusion coefficients to transport momentum, salt, and heat*. A fourth parameter, representing turbulent rotation, would be useful in several contexts.

6.1 Reynolds Stresses and the Gradient Flux Relation

The standard Eulerian equations for fluid motion and transport are averaged in various ways, with the split of the natural fields into "mean" and "variable" parts:

$$V = \overline{V} + \widehat{V} \tag{6.1}$$

Both parts are time- and space-dependent. $\overline{V}$ is the "resolved" part that gets solved in an Eulerian sense; it is smooth and carries with it a truncation error – the effect of finite gridding and resolution of geometry, forcing functions – assumed to be shrinkable to arbitrarily nothing with resolution. This is the classic domain of the numerical analysis of PDEs. $\widehat{V}$ is the residual, "turbulent perturbation," or "unresolved" part. It is subject only to statistical description.

Neither part may be neglected. $\overline{V}$ feeds $\widehat{V}$ through the turbulent cascade of large-scale motions toward smaller scale ones, ultimately transferring the unresolvable energy into $\widehat{V}$. Energy stored in the subgrid decays, transports with the mean flow, and accounts for the Reynolds stresses that are needed for computation of $\overline{V}$.

Averaging in almost any way has an intuitively pleasing feature – by definition, the mean of a perturbation is zero, and so the linear terms in any PDE description average out. But the nonlinear terms do not. For example, the quadratic terms decompose as

$$\overline{V^2} = \overline{V}^2 + \overline{\widehat{V}\widehat{V}} \tag{6.2}$$

$$\overline{V\frac{\partial C}{\partial x}} = \overline{V}\frac{\partial \overline{C}}{\partial x} + \widehat{V}\frac{\partial \widehat{C}}{\partial x} \tag{6.3}$$

The cross terms, for example, $\widehat{V}\overline{V}$, vanish on averaging, by definition. The last terms are the "Reynolds stress" terms, generally the variances and covariances of the perturbations. Any average of nonlinear equations acquires a collection of such terms that must be dealt with. When the lion's share of the motion is accounted for in the mean, then the product terms are small. The reverse is true when the perturbations are large compared with the mean.

For example: averaging the momentum equation over tidal variations one obtains the subtidal momentum balance, with intertidal variability left as a driving force – the Stokes drift, for example. When averaging over large $\widehat{V}$, these terms can be large and significant [286, 326, 492]. A typical current practice is to resolve the tidal motion on an Eulerian grid; simulate it; and store the resolved results, typically in terms of tidal constituents. When needed for transport studies, these fields are then reconstructed. Rather than being stochastic, the transports represented are deterministic. They require care in handling the hydraulics and related data, but do not call for a stochastic model.

Another example is spatial averaging to reduce the dimensions of the equation set. Vertical averaging, lateral averaging, cross-section averaging are all forms of this that have been useful; each generates its own set of Reynolds stress terms. These are familiar matters for all analysts. The information averaged out is **not** computed, so it needs to be properly represented less its impact be neglected.

If we have the simple equation set:

$$\frac{\partial V}{\partial t} + V \cdot \nabla V = 0 \tag{6.4}$$

$$\frac{\partial C}{\partial t} + V \cdot \nabla C = 0 \tag{6.5}$$

averaging the entire set gives

$$\frac{\partial \overline{V}}{\partial t} + \overline{V} \cdot \nabla \overline{V} = -\overline{\widehat{V} \cdot \nabla \widehat{V}} \tag{6.6}$$

$$\frac{\partial \overline{C}}{\partial t} + \overline{V} \cdot \nabla \overline{C} = -\overline{\widehat{V} \cdot \nabla \widehat{C}} \tag{6.7}$$

On the left side we recover the linear part of the governing PDE; the Reynolds stress terms have been moved to the right side. As products, they scale quadratically with the size of the perturbation they are averaging; for small motions, these are not important; for large motions, they can dominate.

6.1.1 The Gradient Flux Relation

More generally, the perturbation fields cannot be described except in terms of their statistics.

A standard approach is to parameterize these in terms of Eddy viscosity in the momentum equation, and their counterpart Eddy diffusivity in the transport equation:

$$\frac{\partial \overline{V}}{\partial t} + \overline{V} \cdot \nabla \overline{V} = \nabla \cdot A \nabla \overline{V} \tag{6.8}$$

$$\frac{\partial \overline{C}}{\partial t} + \overline{V} \cdot \nabla \overline{C} = \nabla \cdot K \nabla \overline{C} \tag{6.9}$$

This is the gradient flux assumption. It achieves the practical need to eliminate the perturbation terms. Theoretically, one reasons that these eddy transport processes depend on significant turbulent fluctuations in the processes under study; and that Reynolds transport is zero when there is no gradient in the mean. The notion of unresolvable turbulence attending high-magnitude motions would lead to an increasing relation between the eddy coefficients (A, K) and $|V|$. The unanswered question is, of course, "what are A and K?" Subsidiary questions include "What is the form of their dependence on the mean flow?"; "Can they be observed?"; and "Can they be modeled successfully?"

The ratio $\frac{A}{K}$ is the Prandtl number Pr, the ratio of momentum eddy diffusivity (eddy viscosity) to thermal diffusivity. In an ideal turbulent fluid, $Pr = 1$; the presence of stratification alters Pr.

This gradient flux relation is generally the gateway for communication between $\overline{V}$ and $\widehat{V}$. The description of mean motion requires parameterization of the Reynolds stress terms and they have classically been represented in this form, sometimes dominating considerations of numerical method. And, resolved shear stress – gradients in $\overline{V}$ – are commonly used to depict the generation of subgrid turbulence. An object of "closure" is to depict the mutual interplay among these two field partitions; a primary focus has been on this gateway, the gradient flux assumption.

Every coastal ocean model has embedded in it, a form of this assumption. In particular, Eddy viscosity/diffusivity is needed to compute Eulerian fields for momentum, salt and heat. If these coefficients are not reported, then the report is incomplete. One can expect to find, with any computed product, either the Eulerian fields A and K themselves, or specific dependencies on other variables. The theoretical bases can be weak, unsatisfying, incomplete; while the basic description of the resolved fields is quite far along, the subgrid description of turbulence remains a frontier problem. Ultimately, both are critical for the practical solution of realistic problems.

Values of the eddy diffusion coefficients A, K vary widely. To begin with, they depend critically on the system under study, its hydrodynamic regime, and the form of averaging implied in the available data. In addition, they depend on the form of the equations being used, the scope of the phenomena they represent, and of course the finite mesh employed. There are some tables of diffusivity values presented in Fischer et al. [153] (chapter 7) for longitudinal dispersion in estuaries; in Schnoor [405] (chapter 2) for streams and lakes; in Martin and McCutcheon [303] (chapter 13) and in Thomann and Mueller [446] (chapter 3) for estuaries. Almost every study of a specific water body will undertake to estimate these coefficients based on a blend of empiricism and prior studies.

Throughout the history of environmental transport modeling, there has been a persistent movement toward refining the basis for these estimates. A description in terms of the statistics of unresolved dynamics, that relies on fundamentals, is universally sought.

Generally the vertical and the horizontal present separate closure considerations, given the vastly different scales of motion. We will cover them separately.

6.2 Vertical Closure

6.2.1 Early Models

In this class of closure, we find a reliance on early turbulence-generation and mixing length themes, blended with empirical evidence for complex water bodies. There are good reviews by Officer [340]; Fischer [152]; Bowden [47]; and Blumberg [38, 39].

In the absence of stratification, these closures rely on classical ideas of shear production of turbulence and a mixing length l:

$$K_z = l^2 \left| \frac{\partial \overline{V}}{\partial z} \right| \tag{6.10}$$

The eddy diffusivity K_z and eddy viscosity A_z would be the same under these conditions. The dependence on turbulent energy is represented as a dependence on its generation in the mean flow. Even when the vertical dimension is not modeled, as is common in tidal flows, the proxy for vertical shear is the $\overline{V}$ itself.

The remaining parameter l needs to be closed, and mixing length ideas reflect a blend of geometry, proximity to boundaries, various types of averaging, and so forth in addition to basic free-stream turbulence. The resort to bulk empiricism is necessary here. Hence many of the reviews cited above will examine the empirical bases established in site-specific studies. The portability of these calibrations is dubious; but of course they reflect some basic underlying ideas, played against specific systems and data.

Density stratification complicates the situation. The Brunt–Väisälä frequency N is a metric of stratification and vertical turbulence generation due to buoyancy:

$$N^2 = -\frac{g}{\rho_0}\frac{\partial \overline{\rho}}{\partial z} \tag{6.11}$$

The gradient Richardson number Ri is the ratio of mixing due to buoyancy and shear:

$$R_i = \frac{N^2}{|d\overline{V}/dz|^2} \tag{6.12}$$

When stable stratification is present, $R_i > 0$ and turbulent mixing is inhibited. The form adopted by many investigators, since at least Munk and Anderson [321], is

$$K_z = K_{zo} \cdot \phi_K(R_i) \tag{6.13}$$

$$A_z = A_{zo} \cdot \phi_A(R_i) \tag{6.14}$$

with the stability functions ϕ decreasing with R_i. Separate dependencies are expected of A and K. Blumberg [39] summarized several in the form

$$\phi(R_i) = \frac{1}{[1+\alpha R_i]^p} \tag{6.15}$$

Many important contributions have used this form; included are Munk and Anderson [321], Bowden and Hamilton [48], Kent and Pritchard [227], Officer [340], and Blumberg [38]. Included are the dependence on shear production of turbulence; the suppression of mixing by stratification; the increase of Prandtl number ($Pr = \frac{A}{K}$) with stratification. Left open is the homogeneous value $K_{zo} = A_{zo}$, which is assumed to depend on mixing length, itself already a surrogate for many things including the overall system configuration.

$$K_{zo} = A_{zo} = f(l^2) \tag{6.16}$$

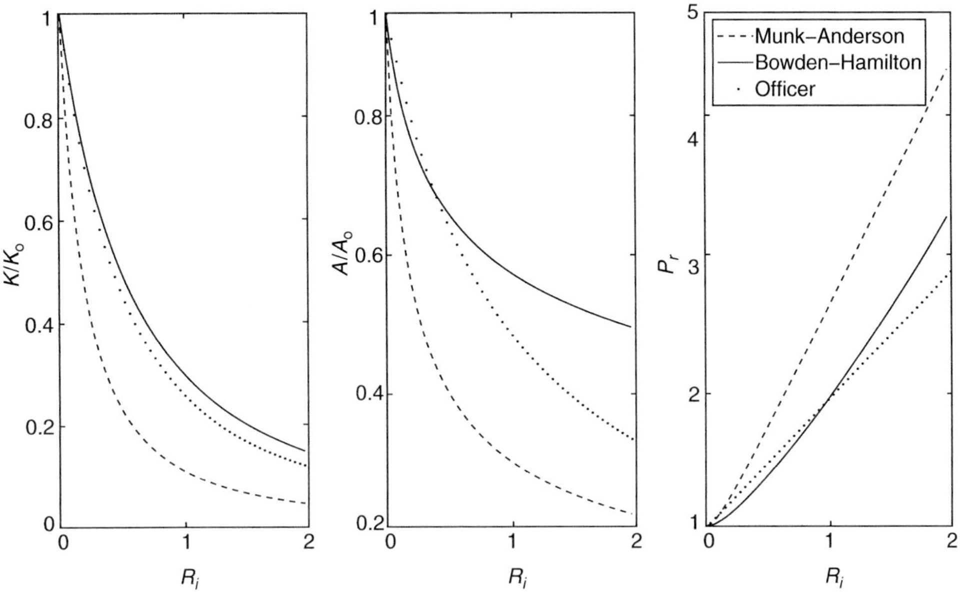

Figure 6.1. Vertical diffusivity K, Viscosity A, and Prandtl number Pr versus Richardson number. —— Munk and Anderson [321]; ··· Officer [340]; – Bowden and Hamilton [48]. A and K are scaled by their unstratified values.

When Ri is large there is no vertical exchange, that is, $K = 0$. When $Ri = 0$, that is, no stratification, mixing under the same shear regime is at a maximum. Absent shear, no turbulence is produced.

Figure 6.1 shows some typical vertical dependencies. These have been confirmed in many specific settings. The diversity of applications and available data would easily explain the variety of parameter values.

This form of closure has been variously described as *semi-empirical* [39], *mixing length* [387], and *mean velocity field* [312]. Implied is the generation of turbulent energy, but there is no accounting for its accumulation; nor is there an agreed-upon approach to the determination of l. The reliance on data is essential, and necessarily site-specific.

6.2.2 Turbulent Kinetic Energy

Second moment models develop the Reynolds stresses in terms of the mean flow $\overline{V}$; and a two-way dynamical interaction between them and $\overline{V}$. A primary variable is q^2, twice the turbulent kinetic energy k at the subgrid scale. It and other necessary subgrid quantities are described and interact dynamically with the mean flow. q^2 has its own dynamic – it is produced, accumulates, decays, and is transported. Accounting for it (and other necessary quantities) is a fundamental goal of turbulence modeling. Excellent reviews are available by Mellor and Herring [312] and Rodi [386, 387].

Mellor and Yamada [313] present a full development of the Reynolds stresses. This basic description invokes a handful of universal coefficients, obtainable experimentally, not site-specific. The analysis admits to ordering of terms based on their departure from anisotropy. This is used as the starting point for a balanced hierarchy of models in terms of precision and computational complexity.

Level 4 is the full model. It requires the solution of 10 differential equations, in addition to those governing the bulk or mean motion; these must be solved simultaneously. These are in "transport equation" form, that is, quantities advecting and diffusing in the mean flow.
Level 3 is a simplification that renders 8 of the 10 into algebraic form. The remaining two are differential equations, describing the transport of q^2 and $\overline{\theta^2}$ (the turbulent variance in temperature).
Level 2 is a further simplification in which the transport of turbulent quantities is entirely removed; the resulting system is wholly algebraic.
Level 1 is a further simplification of Level 2, without significant advantages.

All of these models do not stipulate the closure for l, beyond the requirement that it vary linearly with distance from a solid boundary. As an interim step, this analysis uses the simpler Blackadar [35] formulation (see later), scaled by the vertical integral of q.

Results with all model levels indicate the successful extinguishing of turbulence at $R_i = 0.21$. The Level 2 model (all algebra) is recommended as adequate. Level 3 shows some advantages; Level 4 is impractical.

In a subsequent analysis [311], these authors advance a hybrid of Levels 2 and 3 – Level 2.5 – that became a standard in coastal ocean modeling. Citing the weakness of models for the turbulent length scale l, a new formulation involving the transport of q^2l is advanced. This model has two dynamical quantities (q^2 and q^2l) that are transported in the mean flow. This formulation and that for Level 2 is described in more detail below.

Despite the relative weakness in the description of l, the Level 2.5 and 2 models share the simple structure for the eddy viscosity/diffusivities:

$$A_z = lqS_m \tag{6.17}$$

$$K_z = lqS_h \tag{6.18}$$

$$K_{zq} = lqS_q \tag{6.19}$$

(K_{zq} is the diffusivity for the turbulent quantities.) The coefficients (S_m, S_h, S_q) are algebraic stability functions; they bear a conceptual relation to the stability functions $\phi(R_i)$ defined previously.

There is broad general agreement on the q^2 equation [387]:

$$\frac{\mathcal{D}q^2}{\mathcal{D}t} = P_s + P_b - \epsilon \tag{6.20}$$

where $\frac{\mathcal{D}}{\mathcal{D}t}$ indicates the conventional transport in the macroscopic field, combining local change, advection, *and eddy diffusion*:

$$\frac{\mathcal{D}(\,)}{\mathcal{D}t} \equiv \frac{\partial(\,)}{\partial t} + \overline{V} \cdot \nabla(\,) - \nabla \cdot K_q \nabla(\,) \tag{6.21}$$

Here,

- The left-hand terms are the conventional macroscale transport of q^2.
- The macroscopic eddy diffusivities in the transport operator $\mathcal{D}$ depend on the subgrid variables as in Equations 6.17–6.19.
- On the right-side, the balance of production in the mean flow by shear P_s and by buoyancy P_b, and subgrid dissipation ϵ.
- The macroscopic viscosity $A_z \propto ql$.
- The Prandtl number is the ratio of the stability functions: $Pr = S_m/S_h$.
- The subgrid decay $\epsilon \propto \frac{q^3}{l}$.

These are recognizable conservation statements linking the resolved and subgrid quantities. The Level 2 closure has the left-side transport quantities absent; it balances production with decay, algebraically. Levels 3 and 4 have turbulence transporting in the mean flow.

Generally there is good agreement on the form of the q^2 equation. It expresses a simple conservation property, with sources in $\overline{V}$, sinks in $\widehat{V}$, transported in $\overline{V}$. The macroscopic eddy viscosity/diffusivity are formed from q^2 and partly generate q^2 via shear.

Various studies have revealed difficulties with the initial formulation of the stability functions S. Galperin et al. [166] simplified the formulation for the stability functions by applying "quasi-equilibrium" assumptions that are widely used and circumvent some of these difficulties. Among other things, the quantity G_h is used as the metric of stratification:

$$G_h = \frac{l^2}{q^2} \frac{g}{\rho_o} \frac{\partial \rho}{\partial z} \tag{6.22}$$

This is a modified Richardson number with the macroscopic shear $\frac{\partial \overline{V}}{\partial z}$ replaced by the turbulent ratio q/l. (See, e.g., Deleersnijder and Luyten [110]; Kantha and Clayson [222].)

6.2.3 Turbulent Length Scale

In all of the developments cited, the turbulent length scale l plays a major role. Here there is a recognized need for a second differential equation in l or its surrogates, but less convergence on the form and details.

Algebraic. Blackadar [35] gave what is now the classic algebraic closure for l.

$$l(z) = l_o \left[\frac{\kappa z}{\kappa z + l_o} \right] \tag{6.23}$$

where $\kappa = 0.4$ is the von Karman constant. Following a review of this and other modeling relations [312], the Blackadar formulation was adopted in all the original Mellor/Yamada [313] closure levels, scaled by the vertically averaged turbulence:

$$l_o = \alpha \frac{\int qz\, dz}{\int q\, dz} \tag{6.24}$$

The coefficient α is of order 0.1.

It is common to locate the computational bottom in the log layer, ξ_b above true bottom; and apply a slip condition there in terms of drag coefficient c_d. With z the height above computational bottom, the Blackadar relation is

$$l(z) = l_o \left[\frac{\kappa(z+\xi_b)}{\kappa(z+\xi_b) + l_o} \right] \tag{6.25}$$

Prognostic. In the Level 2.5 model [311] Mellor and Yamada use a second equation for the evolution of $q^2 l$:

$$\frac{\mathcal{D}q^2 l}{\mathcal{D}t} = E_1 l [P_s + P_b] - \frac{q^3}{B_1} \left[1 + E_2 \left(\frac{l}{\kappa l_z} \right)^2 \right] \tag{6.26}$$

E_1, E_2, B_1 are empirical constants; κ is the von Karman constant; and l_z is the geometric proximity to a bounding wall.

The alternative $k - \epsilon$ **model** was originally proposed by Jones and Launder [216], later modified by Launder and Sharma [239], and more significantly by Rodi [386, 387]. The second prognostic equation is for the turbulence dissipation rate:

$$\frac{\mathcal{D}\epsilon}{\mathcal{D}t} = C_{1\epsilon} \frac{\epsilon}{k} [P_s + C_{3\epsilon} P_b] - C_{2\epsilon} \frac{\epsilon^2}{k} \tag{6.27}$$

The similarities with Level 2.5 closure are obvious. Both share the same differential equation for $q^2 \equiv 2k$, and the accompanying consensus for the dissipation rate $\epsilon \propto \frac{q^3}{l}$. This latter property makes ϵ and l proxies for one another. There are some differences related to the experimental basis, and the $q^2 - q^2 l$ form invokes a wall proximity function that is explicitly absent from the $k-\epsilon$ form. Consistently, in the $k-\epsilon$ model, the variable l may be eliminated such that the vertical viscosity is

$$A_z = C_\mu \frac{k^2}{\epsilon} \tag{6.28}$$

which is equivalent to Equation 6.17. (We recall that $q^2 \equiv 2k$.)

A third possibility for the length-scale variable and prognostic equation was advanced with the $k-\omega$ model of Wilcox (1988)[483]. This is extended to geophysical flows by Umlauf et al. [456] who suggest that ω is the optimum "second variable".[1] $\omega \sim \frac{q}{l}$ is the inverse turbulent time scale, or turbulent frequency.

Umlauf and Burchard [455] and Kantha and Carniel [220] suggest a unifying scheme in a generic variable ψ, a power-law combination of q and an additional turbulent length or time, an idea suggested earlier by Mellor and Herring [312]. Kantha [221] extends this in the form $\psi = q^m l^n$. In this useful review Kantha demonstrates (1) that five common two-equation closures are reproduced with the right selection of m and n for the second variable; (2) that they are equivalent; and (3) that reported discrepancies among them are due to the improper modeling of the diffusion term in the length scale equation. An example is the requirement of a wall proximity function in the $q^2 - q^2 l$ formulation.

Generalized Length Scale. As mentioned, Umlauf and Burchard [455] introduced a generic two-equation turbulence closure scheme that can be tuned to behave like several of the traditional schemes. While also introducing some suggested closure parameterizations, their main result was the generic length scale, or GLS, vertical mixing scheme that allowed the flexibility to easily choose and compare different schemes.

The first of the two equations representing transport of TKE is the same as Equation 6.20 with $k = 1/2q^2$. The equation for dissipation is given by

$$\epsilon = c_\mu^{3+p/n} k^{3/2+m/n} \psi^{-1/n} \tag{6.29}$$

where ψ is a generic parameter that is used to establish the turbulence length scale. The equation for ψ is:

$$\frac{D\psi}{Dt} = \frac{\partial}{\partial z}\left(K_\psi \frac{\partial \psi}{\partial z}\right) + \frac{\psi}{k}(c_1 P_s + c_3 P_b - c_2 \epsilon F_{\text{wall}}) \tag{6.30}$$

Coefficients c_1 and c_2 are chosen to be consistent with observations of decaying homogeneous, isotropic turbulence. The parameter c_3 has differing values for stable

[1] Mellor and Herring [312] noted that there is an element of arbitrariness in this choice.

Table 6.1. *GLS Parameters from [469] with Updates*

	MY2.5 $\psi = k^1 l^1$	$k-\epsilon$ $\psi = c_\mu^3 k^{3/2} l^{-1}$	$k-\omega$ $\psi = c_\mu^{-1} k^{1/2} l^{-1}$	Generic $\psi = c_\mu^2 k^1 l^{-2/3}$
p	0.0	3.0	−1.0	2.0
m	1.0	1.5	0.5	1.0
n	1.0	−1.0	−1.0	−0.67
$\sigma_k = \frac{K_M}{K_k}$	1.96	1.0	2.0	0.8
$\sigma_\psi = \frac{K_M}{K_k}$	1.96	1.3	2.0	1.07
c_1	0.9	1.44	0.555	1.0
c_2	0.52	1.92	0.833	1.22
c_3^-	2.5	−0.4	−0.6	0.1
c_3^+	1.0	1.0	1.0	1.0
$k_{\min}$	5.0e−6	7.6e−6	7.6e−6	1.0e−8
$\psi_{\min}$	5.0e−6	1.0e−12	1.0e−12	1.0e−8
c_μ	0.5544	0.5477	0.5477	0.5544

Source: J. Warner, pers. comm. 2013.

(c_3^+) and unstable (c_3^-) stratification. Also:

$$\psi = c_\mu^p k^m l^n \tag{6.31}$$

$$l = c_\mu^3 k^{3/2} \epsilon^{-1} \tag{6.32}$$

Careful choice of the parameters in these equations renders a variety of traditional two-equation turbulence closure models. Warner et al. [469] provide a table with a compilation of values from Umlauf and Burchard [455], Burchard and Bolding [63], and Umlauf et al. [456]. Table 6.1 shows an updated version.

6.3 Vertical Closure Examples

6.3.1 Level 2.5 Formulation

Here we summarize one model, the Mellor–Yamada Level 2.5 closure, with subsequent development as cited. This description is taken from the implementation by Lynch et al. [282, 283]. This is offered as an example of a two-equation closure.

The primitive equations are assumed for the mean or macroscale circulation. Momentum, heat, and salt are conserved. These conservation equations include eddy viscosity and eddy diffusivity terms A and K, representing the subgrid turbulence. We have the conventional primitive-equation variables at the macroscale, plus two variables to represent the turbulence, q^2 and $q^2 l$. These are represented as gridded Eulerian variables. We concentrate on the vertical portion of the closures.

6.3.1.1 Governing Equations

For the motion, the standard 3-D hydrodynamic equations with the conventional Boussinesq and hydrostatic assumptions govern. The subgrid-scale dissipation is represented in eddy viscosity (diffusivity) form, parameterized in terms of stratification plus turbulent kinetic energy and mixing length, both of which evolve at the macroscale.

The equations for the turbulent kinetic energy and mixing length are conventional macroscopic transport equations:

$$\frac{dq^2}{dt} - \frac{\partial}{\partial z}\left(K_{zq}\frac{\partial q^2}{\partial z}\right) = 2\left[A_z\left|\frac{\partial \mathbf{v}}{\partial z}\right|^2 + \frac{g}{\rho_0}K_z\frac{\partial \rho}{\partial z}\right] - 2\left[\frac{q^3}{B_1 l}\right] \tag{6.33}$$

$$\frac{dq^2 l}{dt} - \frac{\partial}{\partial z}\left(K_{zq}\frac{\partial q^2 l}{\partial z}\right) = lE_1\left[A_z\left|\frac{\partial \mathbf{v}}{\partial z}\right|^2 + \frac{g}{\rho_0}K_z\frac{\partial \rho}{\partial z}\right] - lW\left[\frac{q^3}{B_1 l}\right] \tag{6.34}$$

wherein E_1 and B_1 are experimental constants (Mellor and Yamada [311]) and W is a wall proximity function [Blumberg et al.][37, 40]; Table 6.2).

The closure for the vertical turbulent mixing coefficients is

$$A_z = qlS_m \qquad K_z = qlS_h \qquad K_{zq} = qlS_q \tag{6.35}$$

where S_q is a constant and the stability functions S_m and S_h are simple algebraic functions of the local stratification G_h [Galperin et al.][166]:

$$G_h \equiv \frac{l^2}{q^2}\frac{g}{\rho_0}\frac{\partial \rho}{\partial z} \tag{6.36}$$

$$S_m = \frac{(g_2 - g_3 G_h)}{(1. - g_4 G_h)(1. - g_5 G_h)} \tag{6.37}$$

$$S_h = \frac{g_6}{(1. - g_4 G_h)} \tag{6.38}$$

$$S_q = 0.2 \tag{6.39}$$

The constants g_i, introduced in [282] for simplicity, are related in Table 6.2 to the standard constants of Mellor and Yamada [311] derived from experiment. Also from Galperin et al. [166] we enforce an upper bound on the mixing length under stable stratification:

$$l \leq \frac{0.53q}{\sqrt{-\frac{g}{\rho_0}\frac{\partial \rho}{\partial z}}} \tag{6.40}$$

This implies an effective lower limit $G_h \geq -0.28$. Under unstable stratification, G_h is capped at the upper limit $G_h \leq 0.0233$. Among other things, this prevents the computation of negative diffusivity according to Equations 6.37 and 6.38.

Table 6.2. *From [282]. Summary of Constants Associated with the Vertical Closure*

I	II	III
$A_1 = 0.92$	$g_0 \equiv 1. - 6A_1/B_1$	$g_0 = 0.66747$
$A_2 = 0.74$	$g_1 \equiv 6A_1 + B_2$	$g_1 = 15.620$
$B_1 = 16.6$	$g_2 \equiv A_1(g_0 - 3C_1)$	$g_2 = 0.39327$
$B_2 = 10.1$	$g_3 \equiv 3A_1A_2[(B_2 - 3A_2)g_0 - 3C_1g_1]$	$g_3 = 3.0858$
$C_1 = 0.08$	$g_4 \equiv 3A_2g_1$	$g_4 = 34.676$
$E_1 = 1.8$	$g_5 \equiv 9A_1A_2$	$g_5 = 6.1272$
$E_2 = 1.33$	$g_6 \equiv A_2g_0$	$g_6 = 0.49393$
$E_3 = 0.25$		
$\kappa = 0.4$		

Note: Column I contains the Mellor-Yamada [311] experimental constants, plus E_3 from Blumberg et al. [37] and von Karman's constant κ. Column II contains the formulas from Galperin et al. [166] used in Equations (9) herein. Their numerical values are recorded in Column III.

The Prandtl number for this closure is

$$P_r = \frac{A_z}{K_z} = \frac{S_m}{S_h} = \frac{(g_2 - g_3G_h)}{g_6(1 - g_5G_h)} \tag{6.41}$$

Problem 6.1 For a Level 2.5 closure: what is the relation between the dimensionless numbers G_h, the Richardson number number R_i, and the Brunt–Vasala frequency N?

Problem 6.2 For a Level 2.5 closure, plot versus G_h: the stability functions S_m and S_h and the Prandtl number P_r.

6.3.1.2 Vertical Boundary Conditions

Conventional horizontal boundary conditions apply to all transported quantities. Vertical boundary conditions may be posed as follows. (It is important here and throughout to keep the turbulent variables compatible with the mean motion $\overline{V}$).

For the horizontal velocity $\overline{V}$: The shear stress $A_z\frac{\partial \overline{\mathbf{V}}}{\partial z}$ is specified at the surface from data – a conventional Type II or Neumann condition. At the bottom, assuming that the no-slip condition cannot be resolved, a quadratic slip condition is appropriate. With finite bottom velocity $\overline{\mathbf{V}}_b$ and the dimensionless drag coefficient, c_d:

$$A_z \left.\frac{\partial \overline{\mathbf{V}}}{\partial z}\right|_{z=-h} = c_d|\overline{\mathbf{V}}_b|\overline{\mathbf{V}}_b \tag{6.42}$$

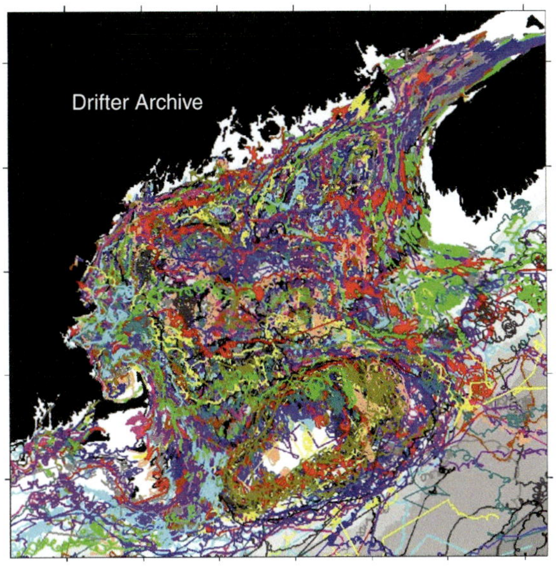

Figure 2.1. Observed drifter tracks off the northeast United States. (Manning et al. [299]). Reprinted with permission.

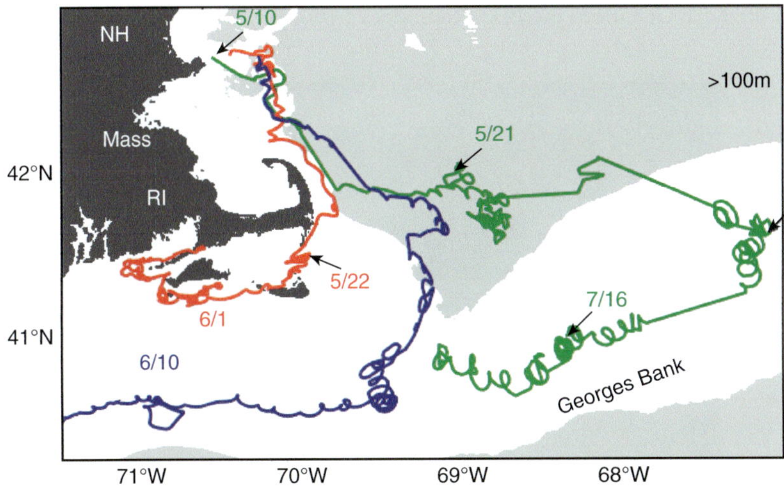

Figure 2.5. Observed branch and hyperbolic points in exit pathways from the Gulf of Maine. Three drifters launched in close proximity to Cape Ann followed very different pathways to Nantucket Sound, Great South Channel, and Georges Bank. Adapted from Anderson et al. [4]. (Reproduced by permission.)

Figure 2.6. HoleySock drifter. Photo of Jack Barth (Oregon State University) preparing for the NEP GLOBEC project. (Reproduced with permission.)

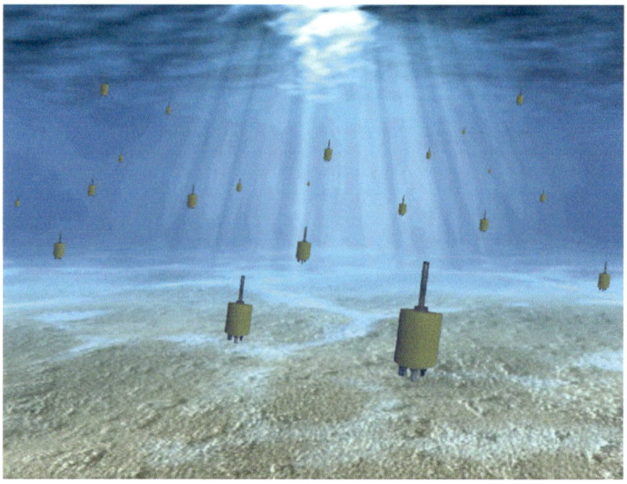

Figure 2.9. Autonomous underwater explorers. Prototypes of these free-floating drogue devices have been built and tested off the coast of Scripps Institute of Oceanography, (Jaffe and Shurgers [208]). They show promise in being an acoustically trackable swarm of Lagrangian-like particles in the ocean. (Image courtesy of J. Jaffee, C. Shurgers, pers. comm.)

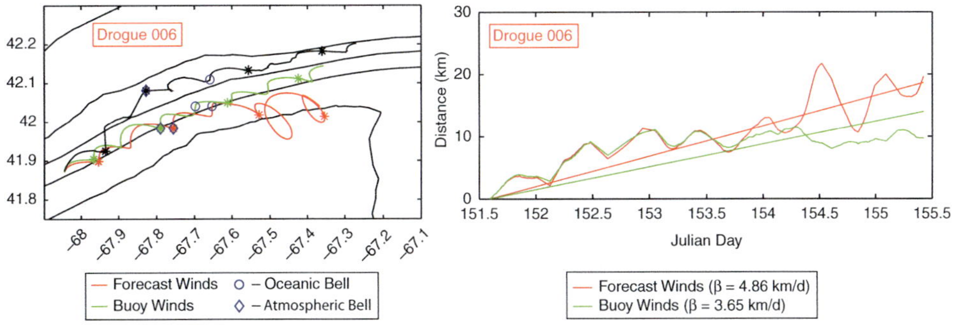

Figure 2.11. Northern Flank, Georges Bank, (From Proehl et al. [366]). (Reproduced with permission.)

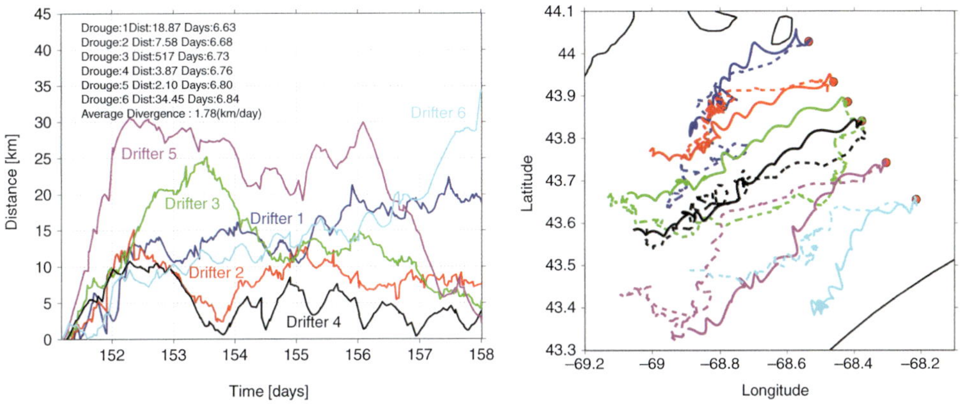

Figure 2.13. Drifter misfit growth (left) and trajectories (right) in the Maine coastal current. The solid lines are modeled; the dash lines, observed. From He et al. [193] (Reproduced permission from the *Journal of Geophysical Research.*)

Figure 2.15. Aboard R/V Oceanus: preparation for drifter launches during the real-time data-assimilative modeling reported by He et al. [193]. Pictured are Ruoying He and Jim Manning (Reproduced with permission.)

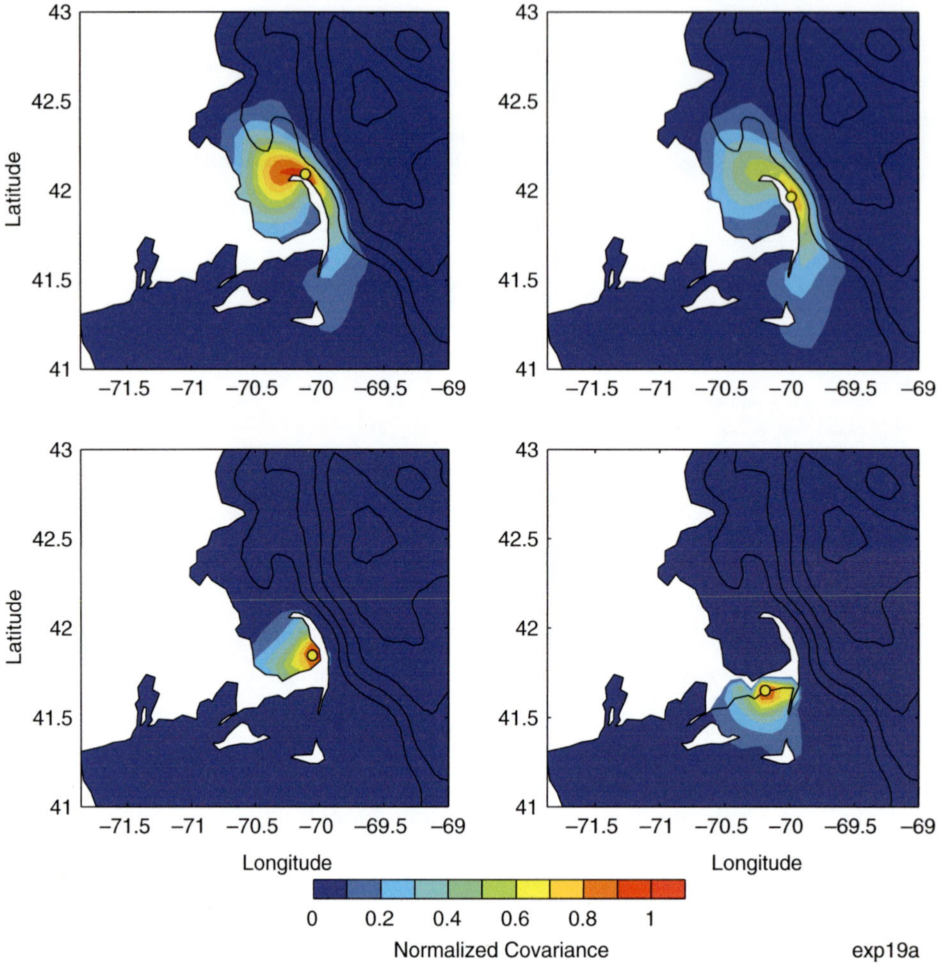

Figure 3.20. Correlogram examples for the PDE-based covariance with landward boundary conditions. Four different correlograms are shown, each centered as indicated. From Lynch and McGillicuddy, [285]. (Reproduced with permission.)

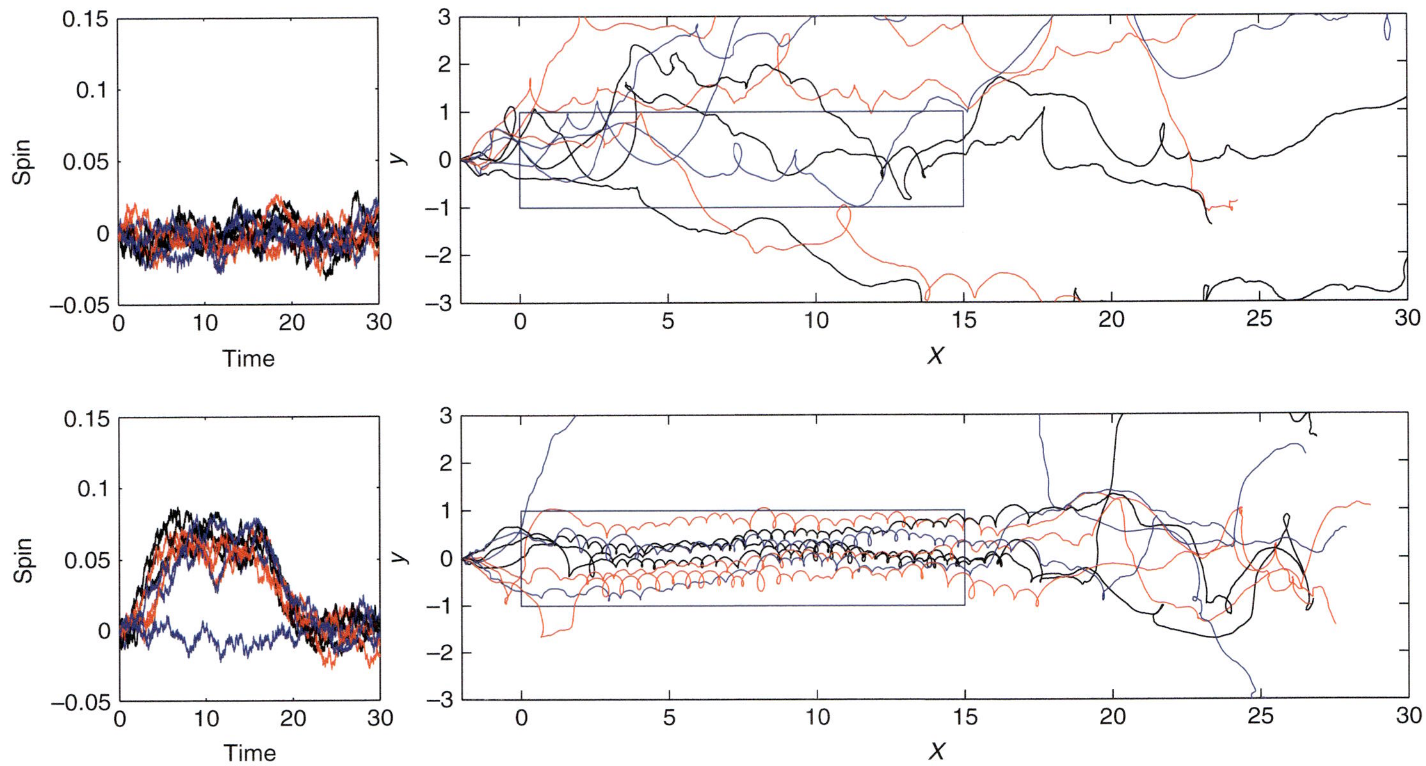

Figure 4.15. As in Figure 4.14, but with full AR1 dynamics for ω_i. σ_ω is everywhere constant. Top: No vorticity generation in the box. Bottom: Constant $\overline{\omega}$ in the box, as before. Particles that walk into the boxed region, are trapped there relative to those that evade it or those in the upper panel.

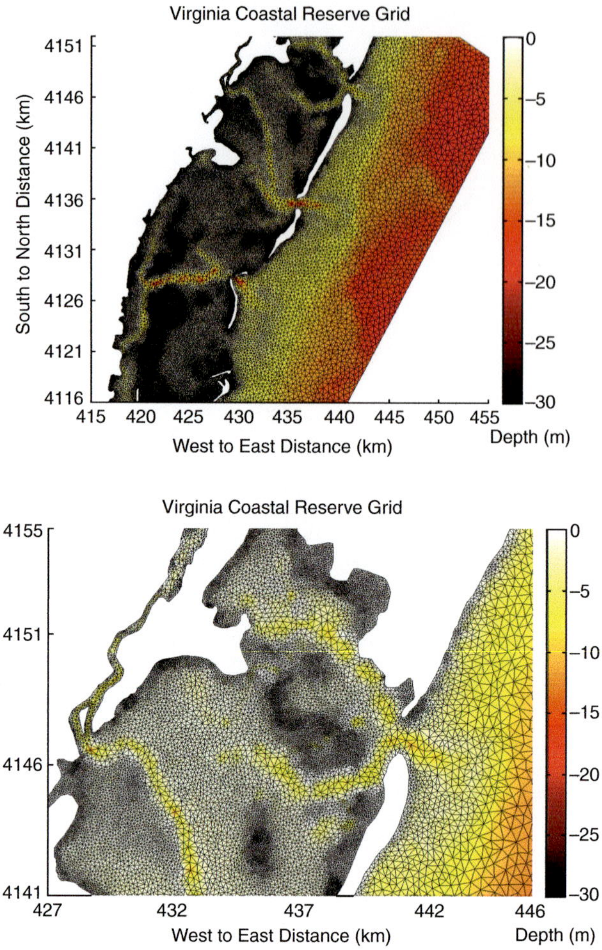

Figure 7.2. Example of a coastal mesh of triangles: Hogg Island Bay Coastal Reserve LTER, VA. As used in Bilgili et al. [32].

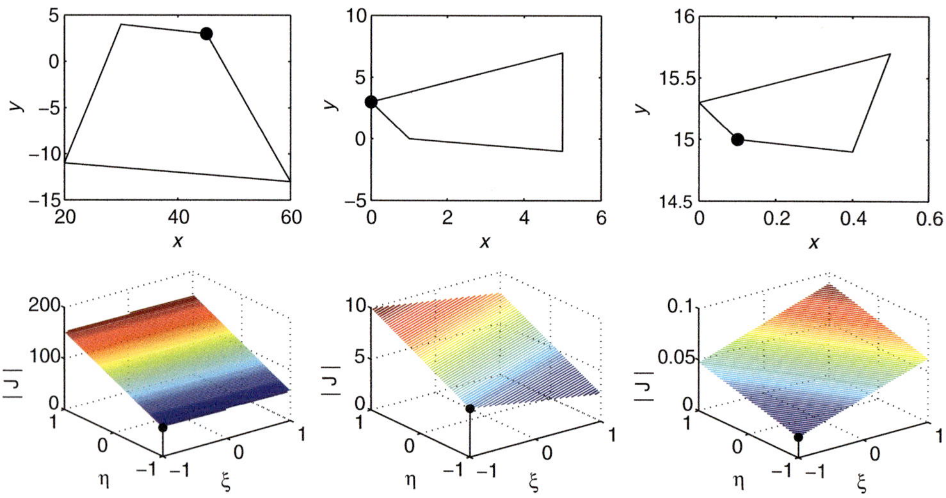

Figure 7.24. Variation of $|J|$ on three elements. Top: Element plotted in global (x, y) coordinates. Bottom: Contours of $|J|$ in the local (ξ, η) coordinates. The dot marks local node 1 and its global counterpart.

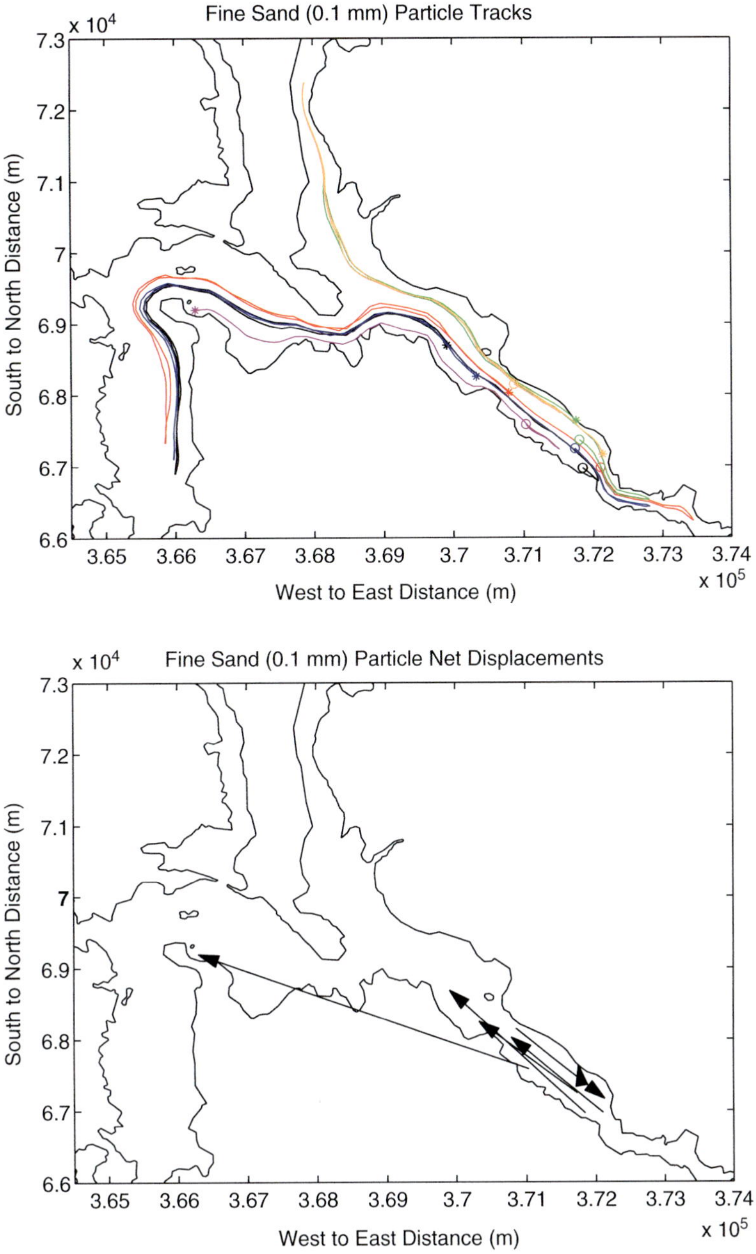

Figure 9.20. Tidal displacement (top) and net displacement (bottom) for fine sand ($D = 0.1$ mm). "o"s mark the starting point and "$*$"s mark the final location. Net displacement arrows are drawn from "o" to "x" for each particle.

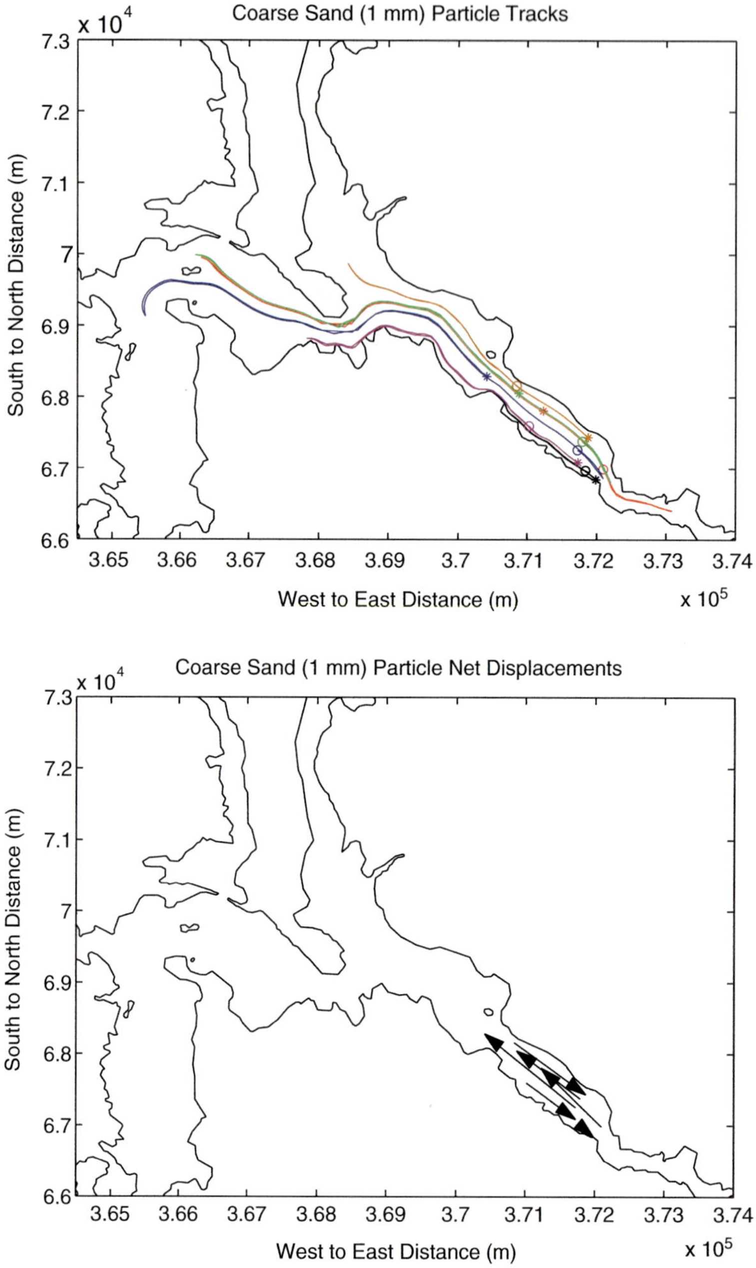

Figure 9.21. Tidal displacement (top) and net displacement (bottom) for coarse sand ($D = 1$ mm). "o"s mark the starting point and "$*$"s mark the final location. Net displacement arrows are drawn from "o" to "x" for each particle.

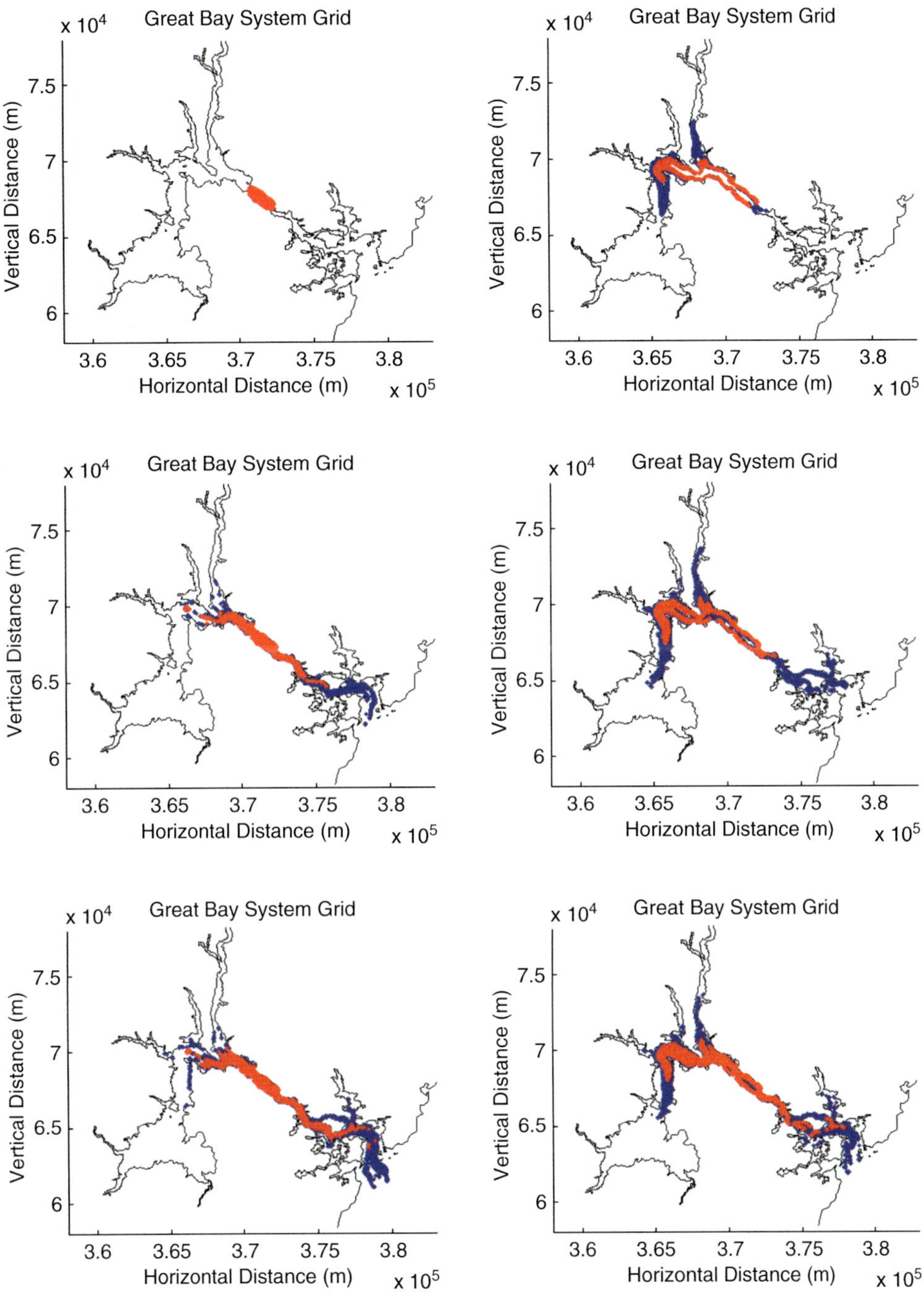

Figure 9.22. Displacement of particles over 2.5 tidal cycles. Fine: blue; coarse: red. Release at low tide (upper left panel). The panels depict subsequent high and low tide points in sequence.

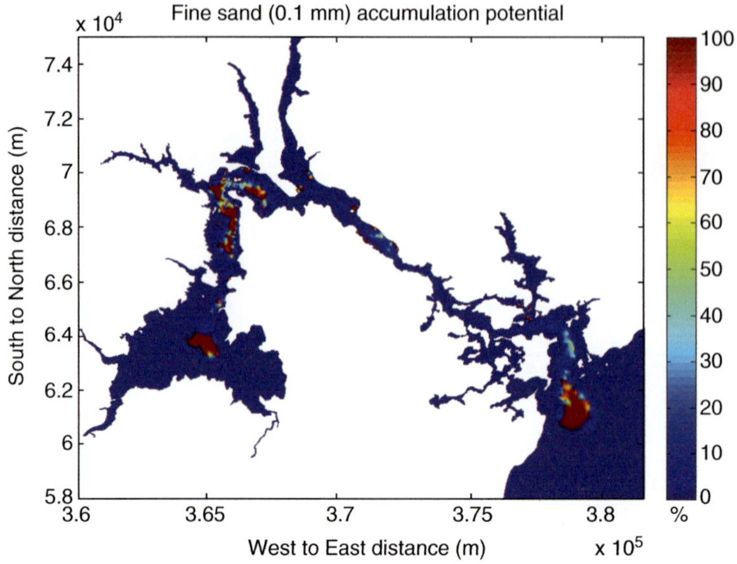

Figure 9.23. Accumulation potential for fine ($D = 0.1$ mm) sediment particles. The color bar gives accumulation potential in percentage terms.

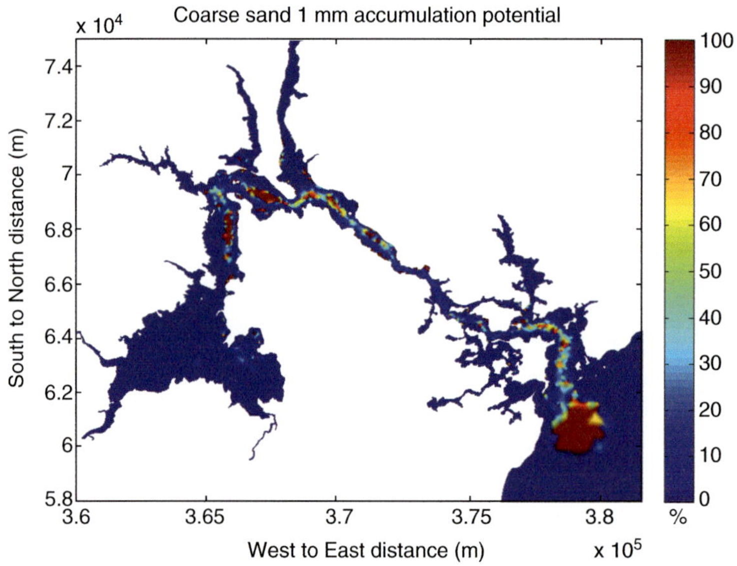

Figure 9.24. Accumulation potential for coarse ($D = 1$ mm) sediment particles. The color bar gives accumulation potential in percentage terms.

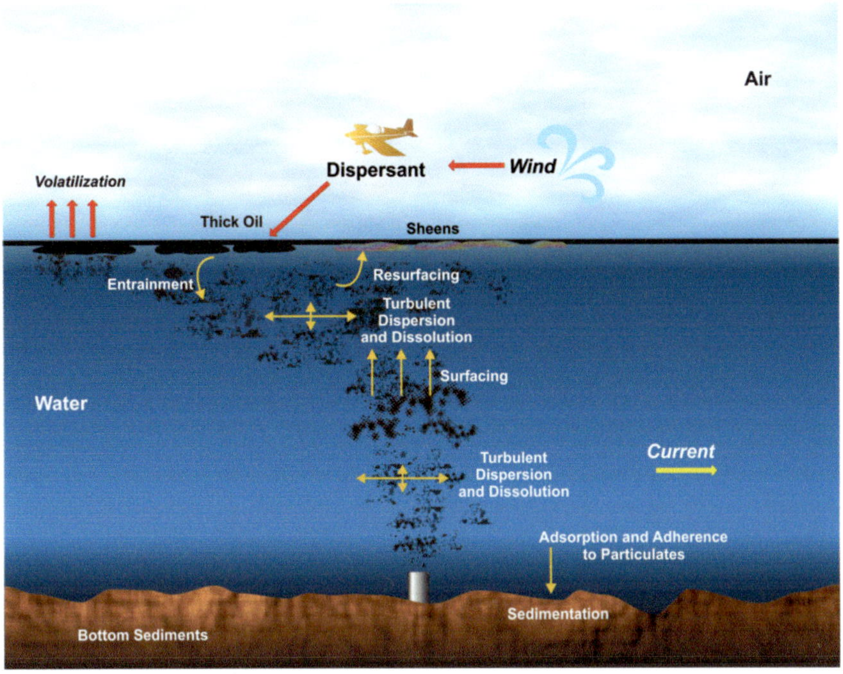

Figure 10.1. Occurrence and processes for marine oil. (Image provided by ASA-RPS, courtesy M. Spaulding.)

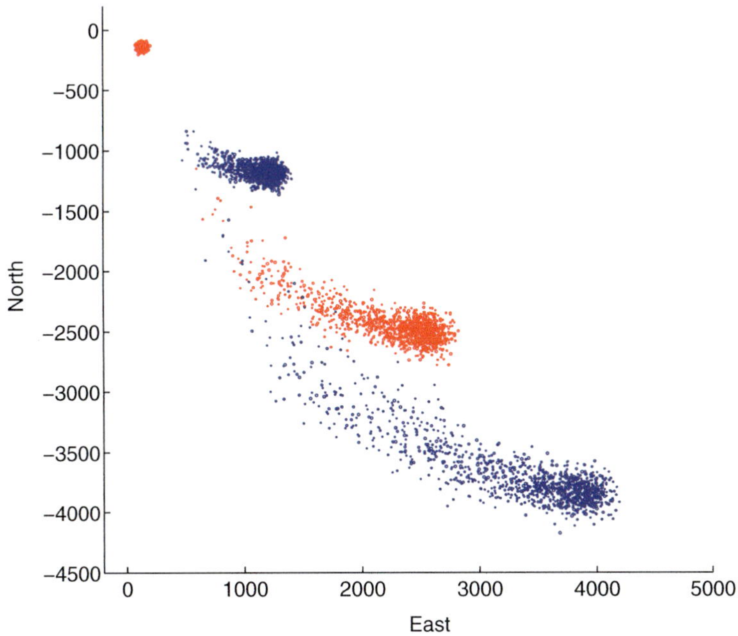

Figure 10.19. As in Figure 10.18: The particle locations at four times, on the same axis.

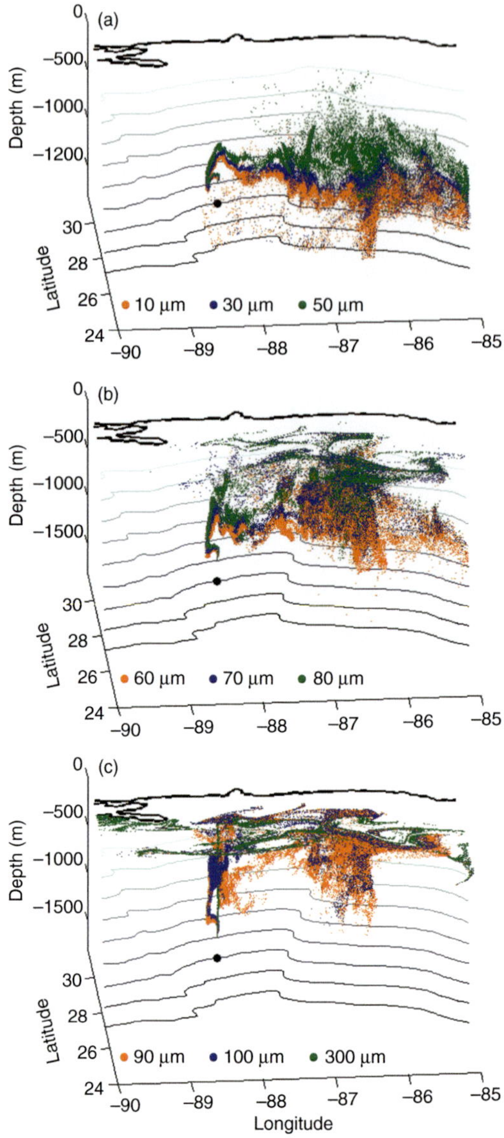

Figure 10.22. 3-D view of simulated particles originating from the Macondo Blowout on June 3, 42 days into the incident. Colors indicate particle diameter. Particles were released at the plume exit height above the well, indicated by the black dot. (From North et al. [334]. Source: E. North, pers. comm. Reproduced with permission.)

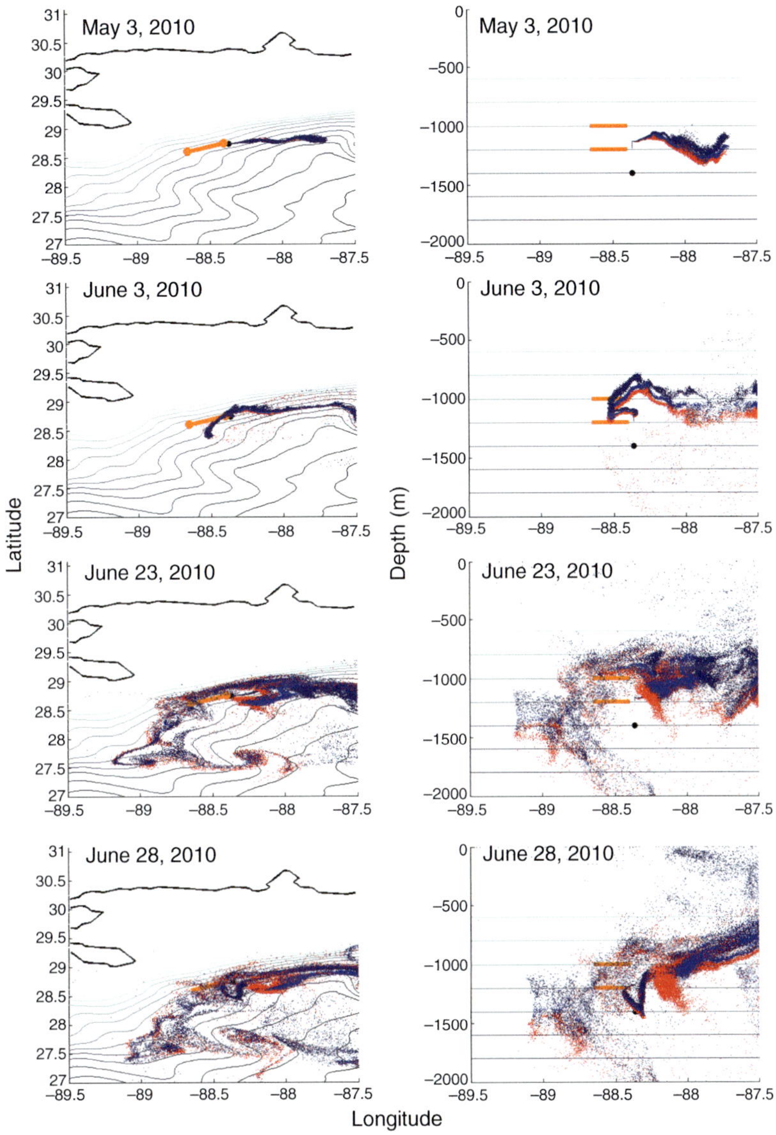

Figure 10.23. Temporal development of the simulated Macondo plume for the smaller particles, $D = [10, 30, 50]$ microns ([red, blue, dark blue]). Particles were released continuously from April 22 to July 15, 2010, at the estimated rates. Evidence of a subsurface plume was reported at the locations marked in orange. (From North et al. [334]. Source: E. North, pers. comm. Reproduced with permission.)

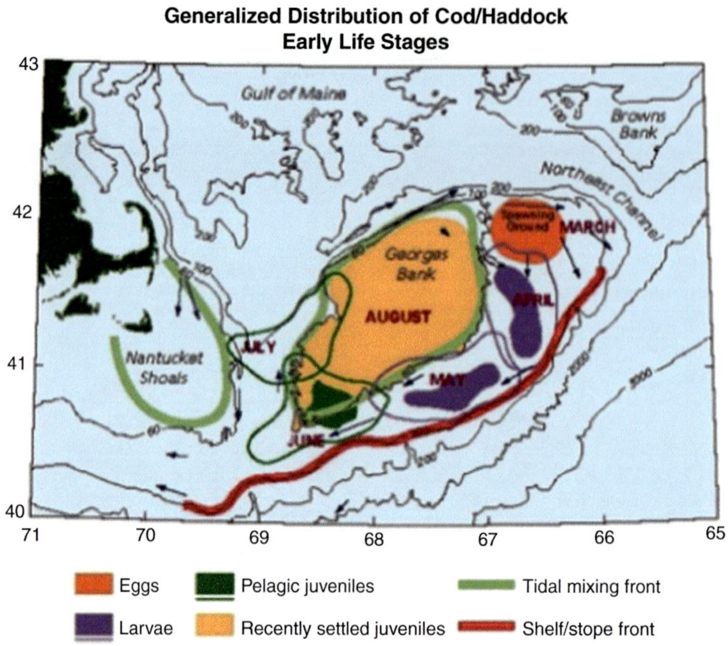

Figure 11.27. Spatial locations relevant to the Lough-Werner IBM for cod on Georges Bank. As it appeared in Lough and Manning [264]. (Original source courtesy of R. G. Lough, pers. comm. 2012.)

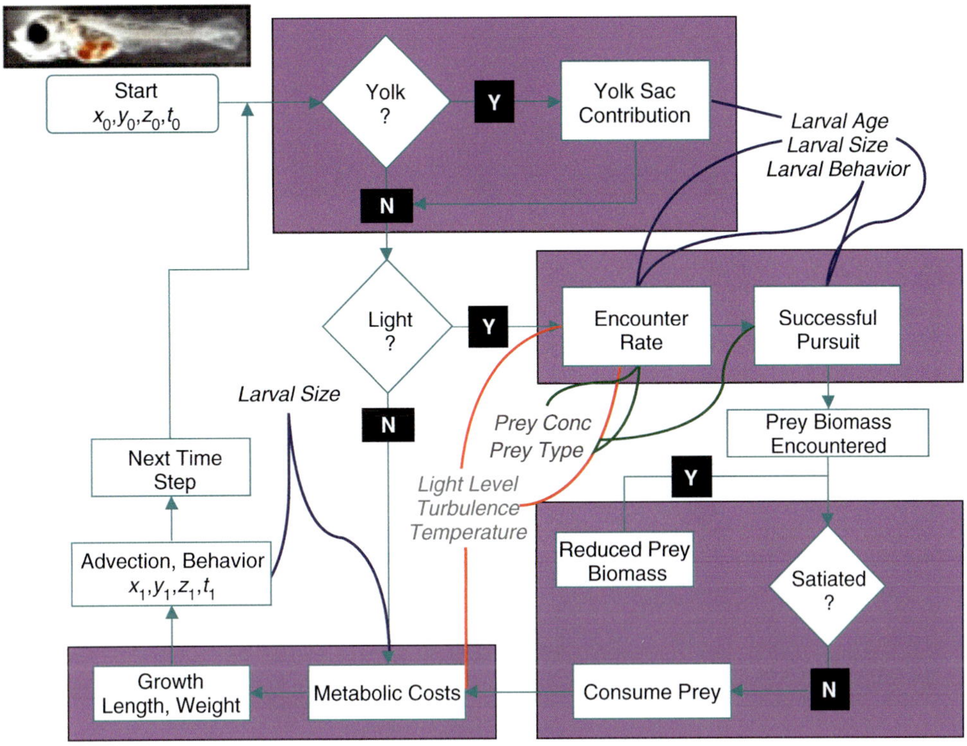

Figure 11.28. The larval fish IBM used by Lough, Broughton, Buckley et al. in [261, 262]. As it appears in Lough et al. [262]. (Original source courtesy R. G. Lough, pers. comm. 2012.)

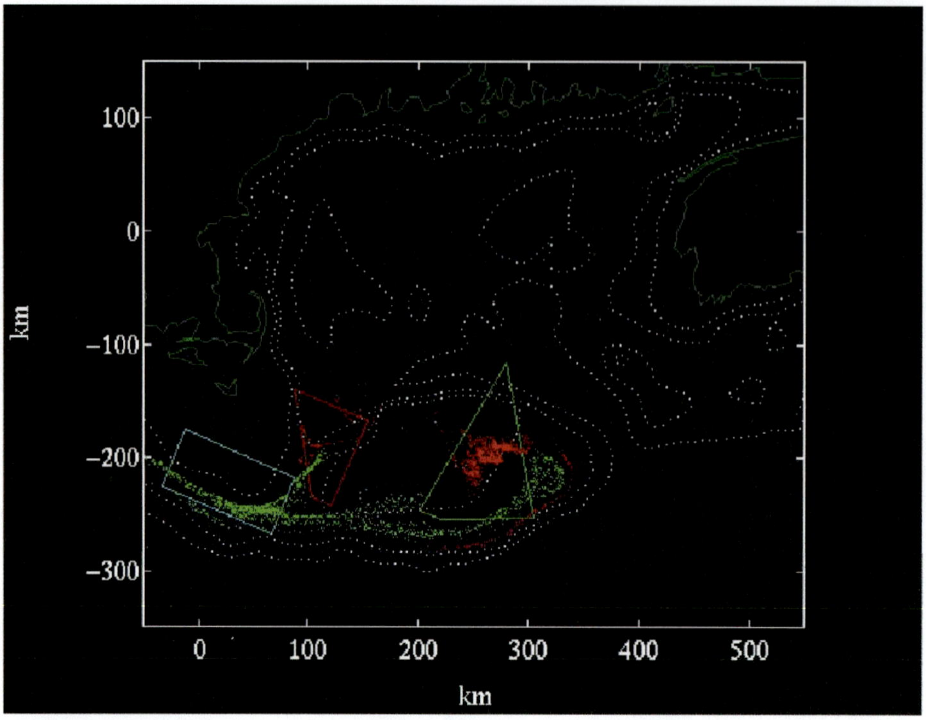

Figure 11.30. Results from the scallop IBM used by Lewis et al. [249]. The three polygons shown are observed adult aggregations. Numerical larvae were spawned there. The dot colors indicate spawning location; the dot locations indicate subsequent settlement. Reflecting management concerns, fishing mortality was reduced in the three adult aggregations. The system is partly self-sustaining, with the westernmost aggregation largely exporting larvae, and the other two, resupplying each other. Results reflect the reduction in fishing mortality in the three adult aggregations.

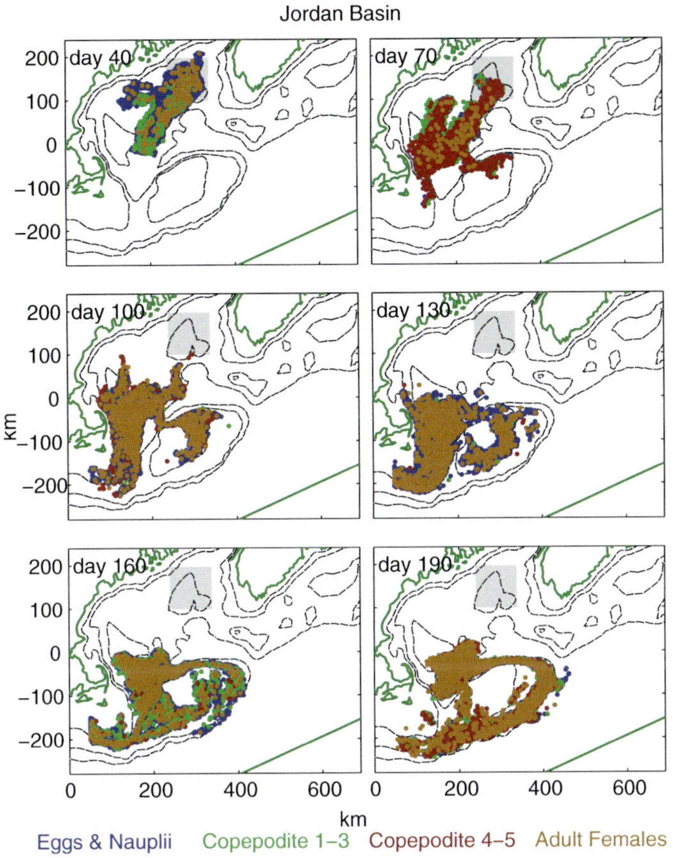

Figure 11.32. *Calanus finmarchicus* simulation results from Miller et al. [315]. Source: NML-98-7 [314]. Time evolution of population structure initiated from diapause in the shaded area on the winter solstice.

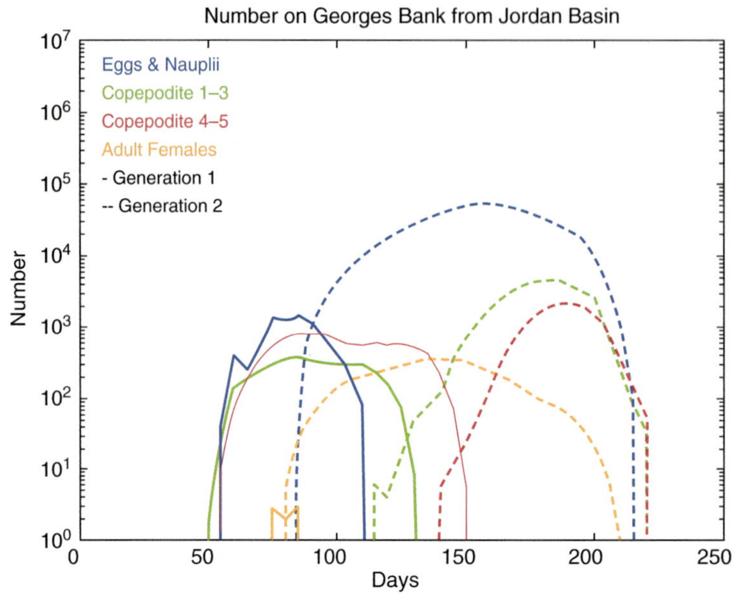

Figure 11.33. *C. finmarchicus* simulation results from Miller et al. [315]. Source: NML-98-7 [314]. Generational structure of population initiated from diapause on the winter solstice in the shaded area of Figures 11.32 and 11.34.

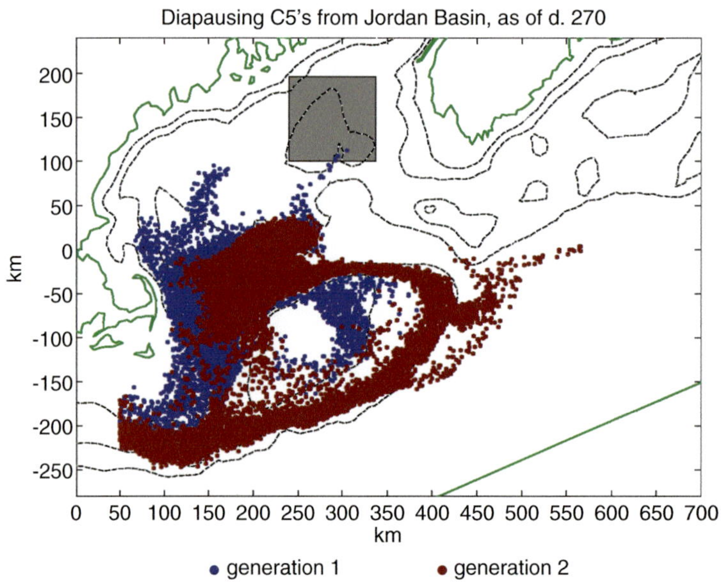

Figure 11.34. *C. finmarchicus* simulation results from Miller et al. [315]. Source: NML-98-7 [314]. Location of diapausing individuals, for a population initiated from diapause on the winder solstice, in the shaded area.

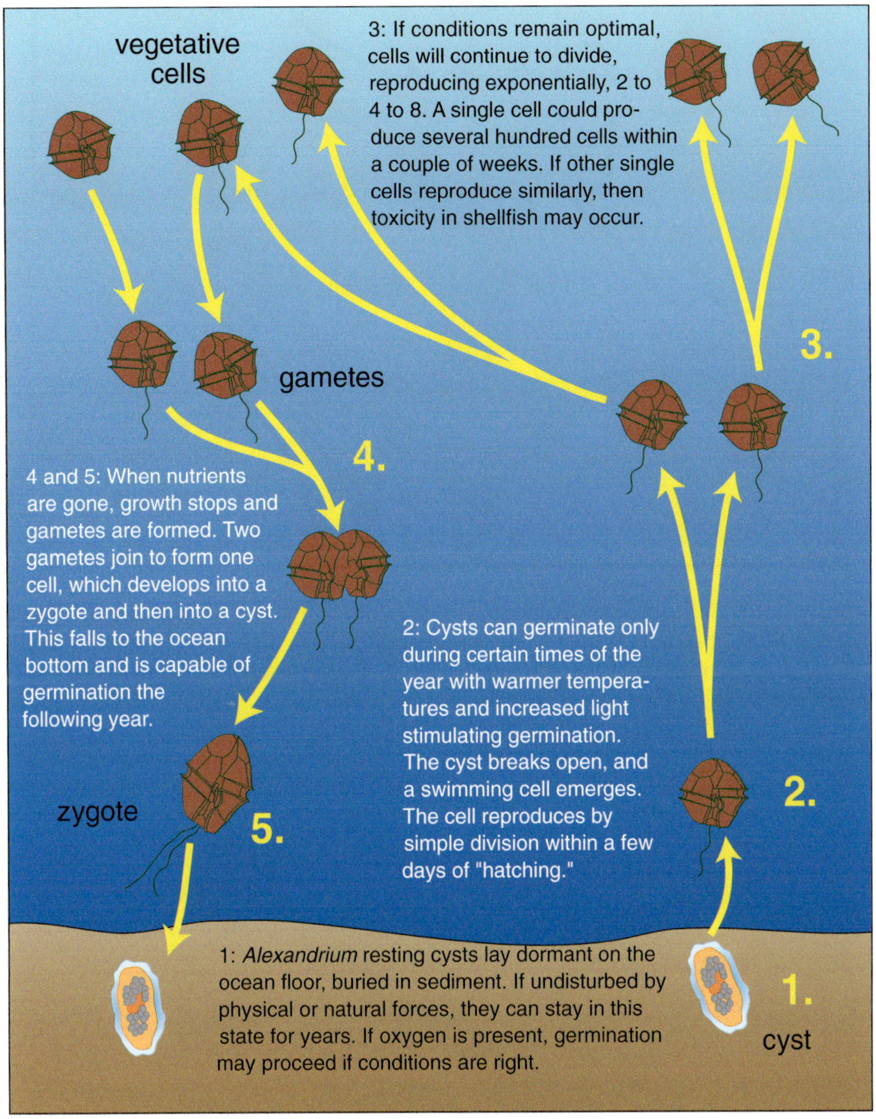

Figure 11.35. Life cycle for individual phytoplankton: *Alexandrium fundyense*. (Image courtesy of D. Anderson, B. Keafer and J. Kleindinst, pers. comm. 2014.)

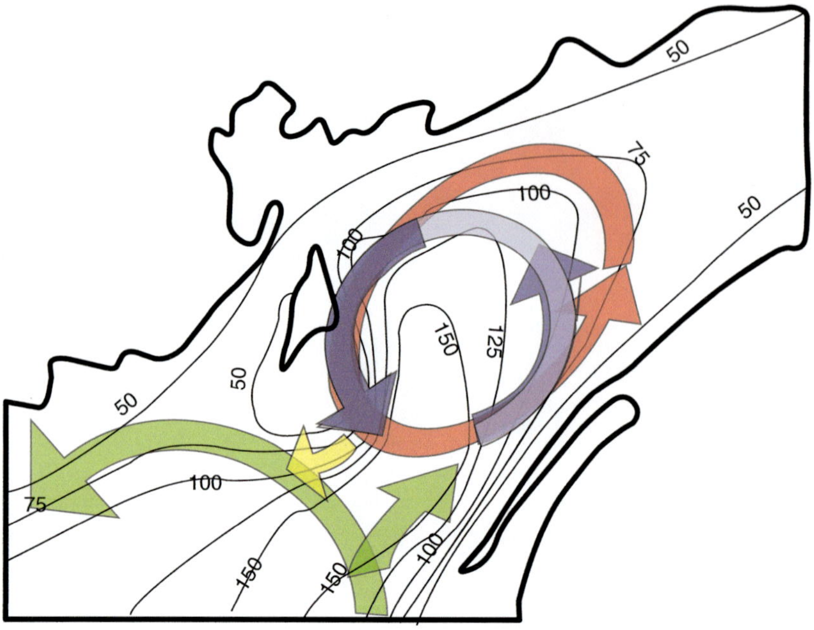

Figure 11.37. Schematic of leaky grye in the Bay of Fundy, as reported by Aretxabaleta et al. [9] and predicted by simulating passive Lagrangian drifters in the climatological flow fields (NML-05-2, [274]. Image from [9], Reproduced with permission.)

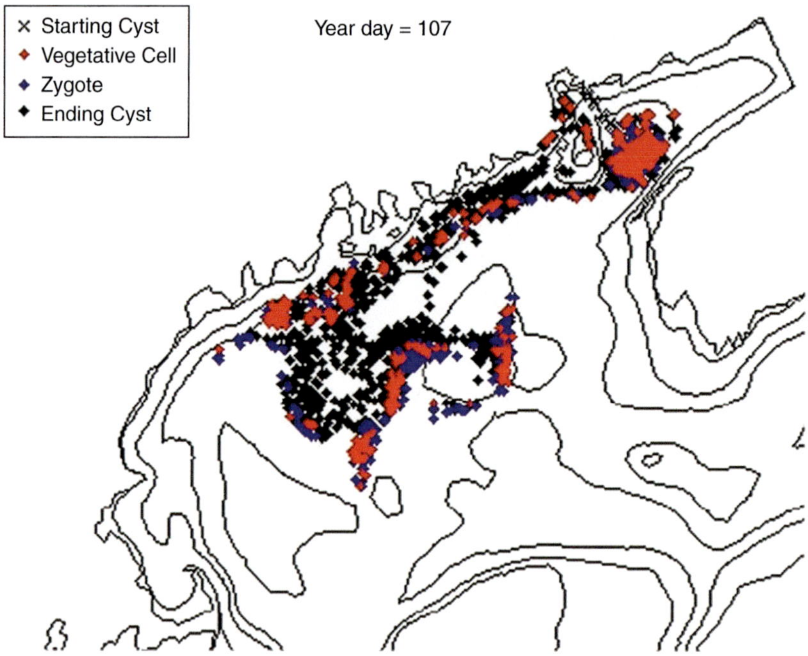

Figure 11.38. *Alexandrium* IBM in the Gulf of Maine. The benthic cysts were seeded in the Bay of Fundy (the northeastern end of the domain) during winter. Excystment was triggered by an endogenous clock. Vegetative cells were transported along coast near the surface, dividing along the way; encountered a major branch point directing part of the flow south; and generally reached nutrient limitation based on climatological records, which triggered the encystment process. Major life stages are indicated by colors.

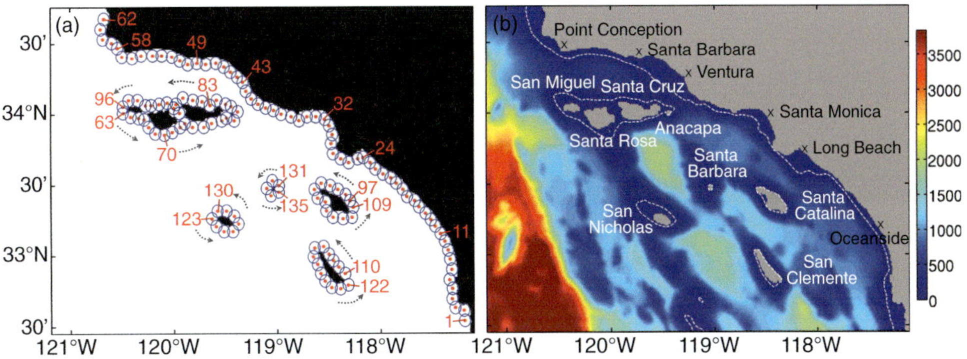

Figure 11.40. (a) Site locations used for the assessment of Lagrangian PDFs. Red dots indicate the center location of each site, with blue circles showing the area covered by each site (with a 5-km radius). The sites are categorized into three regional groups, that is, along the mainland (sites 1–62; from south to north), Northern Islands (sites 63–96), and Southern Islands (sites 97–135). (b) Model bathymetry (color bar in meters). The white dashed line indicates the 100 m isobath. (c) Sample Lagrangian particle trajectories from a site near San Nicholas Island (yellow circle). The blue lines indicate simulated 30-day trajectories, with red dots indicating the endpoints. (From Mitarai et al. [317], with permission. Source courtesy S. Mitarai (pers. comm)).

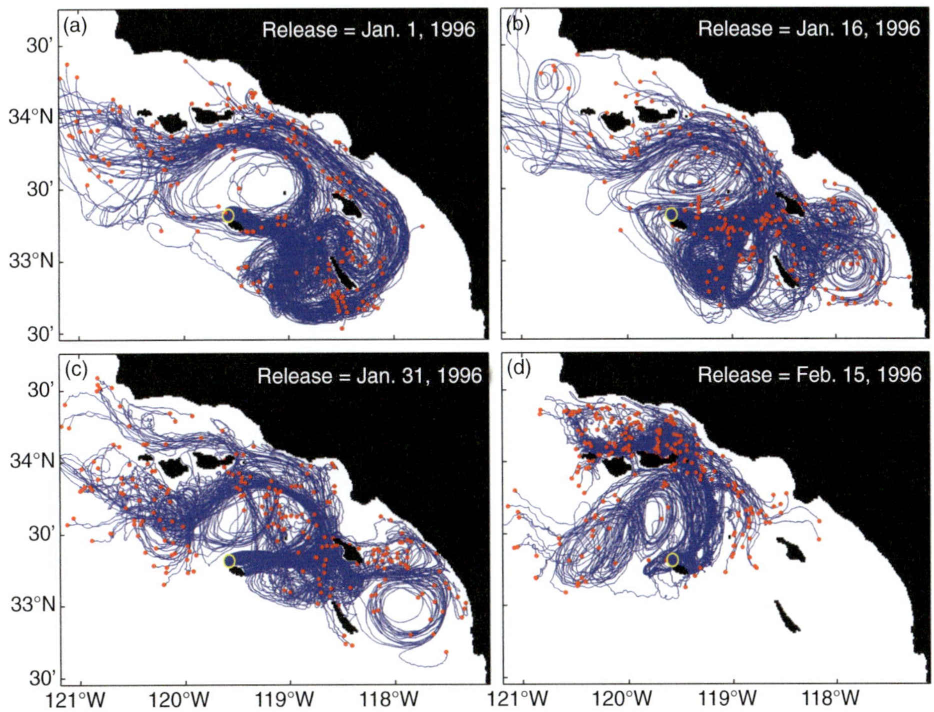

Figure 11.41. Sample Lagrangian particle trajectories from a site near San Nicholas Island (yellow circle). The blue lines indicate simulated 30-day trajectories, with red dots indicating the endpoints. (From Mitarai et al. [317], with permission. Source courtesy S. Mitarai (pers. comm)).

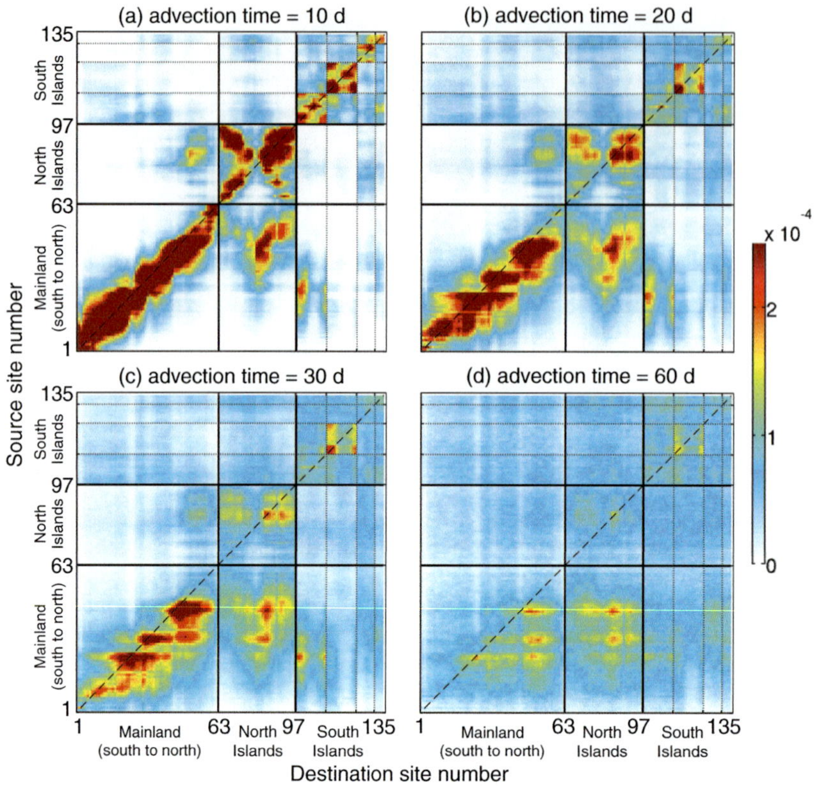

Figure 11.42. Mean coastal connectivity for the advection time of (a) 10 days, (b) 20 days, (c) 30 days, and (d) 60 days. The mean connectivity for a given source site j (horizontal axis) and destination site i (vertical axis) is determined by evaluating Lagrangian PDFs. Connectivity values are normalized so that the summation of the mean connectivity over all possible combinations is one. The black dashed line indicates self connectivity, that is, the source and destination sites are identical. The solid lines define three regions: the mainland (sites 1–62), Northern Islands (sites 63–96), and Southern Islands (sites 97–135). (From Mitarai et al. [317], with permission. Source courtesy S. Mitarai (pers. comm)).

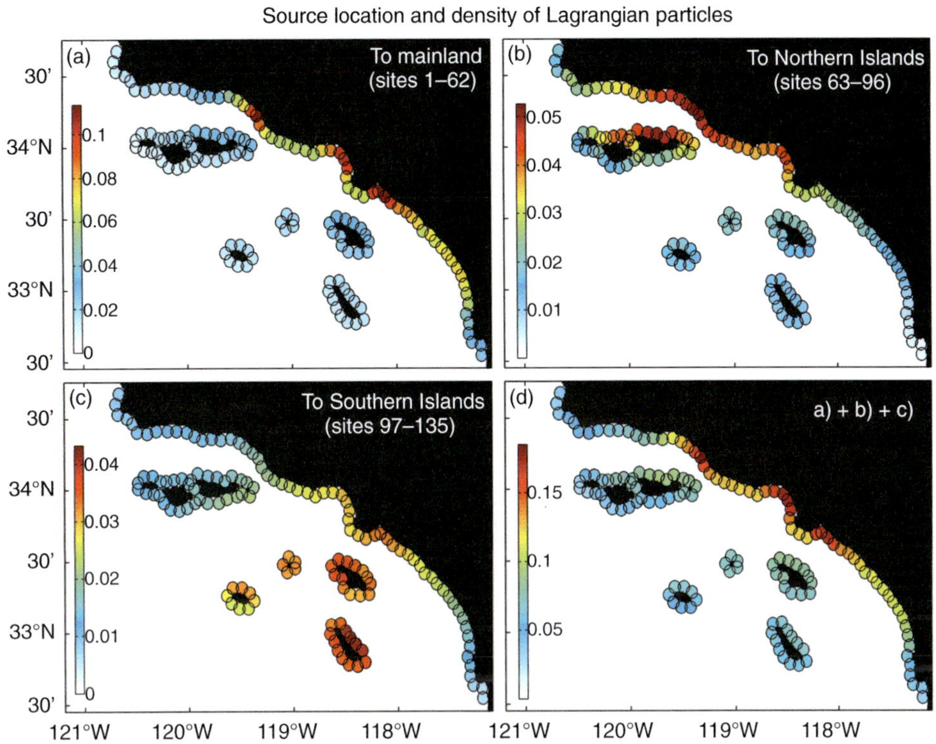

Figure 11.43. The source location and density (source strength) of particles transported to (a) the mainland (sites 1–62), (b) the Northern Islands (sites 63–96), (c) the Southern Islands (sites 97–135), and (d) all sites (sites 1–135) for an advection time of 30 days. For example, a value of 0.1 indicates that 10% of Lagrangian particles that were released from this site were found somewhere along the mainland after 30 days. Note that the color scales differ among the four panels. (From Mitarai et al. [317], with permission. Source courtesy S. Mitarai (pers. comm)).

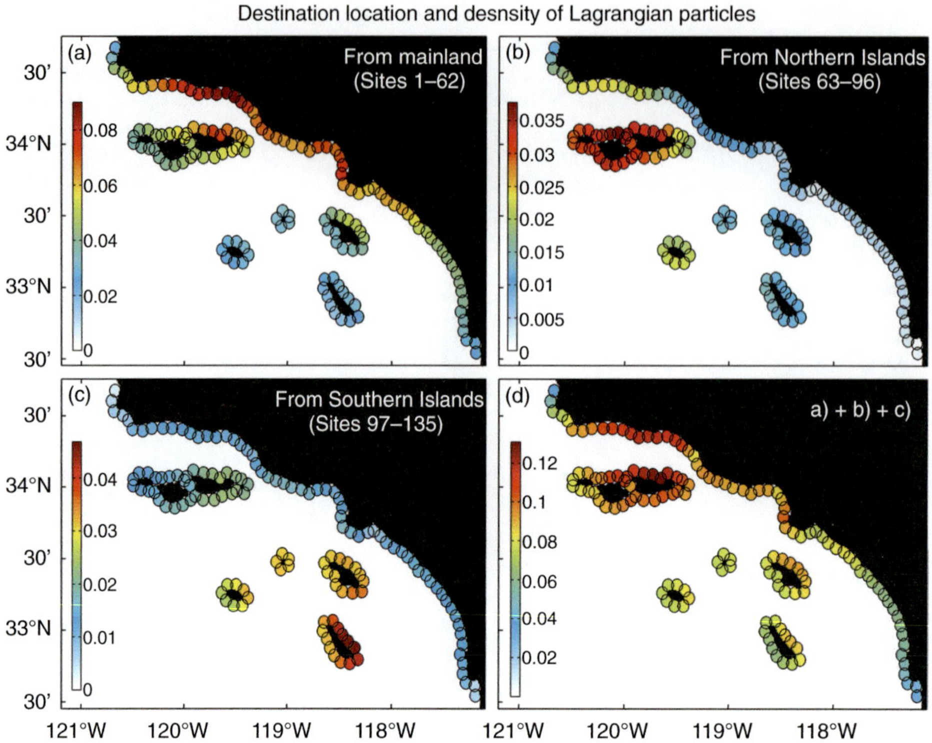

Figure 11.44. The destination location and density (destination attractiveness) of particles transported from sites (a–d) as in Figure 11.43. Advection time 30 days. (From Mitarai et al. [317], with permission. Source courtesy S. Mitarai (pers. comm)).

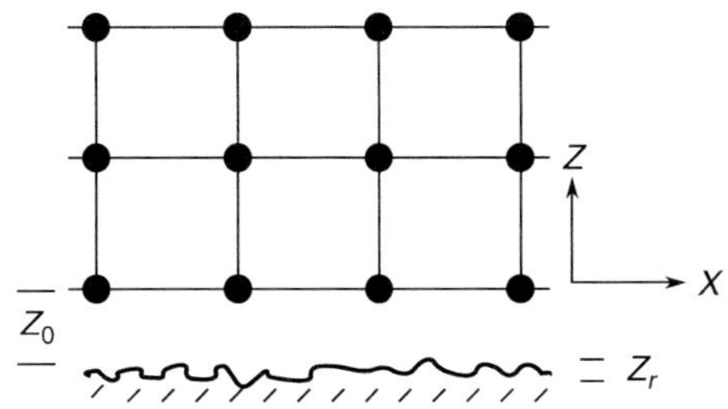

Figure 6.2. The computational bottom. The true bottom is z_0 below; z_r is its roughness height.

The computational bottom is located z_0 above the true sea floor, in the constant stress layer; the value of c_d is set accordingly:

$$c_d = \left[\frac{1}{\kappa}\ln\left(\frac{z_0}{z_r}\right)\right]^{-2} \tag{6.43}$$

with the dimensionless von Karman constant $\kappa = 0.4$, and the parameter z_r the roughness size characterizing the true bottom. Clearly, there is only one dimensionless parameter here that characterizes the bottom: z_0/z_r, the gap between model and true bottom, normalized by the surface roughness[2] (Figure 6.2).

For the turbulent variables: For q^2, Type I or Dirichlet conditions are enforced at surface and bottom:

$$q^2 = B_1^{2/3} u_*^2 \tag{6.44}$$

with the friction velocity $u_*^2 = |A_z \frac{\partial \overline{\mathbf{V}}}{\partial z}| = C_d |\overline{\mathbf{V}}_b|^2$, consistent with the hydrodynamic slip condition at the computational bottom.

The mixing length at the computational bottom is set in accord with the law of the wall:

$$l = \kappa z_0 \tag{6.45}$$

where $\kappa = 0.4$ is von Karman's constant and z_0 the distance between true and computational bottom.

At the free surface, the mixing length is conventionally set to zero; less common is specification of the vertical flux at the surface, $\frac{\partial}{\partial z} q^2 l$. Either is compatible with the diffusion form of the turbulent equations, although we find these unsatisfying in some situations, notably when there is wind-generated turbulence. (The free surface

2 For example, Lynch et al. [275] found an adequate resolution with $c_d = 0.005$, at computational bottom $z_0 = 1$ m above true bottom, roughness height $z_r = 0.0035$ m, and near-bottom resolution $\Delta z = 1$ m for tidally energetic conditions in the Gulf of Maine.

is not a rigid wall.) For generality we enforce the equivalent of 6.45 at a distance z_s from the free surface, and make the hydrodynamic condition there consistent with it.[3] The wall proximity function W is taken from Blumberg et al. [37] and incorporates asymmetry among surface and bottom "walls":

$$W = 1 + E_2 \left[\frac{l}{\kappa d_b} \right]^2 + E_3 \left[\frac{l}{\kappa d_s} \right]^2 \tag{6.46}$$

with $l/\kappa d$ approaching unity as the distance d from either wall vanishes.

6.3.2 Level 2.0 Formulation

If we ignore the left-side or transport side of Equation 6.33, we obtain the production = dissipation balance characteristic of the Level 2 model:

$$\left[A_z \left(\left[\frac{\partial \overline{V}}{\partial z} \right]^2 + \left[\frac{\partial \overline{U}}{\partial z} \right]^2 \right) + \frac{g}{\rho_0} K_z \frac{\partial \overline{\rho}}{\partial z} \right] = \left[\frac{q^3}{B_1 l} \right] \tag{6.47}$$

The left side is the macroscopic production; the right is the microscopic dissipation. Solving for q:

$$q^3 = l B_1 \left[A_z \left(\left[\frac{\partial \overline{V}}{\partial z} \right]^2 + \left[\frac{\partial \overline{U}}{\partial z} \right]^2 \right) + \frac{g}{\rho_0} K_z \frac{\partial \overline{\rho}}{\partial z} \right] \tag{6.48}$$

An l equation is needed. The equivalent treatment of the $q^2 l$ equation leads to the wall function (Equation 6.46) being constant:

$$E_1 = W(z) \tag{6.49}$$

$$= 1 + E_2 \left[\frac{l}{\kappa d_b} \right]^2 + E_3 \left[\frac{l}{\kappa d_s} \right]^2 \tag{6.50}$$

E_1, E_2, E_3 are constants (Table 6.2). If d is a composite "distance" from surface and bottom walls:

$$\frac{1}{d^2} \equiv \frac{E_2}{d_b^2} + \frac{E_3}{d_s^2} \tag{6.51}$$

then the expression is especially simple:

$$l(z) = \kappa d \sqrt{E_1 - 1} \tag{6.52}$$

[3] The surface drag is quadratic in the air-sea velocity difference. In the presence of wind-waves, the drag coefficient would need to be enhanced.

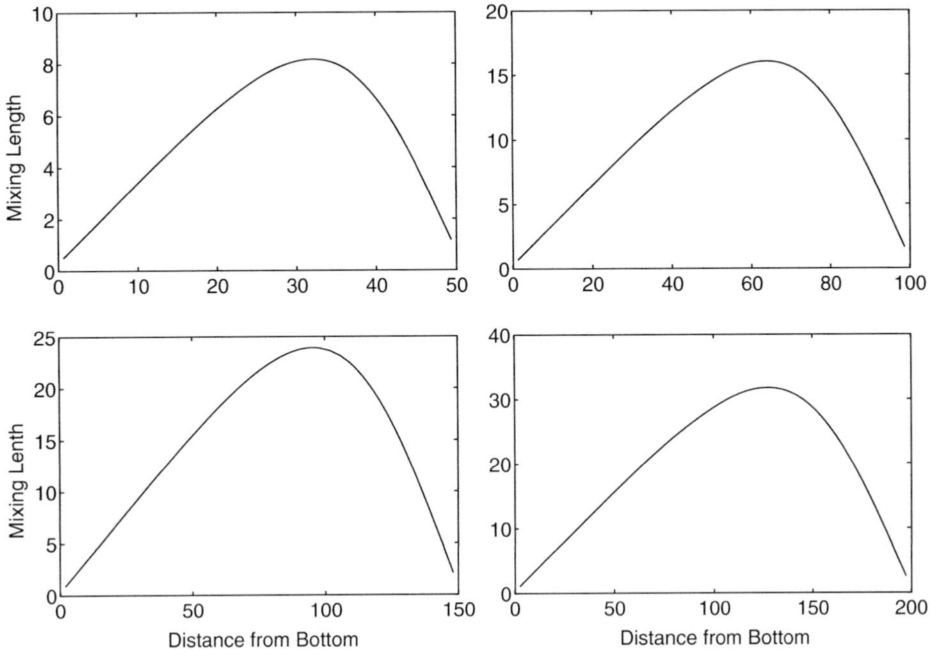

Figure 6.3. Mixing length functions for 4 different depths, using Level 2 closure, according to Equation 6.51. The unresolved boundary layers $[z_s, z_0]$ are 1 m each. For this formulation, l at mid-depth rises monotonously with H.

Depth limits l; for example at mid-depth, $d_b = d_s = H/2$, and $l = 0.14H$. Figure 6.3 shows $l(z)$ for some representative depths. The asymmetry embedded in the wall function is evident; as is the monotone relation between H and l at mid-depth.

Alternately, one can use the Blackadar formulation for l given at Equations 6.23 and 6.24.

The final closure is for the vertical turbulent mixing coefficients

$$A_z = qlS_m \tag{6.53}$$

$$K_z = qlS_h \tag{6.54}$$

where the stability functions S are simple algebraic functions of the local stratification. This would constitute a Level 2 closure.

An alternate in this category might be the $k-\epsilon$ model. The reader is encouraged to pursue this form of closure in Problems 6.10, 6.11, 6.12.

6.3.3 The Point Model

A *point model* would be an z,t representation of momentum, turbulence, and heat dynamics in the vertical, assuming horizontal homogeneity. It would employ tide, rotation, bathymetry, wind, and so forth. characteristic of a particular horizontal point. Accordingly, the vertical turbulence formulation is exercised in all its nonlinear detail; but all horizontal gradients vanish by assumption. Such an implementation would for example constitute a study of vertical resolution; and it reveals the essential computational nature of the vertical dynamics, and the nonlinear interaction among velocity, turbulence, and the diffusivities.

The momentum equation would be simple, with $\frac{d}{dt} = \frac{\partial}{\partial t}$ for all variables:

$$\frac{\partial \mathbf{v}}{\partial t} + \mathbf{f} \times \mathbf{v} - \frac{\partial}{\partial z}\left(A_z \frac{\partial \mathbf{v}}{\partial z}\right) = -g\nabla\zeta(t) \tag{6.55}$$

The right-hand side gravity term is known and barotropic; the left-hand side is quasi-linear in $\mathbf{v}$, assuming the eddy viscosity is known. The resulting diffusion equation is a well-understood computational formulation which works here, with A_z set at its value at the beginning of a time step.

A comparable diffusion equation governs buoyancy distribution:

$$\frac{\partial \rho}{\partial t} - \frac{\partial}{\partial z}\left(K_z \frac{\partial \rho}{\partial z}\right) = 0 \tag{6.56}$$

These diffusion equations are bounded at top and bottom with conventional boundary conditions (BCs) that are supported by exogenous data. For $\mathbf{v}$, wind stress at top (Type II, Poisson) and quadratic bottom slip (Type III).

$$A_z \frac{\partial \mathbf{v}}{\partial z} = \mathbf{W} \qquad \text{surface} \tag{6.57}$$

$$= c_d |\mathbf{v}| \mathbf{v} \qquad \text{bottom} \tag{6.58}$$

For ρ, boundary conditions (BC) are similarly conventional: at top, buoyancy flux given from data as a Type III or radiation condition, with reference density ρ_s (Type III); and no flux (Type II, Neuman) at the bottom.

$$K_z \frac{\partial \rho}{\partial z} = -k(\rho - \rho_s) \qquad \text{surface} \tag{6.59}$$

$$= 0 \qquad \text{bottom} \tag{6.60}$$

For level 2 closure, these are the only *dynamic* variables. The balance of the relations is algebraic, depending ultimately on the two dynamic variables $\mathbf{v}(z,t)$ and $\rho(z,t)$.

- Turbulence q is a function of velocity and stratification, Equation 6.48.
- Mixing length l is given, for example, by the Blackadar formulation, Equations 6.23 and 6.24.

- The diffusivities are the product of ql and the stability functions (S_m, S_h), Equations 6.53 and 6.54.
- The stability functions depend on stratification, $\frac{\partial \rho}{\partial z}$, Equations 6.36–6.38.

Throughout these relations, there are a handful of experimental constants (Table 6.2) whose values are taken to be universal.

Of course, other closures may be substituted as a pathway to determining the turbulent variables. Use of Level 2.5 closure, for example, would involve dynamic equations for q^2 and $q^2 l$.

Point models have been used to advantage in basic studies (e.g., Deleersnijder and Luyten [110]; NML-96-1 [324]; Lynch et al. [275]); and in site-specific studies where a complete hindcast is not available (e.g., Lough et al. [262]). Problems 6.3, 6.5, 6.7, and 6.9, explore these ideas.

These equations are friendlier than they may appear. Below we explore some simple **steady-state** cases.[4]

6.3.4 Steady-State Point Model, Level 2 Closure

To fix ideas, consider the steady state. The hydraulic and transport equations above govern, with all $\frac{\partial}{\partial t} = 0$ and unchanged boundary conditions.

For density, $K_z \frac{\partial \rho}{\partial z}$ is a constant at all depths, zero from the bottom boundary condition; hence from the top BC, ρ is constant at the input value ρ_s; there is no stratification in the steady case.

As a result, the expression for q^2 is simplified to balance dissipation with only shear production:

$$q^4 = \frac{B_1}{S_m} \left[\left[A_z \frac{\partial u}{\partial z} \right]^2 + \left[A_z \frac{\partial v}{\partial z} \right]^2 \right] \tag{6.61}$$

The stability functions S_m, S_h are constants. The diffusivities are as usual $\left[A_z, K_z\right] = ql\,[S_m, S_h]$.

Finally, the turbulent mixing length is the Blackadar–Mellor formulation,

$$l(z) = l_o \left[\frac{\kappa(z+z_0)}{\kappa(z+z_0)+l_o} \right] \qquad l_o = \alpha \frac{\int qz\, dz}{\int q\, dz} \tag{6.62}$$

$l = \kappa z_0$ at the computational bottom ($z = 0$) and grows monotonically with height above bottom. Aloft, l is bounded from above by l_0. The constant $\alpha = \mathcal{O}(0.1)$.

[4] In the steady-state point model, Level 2, Level 2.5, and $k - \epsilon$ models all share the same formulation for q^2; some important points of difference will remain on the formulations for mixing length l.

The momentum equations in simple scalar components are

$$-fv - \frac{d}{dz}\left(A_z \frac{du}{dz}\right) = -g\frac{d\zeta}{dx} \tag{6.63}$$

$$fu - \frac{d}{dz}\left(A_z \frac{dv}{dz}\right) = -g\frac{d\zeta}{dy} \tag{6.64}$$

This plus the velocity boundary conditions concludes the macroscale description for momentum in the steady state.

Below we examine some simple cases, to develop familiarity with some of the features of closure.

6.3.4.1 Wind Only

In the absence of rotation and gravity forcing, the flow is unidirectional, directed by the wind. Let x be the wind direction:

$$-\frac{d}{dz}\left(A_z \frac{du}{dz}\right) = 0 \tag{6.65}$$

$$\text{(surface)}\quad \mathrm{A_z}\frac{\mathrm{du}}{\mathrm{dz}} = \mathrm{W} \qquad A_z\frac{du}{dz} = c_d|u|u \quad \text{(bottom)} \tag{6.66}$$

The solution: $A_z\frac{du}{dz} = W$, constant with depth. The bottom condition gives $\frac{W}{c_d} = u^2$ at the bottom, u directed with the wind.

The turbulent energy is constant throughout the water column:

$$q^4 = \frac{B_1}{S_m}W^2 \tag{6.67}$$

The mixing length scale is $l_0 = \alpha\frac{H}{2}$, and the Blackadar formulation above has $l = \kappa z_0$ at the computational bottom ($z = 0$) and increasing monotonically, bounded by the asymptote l_0. Hence the eddy viscosity A_z has the same shape $l(z)$ scaled by constant qS_m:

$$A_z(z) = q \cdot S_m \cdot l(z) \tag{6.68}$$

$$= \left(\left[B_1 W^2\right]^{1/4} [S_m]^{3/4} \alpha \frac{H}{2}\right) \left[\frac{\kappa(z+z_0)}{\kappa(z+z_0) + \frac{\alpha H}{2}}\right] \tag{6.69}$$

$$= H\sqrt{W}\left([B_1]^{1/4} [S_m]^{3/4} \frac{\alpha}{2}\right) \left[\frac{\kappa(z+z_0)}{\kappa(z+z_0) + \frac{\alpha H}{2}}\right] \tag{6.70}$$

($l_0 = \frac{\alpha H}{2}$.)

Overall, for this simple case,

- The basic vertical shape of A_z is that of the Blackadar $l(z)$.
- A_z scales with depth and wind stress: $H\sqrt{W}$.
- Everything else in the brackets ( ) (Equation 6.70) is constant with depth: $[B_1, S_m, \alpha, \kappa] = [16.6, 0.39327, 0.1, 0.4]$.
- Above the bottom, A_z approaches the approximate asymptote $A_z = H\sqrt{W}\,(.0501)$. (See Problem 6.3 for a refined value in this case.)
- Generally,

$$A_z(z) = H\sqrt{W}\,(.0501)\left[\frac{\kappa(z+z_0)}{\kappa(z+z_0)+l_o}\right] \tag{6.71}$$

The *velocity profile* is given by integrating $A_z\frac{\partial u}{\partial z} = W$ from the computational bottom $z = 0$:

$$\int_0^z du = W\int_0^z \frac{dz}{A_z} \tag{6.72}$$

$$u(z) = \sqrt{W}\left(\frac{1}{\sqrt{c_d}} + \frac{1}{H}\frac{1}{(.0501)}\int_0^z \left[\frac{\kappa(z+z_0)+l_0}{\kappa(z+z_0)}\right]dz\right) \tag{6.73}$$

$$= \sqrt{W}\left(\frac{1}{\sqrt{c_d}} + \frac{1}{H}\frac{1}{(.0501)}\left[z + \frac{l_0}{\kappa}\ln\left(\frac{z+z_0}{z_0}\right)\right]\right) \tag{6.74}$$

We find in this case that velocity is a simple integral of a simple monotone function; it scales as $\sqrt{W}$. (See Figure 6.4.)

The combination of 6.67 and 6.74 show a distinct departure from linear theory wherein A_z is fixed and u scales with W. All the wind energy goes into resolved kinetic energy. When the closure is introduced, both the resolved motion u and the subgrid motion q grow as $\sqrt{W}$. Both are enhanced by W. In both cases the product $A_h\frac{\partial u}{\partial z}$ (the shear stress) scales with the input W.

As mentioned above, to be consistent with the law of the wall, the drag coefficient c_d depends on the distance from true to computational bottom z_0, and on the bottom roughness z_r:

$$c_d = \left[\frac{1}{\kappa}\ln\left(\frac{z_0}{z_r}\right)\right]^{-2} \tag{6.75}$$

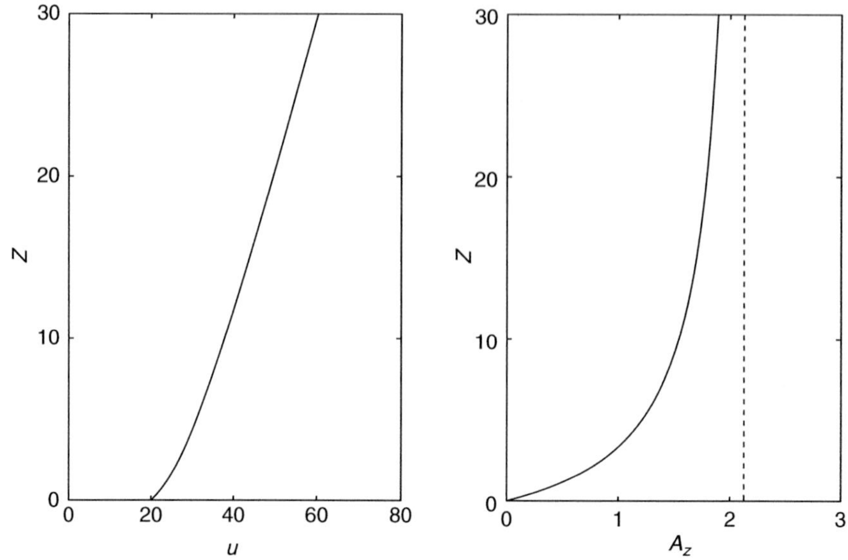

Figure 6.4. Velocity (left) and viscosity (right) profiles for the Steady State Level II closure, Blackadar mixing length, $f = G = 0$. Parameters: $[H, W, c_d, z_0] = [30, 2, .005, 1]$. The asymptotic upper limit of A_z is 2.13, shown as the dashed line.

6.3.4.2 Wind and Gravity

If we add a barotropic pressure gradient $G = g\frac{\partial \zeta}{\partial x}$, we acquire a constant on the right-hand side of the diffusion equation, with the same BC's at top and bottom:

$$-\frac{d}{dz}\left(A_z\frac{du}{dz}\right) = -G \tag{6.76}$$

$$\text{(surface)}\quad \mathrm{A_z}\frac{\mathrm{du}}{\mathrm{dz}} = \mathrm{W} \qquad A_z\frac{du}{dz} = c_d|u|u \quad \text{(bottom)} \tag{6.77}$$

The shear is now linear in height above the computational bottom:

$$A_z\frac{du}{dz} = W - G(H - z) \tag{6.78}$$

and the vertical balance of forces gives the bottom velocity:

$$c_d|u|u = W - GH \tag{6.79}$$

The turbulent energy is linear in the vertical[5]:

$$q^2 = \sqrt{\frac{B_1}{S_m}}\,|W - G(H - z)| \tag{6.80}$$

[5] q^2 must be positive; it can go to zero within the water column.

and the eddy viscosity, assuming Blackadar l:

$$A_z = q \cdot S_m \cdot l(z) \tag{6.81}$$

$$= \left[\sqrt{\frac{B_1}{S_m}} \, |W - G(H - z)| \right]^{\frac{1}{2}} S_m l(z) \tag{6.82}$$

$$= B_1^{\frac{1}{4}} S_m^{\frac{3}{4}} \, |W - G(H - z)|^{\frac{1}{2}} \, l_0 \left[\frac{\kappa(z + z_0)}{\kappa(z + z_0) + l_0} \right] \tag{6.83}$$

As before, we have the integral of l, this time weighted by the variable $q(z)$.

Suppose we normalize the gravity term with $GH = cW$. Then we have

$$A_z = \sqrt{W} \left([B_1]^{\frac{1}{4}} \, [S_m]^{\frac{3}{4}} \, l_0 \right) \left| 1 - c \left(1 - \frac{z}{H} \right) \right|^{\frac{1}{2}} \left[\frac{\kappa(z + z_0)}{\kappa(z + z_0) + l_0} \right] \tag{6.84}$$

This is a natural extension of the $c = 0$ case, Equation 6.70. In Problem 6.4 it is developed further.

6.3.4.3 Rotation and Wind – The Ekman Layer

Here we limit ourselves to the deep ocean with only wind forcing at the surface; there is no gravity term. The surface wind force will be balanced with the Coriolis force acting throughout the depth. For very deep water, the motion will be confined to a finite-depth surface layer, with no bottom motion. The equations:

$$-fv - \frac{d}{dz} \left(A_z \frac{du}{dz} \right) = 0 \tag{6.85}$$

$$fu - \frac{d}{dz} \left(A_z \frac{dv}{dz} \right) = 0 \tag{6.86}$$

The surface boundary conditions:

$$A_z \frac{du}{dz} = W \qquad A_z \frac{dv}{dz} = 0 \tag{6.87}$$

At depth, $u = v = 0$. The coordinate z increases upward, as usual.[6]

[6] We will also use the coordinate d to indicate the depth *below* the free surface; $d = 0$ at the surface and increases downward.

The transport (vertically integrated velocity) is directed perpendicular to the wind (90° to the right in the Northern Hemisphere).

$$-fH\bar{v} = W \tag{6.88}$$

$$fH\bar{u} = 0 \tag{6.89}$$

These properties are unaffected by the form of closure. There is no transport in the direction of W.

When A_z is constant, one obtains the classic Ekman spiral. Some of the standard properties are

- Current speed decays with depth, as $e^{-d/L}$, with $L = \sqrt{\frac{2A_z}{f}}$.
- Velocity veers to the right with depth, with rotation rate $\theta = d/L$, such that one reversal in direction occurs at $d = \pi L$.
- Surface velocity $\mathbf{v}_o$ is at 45° to the wind, downwind to the right in the Northern Hemisphere.
- Surface speed $|\mathbf{v}_o|$ is $\frac{W}{\sqrt{A_z f}}$.

For turbulence closure: A is dependent on shear. The surface condition is $q^2 = W\sqrt{B_1/S_m}$. This is a reasonable first approximation for turbulent energy in the surface layer.

Using Blackadar l: in deep water, a reasonable approximation is $l = l_0$, constant. A reasonable approximation is $l = \beta L = \beta\sqrt{\frac{2A_z}{f}}$, that is, a fraction of the Ekman depth.

Closing these with $A_z = qlS_m$ gives

$$A_z = k_a W \tag{6.90}$$

$$l = k_l\sqrt{W} \tag{6.91}$$

$$L = \frac{k_l}{\beta}\sqrt{W} \tag{6.92}$$

$$q = k_q\sqrt{W} \tag{6.93}$$

$$|\mathbf{v}_o| = \frac{1}{\sqrt{k_a f}}\sqrt{W} \tag{6.94}$$

with the intermediate constants

$$k_q = \left[\frac{B_1}{S_m}\right]^{\frac{1}{4}} \qquad k_a = \left[S_m^2\beta^2\left(\frac{2}{f}\right)k_q^2\right] \qquad k_l = \left[\frac{2k_a\beta^2}{f}\right]^{\frac{1}{2}} \tag{6.95}$$

This is "Approximation A":

- The turbulent velocity q scales as $\sqrt{W}$, in accord with the surface BC.
- The mixing length l and the Ekman depth L both scale with $\sqrt{W}$.

- The viscosity scales with W.
- Increase in W is shared equally among increases in q and in l;
- The surface velocity sales with $\sqrt{W}$.
- Stronger wind means a more energetic, deeper Ekman layer with higher mixing.

The nonlinearity here is hard to manage further, without either more approximation, or simulation.

In the simulation category: we solved the steady point equations iteratively, beginning with reasonable approximations for A_z, and then closing on the nonlinearities iteratively for $\mathbf{v}, q, l$, and A_z. Figure 6.5 shows one such calculation for an initial guess and three subsequent iterations. The turbulent variables were initially held constant through the depth, and develop strong vertical gradients that are stable after only four iterations; most of the change occurs after the very first iteration.[7]

The velocity profiles are remarkably robust – the starting point is close to the final profiles. Evidently the initial calculation with A_z based on W approximates its value in the surface layer quite well. Parameters: $[W, H, f, \alpha, S_m] = [0.5 \times 10^{-4}, 300, 10^{-4}, 0.1, 0.39327]$ (MKS units).[8]

Figure 6.6 shows the converged velocity vectors versus depth for the same simulation. The classic Ekman features are evident.

In Figure 6.7 we show the same result with wind decreased by a factor of 10. The approximate scaling above (Equations 6.90–6.94) is qualitatively verified. Most conspicuous, however, is the nonsensitivity of the computed mixing length, contrary to Equation 6.91. It would appear that the assumptions used in the l approximation are not appropriate here. In detail, the ratios of surface quantities obtained in these two experiments are given in Table 6.3; L by visual inspection is roughly $1/2$; fitting an exponential decay to the top 30 m, L decreases from 57.82 (high wind) to 30.75 (low wind).

If we retrace the steps leading to the scaling (Equations 6.90–6.94), but making l constant, we obtain "Approximation B"[9]:

$$q \sim \sqrt{W} \tag{6.96}$$

$$l \sim \text{constant} \tag{6.97}$$

$$A_z \sim \sqrt{W} \tag{6.98}$$

$$L \sim \sqrt{A_z} \sim W^{1/4} \tag{6.99}$$

$$|\mathbf{v}_o| \sim W^{3/4} \tag{6.100}$$

[7] The initial near-surface approximations are good. q is held constant at the surface, based on the wind BC; A_z is initially qlS_m throughout the depth.

[8] For a typical value: $\rho W = 0.1$ Pa $= 0.1$ kg/m/s^2; $\rho = 10^3$ kg/m^2; $W = 10^{-4}$ m^2/s^2. In shallow water, $W = 0.005 * V_b^2 = .5 \times 10^{-4}$. Estimate: $W = (0.5 \rightarrow 1.0) \times 10^{-4}$.

[9] $q^2 \sim W$ by the surface boundary condition, that is part of both approximations.

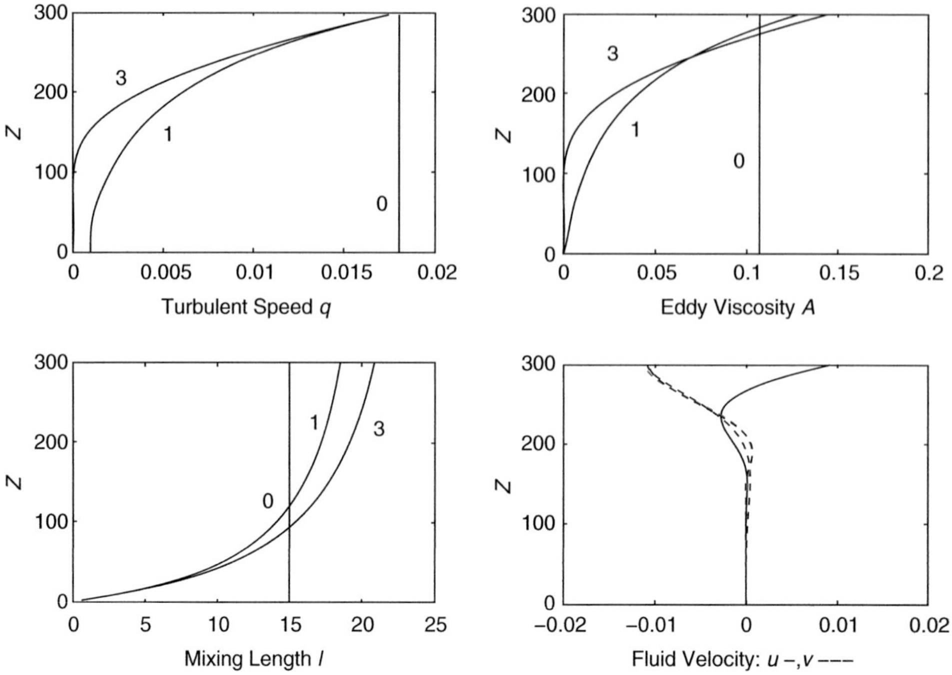

Figure 6.5. Iterative solution of the nonlinear point model showing velocity and the turbulent variables. The turbulent variables develop a strong dependence on depth; the velocity profile is very similar to its starting profile with constant turbulence. Solution number 0 is the starting point; subsequent iterations 1 and 3 are shown. Beyond the third iteration, there was little practical change.

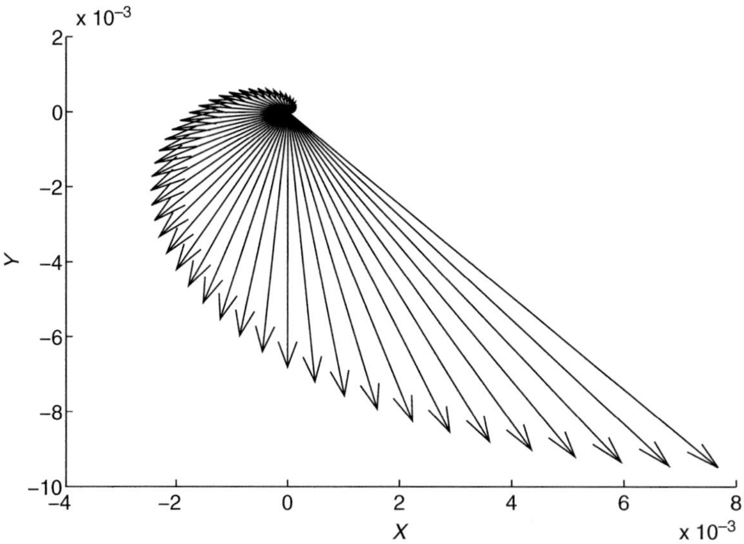

Figure 6.6. Ekman spiral for the converged solution as in Figure 6.5; $W = 0.5 \times 10^{-4}$.

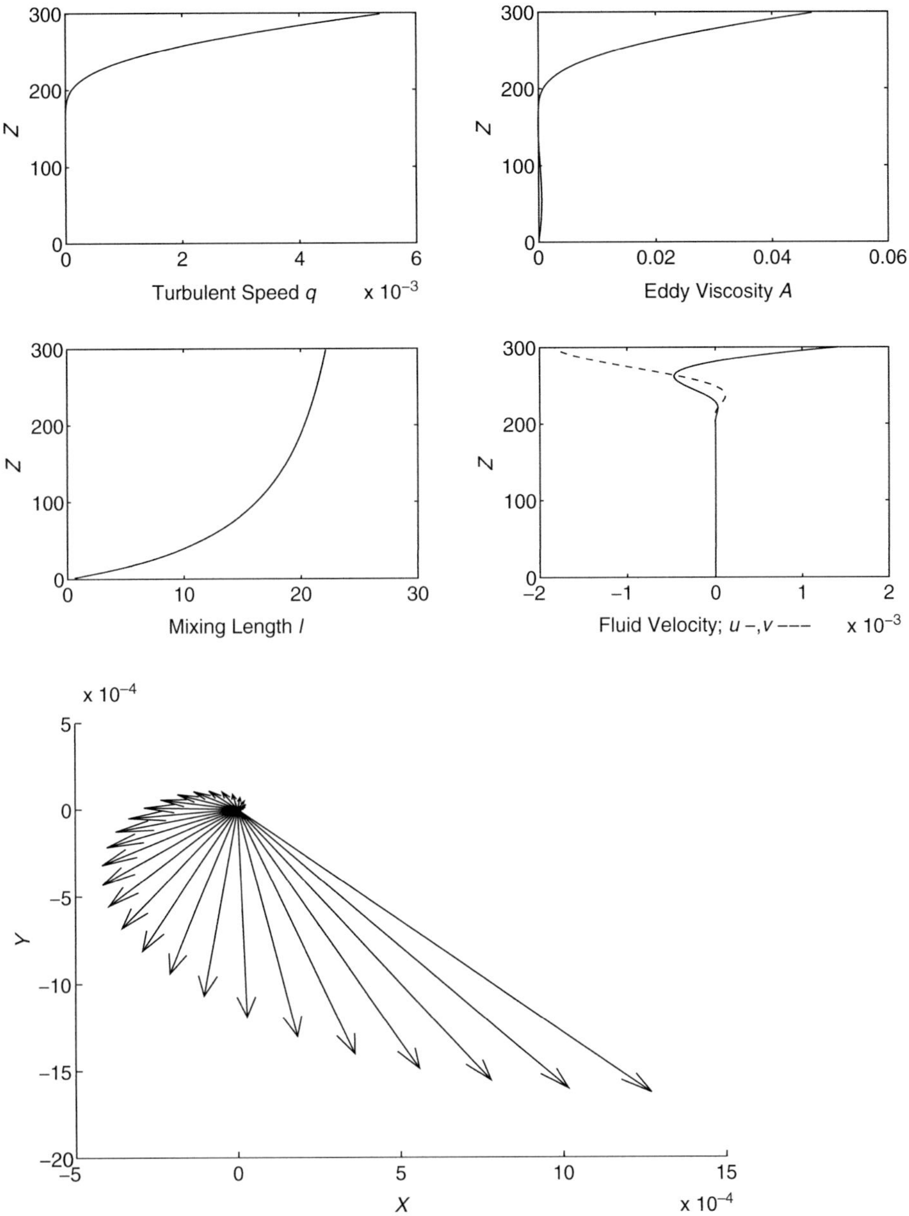

Figure 6.7. The converged solution as in Figures 6.5 and 6.6; but lighter wind, $W = 0.05 \times 10^{-4}$.

This is much closer to the computed result. The mixing length closure generally attracts attention here. There is good consensus for the q closure; less so for l. (See Table 6.3.) The simulation strategy is further explored in Problems 6.4, 6.5, 6.6 and 6.7.

Table 6.3. *Approximation Ratios for the Ekman Problem, with Wind Reduced by a Factor of* 10

	W	q	l	A_z	u	v	L
Linear analysis	0.1	–	–	1.0	0.1	0.1	1.0
Approximation A	0.1	0.316	0.316	0.1	0.316	0.316	0.316
Approximation B	0.1	0.316	1.0	0.316	0.178	0.178	0.562
Nonlinear simulation	0.1	0.31	1.06	0.327	0.1628	0.1698	0.532

Note: Approximation B is clearly superior. Simulation results are plotted in Figures 6.5, 6.6, and 6.7.

6.3.5 Some Implementations

Point models have been widely used; example implementations include COHERENS (Luyten et al. [272], PROBE (Svensson [441]), GOTM (Burchard et al. [61]), NUBBLE (Naimie [324]). For a recent report on GOTM results, see Burchard et al. [62].

Problem 6.3 For a steady-state point problem with wind forcing only, no rotation, and Level 2.0 closure with Blackadar mixing length: express the mixing length l, eddy viscosity A_z, and velocity u at the top of the water column, $z = H$. Compare with the asymptotes mentioned in the text.

Problem 6.4 For a steady-state point problem with wind and gravity forcing, no rotation, and Level 2.0 closure with Blackadar mixing length: *evaluate numerically* the mixing length l and its asymptote l_0; the eddy viscosity A_z, and the velocity u throughout the water column, from the formulae developed in the text. Use the same parameters as in the text example, with $GH = 2W$. Show that when $G = 0$ you recover the correct solution, which is given in closed form.

Problem 6.5 Implement the point model (make a computer simulation) for Level 2 closure described in Section 6.3.3. Assume zero stratification. With it, explore the influence of wind stress, depth, tide. Specifically:

- Show that you can generate an Ekman layer at top and bottom of proper size. At what depth do top and bottom layers intersect? What is a practical criterion for Δz?
- For a unidirectional $M2$ tide and no wind: show the bottom-generated turbulence as a function of time. Under what conditions does the bottom turbulence penetrate the entire water column? Compare with the Simpson–Hunter criterion.
- Repeat, but for a completely rotary $M2$ tide.

Problem 6.6 Using the simulation model from Problem 6.5, examine the approximate Ekman relations developed in Section 6.3.4.3 for a very deep water column. Is the scaling with W qualitatively correct? Quantitatively?

Problem 6.7 As in Problem 6.5, using wind stress but no tide. Add stratification by heating at the surface. Do you obtain a stable mixed layer? How do you characterize its thickness and the dependence of thickness on relevant dimensionless parameters?

Problem 6.8 In the text the Blackadar relation for mixing length is used. Redo the Ekman example using instead the steady-state version of Level 2.5 for l. What do you observe?

Problem 6.9 Modify the description of the Level 2 point model, Section 6.3.3, to incorporate the Level 2.5 closure. What exactly has to change? Specifically, what differential equations govern what variables?

Problem 6.10 For a Level 2 $k-\epsilon$ closure: What are the variables and the governing equations?

Problem 6.11 For a Level 2 $k-\epsilon$ closure: What is the relation between k and the decay ϵ?

Problem 6.12 For a Level 2 $k-\epsilon$ closure: What is the expression for the eddy viscosity?

6.4 Horizontal Closure

By comparison with the vertical closures reviewed in the preceding text, the horizontal closure remains less developed. The basic ideas are not different, and the focus is on an eddy viscosity/diffusivity closure that represents the middleman between resolved mean flow and unresolved subgrid scale turbulence. It is common for practical computations to require viscosity and/or diffusivity in order to function at all; and the blend of computational necessity and fidelity to physical processes can be blurred. As modeling has improved in resolution relative to the various length and time scales involved, there remains the hope that the primary turbulence – and the various nonlinearities in the governing equations that generate it – will be *resolved*, such that the subgrid processes will be less important and the resolved processes will contain more and more of the energy. Nevertheless, this remains a hope, not yet fully realized in the spectrum of practical applications; yet the importance of these remains.[10]

[10] In a recent review, the situation is summarized: "Suffice it to say that modeling horizontal diffusion terms remains rather ad hoc"[219].

The most commonly cited formulation is that by Smagorinsky [421].[11] The momentum formulation is in classic form, in terms of the tension and shearing strains:

$$D_T \equiv \frac{\partial \overline{U}}{\partial x} - \frac{\partial \overline{V}}{\partial y} \tag{6.101}$$

$$D_S \equiv \frac{\partial \overline{V}}{\partial x} + \frac{\partial \overline{U}}{\partial y} \tag{6.102}$$

$$D_e = \sqrt{D_T^2 + D_S^2} \tag{6.103}$$

The deformation field D_e is the metric of resolved horizontal shear. The eddy viscosity is dependent on it and the grid scale Δx:

$$A_h = [A_o \Delta x]^2 D_e \tag{6.104}$$

with $A_o = 0.28$. (A_o values ranged $0.1 - 0.3$, Germano et al. [169]) The macroscopic momentum equations have diffusive right-hand sides representing Reynolds stresses:

$$F_x = \frac{\partial}{\partial x}[A_h D_T] + \frac{\partial}{\partial y}[A_h D_S] \tag{6.105}$$

$$F_y = \frac{\partial}{\partial x}[A_h D_S] - \frac{\partial}{\partial y}[A_h D_T] \tag{6.106}$$

We see that we are recovering a mixing-length type formulation, as in the early vertical models, with horizontal Eddy viscosity given as the product of l^2 multiplying a metric of macroscopic horizontal shear. In this case, the mixing length is the grid spacing.

It is common to assume a unit Prandtl number, that is, $K_h = A_h$ and to represent the non-advective flux of, say, a passive substance (concentration C) in the Laplacian form:

$$\frac{dC}{dt} = \nabla \cdot K_h \nabla C \tag{6.107}$$

Another simple approach is the inclusion of harmonic or biharmonic formulations. The horizontal turbulent viscous terms F_h are specified according to the combination of a number of formulations. A harmonic expression (Gent and McWilliams [168]):

$$F_{h1} = A_h \nabla^2 \vec{V} \tag{6.108}$$

[11] This important early report[421] is lengthy and deals with many things; the passage cited is at p. 105.

a biharmonic formulation (Roberts and Marshall [383]):

$$F_{h2} = B_h \nabla^4 \vec{V} \tag{6.109}$$

and a viscous dissipation depending on the local rate of strain:

$$F_{h3} = \nabla \left(\mu_A T^2 \vec{\nabla}_H \vec{V} \right) \tag{6.110}$$

where A_h, B_h, and μ_A are constant coefficients, $\nabla^2 = \nabla \cdot \nabla_H$ the horizontal Laplace operator, and T^2 is a function of the local rate of strain $\frac{\partial \overline{V}}{\partial x} + \frac{\partial \overline{U}}{\partial y}$.

The biharmonic viscosity is highly dissipative as in discretize form it requires a larger horizontal footprint. Griffies and Hallberg [187] provide examples of applications. Issues with biharmonic mixing have also been widely described (e.g., Delhez and Deleersnijder [111])

The introduction of anisotropy (e.g., topographic, frontal, beta effects) in simple viscosity models becomes a trickier problem and a potential approach (Smith and McWilliams [423]; Smith and Gent [422]) is the inclusion of an arbitrary field of horizontal vectors that specify a preferred direction at each horizontal point in space to break the isotropy.

6.5 The Main Points

- Division of the flow field into grid and subgrid parts is classic, and a necessity with finite (computational) fields. It is a permanent problem in computational fluid dynamics. It is possible that eventually, very high resolution will be capable of resolving most of the important energetic features.
- Eddy viscosity and diffusivity closure is dominant, with the unresolved motions portrayed as macroscopic gradient flux form. These terms are the go-betweens between the macroscale and the microscale. The resolved motions feel the microscale through this closure; the microscale is fed by the macroscale through them.
- Reynolds averaging gives a classic context for this general approach.
- Second moment methods provide closure in terms of the mean Reynolds stresses
- Eulerian fields describe the mean motion $\mathbf{V}$; it is deterministic.
- The subgrid motion is represented by statistical relationships; a stochastic random walk realization is implied.
- The vertical closure has achieved closures based on second-order moments of the flow:

$$K_z = q l S_h \tag{6.111}$$

Level 2, 2.5 and so forth involve two prognostic variables – energetic velocity q, mixing length l, or a combination of the two. These require representation as Eulerian fields alongside the conventional hydrodynamic fields. S_h represents a stability function when density gradients are present.

- The horizontal Reynolds stresses are typically parameterized in mixing length-type models:

$$K_h = l^2 \cdot \left| \frac{\partial \overline{V}}{dx} \right| \tag{6.112}$$

with the mixing length given by the mesh spacing.

- The importance of the stochastic model of the subgrid needs to be constantly kept in mind. The question "What is random walk?" is from this viewpoint, equivalent to the question "What is the model of subgrid motion?"; this adds a stochastic motion to the deterministic mean motion that is the subject of conventional gridded models. It is equivalent to the question "What is the diffusion coefficient?"; that question needs to be taken as seriously as those surrounding the mean motion.
- We have touched only lightly on the applicable boundary conditions. This topic will assume greater importance when one considers the fate of particles of various sorts when transported to a computational boundary, where the boundary layer is almost certainly unresolved.

Part II

Elements

7

Meshes: Interpolation, Navigation, and Fields

Operating with existing gridded fields is the subject of this chapter.

Continuous fields are assumed in almost all coastal ocean discretization methods in operational practice. There are important exceptions, especially for vector fields in certain contexts, or at steep fronts that are essentially unresolvable any other way. But as a baseline, gridded fields will be assumed continuous; if not, then special instructions about their use would need to be appended to them. We briefly visit geometry, interpolation, and navigation on grids.

7.1 The Horizontal Mesh

7.1.1 Triangles

The triangle is the most flexible unit of horizontal discretization. We describe it first, and then describe the quadrilateral. Both present the same options in the vertical direction.

7.1.2 Geometry

Simulation of natural systems requires that we accommodate the geometric needs first – the coastal boundary is maybe the most obvious; a close cousin is the bathymetric realism. The unstructured mesh of triangles achieves much of what is needed here, in the sense of making numerical length scales and orientations "fit" the natural ones.

Examples of triangular meshes appear in Figures 7.1 and 7.2. Resolution matters! In the coastal ocean, the physics and chemistry of human interaction can be quite local; and all data are local and *include* near-field effects. The shelf-coast interactions occur over a wide range of length and time scales, all demanding resolution.[1] Though these meshes have been important, time will render them obsolete due to inadequate resolution by then-current expectations. The truism – *there is never enough resolution!*

These are *unstructured* meshes – all the space is filled with contiguous, non-overlapping triangles of locally appropriate size. Beyond that, there is no predetermined pattern constraining the discretization. Each triangle will be home to all that occurs there – inputs, coefficients, discretization, and results – each is *local.* The whole picture is the union of all the individual triangles. Noteworthy here is the natural ability of such an unstructured mesh to fit the coastal geometry and the bathymetry, and any boundary where important PDE information needs to be specified.

The triangle is the workhorse of horizontal geometry here. Implied is local interpolation within each triangle, with specified continuity between triangles. The usual geometric description uses *linear* interpolation in each triangle, with interpolated functions continuous in value, but discontinuous in slope – "piecewise linearity" with "C^0 continuity" between triangles. It is easy to see that by adjusting the size of triangles locally, this discontinuity in slope can be managed to be "uniform" by making small triangles where the geometry changes fast, and bigger ones elsewhere. Other criteria clearly can be brought to bear here; we discuss these later. And of course, one need not stick with linear elements – entire families of interpolating functions have been well investigated. A basic criterion is always: *resolve the geometry!* Given the nature of coastal ocean simulation, this is a serious matter.

The triangles we describe are 2-D, the (x,y) view of the coastal ocean from an airplane. The vertical dimension has different requirements. It is common to extend the horizontal from surface to seafloor in a structured way; we discuss the vertical later. To begin, we elaborate the 2-D linear triangle and develop its properties. Details are developed more fully in many texts, for example, [279].

It is evident that any discretization based on squares, rectangles, or other quadrilaterals may be reduced to a triangular one by simply dividing each element in half. Figure 7.3 illustrates an obvious example. This would pertain to the common finite difference cells, in Cartesian, polar, or other mapped coordinate system.

7.1.3 Triangle Basics

Fundamental to working with triangles is the local view – the view from within an element. In all the examples shown above, each element is the same from this local view. Globally, any point is always in exactly one element – else the location is

[1] Luettich and Westerink [267]; Lynch et al. [287]; Greenberg et al. [179]; Dupont et al. [121]; Bilgili et al. [32]; Greenberg et al. [178]; Greenberg [180].

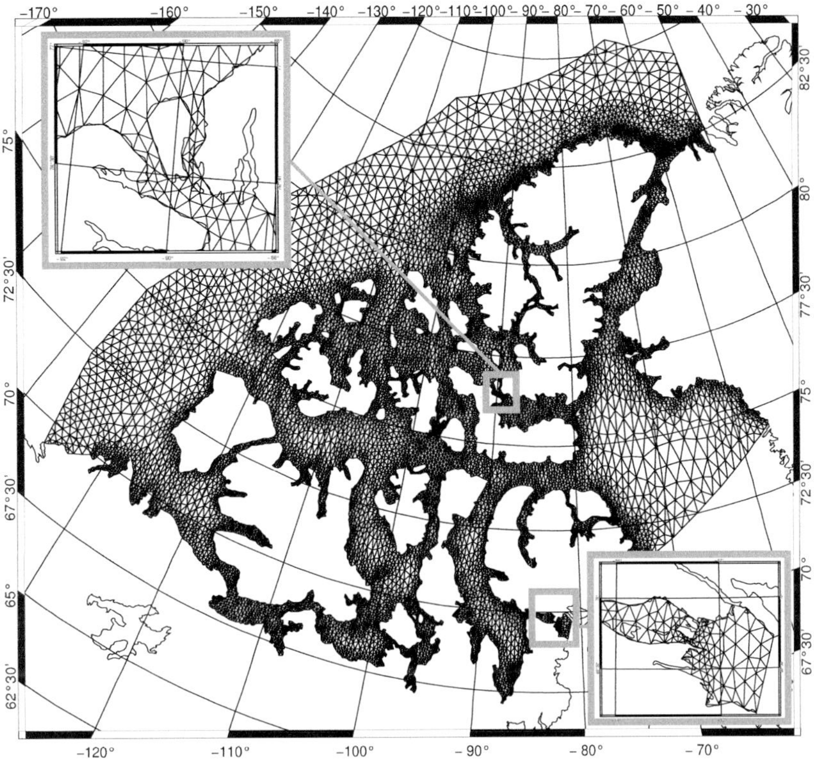

Figure 7.1. Example mesh of triangles: Canadian Arctic Archipelago. (From Greenberg et al. [178]. Reproduced with permission.)

outside the system. Phenomena are described locally; the whole is equal to the sum of its parts.

Figure 7.4 illustrates a single linear triangle. The vertices or *nodes* of this triangle are numbered locally 1 through 3, in a counterclockwise fashion as indicated. Each of the three *sides* is uniquely opposite to one of the nodes. Sides are *directed line segments*, running counterclockwise per convention. Side 1 runs from node 2 to node 3; side 2 runs from node 3 to node 1; and so on. The *area* of this triangle is

$$A = \frac{1}{2}\begin{vmatrix} 1 & x_1 & y_1 \\ 1 & x_2 & y_2 \\ 1 & x_3 & y_3 \end{vmatrix} \tag{7.1}$$

The area will be positive if all the nodes are distinct, and the counterclockwise convention is respected.

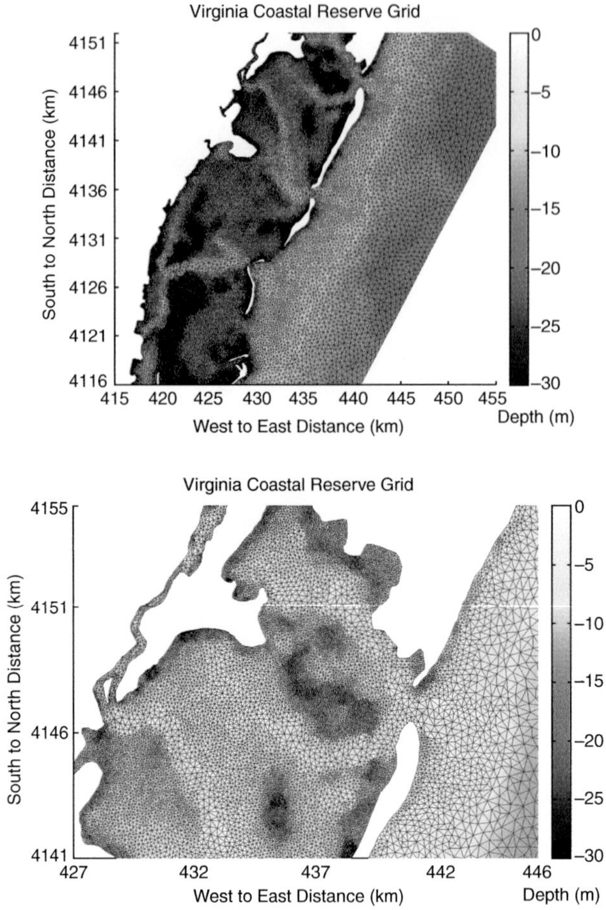

Figure 7.2. Example of a coastal mesh of triangles: Hogg Island Bay Coastal Reserve LTER, VA. As used in Bilgili et al. [32]. For a color version of this figure, please see the color plate section.

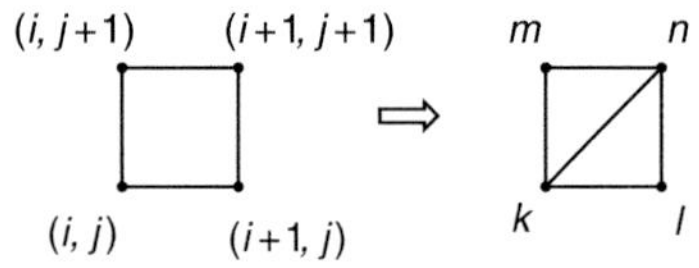

Figure 7.3. A four-sided cell converted to two triangles. The former may use a structured i,j coordinate system intrinsically linked to the Cartesian mesh spacing $(\Delta x, \Delta y)$ or a mapped version of same. The latter uses unstructured node numbering, compatible with irregular, unstructured grids.

Bases. On this triangle, three local functions $\phi_i(x,y)$ are defined. The general linear form is

$$\phi_i(x,y) = a_i + b_i x + c_i y \tag{7.2}$$

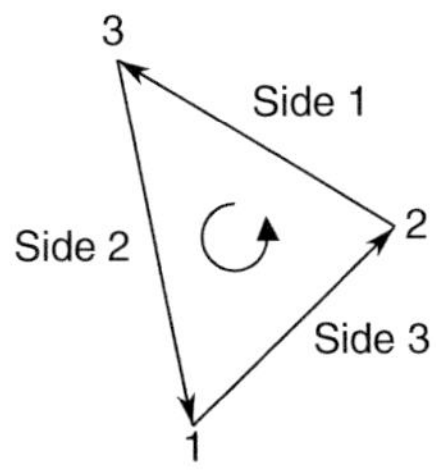

Figure 7.4. The linear triangle. There are three nodes, numbered counterclockwise. Right: There are three sides, each opposite a node: side 1 is opposite node 1, and so forth. The local coordinates ϕ_i indicate proximity to node i; distance from side i.

These are called *bases*, or *interpolants*. It is conventional and very convenient to require these local functions ϕ_i to have unit value at node i, and to vanish at the other two nodes. This gives the three conditions necessary to specify the coefficients (a_i, b_i, c_i) for each of the three ϕ_i. With these conventions, $\phi_i(x,y)$ rises linearly from 0 on side i to 1 at node i (Figure 7.6).[2] For any point within the triangle, $0 \leq \phi_i(x,y) \leq 1$.

An alternate and very intuitive interpretation of ϕ is in terms of *area coordinates*. As in Figure 7.5, for any position (x,y) on the element: indicate area A_i whose base is side i, and A the full element area as above. The ratio $\frac{A_i}{A}$ is linear in (x,y), vanishes along side i, and is unity at node i. Hence, the ϕ are in fact the area coordinates:

$$\phi_i(x,y) = \frac{A_i(x,y)}{A} \tag{7.3}$$

A_i is gotten by permuting the area formula above:[3]

$$A_1(x,y) = \frac{1}{2}\begin{vmatrix} 1 & x & y \\ 1 & x_2 & y_2 \\ 1 & x_3 & y_3 \end{vmatrix} \qquad A_2(x,y) = \frac{1}{2}\begin{vmatrix} 1 & x_1 & y_1 \\ 1 & x & y \\ 1 & x_3 & y_3 \end{vmatrix}$$

$$A_3(x,y) = \frac{1}{2}\begin{vmatrix} 1 & x_1 & y_1 \\ 1 & x_2 & y_2 \\ 1 & x & y \end{vmatrix} \tag{7.4}$$

The ϕ are independent and linear by design; but not orthogonal. By definition we have the property that

$$\sum_{i=1}^{3} \phi_i(x,y) = 1 \tag{7.5}$$

[2] Being linear, ϕ_i vanishing at nodes j and k implies that it vanishes everywhere along the line connecting them.

[3] There are other appealing area formulas, for example, these utilizing cross products.

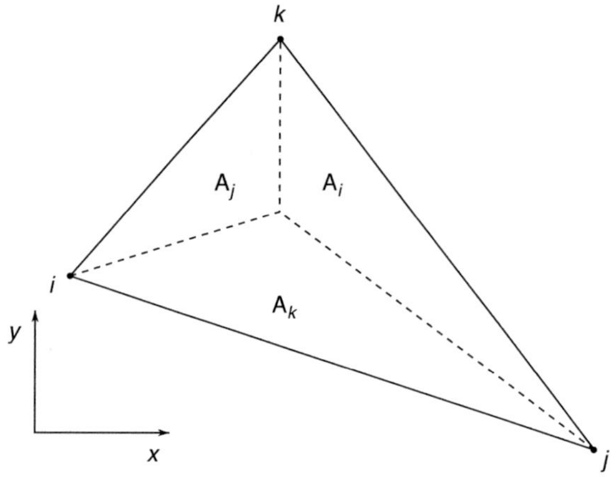

Figure 7.5. Area coordinates.

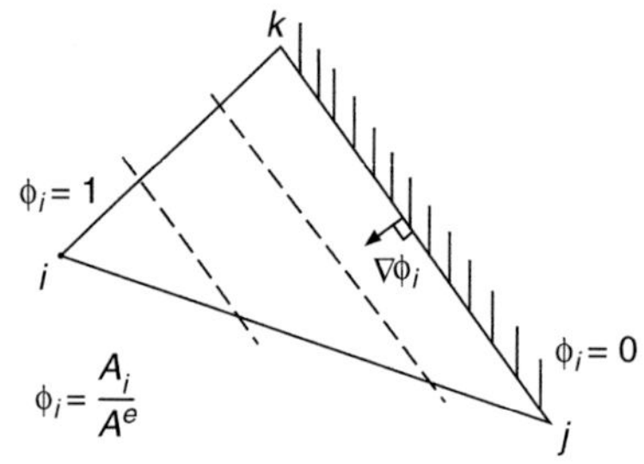

Figure 7.6. Contours of $\phi(x,y)$ are parallel to side i, opposite node i. The gradient $\nabla\phi_i$ is directed toward node i and is perpendicular to side i.

This is evident at each of the nodes or vertices; the functions being linear, the identity holds everywhere. This is particularly evident in terms of area coordinates (Figure 7.5). As a result, any one of the three, say $\phi_k(x,y)$, is constrained everywhere by

$$\phi_k(x,y) = 1 - \phi_i(x,y) - \phi_j(x,y) \tag{7.6}$$

Local Coordinates. The coordinate system natural to the triangle comprises the three linear functions $\phi_i(x,y)$. Any pair of these suffice for a local coordinate system. From inside a triangle, a global position (x,y) is equivalent to a unique local position (ϕ_i, ϕ_j). These local coordinates have the nice "peak bagging" property, that is, ϕ_i is always directed toward node i, ascending from side i where it is zero, and constant along lines parallel to side i (Figure 7.6).

Gradient. ϕ_i is linear in x and y; its gradient is constant. A convenient expression is

$$\frac{\partial\phi_i}{\partial x} = \frac{\Delta y_i}{2A} \qquad \frac{\partial\phi_i}{\partial y} = -\frac{\Delta x_i}{2A} \tag{7.7}$$

The Δ are the displacements along side i:

$$\Delta x_1 = x_2 - x_3 \qquad \Delta y_1 = y_2 - y_3 \tag{7.8}$$

$$\Delta x_2 = x_3 - x_1 \qquad \Delta y_2 = y_3 - y_1 \tag{7.9}$$

$$\Delta x_3 = x_1 - x_2 \qquad \Delta y_3 = y_1 - y_2 \tag{7.10}$$

The displacements have a "from, to" convention as one circulates along the perimeter of the element, counterclockwise. The element area may be expressed in terms of the Δ:

$$A = \sum_1^3 x_i \Delta y_i = -\sum_1^3 y_i \Delta x_i \tag{7.11}$$

The sum of ϕ is everywhere unity: so the sum of their gradients vanishes everywhere.

$$\frac{\partial}{\partial x} \sum_{i=1}^3 \phi_i(x,y) = \sum_{i=1}^3 \frac{\partial}{\partial x} \phi_i(x,y) = 0 \tag{7.12}$$

and similarly for the y-derivatives.

There is a good summary of geometric formulas for triangles in Weisstein [472].

7.1.4 Depth

Geometrically, the vertical dimension is typically represented by the bathymetric function $H(x,y)$, discretized as nodal values H_i, interpolated among them with the bases ϕ:

$$H(x,y) = \sum_1^3 H_i \ \phi_i(x,y) \tag{7.13}$$

The horizontal triangle projects downward from sea surface (the $z = 0$ geoid) to bottom ($z = -H(x,y)$) (Figure 7.7). Normally, H_i is positive; a negative value would indicate a location above mean sea level, periodically wet/dry. Each horizontal triangle represents a 3-D piece of the ocean – a prism with horizontal top plane and sloping bottom plane (Figure 7.8).

The surface plane can be tilted too – by introducing the analogous function $\zeta(x,y)$ for the sea surface elevation above datum. Generally, the top plane is $z(x,y) = 0 + \zeta(x,y)$ and ζ is interpolated like H, among nodal values:

$$\zeta(x,y) = \sum_1^3 \zeta_i \ \phi_i(x,y) \tag{7.14}$$

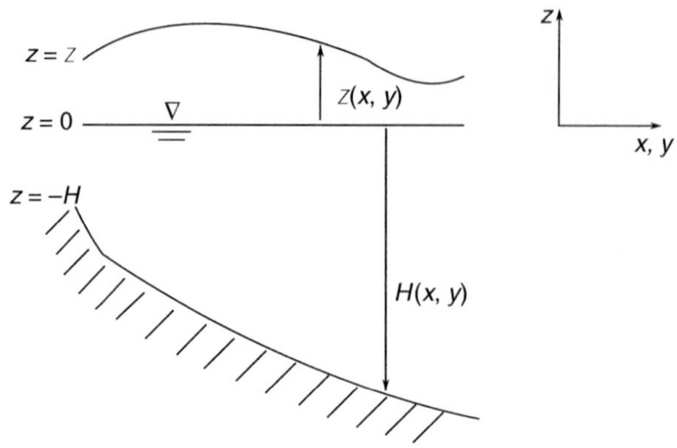

Figure 7.7. Depth conventions. Shown is the bathymetric depth H and the sea surface elevation ζ. Total depth $D = \zeta + H$. z is positive upward; the datum $z = 0$ is at mean sea level.

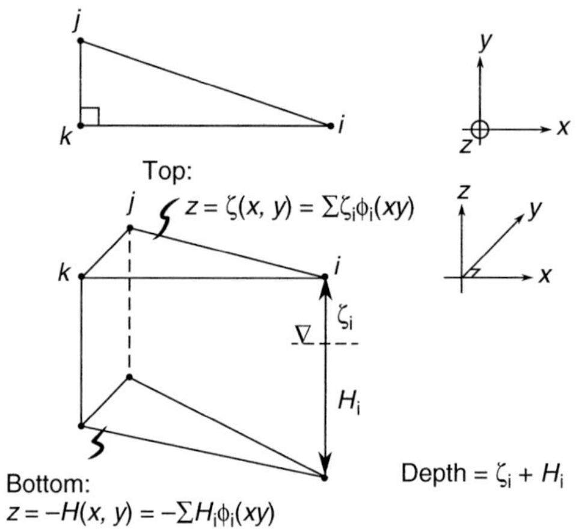

Figure 7.8. Linear triangle and its prism. On the top, a single element is shown in the horizontal plane, with sides i and j at right angles (arbitrarily, for graphical convenience). On the bottom: its projection downward to the bottom, $z = -H$. On the top, its projection upward to the free surface, $Z = \zeta$. The prism has depth $D = \zeta - H$, both parts of which are interpolated horizontally in the nodal basis ϕ. x, y, z constitute a global right-handed coordinate system. z is oriented upward, with $z = 0$ at mean sea level.

And so the full depth of the water column is

$$D(x,y) = \zeta(x,y) + H(x,y) \tag{7.15}$$

Normally ζ is a computed fluid mechanical result, the metric of hydrostatic pressure. Though critical from a dynamical point of view, ζ is kinematically unimportant *except* in the important cases where details of the surface boundary layer are of interest; or where coastal inundation or intertidal flooding is required. In most cases H_i is a constant in the absence of scouring or deposition. Offshore, $\zeta \ll H$.

The full ocean mesh comprises many such horizontal triangles. Each one represents a prism projecting a horizontal triangle vertically from sea surface to bottom. The bathymetric planes in adjoining elements will match by construction – the values of H_i are shared along each triangle side. But their horizontal slope will be constant on a triangle; it will change to a different constant in adjacent triangles. This is "C^o continuity" – the zeroth derivative is continuous globally; higher derivatives are not. The same pertains to the sea surface elevation $\zeta(x,y)$.

7.1.5 Spherical-Polar Coordinates

Greenberg [183] called attention to the weakness of Mercator projections for large meshes used for long-wave simulation, such as barotropic tides and wind response. For example, use of a Cartesian system in the Gulf of Maine was found to distort the tidal resonance in the Bay of Fundy by roughly 6%.

The basic (x,y) coordinate system described here is assumed locally orthogonal, but need *not* be Cartesian. Plane, spherical, or curved surfaces can all be triangulated.

If the horizontal nodes are located in the general spherical-polar system, then there are two paths. In the first, one describes the governing PDEs in this S-P system ahead of time, embedding the S-P transformation in the equations themselves. In the second, one uses a locally defined Mercator projection on each element. An example is Greenberg et al. [180]. Consistency here is critical; but not in itself a special computational difficulty. Additional examples of use of S-P coordinates for oceanic problems include LeProvost et al. [241], Platzman [357], and Westerink et al. [479].

7.1.6 Horizontal Interpolation

Eulerian fields require interpolation within an element if trajectories are to be properly accounted for. Examples include velocity, diffusivity, turbulent energy and mixing length, temperature, salinity, pressure, ocean depth, bottom characteristics, light availability, and so forth. These fields are conventional modeling products. They may be the object of simulation per se; they may be required input fields; or they may be supplementary information, from separate sources. All of these gridded fields require interpolation.

Interpolation on unstructured meshes follows familiar forms. In general, one defines interpolation functions $\psi_k(x,y)$ and interpolates $F(x,y)$ on an element with

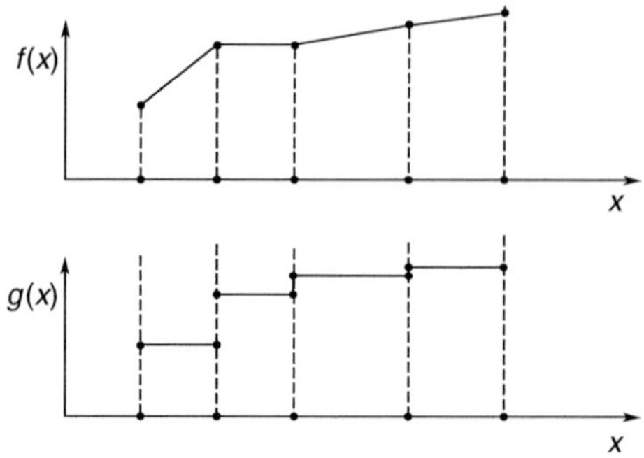

Figure 7.9. Interpolation strategies Top: Piecewise linear (nodal interpolation). Bottom: Piecewise constant (element-wise constants).

known coefficients F_k:

$$F(x,y) = \sum_{1}^{N} F_j \psi_j(x,y) \tag{7.16}$$

It is necessary to be specific in terms of the form of the interpolating functions ψ on the element in question. There are two important cases, illustrated in Figure 7.9. Both are very practical, and amount to simple forms of the general case:

1. Nodal support: The function $F(x,y)$ is expressed in the basis $\phi(x,y)$ defined already for this element:

$$F(x,y) = \sum_{1}^{3} F_i \phi_i(x,y) \tag{7.17}$$

   where F_i is a nodal value or sample of F. This amounts to linear interpolation on the element. All the properties associated with $\phi_i(x,y)$ above, are inherited by the interpolant $F(x,y)$ – specifically, F will be piecewise linear, with C^0 continuity; its gradient will be constant on an element's geometry.
   Figure 7.10 illustrates contours of a scalar with this form of interpolation. C^o continuity guarantees that the contour values match at the boundary with an adjacent element, but the slopes change abruptly.
2. Element support: The function $F(x,y)$ is approximated by a element-wide constant F^e. This is the simplest, applicable where F is assumed smooth compared to the triangle size (changes in F from one element to another are effectively small).

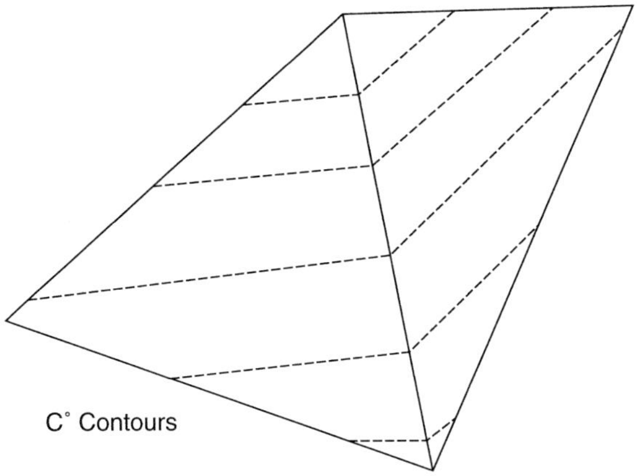

Figure 7.10. Contours for nodal interpolation with linear bases.

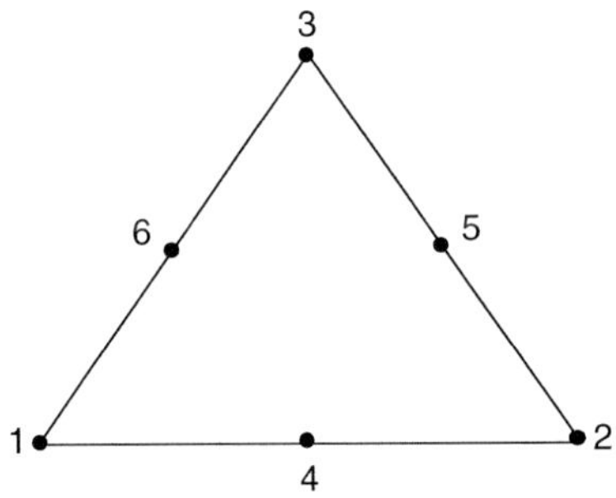

Figure 7.11. Quadratic triangle.

7.1.6.1 Higher Order Interpolation

Entire families of higher order interpolating functions ψ have been worked out and described in contemporary FE texts (for example, [279].) Typical is to use low-order polynomials expressed in the local ϕ coordinates. Functions would be characterized by (a) the polynomial order and (b) their continuity when crossing element boundaries.

For example, quadratic interpolation requires six constants $\psi(x,y) = a + bx + cy + dx^2 + ey^2 + fxy$ and would need six constraints – awkward in global (x,y) coordinates. In local ϕ coordinates, this is simple. There would be six nodes, three at triangle vertices and three midside nodes as in Figure 7.11. Each node would be home to an independent quadratic, in one of two forms:

- For a corner function: $\psi_1 = \phi_1(2\phi_1 - 1)$
- For a midside function: $\psi_4 = 4\phi_1\phi_2$.

These functions are simple. By inspection, each ψ_i is quadratic; has value unity at node i; and 0 at all other nodes. Nodal values of the interpolant at each node are all that is needed.

This family is devised to have the interpolant match at element edges, but not its first derivative. The descriptor "C° elements" describes this important feature – continuity of the interpolant (0th derivative) but discontinuous first and higher derivatives – across element boundaries.

In the polynomial family, higher continuity is possible, for example, C^1 with splines or the Hermite polynomials by arranging their first derivatives to match across element edges.

When a field is interpolated with the same basis as the geometry, ($\psi = \phi$), we have "isoparametric" elements. Quadratic velocity, for example, on linear triangles would be "superparametric."

There are significant families of discontinuous interpolants – locally continuous, simple, differentiable, and integrable – but that have discontinuities at element sides. Special handling is needed to properly account for these.

It is critical to know exactly what the interpolation scheme is intended to be for various fields. The twin levels of local variation (e.g., polynomial level) and interelement continuity are assumed to be properties embedded in gridded fields, whether overtly or by default. When there is no information about this matter, or if fields need to be invented, then a good rule is *keep it as simple as possible*, and consider simple linear interpolation.

7.1.7 Gradient

For a quantity F interpolated with the functions $\psi(x,y)$ (Equation 7.16): straightforward differentiation gives

$$\nabla F(x,y) = \sum_{1}^{N} F_j \nabla \psi_j(x,y) \tag{7.18}$$

If the basis ψ is polynomial, its derivative will be one degree simpler. In the linear case, $\nabla \psi_i$ is a constant on an element; so the gradient of an interpolant $F(x,y)$ will be a constant too, a simple weighted average of nodal values.

If the element support strategy is used, F cannot be differentiated without invoking singular functions. ($\nabla F = 0$ on the interior of the element, and is potentially singular on element boundaries.) Special considerations will need to be specified for individual fields – their structure, continuity, and how they are to be computed. Conversion to nodal values, interpolating them with C^o functions, and then differentiating is likely to be reasonable in many cases. There are several options here, beyond the present scope.

7.1.8 Location and Navigation

If a particle is to move about, it needs to know where it is, and to access all relevant gridded fields. It needs to navigate within and among elements and to interpolate constantly.

7.1.8.1 Global and Local Coordinates

For the linear triangle, global location is always an interpolant among nodal values:

$$x = \sum_{1}^{3} x_i \phi_i \qquad y = \sum_{1}^{3} y_i \phi_i \tag{7.19}$$

So: we can always go between the local and global coordinate systems:

$$(x, y) \Leftrightarrow (\phi_1, \phi_2, \phi_3) \tag{7.20}$$

This is a 1:1 reversible transformation:

- *Global to local:* Knowing (x, y), use the area coordinates (Equation 7.3; Figure 7.5) to find (ϕ_1, ϕ_2, ϕ_3).
- *Local to global:* Knowing (ϕ_1, ϕ_2, ϕ_3), find (x, y) by interpolating among nodal values (Equations 7.19).

7.1.8.2 Is a Particle in an Element?

Given the location of particle m, $(X, Y)^m$, is this particle within a given element?[4] The element basis $\phi(x, y)$ is one of the simple ways to answer to this question. Simply, if all ϕ at the location of m are within the range [0 1], then the answer is yes; otherwise, $(X, Y)^m$ is not within the element.

The area coordinates provide an intuitive geometric interpretation:

$$0 \leq A_i(X, Y)^m \leq A \tag{7.21}$$

Each A_i must pass this test (Figure 7.12).

7.1.8.3 Motion within an Element

Let the global coordinates of particle m be $(X(t), Y(t))^m$; $(\dot{X}, \dot{Y})^m = \mathbf{V}^m$ is its velocity in the global (x, y) coordinate system.[5] Let the local coordinates of that particle be

[4] Throughout, the superscript m indicates "a property of particle m."

[5] Note that we employ the convention: uppercase signifies the Lagrangian location of a particular particle; lowercase, the Eulerian coordinates.

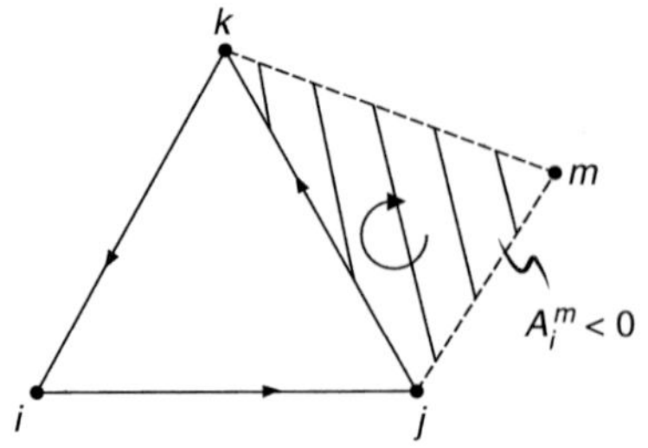

Figure 7.12. The area test for a particle outside an element. The counterclockwise convention is always assumed.

$(\Phi_1, \Phi_2, \Phi_3)^m$. By the chain rule,

$$\frac{d\Phi_i^m}{dt} = \frac{\partial \phi_i}{\partial x}\frac{dX^m}{dt} + \frac{\partial \phi_i}{\partial y}\frac{dY^m}{dt} \tag{7.22}$$

$$= \frac{\partial \phi_i}{\partial x}\dot{X}^m + \frac{\partial \phi_i}{\partial y}\dot{Y}^m \tag{7.23}$$

and so, the local "velocity" components are

$$\frac{d\Phi_i^m}{dt} = \nabla\phi_i \cdot \mathbf{V}^m \tag{7.24}$$

For triangles, $\nabla\phi_i$ are constant on an element.

7.1.8.4 When Does a Particle Leave an Element?

Particle m will exit its element by crossing one of the element sides, where by definition one and only one of the ϕ_i is zero.[6] From a starting position Φ_{io}^m, this will happen when

$$\Delta\Phi_i^m = (0 - \Phi_{io}^m) \tag{7.25}$$

for some i. The time interval Δt_e at which this happens will be

$$\Delta t_e = -\frac{\Phi_{io}^m}{\mathbf{V}^m \cdot \nabla\phi_i} \tag{7.26}$$

The non-negative minimum over i will govern; at that point one of the Φ^m, say Φ_a^m, will be zero; the other two will remain on the interval $[0,1]$; their exit values will be

[6] We are ignoring the degenerate case of a particle colliding exactly with a node, where two of the ϕ vanish.

given by

$$\Phi_a^m = 0 \tag{7.27}$$

$$\Phi_b^m = \Phi_{bo}^m + \Delta t_e \mathbf{V}^m \cdot \nabla \phi_b \tag{7.28}$$

$$\Phi_c^m = \Phi_{co}^m + \Delta t_e \mathbf{V}^m \cdot \nabla \phi_c \tag{7.29}$$

$$= 1 - \Phi_b^m \tag{7.30}$$

The particle will be exiting across side a. If the particle is entering an adjacent element, the values of all Φ_i remain unchanged there, with suitable conversion into the adjacent element's local numbering system.

7.2 Vertical Discretization

7.2.1 Separation of Variables

In the coastal ocean, the z dimension is fundamentally different from the horizontal. It is hydrostatic in all but very special cases; requires elements with extreme aspect ratios; the surface and bottom boundary layers can penetrate the entire water column and dominate the production of turbulence; and so on. As a consequence, a special form of structure is typically used to represent the vertical: typically, a form of separation of variables:

$$F(x,y,z) = \sum F_i(z)\phi_i(x,y) \tag{7.31}$$

The horizontal is assumed to be discretized into triangles; hence the use of the triangle bases $\phi_i(x,y)$. The functions $F_i(z)$ depend on location below node i.

There are two common approaches. The first uses the continuous special functions of mathematical physics to describe the vertical. The second uses simpler, piecewise local vertical functions.

In both cases, one commonly scales the vertical position by the total depth. The most straightforward is the σ transformation, where distance above bottom is normalized by the total water depth $D(x,y) = \zeta(x,y) + H(x,y)$:

$$\sigma = \frac{z+H}{D} \qquad F(x,y,z) = \sum F_i(\sigma)\phi_i(x,y) \tag{7.32}$$

σ increases linearly with z, from $\sigma = 0$ at the bottom to $\sigma = 1$ at the sea surface. σ is always positive, whereas most values of z are negative, underneath mean sea level. Both increase "upward." Figure 7.13 extends Figure 7.7 to indicate this.

Constant-σ surfaces are *not* horizontal. For example, the lowest surface ($\sigma = 0$) parallels the bottom, thus providing a natural local orientation for benthic boundary fluxes.[7]

[7] Care in differentiation is important relative to this point; the normalizing D is a function of (x,y).

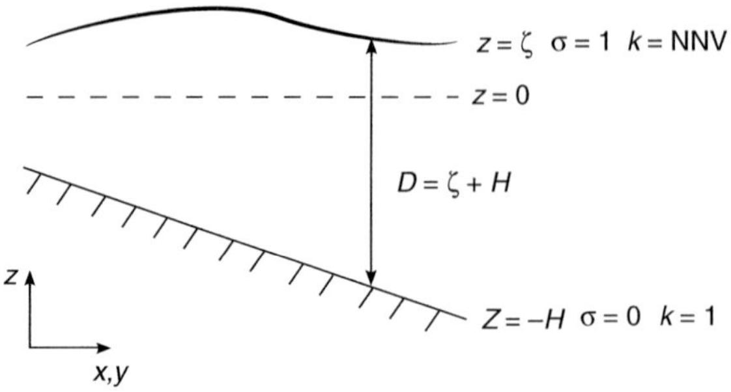

Figure 7.13. Depth conventions for the vertical coordinates z, σ, and k. σ is the local height above bottom, scaled by the local depth; k is the vertical level number. This figure generalizes that used for depth, Figure 7.7.

7.2.2 Special Functions and Global *z*-Interpolation

In the first approach, structure functions are used to describe the continuous $F_i(\sigma)$ throughout the depth. Examples include 1-D Fourier, Chebyshev, Legendre, or other special functions (e.g., Davies [99]; Davies and Xing [101]). The functions used provide the opportunity to embed boundary layers at surface and/or bottom. For example, Davies and Furnes [100] use a special linear-in-z bottom boundary layer; there are many subsequent studies.

7.2.3 Piecewise Local *z*-Interpolation

In the second approach, one defines computational "planes" at fractional depth, k-planes. The prism topography in Figures 7.7 and 7.8, where sea level ζ and sea bed H constitute the top and bottom, is readily generalized to add levels in between, at predefined fractional depths. This is done in Figures 7.13 and 7.14 for the water column beneath a single horizontal triangle, bounded by surfaces k below and $k+1$ above.

UniSigma. The simplest case is that of N_σ levels, equally spaced (including top and bottom surfaces: N_σ nodes, $N_\sigma - 1$ intervals) (see, e.g., Blumberg and Mellor 1987 [40]). In the present terminology, the vertical is discretized into vertical elements. Node locations would be $(x_i, y_i, z_{i,k})$. The vertical index k runs from 1 to N_σ and indicates a computational plane on which σ is constant; the horizontal index i, indicates an (x,y) triangle vertex. $\Delta\sigma = \frac{1}{N_\sigma - 1}$ is everywhere uniform; and Δz is vertically uniform, a fixed fraction of the depth: $\Delta z_i = D_i \Delta\sigma$. The vertical node locations beneath triangle vertex i are

$$z_{i,k} = -H_i + \sigma_k D_i \qquad \sigma_k = \frac{k-1}{N_\sigma - 1} \tag{7.33}$$

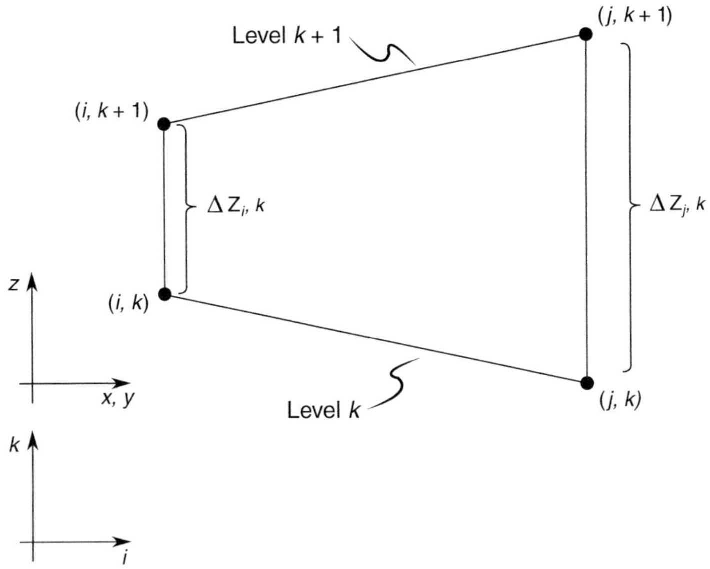

Figure 7.14. A single vertical element, viewed from the side. i and j are horizontal node numbers indicating position in a triangle. k indicates vertical position within that triangle's vertical projection. This figure complements Figure 7.8.

This constitutes a 3-D mesh! In addition to the horizontal information, a single number N_σ is sufficient.

VariSigma. The σ levels need not be equally spaced. For example, extra resolution in surface and bottom boundary layers can be accommodated with small $\Delta\sigma$ adjacent to top and bottom, and bigger $\Delta\sigma$ on the interior. Let surface k be characterized by its fractional depth σ_k. Node locations would be

$$z_{i,k} = -H_i + \sigma_k D_i \tag{7.34}$$

These k-planes would be capable of enhancing resolution in surface and bottom layers. The number of σ levels would remain constant, such that the connectivity of the surface triangles is the same on each level. Here we retain the feature that on any k-plane, $\sigma = \sigma_k$ is a constant; the σ_k are a vertical mesh property; they are the same under each horizontal node.

Mesh generation requires N_σ and a single vector of σ_k, the same fractional depths under each horizontal triangle vertex.

General. The generalization of this is the ϵ transformation defined in Lynch and Werner [276]. This provides quite general spacing of z below each triangle vertex. Importantly, there are still the same number k-planes throughout. This includes both of the above as special cases, but one relaxes the requirement that σ be constant on a k-plane. Required is N_σ values of the fractional depth underneath each vertex i: $\sigma_{i,k}$;

and/or the equivalent specification of the vertical locations $z_{i,k}$ themselves.

$$z_{i,k} = -H_i + \sigma_{i,k} D_i \tag{7.35}$$

The value of this generalization lies in the ability to retain high resolution in, for example, the surface mixed layer, when depth is large. The vertical generation becomes a blend of dimensionless fractional depth considerations, $\Delta\sigma$, and dimensional considerations of Δz.

SinGrid. A popular use of this general idea involves a sinusoidal variation in vertical spacing in surface and bottom boundary layers, expanding in the middle of the water column; but required to be limited to a maximum $\overline{\Delta z}$ at the surface and bottom, irrespective of local depth. As introduced in [275]: the 'SinGrid' transformation is

$$z_k = -H + D\left[\frac{k-1}{N_\sigma - 1}\right] - A\sin\left(2\pi\left[\frac{k-1}{N_\sigma - 1}\right]\right) \tag{7.36}$$

This is illustrated in Figure 7.15. There are N_σ vertical nodes everywhere. The mesh stretches in the center to allow coarser resolution there. The dimensional parameter A is set to achieve the desired bottom and top mesh spacings.[8] It is different under every horizontal node. The result for the layers:

$$\sigma_{i,k} = \frac{k-1}{N_\sigma - 1} - \frac{A_i}{D_i}\sin\left(2\pi\left[\frac{k-1}{N_\sigma - 1}\right]\right) \tag{7.37}$$

Typical in the applications cited is $\overline{\Delta z} = 1$ m to resolve tidal-time bottom stress; and $N_\sigma = 20$ to 40.

Other transformations are desirable and have been used – for example, high resolution in the surface mixed layer (e.g., 1 m in the top 50 m or so; and larger elements below). All produce "structured" meshes in the vertical, and "unstructured" in the horizontal – a single horizontal mesh, projected vertically onto N_σ k-planes.

7.2.4 Vertical Interpolation

With z so discretized, vertical interpolation is linear and local – specifically, in an element bounded above by level $k+1$, below by level k:

$$F(z) = F_k W_k + F_{k+1} W_{k+1} \tag{7.38}$$

$F(z)$ is a simple linear blend of its values sampled on the planes immediately above and below. The W are the two linear basis functions defined over vertical

[8] $A = \left[\frac{1}{\sin\left(\frac{2\pi}{N_\sigma - 1}\right)}\right]\left[\frac{D}{N_\sigma - 1} - \overline{\Delta z}\right]$. When $A \leq 0$ this reverts to UniGrid.

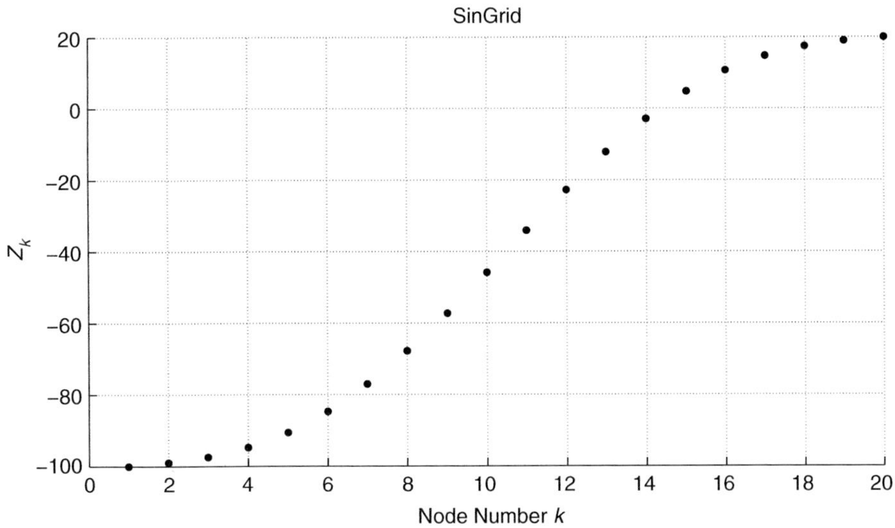

Figure 7.15. Vertical node placement via the SinGrid transformation. Plotted is vertical node coordinate z_k, versus node number k. The bathymetric depth is 100 m; the free surface elevation is $\zeta = 20$ m; there are 20 k-planes. The bottom and top mesh spacing $\overline{\Delta z}$ is 1 m. See [282, 281, 288, 325] for applications to Georges Bank, the New England coast, the Yellow Sea, and the Irish shelf.

element k:

$$W_k(z) = \frac{z_{k+1} - z}{\Delta z_k} \qquad W_{k+1}(z) = \frac{z - z_k}{\Delta z_k} \tag{7.39}$$

$$\Delta z_k = z_{k+1} - z_k \tag{7.40}$$

Notice that the values of z on these k-planes depend on (x,y) – the k-planes bounding an element are not horizontal. Accordingly, we use the notation $W_k(z|x,y)$, that is, $W(z)$ given (x,y). And we understand here that F is sampled immediately above and below, at the same values of (x,y). (F_{k+1} is sampled on the roof immediately above; F_k on the floor immediately below.)

7.3 3-D Location, Interpolation, Navigation

Within each 3-D element: the function $F(x,y,z)$ is to be interpolated among its nodal values $F_{i,k}$, with i indicating horizontal node number (triangle vertex) and k indicating surface k.

7.3.1 Interpolation on a Single Element

Given (x,y,z) is in a particular 3-D element. The local interpolation is illustrated in Figure 7.16. Horizontal nodes i and j are shown, and the two bounding k-surfaces. The interpolation is among the six values $F_{i,k}$ in that element.

Algorithmically: There are many equivalent ways to do this interpolation. A simple way to envision this is to interpolate horizontally on the triangle making up the top and bottom k-levels; then interpolate vertically between them.

1. The elevation of this (x,y) point projected onto the two k-surfaces bounding the 3-D element is

$$z_k = \sum_{i=1}^{3} z_{i,k}\phi_i(x,y) \qquad z_{k+1} = \sum_{i=1}^{3} z_{i,k+1}\phi_i(x,y) \tag{7.41}$$

2. The vertical distance between the planes, at this (x,y) point, is

$$\Delta z_k = z_{k+1} - z_k = \sum_i \Delta z_{i,k}\phi_i \tag{7.42}$$

3. The vertical interpolation functions are

$$\omega_k = \frac{z_{k+1} - z}{\Delta z_k} \qquad \omega_{k+1} = \frac{z - z_k}{\Delta z_k} \tag{7.43}$$

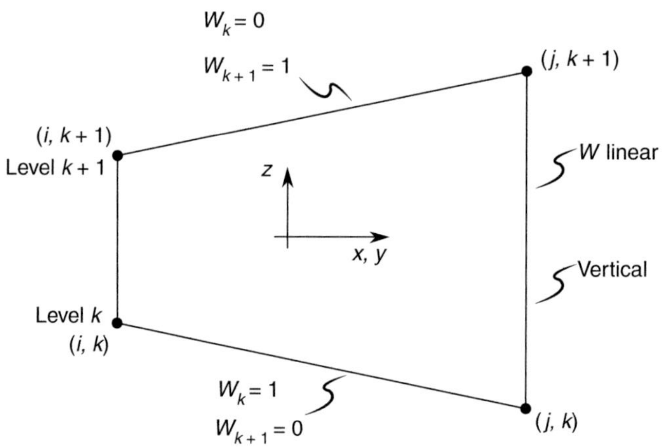

Figure 7.16. The 3-D element illustrated in the x,z plane. Bilinear interpolation involves triangular interpolation on the planes above and below, using the $\phi_i(x,y)$, and linear interpolation between them using the ω. Locally ω_1 is unity over the whole bottom plane; zero over the whole top plane; and linear in z along the sides. The reverse pertains to $\omega_2 = 1 - \omega_1$.

4. So the final interpolated value of F is

$$F(x,y,z) = \omega_k \sum_{i=1}^{3} F_{i,k}\phi_i + \omega_{k+1} \sum_{i=1}^{3} F_{i,k+1}\phi_i \tag{7.44}$$

$$= \sum_{k=1}^{2} \sum_{i=1}^{3} F_{i,k}\ \phi_i(x,y)\ \omega_k(z|x,y) \tag{7.45}$$

Effectively, the 3-D basis is the product of the mesh functions ϕ_i in the horizontal, and the linear ω_k in the vertical. Like in the 2-D case, ω_k are all both between 0 and 1 for points between surfaces k and $k+1$.

The data supporting this particular interpolation are (a) all of the triangular mesh data; (b) the number of k levels N_σ; and (c) the specific values of σ_k that are to be employed beneath each horizontal node, running from $\sigma_1 = 0$ to $\sigma = 1$ at the surface.

7.3.2 Is a Particle in an Element?

This question is prior to local interpolation. The discussion above of locating horizontally within triangles is immediately relevant and unchanged. The question of which 2 k-planes bound a particle in the vertical is an easier one, the default being a 1-D search under a given triangle.

One can do better that an all-out search. The best strategy always is to begin with knowledge of what element last contained a particular particle, before it moved. The checks of basis functions all between 0 and 1 are always relevant and reveal if a particle is still in that element. If a particle crosses a side into another element, or into a transboundary region, that should be noted immediately. (The relevant data structures are explored below.) If a particle remains unlocated, one can begin with a neighbor element search – the neighbors of the most recent known location. If that fails, then a global search is needed.

No particle is ever irretrievably lost!

7.3.3 Motion within an Element

Above we developed the horizontal motion in terms of the rate of change in the local triangle basis $\phi(x,y)$. That remains unchanged in 3-D.

The rate of change in the two vertical bases ω_k and ω_{k+1} (bottom and top) are developed in the same manner.[9] Let Ω_k^m be the value of the local vertical basis function ω_k for particle m. By the chain rule, we have the Eulerian gradient of ω_k and the

[9] We continue the conventions: the superscript m indicates "particle m"; uppercase indicates a Lagrangian property, lowercase, an Eulerian field.

Lagrangian motion $\mathbf{V}^m$ of the particle:

$$\frac{d\Omega_k^m}{dt} = \frac{\partial \omega_k}{\partial x}\frac{dX^m}{dt} + \frac{\partial \omega_k}{\partial y}\frac{dY^m}{dt} + \frac{\partial \omega_k}{\partial z}\frac{dZ^m}{dt} \tag{7.46}$$

(ω changes with x, if the 2-D surface at the element top (z_{k+1}) or bottom (z_k) is sloping.) Differentiation gives

$$\frac{\partial \omega_k}{\partial x} = \frac{1}{\Delta z_k}\left[\omega_{k+1}\frac{\partial z_{k+1}}{\partial x} + \omega_k\frac{\partial z_k}{\partial x}\right] \tag{7.47}$$

$$\frac{\partial \omega_k}{\partial y} = \frac{1}{\Delta z_k}\left[\omega_{k+1}\frac{\partial z_{k+1}}{\partial y} + \omega_k\frac{\partial z_k}{\partial y}\right] \tag{7.48}$$

$$\frac{\partial \omega_k}{\partial z} = -\frac{1}{\Delta z_k} \tag{7.49}$$

Assembling[10] gives

$$\frac{d\Omega_k^m}{dt} = \frac{1}{\Delta z_k}\left(\mathbf{V}^m \cdot \left[\Omega_{k+1}^m \nabla z_{k+1} + \Omega_k^m \nabla z_k\right] - V_z^m\right) \tag{7.50}$$

$$\frac{d\Omega_{k+1}^m}{dt} = -\frac{d\Omega_k^m}{dt} \tag{7.51}$$

The nomenclature looks complicated,[11] but the relation is simple. Vertical motion affects Ω directly – vertical velocity scaled by the element thickness Δz. Horizontal motion adds an extra effect if there is a slope to the top or bottom surface. For example, if a particle is above surface k, and that surface is sloping upward, horizontal motion alone can cause it to exit through the bottom. This is the contribution of the $\mathbf{V}\cdot\nabla z$ terms - $V_x\frac{\partial z}{\partial x} + V_y\frac{\partial z}{\partial y}$

7.3.4 When Does a Particle Leave an Element?

The conditions for a particle exiting an element horizontally were found above (Equation 7.26). They remain unchanged.

For *exit through the top* of an element: vertical motion must reach the element top surface, z_{k+1}, starting from position Z^m. Call this starting distance D_{k+1}^m, the initial *(positive)* distance to the “roof” above particle m. Horizontal particle motion adds an additional vertical distance, at the rate $\mathbf{V}^m \cdot \nabla z_{k+1}$. So the condition for the

[10] The terms in ∇z have only horizontal components, so the advective terms $\mathbf{V}\cdot\nabla$ have no vertical component; they involve only the horizontal components of $\mathbf{V}$.

[11] The superscript m attaches anything to particle m; Ω^m is the value of ω for that particle; z_{k+1} and z_k are the element top and bottom planes; $\Delta z_k = z_{k+1} - z_k$, the element thickness.

time-to-exit through the top is

$$\Delta t_e = \frac{D_{k+1}^m}{\left[V_z^m - \mathbf{V}^m \cdot \nabla z_{k+1}\right]} \tag{7.52}$$

Particle exit motion is relative to the sloping ceiling. If the ceiling z_{k+1} rises as a particle moves horizontally, the vertical motion must overcome that. If the ceiling falls, it assists the vertical motion.

The complementary condition governs if a particle is to *exit the bottom*. Let D_k^m be the initial (*positive*) distance to the bottom, $Z^m - z_k$. The condition for bottom exit is

$$\Delta t_e = \frac{D_k^m}{\left[-V_z^m + \mathbf{V}^m \cdot \nabla z_k\right]} \tag{7.53}$$

This gives two candidates for Δt_e based on vertical exit. Three additional candidates cover the possibility of horizontal exit across the triangle sides. So altogether there are five candidates for Δt_e; the smallest non-negative value will govern.

7.3.5 Summary: An Overlay of Horizontal Meshes

In 3-D, we have a separation of variables strategy, with element-based, piecewise local interpolation. The vertical is separated from the horizontal. Gridded fields will need two indices: i indicating horizontal position in an unstructured mesh; and k indicating vertical position. On surface k, the interpolation of F is as in 2-D, among horizontal elements.

$$F(x,y|k) = \sum_i F_{i,k}\phi_i(x,y) \tag{7.54}$$

The added subscript k indicates position in the vertical. Linear interpolation between the overlying and underlying k-planes handles the vertical variation. The general 3-D interpolant is

$$F(x,y,z) = \sum_k \sum_i F_{i,k}\phi_i(x,y)\omega_k(z|x,y) \tag{7.55}$$

7.4 Meshes

7.4.1 The Union of Elements

All the example meshes shown are the union of many small contiguous triangles representing the horizontal. Vertically, these triangles project between the free surface and bottom. Effectively, each triangle represents a triangular prism. Each is small and simple, and mathematically the same. Their configuration (size, orientation, connections) represents the detailed geometric complexity of the coastal system. Where there is no element, there is no ocean.

7.4.2 Essential Data Structures

The necessary and sufficient data for any mesh are

1. a complete list of nodes and their locations in (x, y)
2. a complete list of depths, by node
3. a complete list of elements indicating which nodes make up that element
4. a list of boundary segments and the type of boundary encountered there

Mesh generation is not easy. There can be mesh errors – elements that overlay each other, negative depths, elements numbered clockwise, orphan nodes or elements, absurd mesh features, and so forth – these are considered elsewhere. Here we assume a valid mesh; and in particular, how to describe it without requiring any built-in, implied structure (like, increasing node number means more eastward location; or, for doubly subscripted schemes, a necessary correspondence between the continuum (x,y) and its discrete representation (i,j)).

Nodes and Elements It is easy to imagine a *list of node locations, bathymetric depths*, and *surface elevations*: x_j, y_j, ζ_i, H_j. There are N nodes; the index j is a unique, global identity for each. Their connections are described effectively by an *incidence list*. Simply, this is a list for each element of the three nodes, in counterclockwise order, for that triangle. This, plus the location of each node, completely specifies the geometry in the global (x, y) system. Figure 7.17 illustrates the incidence list for a single element. Locally, nodes 1, 2, and 3 are actually global nodes IN(1,L), IN(2,L), IN(3,L) for element L. Generally, each element L has a four-integer string: L, IN(1,L), IN(2,L), IN(3,L).

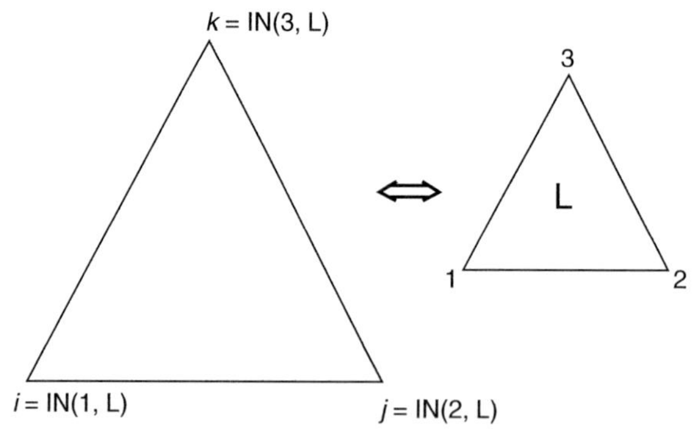

Figure 7.17. The incidence list.

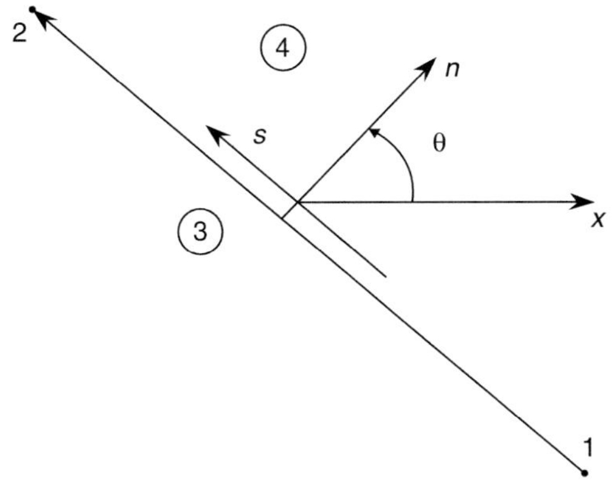

Figure 7.18. The boundary element convention: 'From', 'To', 'Left', 'Right'. Each element is a directed line segment with natural coordinates normal and tangential (n, s). The right-hand convention is assumed: the tangential direction is along the directed boundary element; normal is directed to the right. (n, s, z) is the right-handed coordinate system.

The Boundary Element. Less obvious, but nonetheless essential, is a list of boundary segments and the type of boundary encountered there. The *boundary element* is a useful concept here. It represents a hybrid of known information derivable from the incidence list, plus the extra necessary boundary info – for example, geographical information: land, water, marsh, seawall, etc; or alternately, a PDE-based classification: Dirichlet, Neuman, Radiation. Both are quite context-specific but are absolutely essential mesh descriptors.

A useful form of this list is illustrated in Figure 7.18. There are N_{be} boundary segments, each a directed straight line on the perimeter of the mesh. These are readily identified by considering all sides of all triangles – the interior sides are mentioned in exactly two triangles; the boundary sides are mentioned exactly once. (There is no other possibility; violations indicate an invalid mesh.) For each of the N_{be} boundary elements there are exactly four integers: From, To, Left, Right. The first two indicate the location of the element: from node 1 to node 2, with the node locations set out already in the node list. This is a directed line segment; it divides its left from its right. The remaining two integers identify the left and right regions: easiest is a triangular element number on the "inside," and a transboundary region type on the other. The convention we favor uses a pseudo-element number zero and less for this classification of transboundary regions. It is natural to identify the transboundary with one of a logical set of boundary conditions.

Also indicated in Figure 7.18 is the local orientation angle θ, implied by the boundary element coordinates. It can be convenient to pre-compute θ per boundary element.

Table 7.1. 5 × 5 *Adjacency Matrix for 5-Node, 3-Element Example Mesh in Figure 7.19*

	1	2	3	4	5
1	0	1	0	0	0
2	0	0	3	2	1
3	0	2	0	0	3
4	0	0	2	0	0
5	1	3	0	0	0

Note: $A_{i,j} = L$ indicates that a side of element L runs from node i to node j. There are three nonzero entries per element. Interior sides have $A_{i,j} = A_{j,i}$. The table borders are included for illustration; they indicate the from-to convention: the left border indicates row number, "from node *i*", while the top border indicates column number, "to node *j*". Adjacency entries are element numbers.

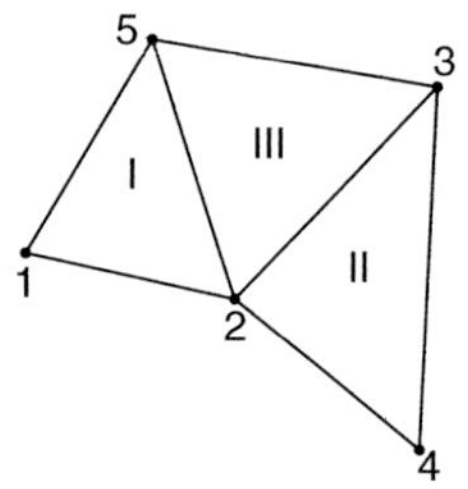

Figure 7.19. Example mesh illustrating the adjacency matrix for locating boundary elements from an incidence list.

It is easy to derive the boundary elements from the incidence list. All element sides are candidates. Most sides are shared by exactly two elements, and are therefore *not* boundary elements. The remaining sides occur only once in the mesh; these are the boundary elements. We can know from this, the element on the inside of this boundary; but not the identity of the transboundary region. For that reason, the boundary element is a hybrid of necessary transboundary information, with subsidiary information derivable from the incidence list.

An example construction is the Adjacency List A. For a single triangle L, all three sides have on their left (when circulating counterclockwise (CCW), element L. Specifically, let $A_{i,j}$ be the element enclosed by a side running from node i to node j. Looping over all elements, we have three entries A that are nonzero. An interior side will occur twice, in opposite directions, and generate symmetry in adjacency: $A_{i,j} = A_{j,i}$. A boundary side, on the other hand, will occur only once, so only one of the entries, say $A_{i,j}$ will be nonzero; while $A_{j,i}$ will remain zero. That segment (i,j) is a boundary element. We have from i, to j, left L; but no information about the right, transboundary region.

This is illustrated by the little mesh in Figure 7.19. The adjacency matrix for this would be as in Table 7.1.

This set of data structures covers the necessary and sufficient information about a mesh. Needless to say, they set up the possibility of organizing many other data items by node, by element, or by boundary element. All can be very useful.

7.4.3 Example

Figure 7.20 shows a mesh of 7 nodes, 5 elements, and 7 boundary elements. The incidence list for this mesh would be as in Table 7.2, and the boundary element incidence list in Table 7.3.

7.4.4 Subsidiary Data Structures

The data structure given above is necessary and sufficient to describe the geometry being modeled. There are many other structures that can be derived from this root.

- Neighbor nodes, by node: a list of nodes directly connected in some element.
- Neighbor elements by node: all elements sharing a given node
- Side neighbor elements, by element: at most three neighbor elements, each sharing one side
- Neighbor elements by element: the side neighbors plus those elements touching only one node
- Boundary elements and conditions – the information necessary to implement boundary conditions should a particle encounter a boundary. Ideally this is connected to the PDE boundary conditions if available.
  - In the example above we used three logical possibilities: -2 (river input), -1(ocean), 0 (land). These would need problem-specific support – for example, along the ocean is there a Dirichlet, Neuman, or Radiation BC? Along the land, is there a slip or no-slip requirement and what is the slip coefficient? And so on.
- Boundary element orientation angles θ. In the defining Figure 7.18, the convention is

$$\sin\theta = \frac{X_1 - X_2}{L} \qquad \cos\theta = -\frac{Y_1 - Y_2}{L} \qquad \tan\theta = -\frac{X_1 - X_2}{Y_1 - Y_2} \tag{7.56}$$

Table 7.2. *Incidence List for Mesh in Figure 7.20*

1	2	5	4
2	2	3	5
3	6	4	5
4	2	4	1
5	7	3	2

Note: There is one line per element; each line is, in order: element, node 1, node 3, node 3.

Table 7.3. *Boundary Element Incidence List for Mesh in Figure 7.20*

1	1	2	4	−2
2	2	7	5	−2
3	7	3	5	0
4	3	5	2	−1
5	5	6	3	−1
6	6	4	3	0
7	4	1	4	0

Note: There is one line per boundary element; each line is, in order: boundary element, node 1 (from), node 2 (to), element on left, element on right. We indicate three types of transboundary regions: land (0), ocean (−1), and riverine (−2).

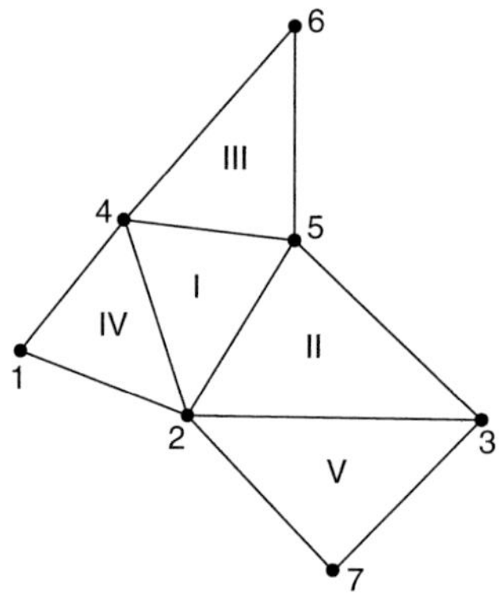

Figure 7.20. Example mesh with 5 elements, 7 nodes, 7 boundary elements. The incidence list is given in Table 7.2 and the boundary element incidence list in Table 7.3. The mesh is surrounded by land, except for (a) the leftmost boundary where it meets a river, and (b) the rightmost boundary where it meets the sea.

Table 7.4. *Neighbor Nodes, by Node, for Mesh in Figure 7.20*

1	2	4			
2	7	3	5	4	1
3	5	2	7		
4	1	2	5	6	
5	6	4	2	3	
6	4	5			
7	3	2			

Note: There is one line per node; each line is, in order: node, followed by all its neighbor nodes in CCW order.

Table 7.5. *Neighbor Elements by Node, for Mesh in Figure 7.20*

1	4			
2	5	2	1	4
3	5	2		
4	4	1	3	
5	2	1	3	
6	3			
7	5			

Note: There is one line per node; each line is, in order: node, neighbor elements in no particular order.

with L the length of the boundary element. Any vector $\mathbf{V}$ may be converted between its (x,y) and n,s expressions in the standard rotations

$$\begin{Bmatrix} V_n \\ V_s \end{Bmatrix} = \begin{bmatrix} \cos\theta & \sin\theta \\ -\sin\theta & \cos\theta \end{bmatrix} \begin{Bmatrix} V_x \\ V_y \end{Bmatrix} \leftrightarrows \begin{Bmatrix} V_x \\ V_y \end{Bmatrix} = \begin{bmatrix} \cos\theta & -\sin\theta \\ \sin\theta & \cos\theta \end{bmatrix} \begin{Bmatrix} V_n \\ V_s \end{Bmatrix} \tag{7.57}$$

By construction, the rotation matrix is nonsingular and orthonormal; and its inverse equals its transpose.

In the example (Figure 7.20): several of these have been constructed as illustrations: Tables 7.4, 7.5, 7.6, and 7.7. Some necessary additional conventions are adopted; perhaps most important is whether CCW ordering may be presumed in any of these lists, and the connection with the transboundary regions. The exercises below invite manipulation of these conventions; as subsidiary structures, they are derivative of the necessary ones so there is freedom to experiment here.

Table 7.6. *Side Neighbor Elements, by Element, for Mesh in Figure 7.20*

1	3	4	2
2	−1	1	5
3	1	−1	0
4	0	0	1
5	2	−2	0

Note: There is one line per element; element number followed by the identity of the element sharing local side 1, 2, or 3. A non-positive entry in this table indicates that side adjoins a transboundary region, as defined in the boundary element list. The ordering is CCW per element convention, and Side 1 is opposite Node 1, etc.

Table 7.7. *Neighbor Elements by Element, for Mesh in Figure 7.20*

1	4	−2	5	2	−1	3	0
2	1	4	−2	5	−1	3	
3	1	2	−1	0	4		
4	0	−2	5	2	1		
5	−2	0	2	1	4		

Note: There is one line per element; element number followed by the identity of all elements that touch a side or vertex; **no order is assumed**. This is a longer list of neighbor elements than the side neighbors; it has variable length. Transboundary regions are again indicated by the three nonpositive integers as defined in the boundary element list.

Problem 7.1 Verify that the subsidiary data structures given in the example are correct as defined.

Problem 7.2 In the example: If a counterclockwise ordering were required in the two neighbor elements, how would the respective data structures change?

Problem 7.3 In the example: Suppose the two landward boundaries needed to be distinguished by 0 (north; seawall) and −10 (south; marsh); and the CCW convention needed to be enforced. Which data structures need to change, and how?

7.4.5 The Galerkin Projection

Eulerian fields must be interpolated – the common ones share either element support (constant on an element) or nodal support (interpolated linearly between nodes).

How to get nodal support for a known function that does not lie in the nodal basis ϕ? Specifically, let $\mathcal{U}(x,y)$ be a known, field, to be approximated by the piecewise linear field $U(x,y) = \sum_i U_i \phi_i(x,y)$. The basis $\phi_i(x,y)$ is known (from the mesh!); the nodal coefficients U_i need to be found.

Clearly, a perfect match is not possible; an approximation is required, one that makes common sense and gets better with resolution. The common procedure would be to make the *local* averages match – averaging in a locally weighted sense. $\phi_j(x,y)$ is a natural choice for a weight function near node j – it is nonzero only in the elements that contain node j, and maximum at node j. The weighted average would be

$$\int\int \left\{ \sum_i U_i \phi_i(x,y) - \mathcal{U} \right\} \phi_j(x,y)\, dx\, dy = 0 \tag{7.58}$$

for each j separately. The usual shorthand employs the inner product notation $< \ >$ for the area integral. So by expansion we acquire the projection

$$\sum_i < \phi_i \phi_j > U_i = < \mathcal{U} \phi_j > \tag{7.59}$$

for each j. The weighting functions are the same as the basis for U – the *Galerkin projection*.

Some properties:

- The quantity on the right-hand side represents a local average of $\mathcal{U}$ in the vicinity of node j. Let that average be $\overline{\mathcal{U}}^j$, defined by

$$< \mathcal{U} \phi_j > \equiv \overline{\mathcal{U}}^j < \phi_j > \tag{7.60}$$

- $\overline{\mathcal{U}}^j$ is allocated among node j and its immediate neighbors according to the coefficients C_{ij}:

$$\sum_i C_{ij} U_i = \overline{\mathcal{U}}^j \qquad C_{ij} = \frac{< \phi_i \phi_j >}{< \phi_j >} \tag{7.61}$$

- The allocation will be total: $\sum_i C_{ij} = 1$ since $\sum \phi_j = 1$.
- The projection has the overall conservation property $< U > = < \mathcal{U} >$.

Globally, the original $\mathcal{U}(x,y)$ is conserved. Locally, it is partitioned completely among node j and its immediate neighbor nodes.

The Least Squares Projection. This projection is based on minimizing the integral $<(U-\mathcal{U})^2>$ by adjusting the U_j. The first-order conditions for a minimum are

$$\left\langle (U-\mathcal{U})\frac{\partial U}{\partial U_j}\right\rangle = 0 \tag{7.62}$$

Since $\frac{\partial U}{\partial U_j} = \phi_j$, this expression becomes *identical* to the Galerkin projection.

The Intuitive Projection. Suppose $\mathcal{U}$ is supported on triangular elements – as element-wise constants $\mathcal{U}_e$ – and we want a projection to the nodes, U_j, for piecewise linear interpolation. The intuitive projection is an area-weighted average of local element values:

$$U_j = \frac{\sum_e A_e \mathcal{U}_e}{\sum_e A_e} \tag{7.63}$$

where the sum is over only those elements that contain node j (Equivalently, these are the neighbor elements for node j, immediately surrounding node j.)

This amounts to the Galerkin projection, with the mass matrix "lumped," that is, the off-diagonals collapsed to the diagonal. It is especially popular when the matrix inversion implied by Galerkin needs to be avoided.

The reverse projection – from nodal support to element constants – is intuitive, $\mathcal{U}_e$ is the local average of the three nodal U_i for the element.

Problem 7.4 The uniform triangular mesh has six equilateral elements surrounding each node, all of the same size. Each node j has six neighbor nodes, and six neighbor elements. Show that the Galerkin projection of element- constant $\mathcal{U}_e$ to node j would be

$$U_j + \frac{1}{6}\sum_{i \neq j} U_i = \frac{1}{3}\sum_e \mathcal{U}_e \tag{7.64}$$

where the sums are over six immediate neighbors (nodes or elements as appropriate).

Problem 7.5 Show that in the case above (uniform triangular mesh), the lumped approximation to the Galerkin projection would be

$$2U_j = \frac{1}{3}\sum_e \mathcal{U}_e \tag{7.65}$$

7.4.6 Mesh Generators

The generation of a mesh of triangles to fill a given space is a classic problem. A central approach is the Delaunay triangulation, in which a given set of points is triangulated and the result has the property that the minimum angle among all the triangles is maximized – that is, this triangulation avoids awkwardly skewed triangles, optimally. The points to be triangulated are presumed already generated. This algorithm is public and widely implemented within almost all comprehensive packages.

The problem of point generation presents great diversity in goal and approach. For oceanic problems, one usually begins with a bathymetric data set – a set of points where H is sampled – and the goal of fitting a function $H(x,y) = \sum_i H_i \phi_i(x,y)$ without knowing the node locations i in advance. There could be a separate data set describing the shoreline; or it could be embedded in the bathymetry data as a line of zero mean depth. With the points generated a Delaunay triangulation would be constructed. Typically one enforces a practical upper limit on the number of nodes and/or elements generated; and in many cases of dense data or multiple samplings, enforces an optimal estimation (objective analysis, optimal interpolation) requirement on the final nodal H in order to deal with uncertainties, instrument precision, or regions of sparse data. Beyond that, there need to be criteria that reflect various goals; these are typically done iteratively, with visual feedback (plotting and graphical inspection of candidate meshes), manual editing, and several criteria experimented with. The problem of mesh generation is quite problem-specific, and remains highly reliant on the skill of the person doing the generation, assisted by advanced tools. (A domain-specific user interface can be very helpful – in this case, the domain of ocean modeling.)

In the point set generation, one finds methods that begin with a very dense set of points and proceed by deletion, and methods that start with a sparse set of points and proceed by insertion.

There are several very good packages. Some are in the public domain and have user groups; others are commercially available. They generally proceed by node point generation, followed by Delaunay triangulation with additional constraints, visual and heuristic inspection, and reiteration, with the goal of making smooth or uniform some mesh properties involving triangle quality, fidelity to the given database, and some aspects of PDE discretization error.

The Delaunay triangulation is fundamental: given a set of mesh points, find a contiguous triangulation that maximizes the minimum angle within any triangle. This is a reliable property, and forms the basis of many 2-D mesh generators. The algorithm is in the public domain and implementations are welltested. (For example, a very good package "Triangle" is widely utilized (Shewchuck: [411],[412],[413]) There are significant complexities in locating the mesh nodes, how to evaluate the bathymetric fit, and what criteria to blend for evaluating mesh quality.

For example, in coastal ocean applications, the governing PDEs generate well-known requirements for resolution. It is not the local mesh spacing (Δx or A) per se that matters – it is their ratio to certain characteristic lengths; and these vary across realistic meshes. This is one of the primary arguments for unstructured meshes in the first place – put the resolution where it is needed. In other words, balance the discretization error across the mesh by varying the element area appropriate to the underlying physics.

Examples include

- The topographic length scale: Where flows encounter steep topography, resolution of $L_T = \frac{H}{|\nabla H|}$ can be critical to simulating both shelf-edge flows and ultimately dominating discretization error. In the present context, this translates to $\frac{\Delta x}{L_t} = \frac{|\Delta H|}{H}$ – the maximum *relative* change in depth over an element – being required to be small uniformly. A typical practical constraint in use is

$$\frac{\Delta H}{H} < 0.3 \tag{7.66}$$

  based on shelfbreak studies (Luettich and Westerink [267]).
- The Peclet length scale: In advective-diffusive transport problems, the ratio $\frac{D}{V}$ sets a length scale requiring resolution if fronts are to be resolved. Accordingly, the numerical ratio $P_e = \frac{V\Delta x}{D}$ is the per-element quality metric; it needs to be small and balanced throughout a mesh – that is, each element needs to keep $P_e < \overline{P_e}$ for assumed speed and diffusivity. The result: all elements have the upper limit on area

$$A < \overline{P_e}\left(\frac{D}{V}\right)^2 \tag{7.67}$$

  As in the topographic length resolution, resolution needs to be balanced uniformly across a mesh.
- The wavelength resolution: For periodic motions, period P: gravity waves have wavelength $\lambda = P\sqrt{gH}$. Resolution requires that $\frac{\Delta x}{\lambda}$ be balanced below some threshold κ (a typical value: $\kappa = 25$ nodes per wavelength). So the meshing requirement in terms of element area is

$$A < \kappa^2 P^2 gH \tag{7.68}$$

  If balanced resolution of a dominant tidal constituent (known P) is needed, we have a Courant-type requirement that the ratio be constrained throughout a mesh:

$$\frac{A}{H} < \kappa^2 P^2 g \tag{7.69}$$

These and other PDE-based considerations are available to the user in mesh generation packages. They complement the strictly geometric criteria such as fidelity to the bathymetry and coastline, avoidance of sharp angles or distorted elements, conformance to selected isobaths, smoothness in the face of noisy or irregular data, and so on. (For an example see the package BatTri [32].)

In the finite element world it is conventional to compute *a posteriori* discretization error metrics, and to rebalance them by adjusting the mesh. One devises a mesh and performs a simulation. The discretization error can then be estimated and the *mesh rebuilt* to avoid unacceptable error patterns, for example, by mesh refinement where error is large. This general idea is important, but beyond this exposition. See the review by Carey [67].

In every case, final mesh judgments are heuristic and involve informed judgment. A skilled person is essential, equipped with a quality user interface and visual feedback, and easy ability to edit a mesh by hand.

7.4.7 Some Mesh Generation Packages

We give examples of some of the mesh creation software with which we are familiar.

BatTri and Resolute are freely available open source packages that share a common back end – the Delaunay triangulation program Triangle of Jonathan Shewchuk. Both BatTri and Resolute prepare data for input to iterate with Triangle to progressively produce nodal density according to different general requirements for resolving current motion based on bottom topography and topographic gradients. The user must supply an appropriate description of bottom depth and model sea and land boundaries. BatTri is run within MATLAB; some of its features are described below. Resolute is composed of a set of perl scripts and C programs. It creates meshes based on wave speed and depth gradients with controls to limit extreme resolution and abrupt resolution changes. Triangle is written in C.

Both of the above packages are essentially command line driven. BatTri has several graphical mesh editing tools included in the package. Resolute can make use of the graphical display program "showme" included in Triangle but has no editing functions based on PDE metrics. Two open source programs, Gmsh and Genesis, that use a graphical interface to create meshes and manually edit them are described below.

Genesis is a based on the package Trigrid, which does not appear to be actively developed or maintained at this time. The mesh generation principles are similar to those of BatTri and Resolute. In addition to generating and editing grids, it can be used for graphically displaying several types of data formatted for the LEGOS[12] models T-UGOm and SYMPHONIE. However, input formats are general enough that much triangle or rectangular data can be visualized.

[12] LEGOS: Laboratoire d'Études en Géophysique et Océanographie Spatiales

Gmsh is an open source package that can "... generate 1D, 2D and 3D simplicial (i.e., using line segments, triangles and tetrahedra) finite element meshes ...". It can be run interactively in a graphical mode or with ASCII text files using Gmsh's own scripting language. Two-dimensional mesh generation is a well-defined subset of its capabilities. Included in its other modes is post processing and display of output.

There are many different formats for ocean model triangular meshes. In the simplest form, the essential properties are (1) the node positions, (2) which nodes comprise which elements, and (3) what depth is associated with each node (or element). If needed, a connectivity list – which elements are adjacent to an element – is easily determined from the node and element lists, as is the outer boundary and any island boundaries (provided all islands are defined by four or more nodes). Programs for conversion between the various formats are easily written and usually easy to find.

7.4.8 Mesh Diagnostics

A valid mesh is presumed; there are many invalidities possible, any of which is enough to render a computation useless or impossible. Generally we have no universal test of validity; one looks heuristically for symptoms. Some diagnostic tests:

- Are all element areas positive?
- Do any element sides cross?
- Are there any orphaned nodes?
- Are there any orphaned elements?
- Are there any out-of range elements?
- Do any elements have poor aspect ratios?
- Do any elements have more than one side constrained by BCs?
- Are any nodes connected to an excessive number of neighbor nodes?
- Are all boundary elements consistently identified among the various lists?
- Are all boundary elements properly connected to relevant and consistent boundary condition information?

This would be relevant to discretizing the plane. The boundary description is the most important here.

There is also the requirement of properly describing the bathymetry, which requires variable resolution in itself. For several reasons one defines a dimensionless metric of resolution based on Δx or δ relative to bathymetric gradients and other length scales.

And then there are the physics-specific considerations. These introduce characteristic lengths that must be resolved; classic numerical analysis would show convergence when that is so and require a mesh length relative to these physical lengths. A measure of discretization length is $\delta = \sqrt{2A}$. Based on this length, and conventional discrete time metrics ($\Delta t, \omega\Delta t$, *etc.*) one can define several dimensionless numbers of numerical importance: Courant no., Peclet no., $\frac{\Delta H}{H}$, and so on. Bilgili et al. describe these in the context of the embedded checks and criteria in the mesh generator BatTri [32].

Other measures might include the three altitudes of each triangle. If Δs_i is a side length, $\delta = 2A/\Delta s_i$ gives three directional metrics; imbalance in any triangle indicates that the triangle is skewed.

The list of refinement requirements available in the BatTri package is instructive.

1. An absolute maximum area
2. Maxima on various per-element combinations of H, $|\Delta H|$ (the difference in H on an element), and $|\nabla H|$, including
   (a) Resolution of the topographic length scale $\frac{H}{\Delta H}$
   (b) Max. bathymetric slope
3. The CFL requirement in three forms
   (a) Advective speed for a given time step
   (b) Wave propagation speed: wavelength resolution for a given time step
   (c) Wave propagation speed: wavelength resolution for harmonic representation at given tidal frequency
4. Peclet refinement for transport problems
5. Avoid unnecessary distortions by spring relaxation

7.4.9 Graphics

Plotting on triangular meshes is straightforward.

Vector plots are facilitated by simply plotting the unstructured bundle of vectors. Rather than default to a uniform grid of vectors, the more general idea is to make each $\mathbf{V}_i$ originate at its node position $\mathbf{X}_i$. This is not difficult. One could automatically generate an initial scaling factor based on the maximum length $\mathbf{V}_i$, and a mesh scale based on the mean or minimum element "length" $\sqrt{2A}$; or to some fraction of the domain size, say 10%; and then like always, iterate for visual quality. Fundamentally, the plotting of vectors is not an issue.

Scalar contour plotting might appear more problematic, but is in fact simple. The standard strategy is to plot contours, element by element. Everything that is needed for a single-element contouring is available in that element. Doing this for all elements sequentially, in any order, assembles the entire picture. The continuity built into the interpolating functions will provide an easy clear check on the adequacy of the triangle size relative to computed gradients – for example, for C^o elements, contours or isolines will change slope at element boundaries, and if one is faithful in rendering the local interpolants, these mesh boundaries will be visible even if the mesh itself is not plotted per se. In fact, scalar plots of metrics of mesh quality are a common form of visual feedback during mesh generation itself.

Line Plots – for example, particle trajectories – may be similarly built up element-by element, one element at a time, and present no special problem.

Graphics capabilities for mesh generation and pre-processing are often built into finite element packages. Together with the Galerkin projection of mesh diagnostics, they are an essential visual feature of iterative mesh generation, *a priori*. ("pre-processing"). The same graphic routines serve the needs of *a posteriori* examination of results – whether one is looking at solutions or at metrics of mesh quality and discretization error ("post-processing").

One can rely on these simple rules in interpreting what is being plotted, supplementing it, tailoring a package to one's use. It is neither necessary nor desirable to be imprecise at the graphics level. Having been careful about how various fields are interpolated within elements, one can and should use these functions to display fields. They are by definition simple, local, and rely on element size being acceptably small. If elements are too coarse, this strategy will certainly reveal that, rather than smooth it over.

7.5 Quadrilateral Elements

The four-sided element is the natural complement to the triangle. There may be entire meshes of quadrilaterals, unstructured or structured; or there may be mixed meshes with both triangles and quads. Generally, one can divide each quadrilateral into two triangles; but it may be necessary or convenient to work directly with the quads themselves.

Quadrilaterals add a small complexity, as the local interpolation can be ambiguous. But this is readily overcome at the very start by defining the interpolation in a systematic and computable manner.

There are many instances of quadrilateral meshes. Figure 7.21 shows a boundary-fitted mesh of quads.

In all cases, the generic questions are

- How to interpolate
- How to navigate
- How to locate
- How to generate random points with given intensity
- How to project points into a density field

All of these matters need to be unambiguous, local, and simple to deal with. The first three are dealt with below; the latter two are addressed in Chapter 8.

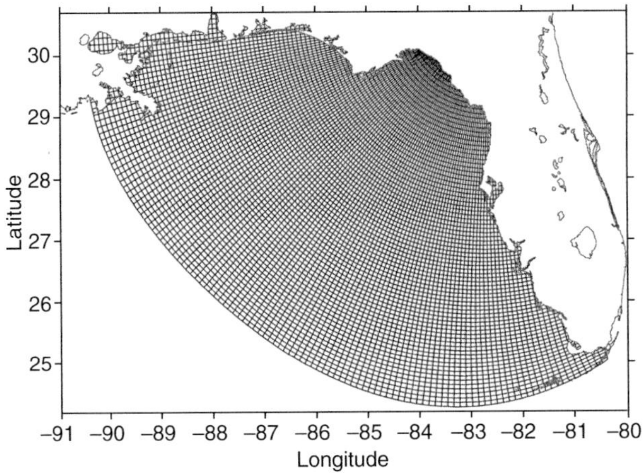

Figure 7.21. Example mesh used by He and Weisberg [194] ((c) American Meteorological Society. Reproduced with permission.)

7.5.1 Local Coordinates and Interpolation

The general quadrilateral can be mapped to an idealized one, expressed in a local coordinate system (ξ,η) (Figure 7.22). By convention, the local nodes are numbered counterclockwise as in the triangle case; this same counterclockwise convention is required in the incidence list (correspondence between local and global node number for a given element).

The simplest interpolants are bilinear in the local system:

$$\phi_i = \frac{1}{4}(1+\xi_i\xi)(1+\eta_i\eta) \tag{7.70}$$

with $\xi_i = \pm 1$ the value of ξ at local node i; and likewise for η_i (Table 7.8). These local interpolants are linear along each side[13], and amount to a formalization of the intuitive mapping shown in Figure 7.23. Everywhere, $\sum \phi_i = 1$, guaranteeing that a constant can be faithfully "interpolated."

The local derivatives are linear in the transverse direction:

$$\frac{\partial \phi_i}{\partial \xi} = \frac{\xi_i}{4}(1+\eta_i\eta) \qquad \frac{\partial \phi_i}{\partial \eta} = \frac{\eta_i}{4}(1+\xi_i\xi) \tag{7.71}$$

The mapping uses the same basis ϕ. This is the "isoparametric mapping" for this element. It guarantees a 1:1 correspondence – each point in the local element has a

[13] There is a quadratic variation generally due to the product term $\xi\eta$.

Table 7.8. *Local Nodal Coordinates for the Bilinear Element, Figure 7.22*

i	ξ_i	η_i
1	−1	−1
2	1	−1
3	1	1
4	−1	1

Note: Node numbering is in CCW order.

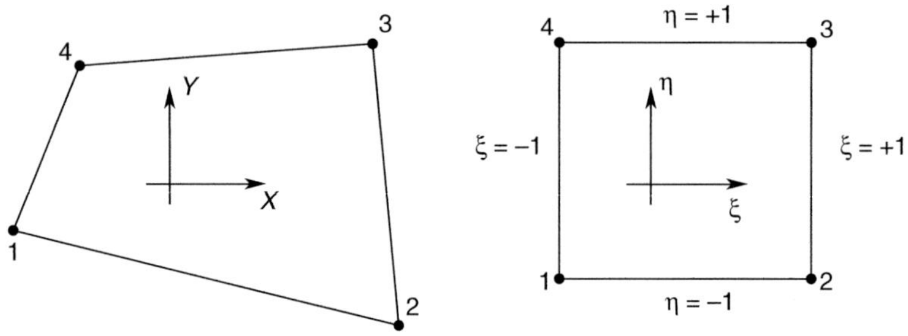

Figure 7.22. Global (x, y) and local (ξ, η) coordinate systems for a quadrilateral element. The CCW convention (right-hand rule) is assumed for both.

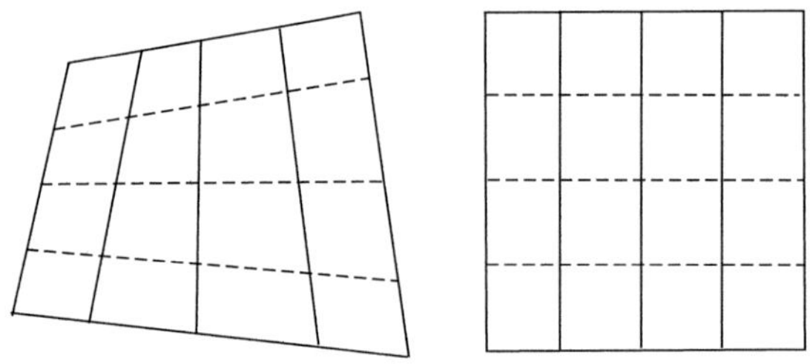

Figure 7.23. Lines of constant ξ, η in local (right) and global (left) coordinates.

unique partner in the global element.

$$x = \sum_{i=1}^{4} x_i \phi_i(\xi, \eta) \qquad y = \sum_{i=1}^{4} y_i \phi_i(\xi, \eta) \tag{7.72}$$

For this element, straight lines locally map into straight lines globally, everywhere in an element (as in Figure 7.23).

7.5.2 Jacobian

The transformation between the two coordinate systems is summarized by the Jacobi matrix:

$$[J] = \begin{bmatrix} \dfrac{\partial x}{\partial \xi} & \dfrac{\partial y}{\partial \xi} \\ \dfrac{\partial x}{\partial \eta} & \dfrac{\partial y}{\partial \eta} \end{bmatrix} \tag{7.73}$$

$[J]$ will generally vary with position within an element. Its entries are the derivatives of the coordinate positions; combining Equations 7.71 and 7.72, they are

$$\frac{\partial x}{\partial \xi} = \sum_{i=1}^{4} x_i \frac{\xi_i}{4}(1+\eta_i\eta) \qquad \frac{\partial x}{\partial \eta} = \sum_{i=1}^{4} x_i \frac{\eta_i}{4}(1+\xi_i\xi) \tag{7.74}$$

and similarly for the y-derivatives. So at any point (ξ, η), $[J]$ is a simple matter to compute, given only the element's corner locations x_i, y_i.

For a field $f(x)$, its gradient in the local and global coordinate systems are related by the chain rule:

$$\begin{Bmatrix} \dfrac{\partial f}{\partial \xi} \\ \dfrac{\partial f}{\partial \eta} \end{Bmatrix} = [J] \begin{Bmatrix} \dfrac{\partial f}{\partial x} \\ \dfrac{\partial f}{\partial y} \end{Bmatrix} \qquad \begin{Bmatrix} \dfrac{\partial f}{\partial x} \\ \dfrac{\partial f}{\partial y} \end{Bmatrix} = [J]^{-1} \begin{Bmatrix} \dfrac{\partial f}{\partial \xi} \\ \dfrac{\partial f}{\partial \eta} \end{Bmatrix} \tag{7.75}$$

For f interpolated from its corner values f_i, the result is analogous to Equation 7.74:

$$\frac{\partial f}{\partial \xi} = \sum_{i=1}^{4} f_i \frac{\xi_i}{4}(1+\eta_i\eta) \qquad \frac{\partial f}{\partial \eta} = \sum_{i=1}^{4} f_i \frac{\eta_i}{4}(1+\xi_i\xi) \tag{7.76}$$

Displacement $(\Delta x, \Delta y)$ maps to local displacement $(\Delta\xi, \Delta\eta)$. Again by the chain rule,

$$\{\Delta x, \Delta y\}\,[J]^{-1} = \{\Delta\xi, \Delta\eta\} \tag{7.77}$$

If a particle has motion U, V in the (x, y) system, its local motion will be

$$\{U, V\}\,[J]^{-1} = \left\{\frac{d\xi}{dt}, \frac{d\eta}{dt}\right\} \tag{7.78}$$

An important feature of this transformation is the deformation required from the locally square element to the more general quadrilateral in x, y. An increment of area

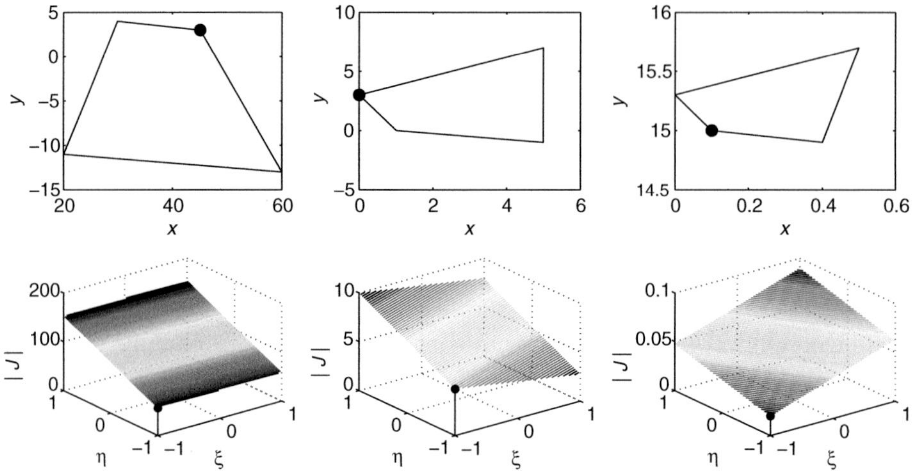

Figure 7.24. Variation of $|J|$ on three elements. Top: Element plotted in global (x,y) coordinates. Bottom: Contours of $|J|$ in the local (ξ,η) coordinates. The dot marks local node 1 and its global counterpart. For a color version of this figure, please see the color plate section.

$d\xi d\eta$ maps to a deformed increment $dx\ dy$ according the Jacobi determinant:

$$dxdy = |J|d\xi d\eta \tag{7.79}$$

This will be very important in integrating any function. Integrals in the form $\int f(x,y)\ dx\ dy$ are typically expressed in local coordinates $\int f(\xi,\eta)|J|d\xi\ d\eta$ and done numerically, by a suitable quadrature rule. Failure to include the $|J|$ term would be serious, amounting to integrating over local space, a distortion of the global. Needless to say $|J|$ has units, the ratio of local area to the global. Even for perfectly square elements, this factor is automatically accounted for in the general formulation here.

$|J|$ is also important in transferring probability density functions f between local and global coordinates. The thing conserved between the two is $fdx\ dy = f|J|d\xi\ d\eta$. Consider, for example, the problem of generating random numbers on a quadrilateral. It is an easy matter to generate local coordinates $(\xi,\eta)_k$ by invoking the uniform deviate twice. The result will be uniform locally, but *not globally*. A simple strategy is to reject some of these particles in the "compressed" areas. The rejection criterion is $|J|$. This is taken up in Chapter 8.

Figure 7.24 illustrates the large variation of $|J|$ for some simple elements.

The Jacobi determinant will be bilinear for these elements. Therefore it can be interpolated to any ξ,η location among its corner values $|J|_i$ and the basis ϕ. The corner values are properties of the mesh and can be pre-computed once, then stored.

There are many formulae for these geometric calculations. For example, a particularly simple formula for the nodal Jacobian[14] is

$$|J|_i = \frac{1}{4}\left[(x_{i+1}-x_i)(y_{i-1}-y_i)-(y_{i+1}-y_i)(x_{i-1}-x_i)\right] \tag{7.80}$$

(As usual, the CCW numbering convention pertains.) This can be pre-computed for the purposes of scattering particles.[15] The general isoparametric theory given here is capable of all necessary manipulations consistently. It is the standard finite element way of handling quadrilaterals, widely implemented.

For the area of a quadrilateral, one of the several approaches is to evaluate the cross product of its two diagonals, $\mathbf{d}_1$ and $\mathbf{d}_2$. The area is

$$A_e = \frac{1}{2}|\mathbf{d}_1 \times \mathbf{d}_2| \tag{7.81}$$

The mean Jacobian determinant can be obtained from Equation 7.79:

$$\overline{|J|} = \frac{A_e}{4} \tag{7.82}$$

Finally, if everything is to be done locally, one element at a time, a potentially limiting step is discovering "what element is a particle in?" So the best specification of a particle's position includes the element number plus the local address. It is good practice to never leave a particle without knowing its element. So if a particle will exit an element in a time step, always know where it exited, and where it ends up. Some elementary bookkeeping at this point is extremely important regarding efficiency.

7.5.3 Locating

To find which element contains a "lost" particle at (X,Y), there are many useful formulae utilizing angles, lengths, or line crossings. One simple way is to compute the area of the four subtriangles made between (X,Y) and the sides of an element, always retaining the CCW convention for the element sides. A particle inside a quadrilateral will have all these four areas nonnegative; otherwise, the particle is outside. These types of tests require searches among many elements; needless to say, knowing which element that particle was recently seen in, is very valuable – it reduces the search list to neighbor elements, or neighbors of elements, and so on.

Mapping a point $(\xi,\eta)_k$ into physical space $(X,Y)_k$ is natural and simple, Equation 7.72. The reverse is a little less so; the isoparametric mapping is still unique, but

14 $|J|_i$ pertains to a particular element; generally, this quantity will be discontinuous across element boundaries without suitable restrictions.

15 For scattering on an element, one needs the element maximum of the nodal $|J|_i$ for the rejection criterion – this is an element property derived from the basis element geometry.

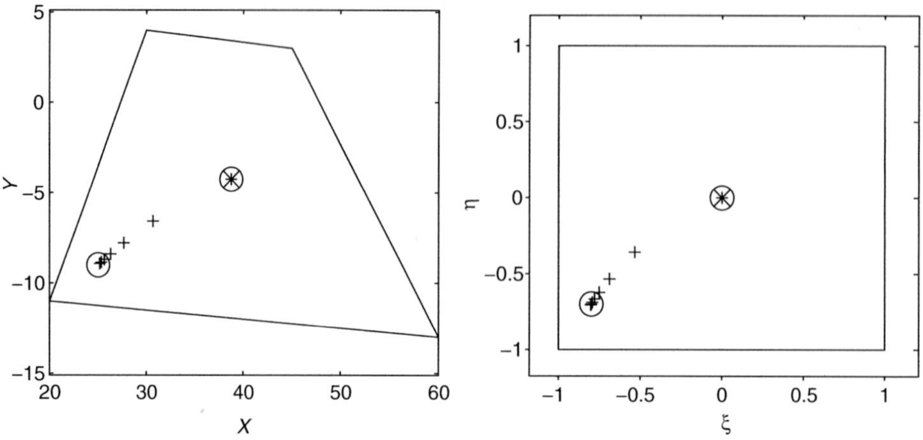

Figure 7.25. Iteration to compute (ξ, η) corresponding to a given location (x, y). The iteration started at the center (bull's eye) and proceeded by Newton–Rhapson iteration to the end point (circle). A relaxation factor $\frac{1}{2}$ was used to avoid getting outside the element. $|J|$ varied from an initial value of 105 to 140 at the endpoint. Curvature in the displacement is visible but very small.

nonlinear in this "reverse" direction. An iteration is needed, and the Jacobi relation (Equation 7.77) can easily be adapted to this task. Figure 7.25 gives an example. The initial estimate was the center of the element; the iteration was relaxed by a factor of $\frac{1}{2}$ to expose the process; generally, one needs to guard against local coordinates going out of range, so a little care is needed to make a procedure reliable.

8

Particles and Fields

8.1 Introduction

There is a discrete set of particles at locations $\mathbf{X}_k$ with an implied/equivalent density field $\rho(\mathbf{x})$. The $\mathbf{X}_k$ represent singularities in the mass field. ρ is assumed to be expressed in a smooth basis with parameters ρ_j, on an unstructured mesh. The correspondence sought is between the two discrete sets

$$\mathbf{X}_k \Longleftrightarrow \rho_j \tag{8.1}$$

along with practical procedures for going between them.

The context is particle simulation. On input, we need to be prepared to generate particle sets from either hypothesis or from data products. On output, we need to interpret simulated particle outcomes. During simulation, we need the capability of generating a field as middleman for particle interactions, as described in Chapter 1.

Colloquially, generating random positions from a density function is "scattering." It requires random number generation on unstructured grids. Constructing a density field from a set of particles on an unstructured grid, is "gathering." The gathering procedure can be formalized as the *projection* of the particle positions into a continuous function space ϕ.

Notation:

$\rho(\mathbf{x})$ is the density.
$\nu(\mathbf{x})$ is the particle *intensity*, number of particles per unit area.
$\mathbf{X}_j$ is a specific location in $\mathbf{x}$, the position of the jth particle.
M is the mass of an individual particle.
$\phi_i(\mathbf{x})$ is the ith basis function, centered at node i.
$\delta_j \equiv \delta(\mathbf{x}-\mathbf{X_j})$ is the Dirac delta function, singular at particle position $\mathbf{X_j}$.
A_e is the area of element e.
$< \; >$ indicates spatial integration; $< \; >_e$ is the integration over one element only.

8.2 Scattering among Elements

There will be N particles distributed over all elements. For a given $\rho(\mathbf{x})$ the mass balance requirement is

$$N = \frac{<\rho>}{M} \tag{8.2}$$

assuming constant M. This will be good to $\mathcal{O}(1)$, since N must be an integer. For large N, we can safely ignore this rounding effect.[1]

The N particles will be distributed among elements according to intensity $\nu(\mathbf{x}) \equiv \rho(\mathbf{x})/M$. This is a classic problem. For a particle, the probability of ending on element e is $p_e = \frac{<\nu>_e}{<\nu>}$. This guarantees that $\sum p_e = 1$. Let the particle sum on element e be N_e. There are many combinations of particles that yield the same set of N_e. The multinomial distribution accounts for this:

$$\{N_e\} \sim \mathcal{M}(N, \mathbf{p}) \tag{8.3}$$

This is the generalization of the binomial distribution. Properties:

$$\overline{N_e} = Np_e \qquad \sigma^2_{N_e} = Np_e(1-p_e) \tag{8.4}$$

$$Cov(N_d N_e) = -Np_d p_e \quad (d \neq e) \tag{8.5}$$

These are generally intuitive; the covariance among elements is a consequence of the zero-sum constraint on their variations.

Covariance and Resolution. In scaled form:

$$\frac{\sigma^2_{N_e}}{\overline{N_e}\,\overline{N_e}} = \frac{(1-p_e)}{Np_e} = \frac{(1-p_e)}{\overline{N_e}} \tag{8.6}$$

$$\frac{Cov(N_d N_e)}{\sigma^2_{N_e}} = -\frac{p_d}{(1-p_e)} \qquad (d \neq e) \tag{8.7}$$

Each p_e scales inversely with the number of elements E. As E is increases, the off-diagonal Cov terms scale as σ^2/E. Increasing mesh resolution drives σ^2 to up to $\overline{N_e}$; and decreases individual covariance by spreading it over more elements.

Figure 8.1 is a 1-D sample distribution of N_e for a linear $\nu(x)$. Shown is the expected distribution and a one-sigma envelope: $\overline{N_e} \pm \sigma_e$ versus e.

As an alternative to the multinomial, one could employ a rejection strategy. For E elements and p_e as defined above:

[1] This pertains to the global mass balance; the graininess of N becomes an issue for individual elements when the particles per element are few – either due to low ρ or low Δx.

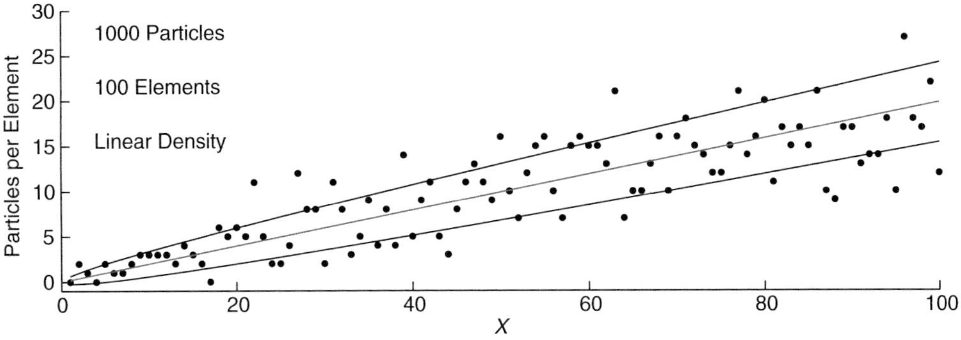

Figure 8.1. One multinomial sample of N_e. The horizontal axis X is divided into equal-length elements. The dots are the individual element counts N_e (number of particles allocated to element e). The lines indicate the expected $\overline{N_e} \pm \sigma_e$, Equation 8.4.

1. Generate a discrete uniform deviate $e \sim \mathcal{U}_D(1,E)$. This is a trial assignment of one particle to element e.
2. Generate a rejection deviate $\eta \sim \mathcal{U}$.
3. Accept this assignment if $\eta \leq p_e$.
4. Otherwise, reject the assignment and repeat.
5. Continue until N assignments have been accepted.

This is a familiar rejection algorithm, useful if the multinomial is not available. Note the use of the discrete distribution $\mathcal{U}_D$ for the element assignment, and the continuous $\mathcal{U}$ for the rejection. Steps 2 and 3 could be further consolidated by using the binary Bernoulli deviate $i \sim \mathcal{B}e(p_e)$; reject if $i = 0$.

8.3 Scattering within an Element

Given N_e, we proceed to scatter particles within each element separately. The simplest distribution is uniform on an element; nonuniform distributions can always proceed from this by rejection (Section 3.3).

On rectangles, uniformity is easily realized as the 1-D uniform distribution in two independent directions, each mapped to local $[0,1]$ coordinates. Figure 8.2 illustrates and this does not need any special description; independent 1-D uniform deviates are intuitive and work. Parallelograms also work in the same way.

The more general quadrilateral, with nonparallel sides, cannot be treated in this way and some care is needed. Simple ideas about 1-D uniform generation along two dimensions end up concentrating particles inadvertently in areas where local coordinates converge (Figure 8.3).

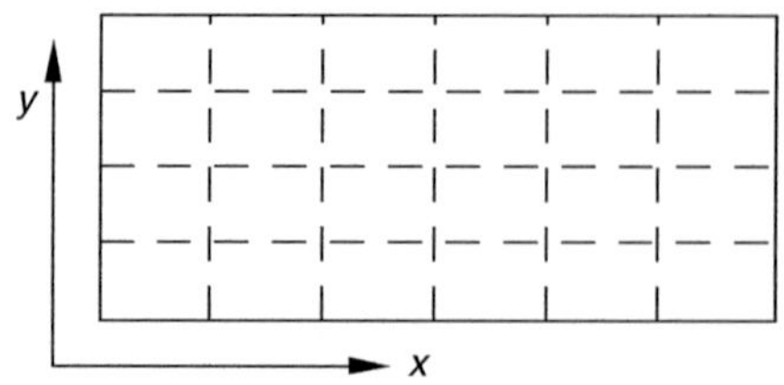

Figure 8.2. A rectangle. Uniform deviates in both directions achieve the 2-D uniform distribution.

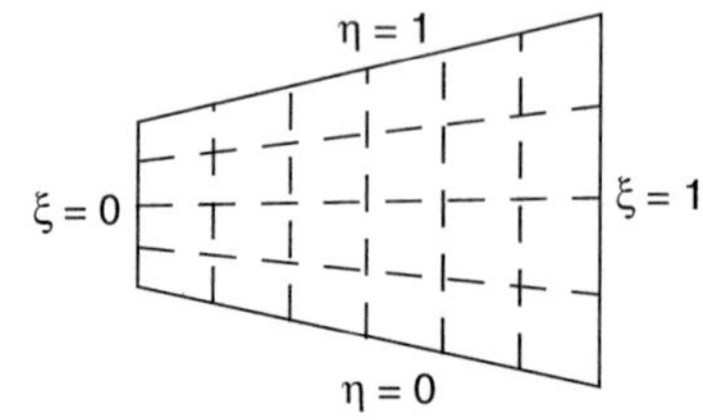

Figure 8.3. A quadrilateral with intuitive local coordinates ξ, η. The dashed lines are lines of constant ξ or η. They converge, and uniform distribution along both local directions cannot achieve uniform 2-D distribution globally.

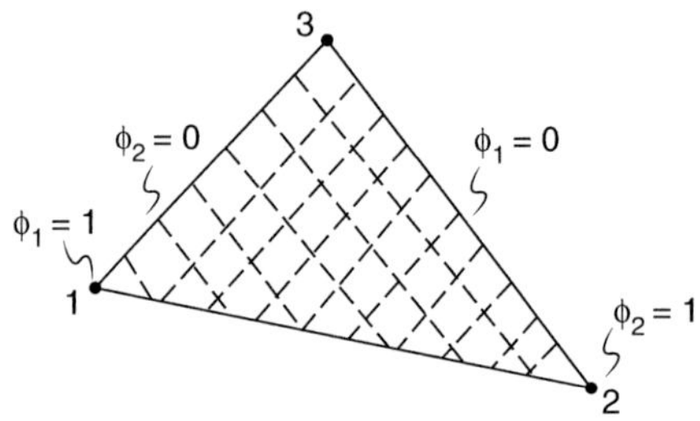

Figure 8.4. A triangle with local coordinates ϕ_1, ϕ_2. The dashed lines are lines of constant ϕ. The coordinates are independent and map the entire triangle without ambiguity. But uniform distribution along these lines produces a convergence of points.

In the case of the triangle, independent local coordinates (the "area coordinates" ϕ_i) are in common use; but they do not map to $[0,1]$ independently; the constraint $\phi_1 + \phi_2 \leq 1$ is necessary or else the location falls outside a given triangle (Figure 8.4).

Both of these elements are quite simple, and the following procedures are structured to avoid the unintended effects associated with unstructured scattering. Between them they are adequate for most 2-D meshes. 3-D elements are the natural extensions.

8.3.1 Triangles

We will generate the positions of particles in the local coordinates. As introduced in Chapter 7, these are the three linear bases ϕ_i. The property $0 \leq \phi_i \leq 1$ holds for each

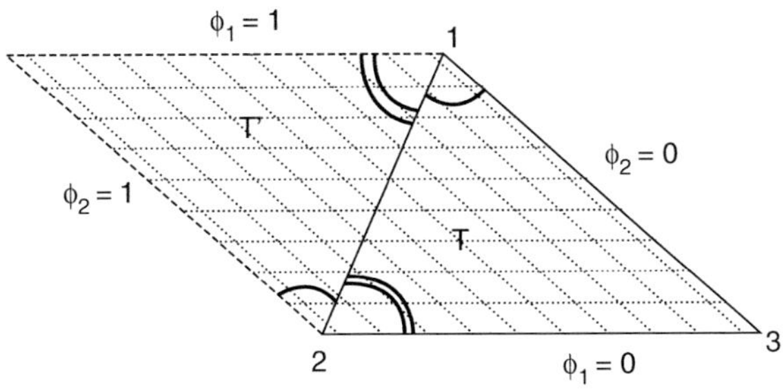

Figure 8.5. Generation of uniform distribution on triangle T and its mirror image T'.

of the three. Only two of these are independent; the constraint $\sum \phi_i = 1$ always holds, everywhere, so $\phi_3 = 1 - \phi_1 - \phi_2$ by definition. To be precise, we indicate the specific numerical location of the kth particle as ϕ_{ik}.

Hence, we generate valid pairs of the local coordinates (ϕ_{1k}, ϕ_{2k}) for a particular particle. Once the local coordinates are known, the global location $\mathbf{X}_k$ of each particle can be interpolated whenever needed from the nodal values $\mathbf{x}$ for that element:

$$\mathbf{X}_k = \mathbf{x}_1\phi_{1k} + \mathbf{x}_2\phi_{2k} + \mathbf{x}_3(1 - \phi_{1k} - \phi_{2k}) \tag{8.8}$$

The $\mathbf{x}_i$ are constant and already assumed part of the mesh database.

8.3.1.1 *Uniform Distribution on a Triangle*

This is the most basic operation, the triangle equivalent of the uniform deviate. Let n be the number of points to be generated.

Begin by supplementing the triangle T to a parallelogram by constructing T' as in Figure 8.5. The two triangles are congruent. Fill the 2-D space with n independent pairs $\phi_1 \sim \mathcal{U}[0,1]$, $\phi_2 \sim \mathcal{U}[0,1]$. Effectively, one has generated two useful sets of locations, half of them in T and the other half in T'. If $\phi_1 + \phi_2 \leq 1$, then the point (ϕ_1, ϕ_2) is in T; assign that location to particle k. If $\phi_1 + \phi_2 \geq 1$, then the point lies in T'; use its reflected presence in T: $(1 - \phi_1, 1 - \phi_2)$ for particle k. This achieves uniform distribution on any triangle by recourse only to the uniform deviate and a reflection rule. Everything is done in local coordinates. All random points generated are used. If n points are needed, then $2n$ independent uniform deviates are needed and the survival rate is 100%.

An alternative is to simply reject the points generated in T', by the same criterion; keep the point if $\phi_1 + \phi_2 < 1$, otherwise discard the point. In this case, to get n points, $4n$ independent uniform deviates are needed. On average, the survival rate is 50% – half of the particles generated are rejected. The comparison is unfavorable but not overwhelming; other factors are likely to govern the choice between these two equivalent methods.

8.3.1.2 Linear Distribution on a Triangle

The desired particle distribution $\rho(\mathbf{x})$ is given – on an element, the nodal densities ρ_j, with linear interpolation among them. We will generate N_e particles and require only the uniform deviate. The method relies on rejection. (The rejection method is quite general and discussed in Section 3.3.)

$\overline{\rho_e}$ is the mean density on an element. The highest density on this element is ρ^*:

$$\overline{\rho_e} = \frac{1}{3}(\rho_1 + \rho_2 + \rho_3) \qquad \rho^* = max(\rho_1, \rho_2, \rho_3) \tag{8.9}$$

1. Generate a particle from a *uniform* distribution on the element, as described previously. Call this candidate particle k.
2. Find the density at that location: $\rho_k = \rho_1\phi_{1k} + \rho_2\phi_{2k} + \rho_3\phi_{3k}$
3. Generate a rejection criterion $u_r \sim \mathcal{U}[0,1]$. If $u_r > \rho_k/\rho^*$, then reject this particle; otherwise, accept it.

Repeat steps 1–3 until N_e points have been accepted. This gives the desired linear distribution for $\sum_e \rho_j\phi_j(\mathbf{x})$ on this element. The fraction of final acceptances will be, on average, $\overline{\rho_e}/\rho^*$ for the element.

If n particles are to survive the final rejection: a uniform distribution of $n_u = n\frac{\rho^*}{\overline{\rho_e}}$ is needed. If the uniform distribution is itself achieved by rejection, $4n_u$ random deviates are required, plus one each for the final selection. Hence, to achieve n particles, we would need $5n\frac{\rho^*}{\overline{\rho_e}}$ random deviates; and the overall survival rate would average $\frac{1}{2}\frac{\overline{\rho_e}}{\rho^*}$.

8.3.1.3 Examples

As a demonstration, we generated finite ensembles of particles for the constant and linear cases; measured their moments; and compared them to the analytic values. (The analytic moments are readily computed using standard triangle procedures, and given in Appendix H, Table H.1.)

Uniform Distribution. Ensembles of 400 and 1600 particles were generated on a triangle and illustrated in Figure 8.6. The distributions are visually correct. The moments of the positions are given in Table 8.1.

Linear Distribution. The procedure was repeated for a triangle with nodal densities [1, 0, 0]. The result for $n = 400$ and 1600 appear in Figure 8.7. Again, the distribution is visually appealing. The measured moments appear in Table 8.2. In this case, the distribution is not symmetric, so the various moments depend on node number. The right (exact) side of Table 8.2 is expanded accordingly.

Table 8.1. *Properties of the Uniform Distributions Shown in Figure 8.6*

0.3394	0.3064	0.3542	μ	0.3333
0.0603	0.0565	0.0620	σ^2	0.0556
−0.4696	−0.4920	−0.5376	r	−0.5000
0.6227	0.7320	0.5098	sk	0.5657
0.3366	0.3275	0.3359	μ	0.3333
0.0575	0.0546	0.0546	σ^2	0.0556
−0.5126	−0.4737	−0.5133	r	−0.5000
0.5189	0.5968	0.5734	sk	0.5657

Note: Properties obtained are on the left, by node. On the right are the analytic values (Appendix H, Table H.1), which are independent of node number for the uniform distribution. The quantities r represent the three off-diagonal entries in the correlation matrix r_{ij}.Top: $n = 400$. Bottom: $n = 1600$.

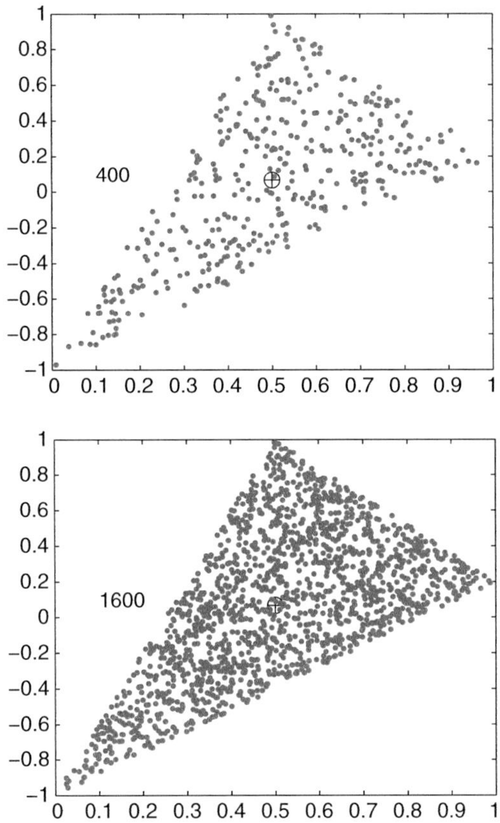

Figure 8.6. Uniform distribution. The ensemble size is shown. The bull's eye marks the centroid of the triangle for reference.

Table 8.2. *Properties of the Linear Distributions Shown in Figure 8.7*

0.4882	0.2590	0.2528	μ	0.5000	0.2500	0.2500
0.0510	0.0393	0.0369	σ^2	0.0500	0.0375	0.0375
−0.5965	−0.3307	−0.5602	r	−0.5774	−0.3333	−0.5774
0.1383	0.8536	0.8178	*sk*	0	0.8607	0.8607
0.5078	0.2527	0.2395	μ	0.5000	0.2500	0.2500
0.0520	0.0385	0.0375	σ^2	0.0500	0.0375	0.0375
−0.5928	−0.3160	−0.5767	r	−0.5774	−0.3333	−0.5774
−0.0927	0.8691	0.9919	*sk*	0	0.8607	0.8607

Note: Actual moments are on the left, by node. On the right are the analytic values, by node number (Appendix H, Table H.1). The quantities r represent the three off-diagonal entries in the correlation matrix r_{ij}. Top: $n = 400$. Bottom: $n = 1600$.

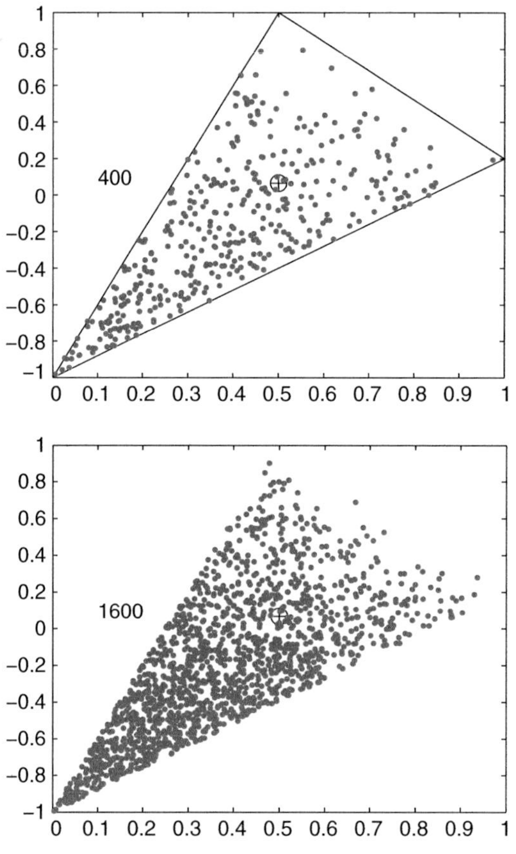

Figure 8.7. Linear distribution, rejection method. $v_j = [1,\ 0,\ 0\ ]$. The ensemble size following rejection is shown. Rejection rate 0.8342 (left), 0.8323 (right).

8.3.2 Quadrilaterals

For these elements, the issues are complicated by the local-global transformation as reviewed in Section 7.5. The Jacobi transformation plays a prominent role. Real, nonsquare elements are mapped into local squares; so uniform distribution locally will be distorted globally. $|J|$ is the metric of distortion. If f is a probability density function, the probability of a particle occurrence on an incremental parcel is $fdxdy$; in local space this is $f|J|d\xi\ d\eta$. This is the gateway for a rejection-based generation strategy on a quadrilateral.

Specifically, to generate a uniform distribution:

1. Start with the uniform 2-D distribution in local space.
2. For each particle, generate a rejection criterion $R = \frac{|J|}{|J|_m}$ with $|J|_m$ the largest nodal value in the element
3. Generate $\epsilon \sim \mathcal{U}$.
4. Accept the particle if $\epsilon < R$; otherwise reject the particle.

This will avoid the artificial concentration of particles where $|J|$ is low, and achieve a uniform distribution in (x, y).

An example appears in Figure 8.8. 899 particles were generated locally and mapped to the (x, y) space. The plot of these shows the expected artificial concentration at the narrow top of the element. 299 particles were rejected, and as the figure shows, the rejects favor the top. The survivors (approximately 2/3) are uniformly distributed. For each particle there were three random deviates needed (two positions and one rejection test) ; per survivor, 4.5 deviates.

If variable intensity ν is wanted on a quadrilateral, this procedure generalizes with the rejection based on $\nu|J|$.

8.4 Projections: The Density of a Set of Particles

Given a set of particle locations and masses (X_k, M_k), what is the equivalent density field?

The "density" of a single particle of mass M_k is a Dirac delta function $M_k\delta_k$. (δ_k is the Dirac singularity at location k.) Superposition of a set of n particles gives the composite density. In the continuum, the approximation is in terms of the nonsingular bases $\phi_i(\mathbf{x})$, weighted by nodal or elemental values ρ_i. These two cannot be "equal", but we seek their weak equality:

$$\sum_{k=1} M_k\delta_k \Longrightarrow \sum_i \rho_i\phi_i(\mathbf{x}) \tag{8.10}$$

Essentially, we seek a principled accommodation of incompatible rivals. This is the "projection" from the singular basis δ into the smooth basis ϕ. Desirable attributes include

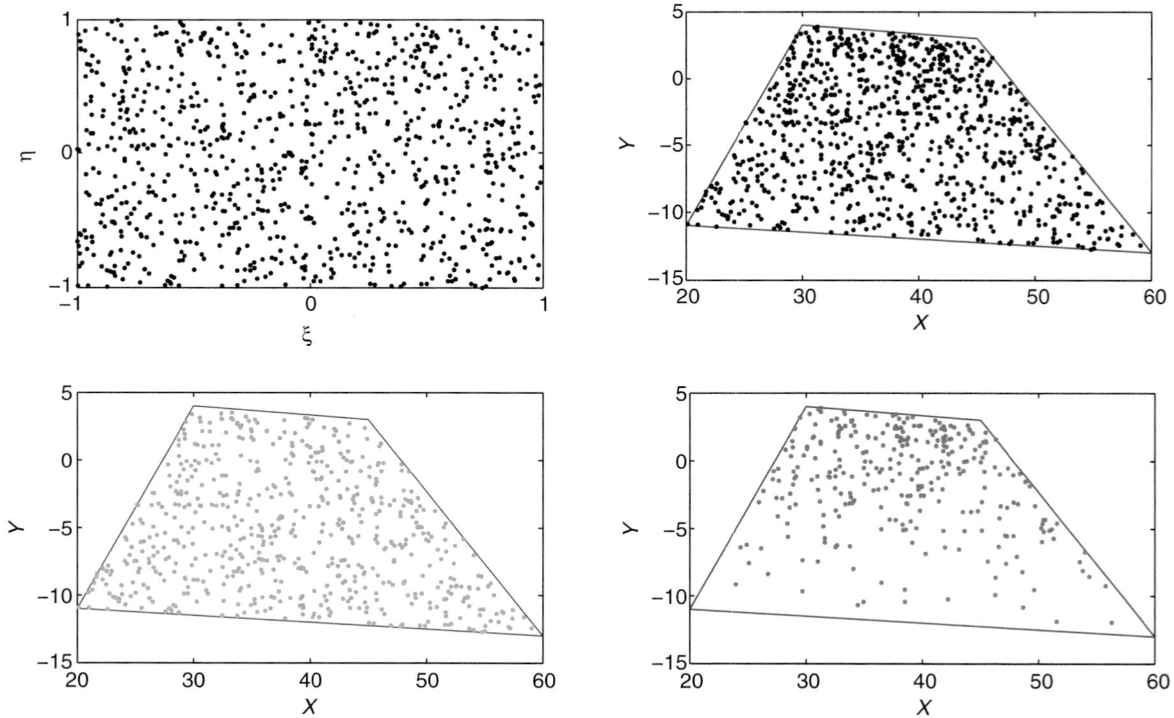

Figure 8.8. Uniform distribution of 600 particles on a quadrilateral. Top left: All 899 particles, uniformly generated in local space. Top right: All particles in x, y space, artificially concentrated near the narrow top; Bottom left: 600 particles that survived the Jacobian-based rejection; Bottom right: The 299 rejects.

- "Compactness" – responsive to local variations, hence limiting the reach of the bases to local support; avoiding global interpolation and large-scale averaging
- Smooth – avoiding unnecessary discontinuities in ρ
- Efficiency – have procedures emphasize one computational element or subvolume at a time
- Mass conserving
- Robust when there are low numbers of particles
- Non-negativity of the ρ_i
- The practical importance of keeping the computational load scaling linearly with N.

Absence of any of these features can become serious problems in large simulations, and require additional safeguards.

Two general classes of projections will be described. The fundamental difference lies in the target basis: a fixed mesh or lattice of points; or a grid associated with the particle positions themselves.

8.4.1 Simple Fixed Mesh Projections

Here we consider projecting particles onto a given (unstructured) mesh, presumably that already in use in a simulation.

By far the simplest approach is to add up the particle mass in a given computational element, and compute an average density over its area or volume. The result is a density field that is piecewise constant – a different constant density ρ_e in each element. If $\mathcal{V}_e$ is the element volume, then we have

$$\rho_e = \frac{\sum M_k}{\mathcal{V}_e} \tag{8.11}$$

with the sum over a single element only. One cannot differentiate this field; but all particles are accounted for, and included in exactly one ρ_e. These "element-based densities" are always non-negative. This is a common procedure, whether formulated that way or not; it requires little beyond intuitive coding and forms a baseline for comparison.

In fact, this projection is the generalization of the *histogram*, implemented on a mesh of variable-size elements. It is easily made, easily displayed and interpreted.

If one were to describe this in functional terms, one would refer to projecting particles onto a "stairstep" basis $\phi_e(\mathbf{x})$ by element averaging. The basis ϕ_e would be binary, 1 on its element and zero on all others. The element volume is $\int \phi_e$. So formally, this projection can be stated in terms of global sums and integrals:

$$\rho_e = \frac{\sum M_k \phi_e}{\int \phi_e} \qquad \rho(\mathbf{x}) = \sum \rho_e \phi_e(\mathbf{x}) \tag{8.12}$$

with piecewise constant ϕ_e. The ρ_e are "element averages."

Beyond this, projection requires specificity about the basis of interpolation. It is common to employ linear interpolation among local (nearest-neighbor) mesh points; formally, piecewise linear bases that match in value at element boundaries. On unstructured meshes, these are extremely common and in wide use – "C_0 linear bases with triangular elements". $\phi_i = 1$ at node i, zero at all other nodes, varying linearly among nodes. All expositions of unstructured mesh discretization describe these bases and elements; they form a baseline of simplicity for continuous interpolation.

Following the form above (8.12), a comparable "nodal average" is

$$\rho_i = \frac{\sum M_k \phi_{i|k}}{\int \phi_i} \qquad \rho(\mathbf{x}) = \sum \rho_i \phi_i(\mathbf{x}) \tag{8.13}$$

where $\phi_{i|k}$ is "the value of ϕ_i where particle k is located" – the consequence of integrating the singular function δ_k. This is the same statement as for the ρ_e; but here the ϕ_i are linear, continuous.[2] Particles will be apportioned among nodes in accordance with their proximity; and provides linear interpolation between nodes. Figure 8.9 illustrates this. Specifically, in any element: a fraction of each particle will be apportioned to each participating node according to nodal proximity; particles have no non-local influence, that is, no influence in elements not sharing a side. The ρ_i favor particles located close to a node, within elements; for the ρ_e, all particles in an element have equal weight at each node. Like the ρ_e projection, this accounts for all particles in the system, but with greater fidelity to location within elements, and continuous interpolation built in. By inspection of Equation 8.13, ρ_i is always non-negative.

Either of these projections is workable on unstructured meshes. Desirable properties include non-negativity, mass conservation, compactness, intuitiveness, and efficiency of implementation (one element at a time).

Figure 8.10 illustrates these two projections. The nodal ρ_i results are sensitive to the within-element particle distribution, while the ρ_e are not.

This example is intentionally under-resolved – only 6 particles on 10 elements – and reveals some distinguishing differences between the 2 projections. The nodal projection is more prone to "wiggles" – caused by unresolvable patterns in the particle

[2] The generalization could be pushed further by defining the "nodal volume" $\mathcal{V}_i = \int \phi_i$.

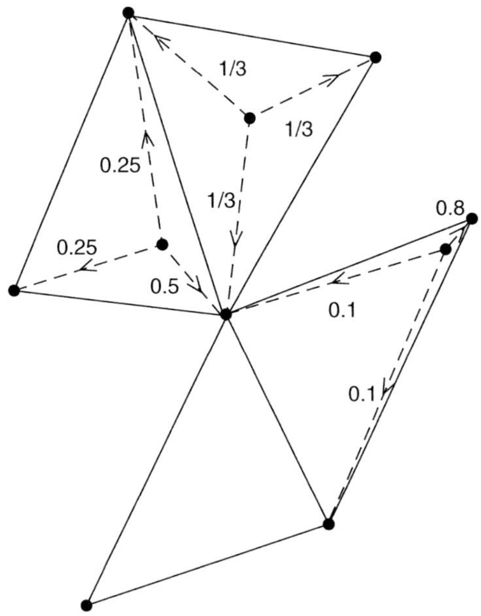

Figure 8.9. Three particles in four elements: relative contributions to ρ_i, Equation 8.13.

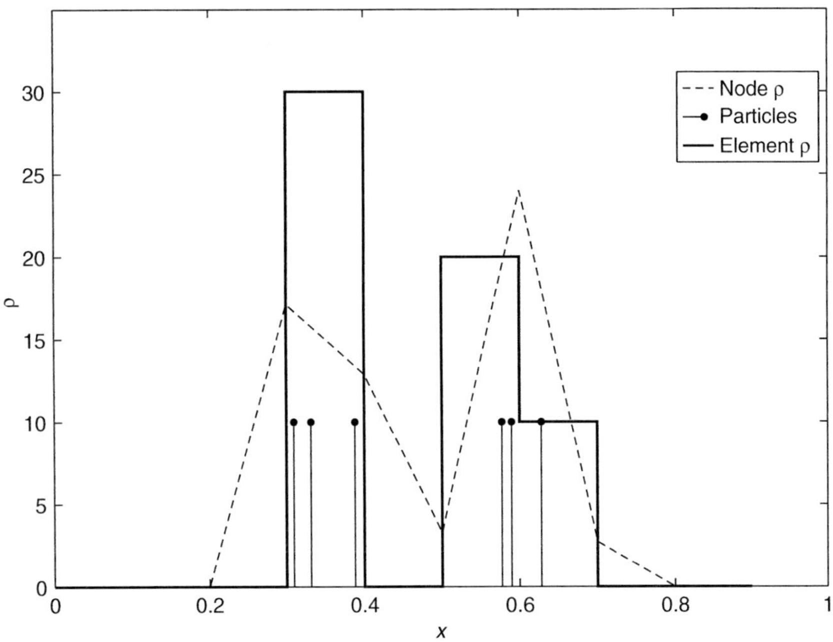

Figure 8.10. Example of 1-D projections of 6 equal-mass particles on 10 equal elements. Linear elements are used for the nodal projection. The element projection has the simplest piecewise-constant bases.

distribution. While the element projection annihilates all intra-element variations, the nodal projection attempts to portray them, at the risk of amplifying "noise."

Perhaps the most serious feature of both of these is their performance at low particle intensity. This inevitably occurs with high-resolution meshes and fixed overall particle counts – there will be fewer particles per element, in the limit a one-particle element surrounded by empty elements. The resultant element density would be huge, $\rho_e = \frac{M}{\mathcal{V}_e}$ in an element, surrounded by zero density in the immediate neighbor elements – approaching a small set of isolated singularities. For example, in Figure 8.10, the same particles on a more refined mesh would produce higher and thinner "chimneys." The nodal projection shares this phenomenon.

Both projections share an equivalent expression, useful for analytic purposes:

$$\rho_e < \phi_e > = \sum_k M_k \phi_{e|k} \tag{8.14}$$

$$\rho_j < \phi_j > = \sum_k M_k \phi_{j|k} \tag{8.15}$$

$< \ >$ indicates integration over the computational grid; $\phi_{e|k}$ means "the value of ϕ_e at particle point k, and analogously for $\phi_{j|k}$ Mathematically the sum can be over all particles here; ϕ is zero outside its home element(s), and does the selecting.

Both have the same right-side form, the particle aggregations T_i:

$$T_i \equiv \sum_k M_k \phi_{i|k} \tag{8.16}$$

These are aggregations of particle mass in elements where ϕ_i is nonzero, and weighted by the size of ϕ_i. Put another way, T_i is the total particle mass sampled by ϕ_i.

An additional projection follows, formulated as a generalized least squares problem.

8.4.2 The Least Squares Projection

This is another fixed-mesh projection. The particles are given; the mesh and interpolants $\phi_i(\mathbf{x})$ are given; the ρ_i are to be found.

Define the misfit ϵ:

$$\epsilon(\mathbf{x}) \equiv \sum_j \rho_j \phi_j(\mathbf{x}) - \sum_k M_k \delta_k \tag{8.17}$$

$\epsilon(\mathbf{x})$ is a smooth field with removable singularities superposed at all particle locations. The integral measure of misfit size is $\mathcal{E}$:

$$\mathcal{E} = \langle \epsilon^2 \rangle \tag{8.18}$$

The roughness of the interpolant is the vector $\nabla\rho$. Define the metric of roughness $\mathcal{R}$ as the integral

$$\mathcal{R} = \langle \nabla\rho \cdot \nabla\rho \rangle = \sum_i \sum_j \rho_i \langle \nabla\phi_i \cdot \nabla\phi_j \rangle \rho_j \tag{8.19}$$

The composite metric blends misfit and roughness, with scaling length L:

$$\Omega = \mathcal{E} + L^2\mathcal{R} \tag{8.20}$$

The ρ_i are to be found by minimizing Ω. Differentiation gives

$$\frac{\partial\Omega}{\partial\rho_i} = 2\langle \epsilon\phi_i \rangle + L^2\frac{\partial\mathcal{R}}{\partial\rho_i} \tag{8.21}$$

These must vanish for all i. The result:

$$\sum_j \langle \phi_i\phi_j \rangle \rho_j + L^2 \sum_j \langle \nabla\phi_j \cdot \nabla\phi_i \rangle \rho_j = \sum_k M_k \langle \delta_k \phi_i \rangle \tag{8.22}$$

$$= \sum_k M_k \phi_{i|k} \tag{8.23}$$

where $\phi_{i|k}$ is the basis function ϕ_i evaluated at particle point k as before.

This is the generalized least squares (GLS) problem, with the $\mathcal{R}$ term representing "regularization" or "penalization." It gives priority to smooth solutions. The right-hand side of Equation 8.23 is what we have called T_i in the two simpler projections: "the total particle mass as sampled by ϕ_i." We will continue to use that nomenclature.

The discrete form 8.22 is recognizable as a standard Galerkin rendering of the PDE

$$\rho - \nabla \cdot L^2 \nabla\rho = \sum_k M_k \delta_k \tag{8.24}$$

with natural boundary conditions. It is a deterministic differential equation with single parameter L, forced by the stochastic particle locations.

L is not arbitrary; it represents physical processes. It is assumed known in advance – the *decorrelation length* of the density field.[3] The estimated nodal densities ρ_i will

[3] Specifically, for 1-D points separated by distance r,

$$Cor_{\rho\rho} = \left(1 + \frac{r}{L}\right)e^{-\frac{r}{L}} \tag{8.25}$$

and the 2-D version is analogous. (See Section 3.5.4.)

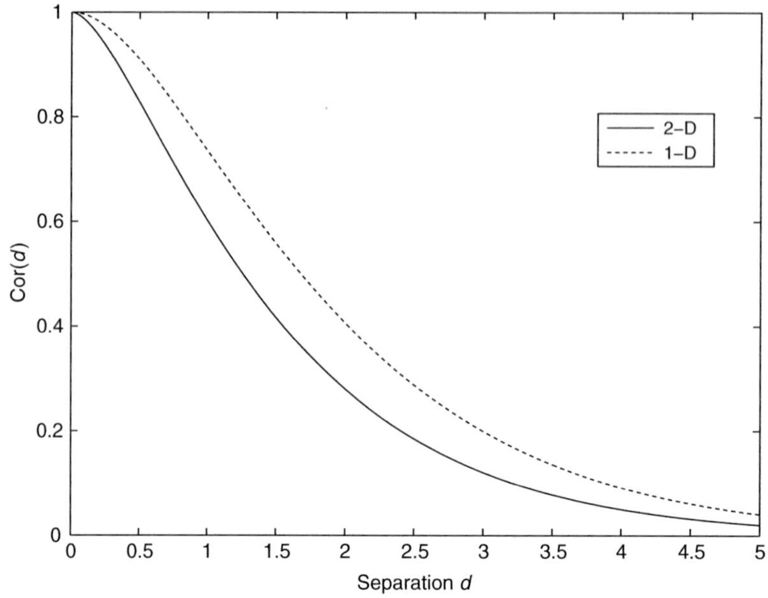

Figure 8.11. Separation-based Correlograms for the GLS projection of density. d is the separation distance, scaled by the GLS parameter L. This is the analytic limit of high resolution, with boundaries far removed. The mathematical forms are given in Equations 3.232, and 3.233.

have this statistical property. For the limiting analytical case, far from boundaries, the autocorrelation of ρ will depend on the separation distance, scaled by L, as in Figure 8.11. The two forms are basically similar; the 2-D correlation falls off faster with distance, being about 2/3 of the 1-D form in the mid-range of separation.

The GLS projection has a single physical parameter L; its assembly and solution on finite meshes are standard and well understood. In matrix form:

$$\left[G_{ij}\right]\{\rho_j\} = \{T_i\} \tag{8.26}$$

with

$$G_{ij} = \langle\phi_i\phi_j\rangle + L^2\langle\nabla\phi_i\cdot\nabla\phi_j\rangle \tag{8.27}$$

$$T_i = \sum_k M_k\phi_{i|k} \tag{8.28}$$

The other two projections have the same form,

$$<\phi_e>\rho_e = T_e \qquad \text{Element} \tag{8.29}$$

$$<\phi_i>\rho_i = T_i \qquad \text{Node} \tag{8.30}$$

$$\sum_j G_{ij}\rho_j = T_i \qquad \text{GLS} \tag{8.31}$$

GLS is a generalization of the Galerkin projection (Section 7.4.5) which is recovered when $L = 0$.

8.4.3 Mass Conservation

These projections all have a global mass conservation property – all the mass in the particles is captured in the projected density field. To see this, we need to exploit the basic property of the basis, $\sum_i \phi_i(\mathbf{x}) = 1$; and its consequence, $\sum_i \nabla\phi_i(\mathbf{x}) = 0$.[4] These are true everywhere.

Plainly, the right-hand quantities T_j sum to the total mass of particles

$$\sum_j T_j = \sum_k \sum_j M_k \phi_{jk} = \sum_k M_k \tag{8.32}$$

No particles are lost; all contribute fully.

For the left-hand side of the projections, summing the relations simply gives $< \sum \rho_i \phi_i > \equiv < \rho >$; so in each case, the global mass conservation property holds:

$$< \rho > = \sum_k M_k \tag{8.33}$$

The projected mass is exactly equal to the sum of the particle masses present. This property is exact despite a finite mesh and finite ensemble size. It is true of each of the three projections. They will differ in detail but not in the aggregate.

8.4.4 Example: Particles on a Mesh

For a simple test, n particles were generated on a 1-D mesh of equal length, linear elements. All particles were the same size; the particle intensity $\nu(\mathbf{x})$ was a sigmoid curve illustrated in Figure 8.12. The particle points were generated by the uniform deviate, and the rejection method[5] was used to enforce their distribution.

In Figure 8.13 there are 12 realizations of the same projections, illustrating the variability in outcomes. The variability is largest where the intensity is largest, with almost no variability where the intensity $\to 0$. The nodal and element projections have comparable variability, though qualitatively different. The GLS does the best of the three.

[4] It is very common interpolation practice to require the bases to sum to unity *everywhere* – otherwise a constant cannot be interpolated properly!

[5] The general rejection method is illustrated at Figure 3.8 and the accompanying discussion.

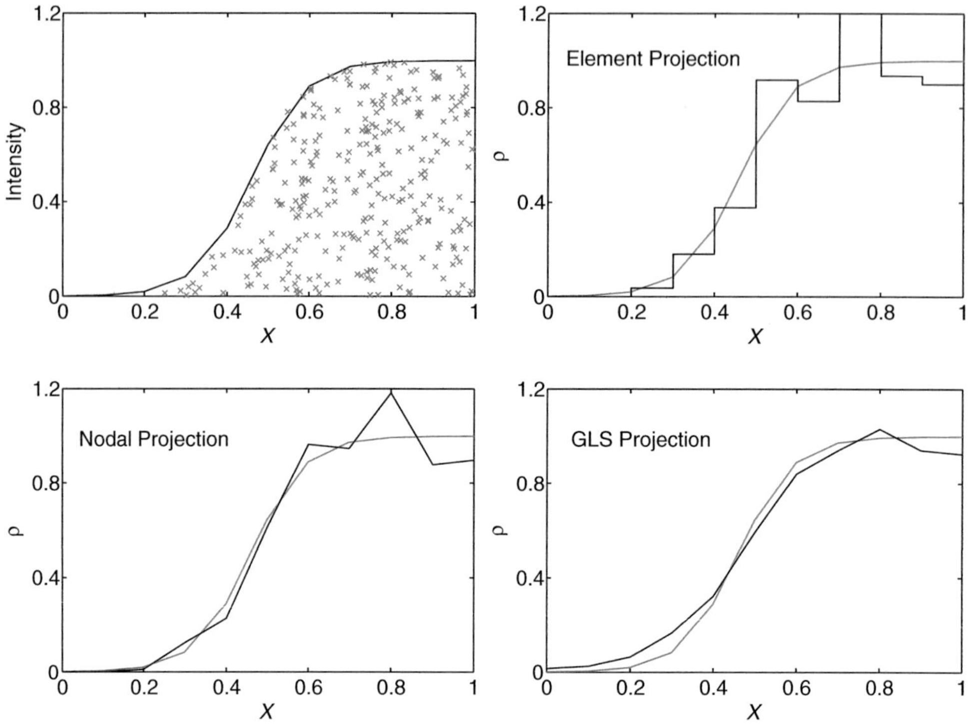

Figure 8.12. Projections of 300 particles by three methods; 10 equally length elements. The particle intensity varies with x and appears in each panel. The particles generated are plotted in the upper left panel, with each particle's rejection criterion plotted as the vertical coordinate. The three projections used the identical set of particles shown. For the GLS projection, $\frac{L}{\Delta x} = 1$.

Based on these and similar tests, the GLS projection is the most durable. The free parameter offers some flexibility in the decorrelation length scale; for well-resolved situations, this is physical; while for coarse resolution, one needs to boost L artificially to keep $\frac{\Delta x}{L} < \mathcal{O}(1)$.

The nodal projection emerges as a reasonable second; and the element projection, the weakest. Both of these exhibit comparable "noise," but the nodal projection avoids the stair-step feature built into the element projection.

All three are workable, and amenable to further development. Only the GLS survives for low particle resolution (low particle count or high mesh resolution). This is described in Section 8.4.5.

8.4.5 Convergence – The Small Δx Problem

For a fixed particle count: as mesh resolution improves, the nodal and elemental projections degrade. Inevitably, the element density will be based on a small number of

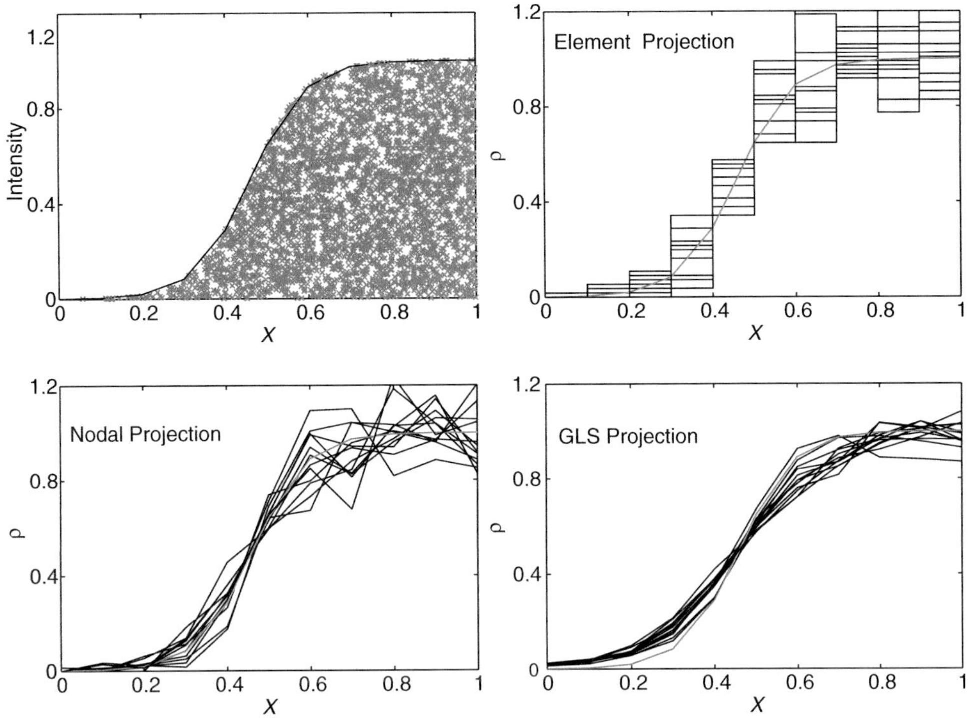

Figure 8.13. As in Figure 8.12, but 12 different realizations are shown. Each projection used 300 particles; all 3600 particles are plotted together in the upper right panel.

fixed-mass particles in a very small element; N will ultimately be confined to very small integers, including zero. And the occurrence of a massless element becomes more common. In such cases, ρ_e can be huge surrounded by many near neighbors with $\rho_e = 0$ – very rough and spikey, approaching the reproduction of the Dirac distributions in the data. Similarly for ρ_i. These projections take no notice of very large local gradients, having no built-in assumptions about correlation length for ρ. The GLS overcomes this problem from the outset.

This phenomenon will occur wherever the local number of particles per element is low. It the node or element projections are used, it will have to be overcome after it occurs. Rather than deal with it in an ad hoc way, it is better to prevent its occurrence in the first place. So on this argument, GLS is superior. It requires the a prior estimate of the decorrelation scale L.

As an example, Figure 8.14 shows some 1-D projections with low particle abundance per element. The GLS L in this case is set at the lower limit $L_0 = 0.1$, instead of $L = \Delta x$ as in the coarser case.

Recommended is a rule that puts a physically inspired L for high resolution, but L proportional to the local Δx for coarse resolution.

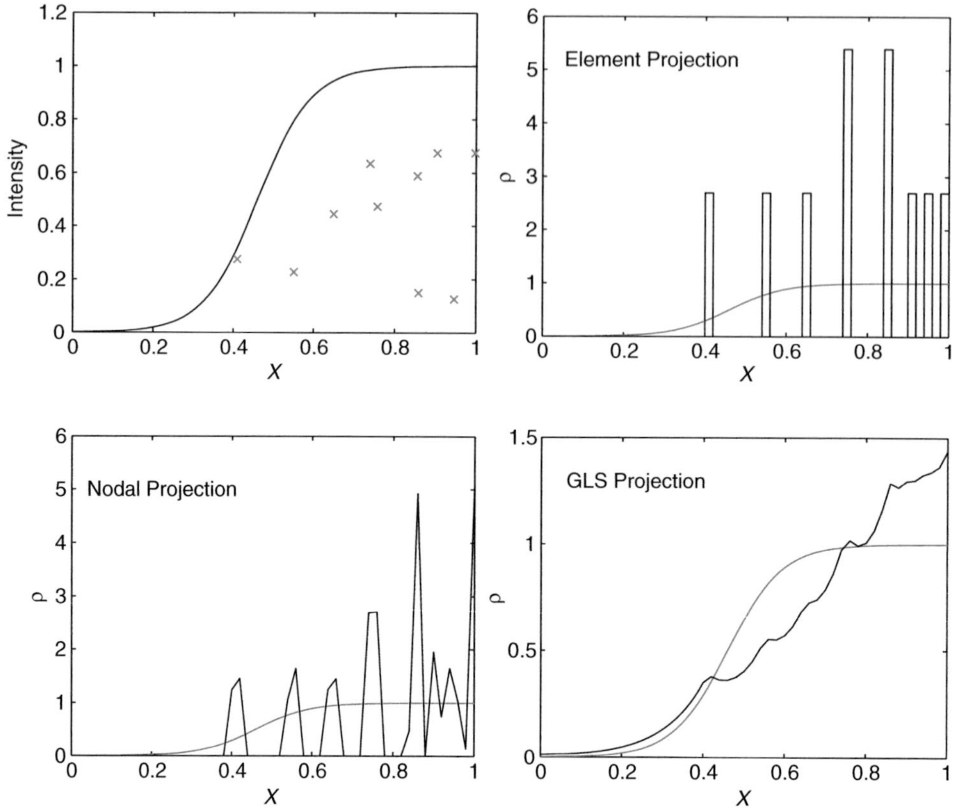

Figure 8.14. Sparse projections. As in Figure 8.12, but there are 50 elements, 10 particles. For GLS, $L = 0.1$. (Note that the vertical axes have been expanded.)

8.4.6 Kernel Methods

Kernel methods seek the estimation of $\rho(\mathbf{x})$ at every point, in terms of predefined kernel functions K. Rather than being associated with the predetermined nodes of a mesh, the kernels are associated with the *particle* locations $\mathbf{X}_k$. Each kernel has potentially global support, and will be smooth and differentiable. General properties:

- Centered at X_k
- A function of a pseudo-distance from particle k: $r_k = \frac{|\mathbf{x}-\mathbf{X}_k|}{L}$
- Normalized, $< K >= 1$

With these features, we have for the kth kernel, $K(r_k)$. There is a single parameter, L which specifies the distance scaling. The specific functional form can be quite general. A simple example is the "box kernel" – uniform within L of point k; and zero otherwise.

The density estimate:

$$\rho(\mathbf{x}) = \sum_{k=1}^{N} K\left(\frac{|\mathbf{x} - \mathbf{X}_k|}{L}\right) M_k \tag{8.34}$$

where N is the number of particles. ρ is defined everywhere and inherits the properties of the kernels. The "box" and Gaussian kernels are common.

Essentially, each particle is spread over all possible values of $\mathbf{x}$, according to its kernel strength there. At each point $\mathbf{x}$, the projection ρ is a weighted sum of particle masses M_k, the weights being the strength of K_k at that point. All data have global influence; the normalization property of the kernels guarantees mass conservation.

Figure 8.15 shows the box kernel projection for the same 1-D problem used above, Figures 8.12 and 8.13. There are 10 equal linear elements, and 25 particles generated with the same intensity. The kernel used relatively coarse $L = \Delta x$, the same as

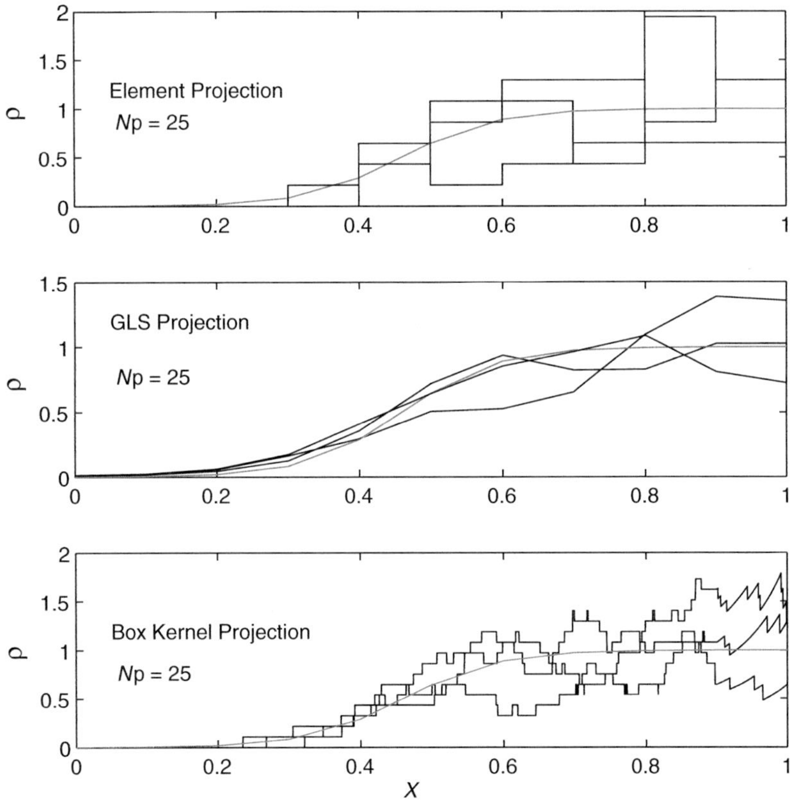

Figure 8.15. Test case as in Figure 8.12. There are 10 elements; 25 particles were generated for each of three tests per panel. For the GLS and kernel projections, $L = 0.1$.

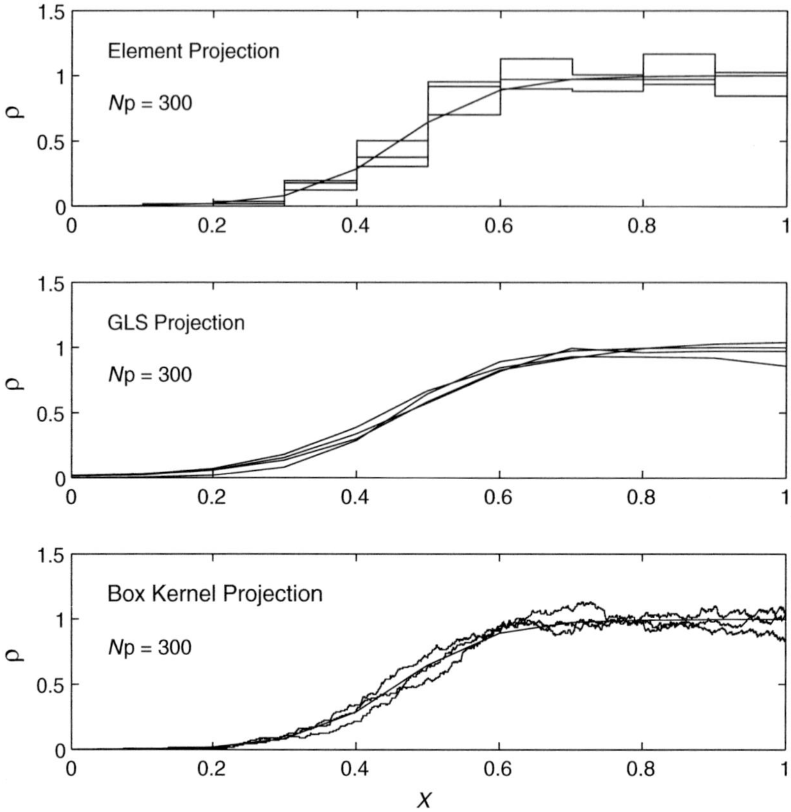

Figure 8.16. As in Figure 8.15, but there are 300 particles per test.

the length scale in the element (histogram) and GLS projections. But the kernel projection was sampled densely (500 samples over the length) and reveals substantial variability with x that is not available to the other projections. The GLS projection appears as a smoothed but continuous form of the kernel projection.

For a given dataset, all three projections are attempting the same thing. The basic assumptions in each are evident. Significantly, each shows comparable variability across several repeated runs. Figure 8.16 shows the results for 300 particles; the same comments pertain.

As described, there is one parameter L, variously called the window width, smoothing parameter, bandwidth, and so on. The kernel approach puts the kernels where the particles are, thereby adapting the interpolating functions to the data. As a result, every projection has to be done again in its entirety. The GLS projection, on the other hand, always involves the same matrices, which can be pre-computed and factored. Both result in the global influence of every particle, and conserve mass. Both have one parameter, related to the assumed decorrelation length of ρ.

Extensions of kernel methods are numerous. For example, the pseudo-distance can have more than one length scale for anisotropy; or be spatially variable. Some variants use a "nearest neighbor" strategy in which the summations in Equation 8.34 are restricted to the n particles nearest to an estimation point $\mathbf{x}$. In all cases, kernels must be adjusted to the extent they spill over model boundaries, to secure the proper normalization. There are hybrids. For example, Equation 8.34 could be used to estimate ρ at fixed points or nodes on a computational mesh. With suitable normalization, these could then be interpolated with a fixed basis ϕ.

Implementation of kernel methods have the feature that the location of the kernel functions (the interpolating functions) are not established a priori, but accompany the particles. One needs to perform integration, interpolation, and so on for each particle, whenever a projection is needed. In addition, the kernels require dealing always with their boundary overlap. Some algorithmic creativity is needed to avoid the potentially intense computational demands. The hybrid suggested avoids some of these.

Kernels were used by Zhang et al. [500] for oil spill modeling; and by Edwards et al. [127], Mitarai et al. [317], and Li et al. [251] for larval transport studies.

There are fascinating connections among all the methods mentioned and optimal estimation; that requires a more lengthy presentation.

Part III

Applications

9

Noncohesive Sediment – Dense Particles

9.1 Introduction

Sediments are dense particles. In still water they sink to the bottom. The introduction of bottom turbulence enables particle motion. At low turbulence, particles will bump or roll along the bottom. Higher turbulence will lift particles off the bottom. Once lifted, they are advected in the fluid flow, and dispersed by 3-D turbulent mixing. Near bottom, both turbulent diffusivity and horizontal advection increase with height; higher particles will move faster with larger random walks.

Bottom turbulence will have a sorting effect, with lighter particles moving higher and faster; and more often. The sorting will result in greater horizontal displacement for particles that fly high and long. When the shear is removed, the turbulence decays and particles sink.

We consider a single noncohesive particle and describe its motion. It is assumed that a 3-D circulation field is available, providing velocity, bottom stress, and eddy viscosity. Consistent with common practice, the computational bottom is slightly above the true bottom, with the fluid slipping horizontally above an idealized boundary layer, wherein the law of the wall governs. As particle motion is initiated by turbulence in this layer, it will be necessary to follow particle motions within and across it.

We will present the 3-D case first. Then, we will turn to the additional simplifications/assumptions necessary to deal with the vertically averaged case. The analysis introduced in Section 5.4 is developed more fully here for the bottom boundary.

Sediment dynamics is a mature field, and the coupling with hydraulic models is well established. The use of 3-D particle tracking is relatively new; comprehensive implementations include the PTM (MacDonald et al. [291]), PSED (Davies [102]) and SandTrack (Soulsby et al. [428, 429]).

There will be several notational conventions. They will be introduced as needed; a summary is included in Section 9.15

9.2 Three States: P, M, B

In Figure 9.1 we illustrate the general arrangement of things. An individual particle will be in one of three states:

P: The pelagic or oceanic state. The Coastal Ocean Model is the gridded area at the top; it provides fluid velocity and bottom stress. Beneath it is a log layer, where analytic solution extends vertically between the computational bottom and the true bottom. In both, sediment particles move and disperse with the fluid, with negative buoyancy and ultimately bottom encounters.

M: The mixing state. Beneath true bottom is the sediment mixing zone, the surface sediments that are in active exchange with the overlying water column. In this zone, horizontal motion is arrested. There is vertical mixing of particles, driven by the surface stress. Surface forms (ripples, bed forms) evolve and affect the hydraulic stress in the log layer. When mixtures of sediment particles are present, smaller particles can "hide" among larger ones, limiting their potential for motion. Turbulence can reverse this.

B: The burial state. Beneath this interfacial mixing zone, vertical motion reflects a blend of net sedimentation and compaction, and particles are generally not in communication with the overlying water column. Vertical mixing is limited,

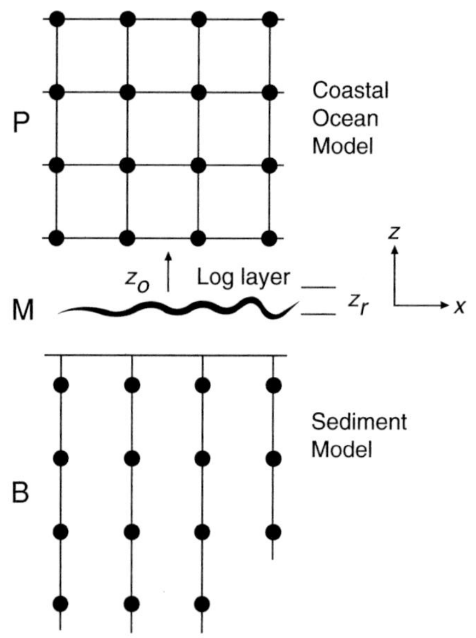

Figure 9.1. The Coastal Ocean Model, the assumed log layer, the surficial sediment mixing zone, and the deeper sedimentary regime, represented by a 1-D (vertical) model under each horizontal node. Bottom boundary layer conventions – the true bottom at $z = 0$, with roughness height $\pm z_r$; the computational bottom at $z = z_0$; and the intervening log layer.

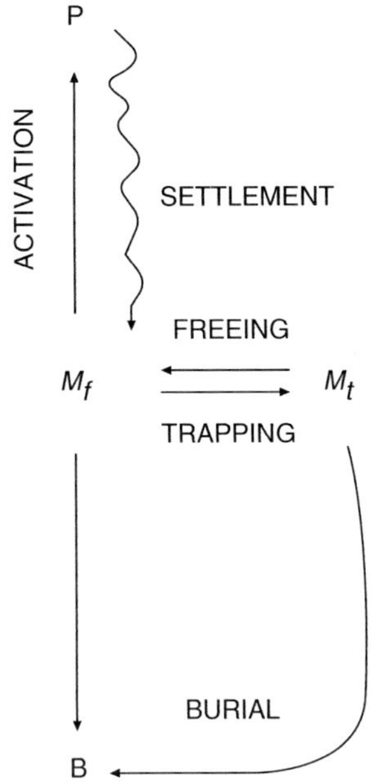

Figure 9.2. Three particle states: pelagic, buried, and the mixing zone. M includes two reversible substates, trapped M_t and free-to-move M_f. Particle exchange between P and M is reversible and the primary concern here.

> except for episodes of very high surface stress, as in unusual storms. Accordingly, longer-lived chemicals adsorbed onto particles are sequestered, subject to episodic remixing upward or in periods of net scouring, as buried particles are brought into the mixed layer.

Figure 9.2 shows an idealization in which a particle is in one of these: the buried state, B; the pelagic state P; and the state of activation, exchange, and mixing, M. The P state includes the Coastal Ocean Model and the log layer. The M state includes the two substates for trapped (M_t) particles and those free to move (M_f).

The major fluxes between them are illustrated as the entrainment, settlement, hiding, exposure, and burial fluxes. Any particle must be in only one state; all the state transition probabilities need description. If there are N particles, then an Eulerian rate $\alpha = \frac{1}{N}\frac{dN}{dt}$ is equivalent to the individual particle's transition probability $\alpha\Delta t$. This is used extensively in other chapters; it is utilized here.

Particles in P move horizontally and vertically. There is entrainment from M; and settling to M. Particles in the B and M states are horizontally immobilized on the bottom. Those in M may be entrained; those in B are isolated from M and P, unless a significant model of the B processes is available.

The P zone includes the log layer beneath the mesh – it is part of a Coastal Ocean Model, and its properties are built into the bottom stress law therein. It will be necessary to account for motion analytically there, matched to the bottom of the computational mesh.

A full sediment model would account for all transport and biogeochemical phenomena beneath the ocean bottom. Such models are emergent, the analogs of Coastal Ocean Models. Our primary goal is to describe the fluxes between the two. Accordingly, we focus on the M and P states, the exchanges between them, and the particle motion that would occur while in the P state.

Our basic goal in this chapter is to enable particles to move among the states; and to account for their motion while in P.

9.3 Sediment Particles

We will use three parameters to describe a particle:

D: the grain size
ρ_s: the particle density
w_s: the particle's settling velocity in water
s: the dimensionless density – normalized by that of water, $\frac{\rho_s}{\rho_w}$

Krumbein's Φ is a metric of variability among particles (Krumbein and Sloss [234]) relative to a reference D_r:

$$\Phi \equiv -\log_2 \frac{D}{D_r} \iff D = D_r 2^{-\Phi} \tag{9.1}$$

The distribution of Φ characterizes a sediment. For instance, a particle set of coarse sand, whose Φ scale is uniformly distributed within the range $[a,b]$, can be generated as

$$D = D_{\max} \cdot 2^{-(b-a)\epsilon} \qquad \epsilon \sim \mathcal{U}(0,1) \tag{9.2}$$

ϵ is the uniform deviate. $D_{\min}/D_{\max} = 2^{-(b-a)}$, so an appealing form of this *loguniform* distribution is

$$\frac{D}{D_{\max}} = \left[\frac{D_{\min}}{D_{\max}}\right]^{\epsilon} \tag{9.3}$$

Alternatively, let Φ be Gaussian with variance σ^2. Then D will be *lognormally* distributed:

$$D = D_{50} \cdot 2^{-\sigma\eta} \qquad \eta \sim \mathcal{G}(0,1) \tag{9.4}$$

where η is the normal deviate (unit variance), D_{50} is the median grain size, and σ^2 is the variance of Φ. The Φ scales for various sediment types are available in the literature (e.g., Dean and Dalrymple [106], Chamley [73]).

9.4 Settling Velocity

Settling velocity has received much attention in the sediment context. The major dimensionless groups are the settling Reynolds number and the dimensionless particle size:

$$R_s \equiv \frac{w_s D}{\nu} \qquad D_* \equiv D\left[\frac{(s-1)g}{\nu^2}\right]^{1/3} \tag{9.5}$$

(ν is the kinematic viscosity of water). As a point of departure, Stokes' law governs a perfect sphere in a laminar fluid:

$$R_s = \frac{D_*^3}{18} \tag{9.6}$$

Many investigators have reported empirical formulae for the settling velocity of natural sediments. A general formula for field situations is not available; but there is a range of formulae, generally developed for specific data sets (Hallermeier [189]; Van Rijn [460]; Soulsby [427]; Cheng [81]; Ahrens [3]; Jimenez and Madsen [210]; Lin et al. [252]; Camenen [65]; Zhiyao et al. [502]). Here, we give three examples.

The first example is Soulsby's [427] formula derived for irregular grains:

$$w_s = \frac{\nu}{D}\left[\left(10.36^2 + 1.049 D_*^3\right)^{1/2} - 10.36\right] \tag{9.7}$$

Soulsby reports that 66% of the w_s predictions obtained using Equation 9.7 lie within 10% of the observations while 90% lie within 20%.

The second example is the Jimenez-Madsen [210] formula, which is a variation of the formula by Dietrich [116]:

$$w_s = \frac{\nu}{D_n}\left[4S_*\left(A + \frac{B}{S_*}\right)^{-1}\right] \tag{9.8}$$

A and B are constants depending on particle shape and roundness; S_* is a dimensionless particle size parameter given in terms of the nominal particle diameter (i.e., diameter of a sphere that has the same volume as the natural sediment particle)

$$S_* = \frac{1}{4}\left[\frac{(s-1)gD_n^3}{\nu^2}\right]^{1/2} \tag{9.9}$$

(S_* is a replacement for D_* here; $D_*^3 \equiv 16S_*^2$ except for the introduction of D_n.) Jimenez and Madsen [210] suggest using values of 0.954 and 5.12 for A and B

Table 9.1. *Typical Values for Noncohesive Sediment Particles*

Sediment Type	Size Range D (mm)	Settling Velocity Range w_s, m/s
Very fine sand	0.06–0.125	0.002–0.008
Fine sand	0.125–0.250	0.008–0.028
Medium sand	0.250–0.50	0.028–0.066
Coarse sand	0.5–1.0	0.066–0.110
Very coarse sand	1–2	0.11–0.17
Very fine gravel	2–4	0.17–0.25
Fine gravel	4–8	0.25–0.36
Medium gravel	8–16	0.36–0.50

Note: In practice sediment calculations treat sands to be composed of quartz (2650 kg/m^3). If sand is composed of seashell material, density would be 2400 kg/m^3.

respectively. They report that although these values have been obtained for sediment grain diameters between 0.063 mm and 1 mm, they are applicable within the entire sand range, up to 2 mm.

The third example comes from Ferguson and Church [143]:

$$w_s = \frac{\nu}{D}\left[\frac{D_*^3}{C_1 + [0.75C_2D_*^3]^{1/2}}\right] \tag{9.10}$$

Here C_1 and C_2 are key constants that can be taken as 18 and 0.4 respectively for smooth spheres. However, Ferguson and Church [143] state that because these values provide poor fit for natural grains, the nominal diameter, D_n, should be used instead of D in Equation 9.10 with C_1 and C_2 values of 20 and 1.1 for a root mean square error of 6%. They also state that for natural sands and when the sieve diameter D is used, C_1 and C_2 can be set to 18 and 1 with an RMS error of 16%. Equation 9.10 applies to the entire range of flows and converges to Stokes' law for fine grains.

For large concentrations of sediments (volumetric concentration C greater than 0.05), the settling velocity will be diminished due to increased drag (Richardson and Zaki [375], Soulsby, [427]). The reduced settling velocity is w_{sc}:

$$w_{sc} = \frac{\nu}{D}\left[\left(10.36^2 + 1.049(1-C)^{4.7}D_*^3\right)^{1/2} - 10.36\right] \tag{9.11}$$

This reduces to Equation 9.7 when $C = 0$. For concentrations around 0.3, the reduction in the clear water settling velocity can be up to 80% according to Baldock et al. [15].

Table 9.1 gives some typical values.

9.5 Bottom Boundary Layer

As in Figure 9.1, beneath the computational grid there is an unresolved layer z_o with resting particles at its bottom. The stress is constant in this layer; the horizontal velocity is logarithmic, vanishing at the bottom. At $z = z_o$ the gridded velocity takes over; velocity and shear match there, by virtue of the slip condition used in the gridded product.

The important features of the log layer are discussed in Sections 5.3 and 6.3.1.2. In summary form:

- τ is constant.
- $u_* \equiv \sqrt{\frac{\tau}{\rho}}$, the friction velocity.
- $u(z) = \frac{u_*}{\kappa} \log\left(\frac{z}{z_r}\right)$.
- z_r is the roughness height; $u = 0$ there.
- $q = \frac{u_*}{\kappa}$ is constant.
- $l = \kappa z$.
- $A_z = \kappa u_* z$ is the vertical eddy viscosity.
- $c_d \equiv \frac{\tau}{\rho |u|^2} = \left[\frac{1}{\kappa} \ln\left(\frac{z}{z_r}\right)\right]^{-2}$.

The ratio of stress to kinetic energy, c_d, is a function of z through the log layer. Enforcing this at the model bottom in the form of a slip condition, $A_z \frac{\partial u}{\partial z} = c_d u|u|$, enforces continuity with the underlying log layer.

There is wide agreement (e.g., Soulsby [427]) on the important contribution of wave-induced turbulence at the bottom. In some cases this can be dominant. The relations reviewed below generally invoke the *total* bottom stress, including the effect of surface waves. Available gridded products that supply the physical fields may or may not have this component; if not, then it needs to be added.

Typical horizontal grid lengths are too large to resolve potentially important bedform variations. The parameters z_r and z_0 must effectively bear the representation of these unresolved variations. These considerations are prior to the hydraulic calculations, so for consistency we assume that the two parameters are available from the Coastal Ocean Model.[1]

9.6 Entrainment

For entrainment of resting particles into the fluid, one needs a threshold and a rate. In addition, an entrained particle needs to be initialized in the overlying water column – an initial height.

[1] The excellent textbooks by Soulsby [427] and Fredsoe and Deigaard [161] treat this matter in detail.

9.6.1 The Threshold of Motion – The Shields Parameter

Motion initiation is commonly described in terms of the fluid shear stress τ at the bottom. For an isolated particle, the Shields parameter Θ expresses the ratio of vertical (buoyancy) and horizontal (shear) forces acting on a particle:

$$\Theta \equiv \frac{\tau}{\Delta \rho g D} = \frac{u_*^2}{(s-1)gD} \tag{9.12}$$

The threshold or critical stress τ_{cr} (Θ_{cr}) is the level above which particle motion is possible (Shields [414]). There is considerable experimental evidence linking Θ_{cr} to the bed Reynolds number

$$R_* \equiv \frac{u_* D}{\nu} \tag{9.13}$$

with ν the kinematic viscosity of water (Shields [414]). The characteristic velocity here is the friction velocity, constant throughout the log layer. (Vanoni [461] provides an overview of early, seminal work.)

Current work is often stated in terms of a third dimensionless variable, the dimensionless grain size D_*:

$$D_* \equiv \left[\frac{(s-1)g}{\nu^2}\right]^{1/3} D \tag{9.14}$$

This is not independent of the others:

$$R_*^2 = \Theta D_*^3 \tag{9.15}$$

The relation reported by Soulsby and Whitehouse [430] is

$$\Theta_{cr} = \frac{0.3}{1+1.2D_*} + 0.055\left[1 - e^{-0.02D_*}\right] \tag{9.16}$$

This is corrected for silt and fine sands ($2 < D_* < 3$) and fits the threshold data for combined waves and current. Similar is the relation proposed by Davies [102]:

$$\Theta_{cr} = \frac{0.23}{1+1.2D_*} + 0.029\left[1 - e^{-0.04D_*}\right] \tag{9.17}$$

which is fine tuned to better fit the data for currents.

Intuitively, for any $\Theta \leq \Theta_{cr}$, a bottom sediment particle is immobile. Once the critical value is exceeded, particle motion is possible.

9.6.2 Entrainment Rate

All identical particles cannot be entrained simultaneously. Various factors will limit this rate. Among these are the ambient bed condition surrounding a particle.

If E is the sediment entrainment rate, mass/time/area, then the relevant dimensionless group is (Einstein [131]):

$$\phi_p = \frac{E}{\rho_s \sqrt{(s-1)gD}} \tag{9.18}$$

For each individual particle, let the probability of entrainment in Δt be $r_e \Delta t$. If N immobile particles at the bed have achieved Θ_{cr}, each with individual mass M, then $r_e = -\frac{1}{N}\frac{dN}{dt}$ will be their collective mean rate of entrainment.[2] We seek formulations for r_e, which has dimensions of inverse time. In specific experiments, it will be proportional to E/ρ_s and some geometric factors.

Van Rijn's [459] parametric pick-up function is a commonly used formulation for entrainment:

$$\phi_p = 0.00033 \left[\frac{\Theta - \Theta_{cr}}{\Theta_{cr}}\right]^{1.5} [D_*]^{0.3} \qquad \Theta > \Theta_{cr} \tag{9.19}$$

$$= 0 \qquad \Theta < \Theta_{cr} \tag{9.20}$$

The equivalent flux in terms of particles/time/area will be $\frac{E}{\rho_s f D}$, where f is a dimensionless geometric factor of order unity; henceforth $f = 1$ is assumed.[3] Thus the rate of particle entrainment is

$$r_e = 0.00033 \left[\frac{\Theta - \Theta_{cr}}{\Theta_{cr}}\right]^{1.5} [D_*]^{0.3} \sqrt{\frac{(s-1)g}{D}} \qquad \Theta > \Theta_{cr} \tag{9.21}$$

$$= 0 \qquad \Theta < \Theta_{cr} \tag{9.22}$$

Nielsen [328] reports that Equation 9.19 is valid for a Θ range of [0.06, 1] and a D range of [0.13, 1.5] mm for steady flows; and presents a modified form for unsteady flows.

For an example: consider a particle of medium sand, $[D, s] = [10^{-3}$ m , 2.65], giving $D_* = 25.3$. The entrainment rate would be

$$r_e = 0.11 \left[\frac{\Theta - \Theta_{cr}}{\Theta_{cr}}\right]^{1.5} \tag{9.23}$$

[2] For large Δt, the entrainment probability is $1 - e^{-r_e \Delta t}$.

[3] fD = ratio of particle volume to cross-sectional area; for spheres $f = 2/3$; a similar correction is potentially needed to account for the packing arrangement of particles in the bed; the peak packing factor is 0.91 (2-D) and 0.74 (3-D) for solid packing; under mixing, this would rise to unity.

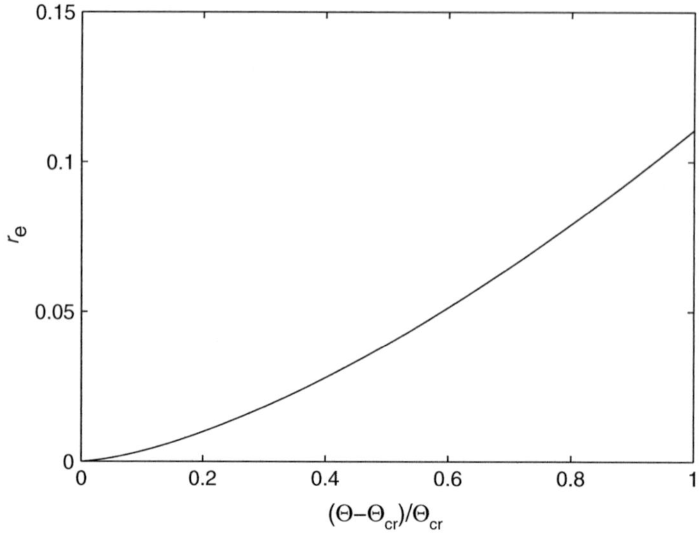

Figure 9.3. Entrainment rate (sec^{-1}) versus flow rate, for the example in Equation 9.23. $\overline{u}_{*cr}=0.0225$.

(Figure 9.3). Equation 9.16 gives $u_{*cr}=0.0225$ m/s; equivalently, in a well-mixed channel[4], $\overline{u}_{cr}=0.45$ m/s. This is a brisk flow; yet the entrainment rate is modest; approaching 10% per second when $\overline{u}=0.6$ m/s. At that point, the mean waiting time would be 10 seconds for a bottom particle to entrain. Lighter flows could have waiting times of order 50 seconds when the stress is only 20% above critical.

Soulsby et al. [429] introduce another distinction – that of trapped versus free particles. Free particles may entrain; trapped particles are hidden or caught among other particles in the bed, therefore not free to move without an enabling transition, keyed to fluid turbulence. We represent this in Figure 9.2 by the two substates M_f and M_t comprising M. If α is the rate of trapping of free particles; and β is the rate of freeing of trapped particles, the formulation is

$$\alpha=\alpha^*\left[1-e^{-\frac{(\Theta-\Theta_{cr})}{\Theta_s}}\right] \qquad (\Theta>\Theta_{cr}) \tag{9.24}$$

$$\beta=\frac{\gamma}{1-\gamma}\alpha \tag{9.25}$$

Both are zero for $\Theta\le\Theta_{cr}$. There are three parameters: α^* is the maximum α; γ is the equilibrium fraction of M_f to (M_f+M_t) particles for a given site; Θ_s is a scaling parameter. The following values are suggested based on sensitivity tests: $\Theta_s=0.1$, $\gamma=0.1$ and $\alpha^*=1.7\times10^{-7}$ s^{-1} [429].

[4] $c_d=0.0025$.

These are slow processes. α^* is the upper limit of the hiding rate; it is of order 1% per day. The exposure rate β is slower by a factor of 10 for this γ.
Both rates represent transfers of particles within M. Initiation by Θ_{cr} is necessary in either case. Only the M_f particles are eligible for entrainment. The trapping/free classification is analogous to the hiding/exposed classification used by others, although varying contexts can make this equivalence problematic.[5]

The bulk rates r_e, α and β apply to aggregates of particles. From the particle point of view, each individual moves among states M_f, M_t, and P with state transition probabilities $p = r\Delta t$, process r being r_e, α, or β as appropriate. The bulk property emerges, on average.[6]

9.6.3 Initial Vertical Position – Entrainment Lift

For an individual particle, the Shields criterion gives the threshold of motion, $\Theta \geq \Theta_{cr}$. Beyond that threshold, the probability of entrainment is $r_e \Delta t$; equivalently, the rate of particle entrainment is r_e.

This is enough for a particle to leave the M state. Entering the P state, however, requires that a particle acquire an initial position from which random walking will begin.

The simplest approach is to initialize $z = z_r$, that is, lifted one roughness height into the log layer. The subsequent random walking will then move the particle from the bottom according to the pseudo-velocity $\frac{\partial A_z}{\partial z}$, minus the settling velocity w_s; plus a random component. But the combination will drive some particles back to the M zone immediately. This would be inconsistent with the prior that the particle is being entrained. Essentially, this option (immediate reversal of entrainment) has already been accounted for in the entrainment probability.

So what is needed is a proper distribution of initial z, given that a particle is being entrained. The time step is spent (a) changing states and (b) lifting a particle to its initial position. The random walk begins in subsequent time steps. Among the requirements: the initial lift must be non-negative; and it should be sensitive to turbulence and the time step. The latter is satisfied by relating the lift distribution variance to $A_z \Delta t$ as already done in random walking. For the non-negativity, among the simplest one-parameter options is the half-normal – where one considers only the positive side of the Gaussian. Operationally, this is a simple matter: with Z the initial lift,

$$Z = S|\epsilon| \qquad \epsilon \sim \mathcal{G}(0,1) \tag{9.26}$$

[5] One can interpret the trapping process as burial, though as used here, "burial" is reserved for a slower process.

[6] There is a range of critical shear stresses that would entrain a sediment particle into the flow field (Einstein [131], Cheng and Chiew [82], Papanicolaou et al. [349], Ma et al. [290]).

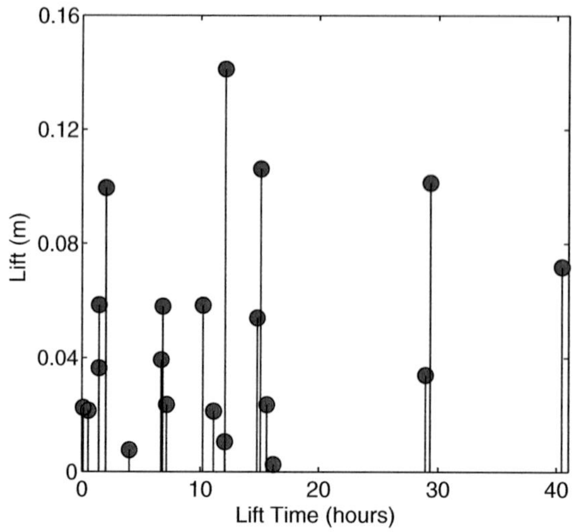

Figure 9.4. Lift events for 20 particles; $r_e = 10\%$ per hour. The half-normal distribution is used for the lift height.

This will have the properties[7]

$$\overline{Z} = S\sqrt{\frac{2}{\pi}} \qquad \sigma_z = S\sqrt{1-\frac{2}{\pi}} \qquad Z_{50} = S\left(0.477\sqrt{2}\right) \tag{9.27}$$

There is a single parameter S that scales the distribution and all three properties. If we set S such that $\sigma_z^2 = 2A_z\Delta t$, then we achieve a credible dependence on turbulence and Δt, qualitatively compatible with the onset of a dispersive random walk.

$$S = \sqrt{\frac{2\kappa u_* z_r \Delta t}{1-2/\pi}} \tag{9.28}$$

(Here we have set A_z at the particle's origin, $\kappa u_* z_r$.)

In Figure 9.4 we implemented this scheme – there are 20 particles in M_f that have achieved Θ_{cr}; the entrainment rate is $r_e = 0.1$ per hour. The individual lifts are plotted versus time.

Another simple non-negative lift distribution could be the uniform distribution with range $[0, S]$:

$$Z = S\epsilon \quad \epsilon \sim \mathcal{U}(0,1) \tag{9.29}$$

[7] Z_{50} is the median; $0.477 = \mathrm{erf}^{-1}(1/2)$.

Unlike the Gaussian, this distribution would never produce extremes beyond $Z = S$. Its properties:

$$\overline{Z} = \frac{S}{2} \qquad \sigma_z = \frac{S}{\sqrt{12}} \qquad Z_{50} = \frac{S}{2} \tag{9.30}$$

If as above we require $\sigma_z^2 = 2A_z\Delta t$, then the result is:

$$S = \sqrt{24\kappa u_* z_r \Delta t} \tag{9.31}$$

The mean lift would be 31% higher in the uniform case than in the half-Gaussian; and the median, 55% higher. The extremes are eliminated; the central part of the distribution is strengthened and flattened.

The exponential distribution is another candidate with only one parameter. It is explored in Problem 9.1

There are several other useful non-negative distributions, including the lognormal, the loglogistic, the Erlang/gamma family, and the Weibull. Each of these has two parameters, roughly corresponding to a shape or mean, and a length scale.

9.7 Vertical Motion: The Rouse Number and z_e

Once activated, a particle begins a random walk, advecting with the fluid velocity. The horizontal aspects of particle motion are covered elsewhere; no special considerations are added here. But the vertical transport deserves special attention. Both advective velocity and eddy viscosity are strongly depth-dependent in the log layer; notably, eddy viscosity A_z will increase from zero, linearly, as a particle rises from the bottom.

A particle random-walking will experience a pseudo-velocity $\frac{dA_z}{dz}$ (Section 4.3.4), such that the combined vertical velocity for a particle will be $w - w_s + \frac{dA_z}{dz}$. The first term is the fluid motion itself; the second represents sinking; and the third is the repulsive effect of an impermeable boundary, as described in Section 5.4. In the bottom boundary layer, w will vanish; and $A_z = \kappa u_* z$, increases upward. So a particle's vertical velocity will be $-w_s + \kappa u_*$. The ratio of these two is the dimensionless Rouse parameter P:

$$P \equiv \frac{w_s}{\kappa u_*} \tag{9.32}$$

So in the constant-stress layer, the vertical particle velocity W will be

$$W = \kappa u_* [1 - P] \tag{9.33}$$

When $P > 1$, the sinking effect is stronger and the particle will be pinned against the bottom; occasional random kicks will produce upward motion, but $z > 0$ is not

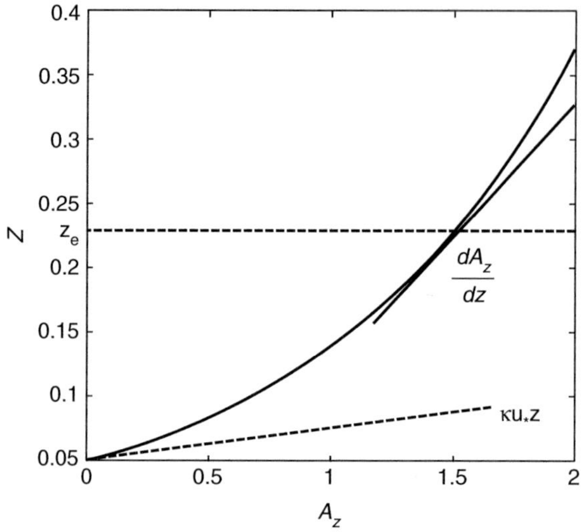

Figure 9.5. The equilibrium height z_e where $\frac{dA}{dz} = w_s < \kappa u_*$.

sustainable and the particle will fall back to the bottom. Horizontal motion during flight will be brief and small.

When $P < 1$, the sinking effect is not dominant everywhere; there will be a neutral or "equilibrium" point z_e around which a particle will hover by random walk:

$$w_s = \left. \frac{dA_z}{dz} \right|_{z_e} \tag{9.34}$$

(Figure 9.5). Above z_e, particles will spread out, the random-walk excursions $\sqrt{2A_z \Delta t}$ being larger, but on average will sink back to z_e; below z_e, smaller particle excursions but rising on average. Collisions with the bottom will limit vertical travel, adding to the concentrating effect below z_e.

As A_z generally scales with u_*, larger bottom stress will result in higher z_e. If A_z peaks with z, then z_e will occur below the peak.

9.8 Profiles

Many particles in random walk will exhibit variability around a stochastic steady-state profile.[8] If $C(z)$ is the time-averaged particle concentration,[9] then the required balance is

$$-w_s C = A_z \frac{dC}{dz} \iff \frac{1}{C}\frac{dC}{dz} = -\frac{w_s}{A_z} \tag{9.35}$$

[8] We assume the hydraulics are not changing.

[9] Alternately, $C(z)$ is the probability density for a single particle.

C will always decline with height. The local rate of decay is $\frac{-w_s}{A_z}$. The general solution to Equation 9.35 is

$$\ln\left(\frac{C}{C_0}\right) = -w_s \int_{z_0} \frac{dz}{A_z} = -P \int_{z_0} \frac{dz}{A_z/\kappa u_*} \tag{9.36}$$

(A_z scales with u_* to meet the bottom boundary condition; so $A_z/\kappa u_*$ is a convenient normalization.) Much will depend on the profile $A_z(z)$.

If we confine attention to the log layer itself, we have $A_z/\kappa u_* = z$, so the result is simple:

$$\ln\left(\frac{C}{C_o}\right) = -P \ln\left(\frac{z}{z_0}\right) \Longrightarrow \frac{C}{C_0} = \left(\frac{z}{z_0}\right)^{-P} \tag{9.37}$$

C decays geometrically with z; the decay is larger for larger P. This expresses a near-bottom property that endures as the A_z profile becomes more realistic – but unbounded growth in A_z is not possible, and the profiles away from the log layer need to recognize this.

Case 1: Deep Boundary Layer For example, consider the case

$$A_z(z) = \kappa u_* \left[z e^{-\alpha z}\right] \tag{9.38}$$

$A_z(z)$ matches the wall boundary condition at bottom, with free decay parameter α. It peaks above the bottom, then trails off. At the top, there is no mixing, effectively the bottom boundary layer is isolated from any surface stress.

The profile could be representative of a bottom boundary layer in deep water, with tidal flow but no wind. It is possible to obtain an exact solution:

$$\ln\left(\frac{C}{C_0}\right) = P \cdot [E_1(-\alpha z) - E_1(-\alpha z_0)] \tag{9.39}$$

$$\frac{C}{C_0} = \left[\frac{e^{E_1(-\alpha z)}}{e^{E_1(-\alpha z_0)}}\right]^P \tag{9.40}$$

with the special function E_1 the exponential integral; and $C(z_0) = C_0$.

This case is illustrated in Figure 9.6. All profiles originate from the bottom and decrease monotonically with height. Low P allows significant C above the point of maximum A_z; increasing P concentrates the profile nearer the bottom.

Case 2: Shallow, Well-mixed For a second example, consider the logistic profile for A_z,

$$A_z(z) = \kappa u_* z \left(1 - \frac{z}{H}\right) \tag{9.41}$$

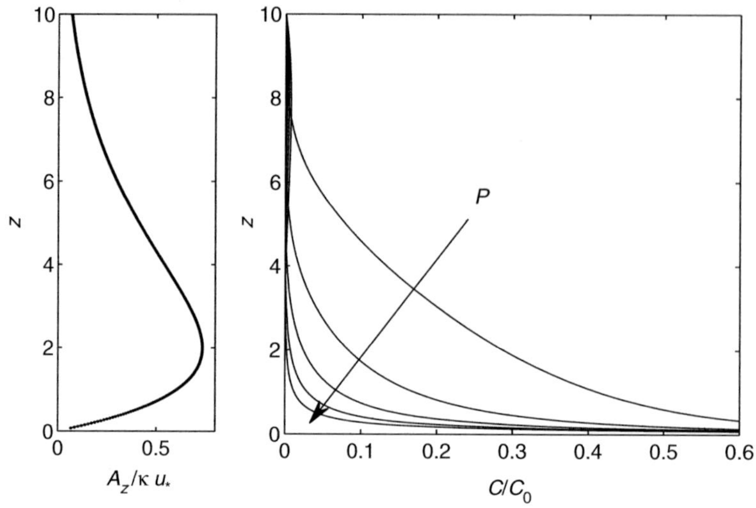

Figure 9.6. Case 1: Deep Boundary Layer. Profiles for $A_z(z)$ and $C(z)$, for $P = [0.25, 0.50, 0.75, 1.0, 1.25]$. $\alpha = 0.5$; $z_0 = 0.05$. The full depth is not shown; the surface is far removed from the bottom boundary layer, $H > \frac{7}{\alpha}$.

and illustrated in Figure 9.7. This is reminiscent of shallow water ($H = 10$ m), with surface and bottom stress the same. (For example, wind balancing bottom stress and both boundary layers overlapping). The law of the wall is enforced at both top and bottom. The C profile in closed form:

$$\frac{C}{C_0} = \left[\frac{(H-z)z_0}{(H-z_0)z}\right]^P \tag{9.42}$$

Near bottom, the Case 2 profiles are very similar to Case 1. Higher in the water column, the mixing from both boundaries overlaps, sustaining higher C there. The case $P = 0.25$ is the most obvious; in that case, C has a sigmoid shape, both top and bottom being nearly the same. (There is still asymmetry in this low-P case, however, as w_s biases the profiles downward and the source of all particles is at the bottom.) These are the classic Rouse profiles.

MacDonald et al. [291] and Davies [102] offer regression relations for the centroid z_c for the (Case 2) Rouse profiles. As can be seen in Figure 9.8, z_e is consistently below z_c, following the same trends. z_e is zero for $P > 1$, while z_c is small but positive.

The overall trends in these two cases are qualitatively the same: uniformly decreasing C with z; higher P results in lower, more bottom-trapped profiles. Both share the feature that C is unbounded at the bottom, and vanishes at the surface (asymptotically for Case 1, abruptly for Case 2 as the wall is encountered.

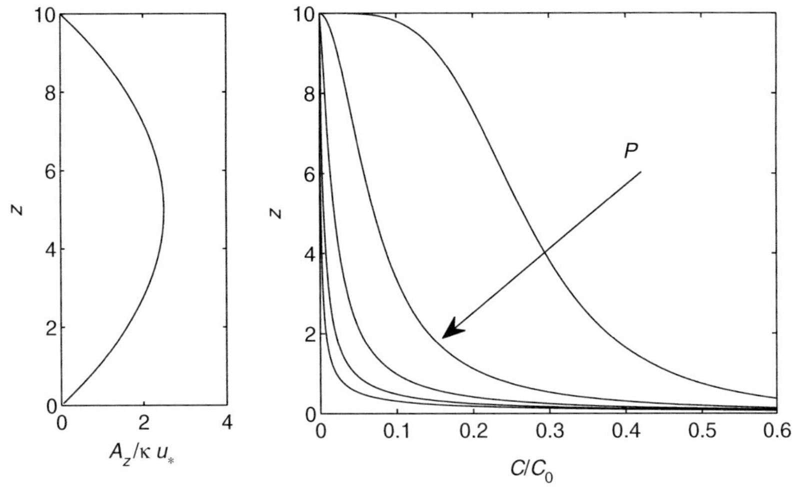

Figure 9.7. Case 2: Well-mixed. Profiles for $A_z(z)$ and $C(z)$, for $P = [0.25, 0.50, 0.75, 1.0, 1.25]$. The surface boundary layer is evident here for low P; the bottom layer resembles Case 1. $H = 10$; $z_0 = 0.05$;

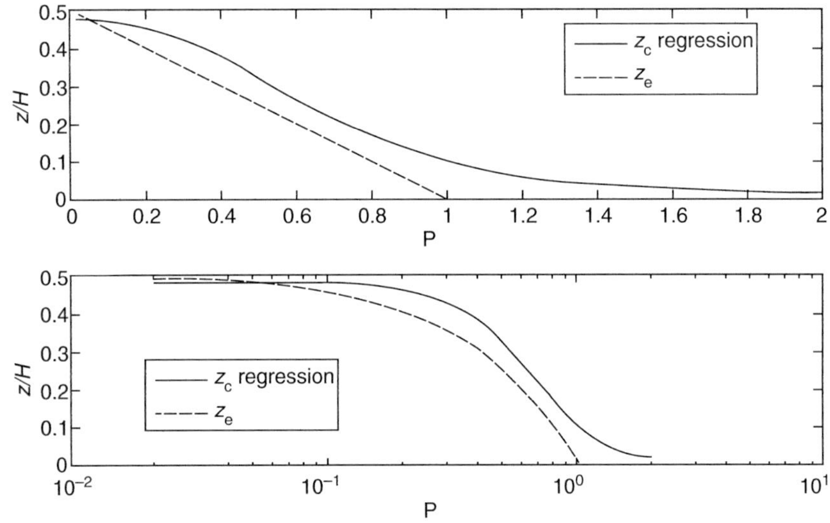

Figure 9.8. Case 2: Well-mixed. The centroid z_c of the Rouse profile, from the regression [291]; and z_e; versus P.

9.9 Flight Simulations

As examples, we flew some particles vertically, using AR0 with Gaussian random numbers, and time stepping as in Visser [464] to account for variable A_z. When particles collide with the top or bottom boundary, their motion is arrested. Surface particles continue walking in the next timestep; bottom particles are absorbed into the mixing zone, and subject to the reentrainment process in subsequent timesteps.

Table 9.2. *Particle Parameters for the Flight Simulations*

D	s	D_*	w_s	u_{*cr}	$\overline{u}_{cr}$	P	u_*	r_e
0.0001	2.65	2.53	0.0079	0.0112	0.2233	0.5	0.0385	6.286
						1.0	0.0192	0.4842
						1.5	0.0128	0.03616
0.001	2.65	25.28	0.12	0.0225	0.4507	0.5	0.5868	1900
						1.0	0.2934	242
						1.5	0.1956	71

Note: w_s from 9.7; u_{*cr} from Equation 9.16; r_e from Equation 9.21. $\overline{u}_{cr}$ is an approximation for well-mixed flow, $c_d = 0.0025$. The examples given use $D = 0.0001$. Problem 9.9 explores the $D = 0.001$ case.

Particles are initiated at the bottom; Θ is constant, above Θ_{cr}. We use the two representative profiles for $A_z(z)$ introduced above.

Simulations

Case 1: Deep Boundary Layer. For $P = 1/2$: simulations were made with the Case 1 viscosity, for $D = 0.0001$, $s = 2.65$. Parameters are listed in Table 9.2. Hydraulically, the bottom condition $\frac{dA_z}{dz} = \kappa u_*$ is always enforced. The bottom condition on particles in M used the half-normal launch height, with entrainment rate r_e as in Van Rijn. There is no surface influence; the depth here is effectively infinite. Figure 9.9 shows the viscosity profile and 11 random walks, for $P = 0.5$; z_e is plotted for reference in each case. A_z peaks at $z = 2$; while particles sometimes venture high relative to this peak, the trajectories are all contained under $z \simeq 5$.

Despite its appearance, z_e is a good estimate of the median z. Figure 9.10 shows for each of several equivalent flights, the median z for each flight since inception. The flights are statistically equivalent to those in Figure 9.9. The time axis is longer; the vertical is shrunk as the medians are much less variable.

Figure 9.11 depicts all points of several random walks – a "cloud" version of Figure 9.9, representing statistically identical random walks.

Case 2: Well-mixed. This same set of calculations was repeated with the A_z profile changed to reflect the presence of surface and bottom stress, in shallow water. All other parameters were unchanged. Results appear in Figures 9.12, 9.13, and 9.14. The particles fly higher, achieving closer to a well-mixed condition, but are still biased toward the bottom.

Problem 9.1 If the entrainment lift is exponentially distributed: (1) Find its mean and variance in terms of the exponential rate parameter. Relate this to the ambient turbulence and the time step as in the half-normal or uniform cases. (2) Compare your results to those in the text for the lift mean and variance for half-normal and uniform distributions.

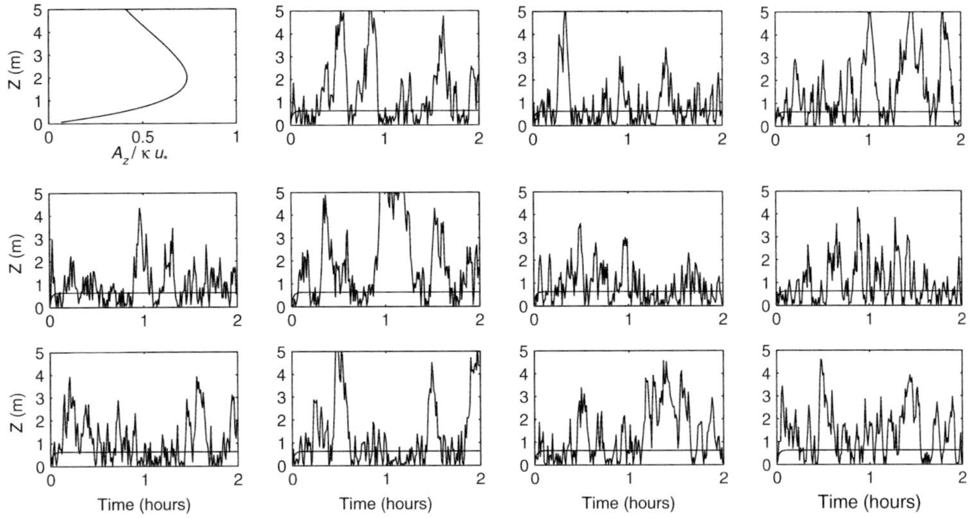

Figure 9.9. Case 1: Deep Boundary Layer. Individual random walks for $P = 0.5$. The smooth line in each case is z_e. The eddy viscosity profile is illustrated in the upper left panel.

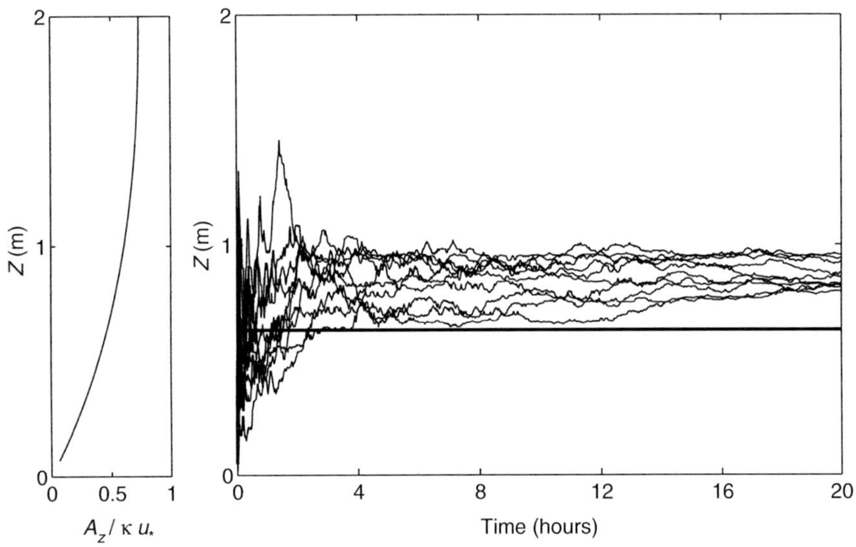

Figure 9.10. Case 1: Deep Boundary Layer. Running median z for individual random walks, $P = 0.5$. As in Figure 9.9, the smooth line is z_e. The flights are statistically the same as in Figure 9.9. The duration is longer here to allow the median to overcome the initial condition; and the vertical axis is shrunk as the medians are smoother.

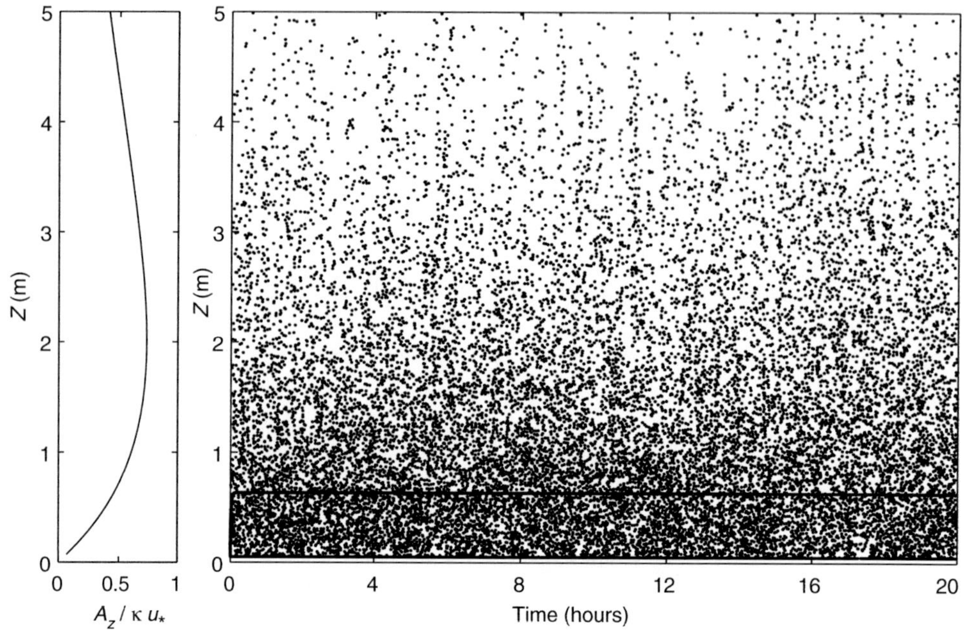

Figure 9.11. Case 1: Deep Boundary Layer. Cloud depiction of random walks as in Figures 9.9 and 9.10. Each point of each of several random walks is plotted independently.

Problem 9.2 Find the maximum A_z for the Case 1 profile, Equation 9.38; and the height at which it occurs. What is their dependence on the bottom stress?

Problem 9.3 For Case 1, find and plot the relation between z_e and P.

Problem 9.4 (1) For the A_z profile in Case 1: find the peak of A_z and its location z_p. (2) For $z > z_\infty$, A_z diminishes to below 2%. What is z_∞?

Problem 9.5 a) For Case 2: (1) What is the peak value of A_z? (2) Find and plot the relation between z_e and P.

Problem 9.6 Develop Case 3 where $A_z = A_o(1 - e^{-rz})$, with parameter r. (1) What must A_o be to satisfy the bottom boundary condition? (2) Find and plot the relation between z_e and P for this case. (3) Find and plot the profile C/C_0. (4) Interpret the results – are they consistent with the discussion in the text? How do they differ from Cases 1 and 2?

Problem 9.7 Same as Problem 9.6, but use Case 4: $A_z = A_o\left[\frac{z}{z+z_h}\right]$, with the half-saturation parameter z_h.

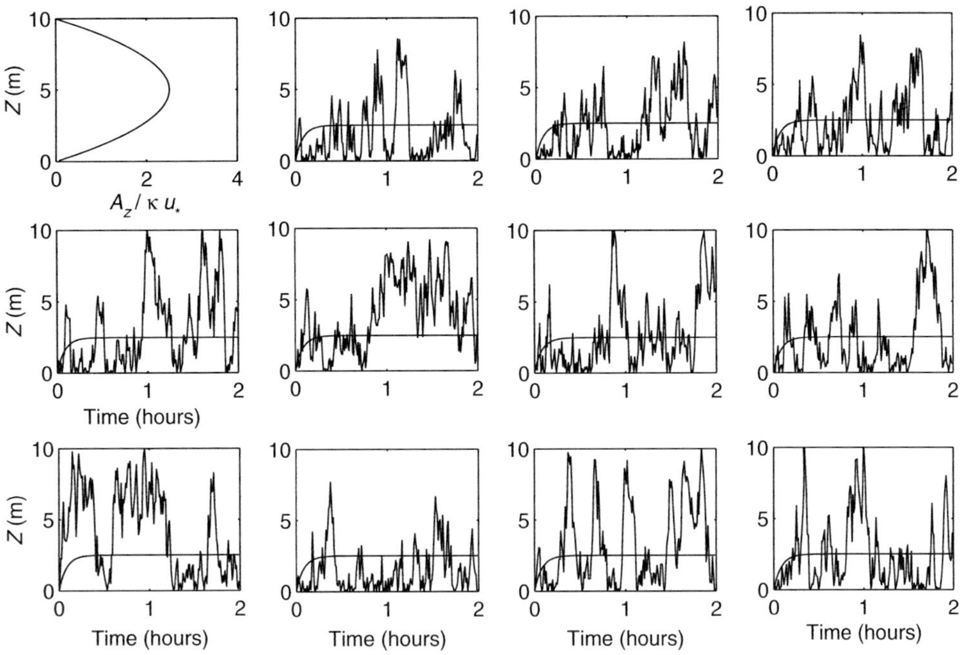

Figure 9.12. Case 2: Well-mixed: Individual random walks for $P = 0.5$. The smooth line in each case is z_e. The eddy viscosity profile is illustrated in the upper left panel. The depth is 10 m, the full vertical axis.

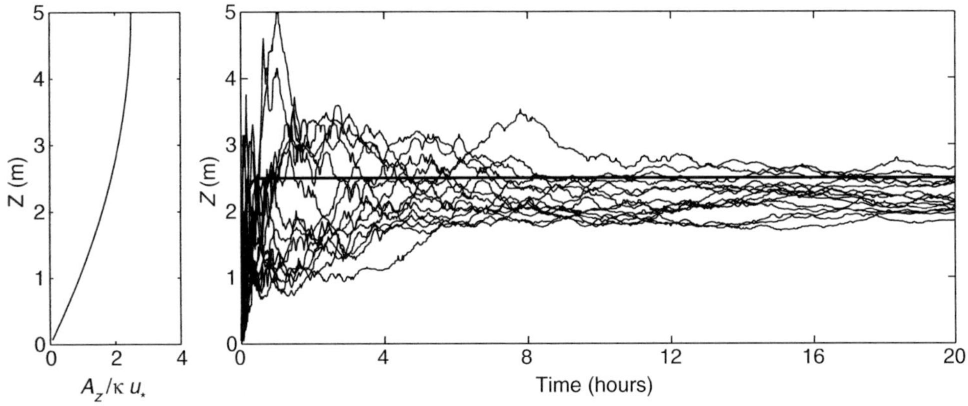

Figure 9.13. Case 2: Well-mixed: Running median z for individual random walks, $P = 0.5$. As in Figure 9.12, the smooth line is z_e. The flights are statistically the same as in Figure 9.12; the duration is longer; the vertical axis stops at half-depth.

Problem 9.8 Redo the flight simulation examples shown, $D = 0.0001$, $P = 0.5$; but with w_s reduced by a factor of 10. Explain your results and why/how they differ from the examples shown.

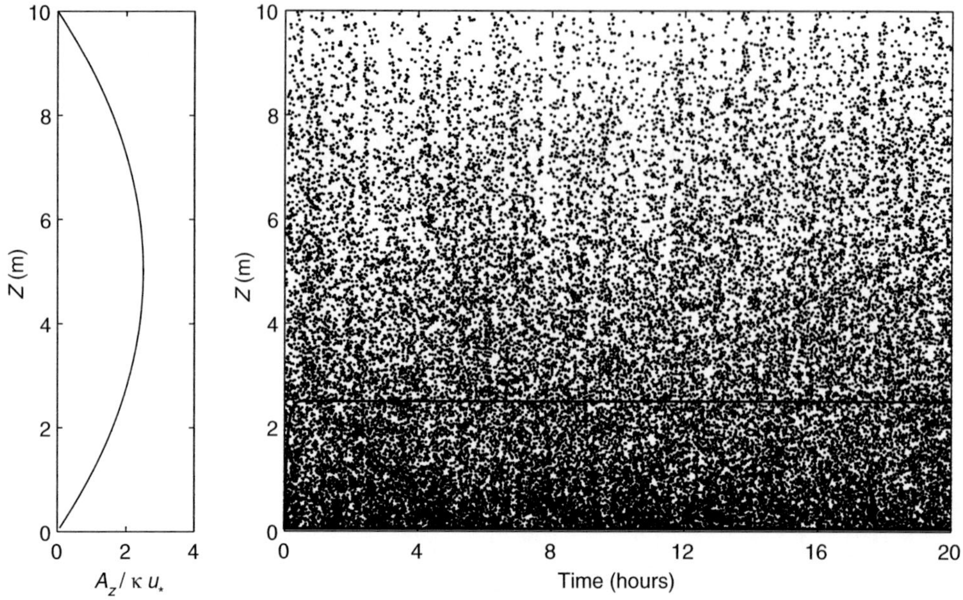

Figure 9.14. Case 2: Well-mixed: Cloud depiction of random walks as in Figures 9.12 and 9.13 . Each point of each of several random walks is plotted independently.

Problem 9.9 Redo the flight examples shown, but use $D = 0.001$ and the parameters in Table 9.2 for $P = 0.5$. Explain your results and why/how they differ from the examples shown.

9.10 Saltation

Saltation refers to particles rolling or bumping along the bottom. This occurs at high P; it is not a property of the particle alone, but of the particle under stress. At low stress, even the lighter particles may saltate. So, saltation should be a simulation outcome, not a prior assumption. The Rouse number P is key.

In the above simulations, a pelagic particle enters the M state freely when it hits bottom, $z < z_r$; it is then a candidate for reentrainment. For low P, particles spend much of their time aloft, so the reentrainment events are not dominant; and the entrainment rate is very high in these simulations, essentially instantaneous. For high P, the reentrainment rate becomes more important. As above, we will retain $D = 0.0001$.

Figure 9.15 shows $P = 1.5$ trajectories. Particles stay near the bottom, and are characterized by less frequent reentrainment events, followed by re-sinking via random walk. The lift variability and its extrema are important. The sinking effect makes for brief walks. The higher P keeps particles closer to the bed, and somewhat less frequent reentrainment. This case is depicted in "cloud" format in Figure 9.16.

Qualitatively, these saltating walks resemble the low-P outcomes, having been generated in the same fashion. The episodic nature of the separated reentrainment events is enhanced. These features require no special treatment, they are natural outcomes of a single basic set of particle-based relations – state transitions and random walks.

Problem 9.10 Redo the saltating flight examples shown for $P = 1.5$, but reduce w_s by a factor of 10. Explain your results and why/how they differ from the examples shown.

Problem 9.11 Redo the saltating flight examples shown, but use $P = 3$. Explain your results and why/how they differ from the examples shown.

9.11 Burial

An individual sediment particle is considered buried once it makes the transition from the mixing phase into the benthic phase (Figure 9.2). This phase is isolated from the water column. That burial is completely permanent is unlikely; surely some buried particles can become available for activation under the right conditions, and the recovery of buried sediments is a key concern.

Physically, benthic sediment mixing and compaction are responsible for vertical particle transport in this phase. A full accounting for them is beyond the present scope; effectively a vertical transport model is needed, underlying every horizontal node or element. At the minimum, the burial process needs to be operative for a particle in M; and its reverse, the reentry from B to M. Accounting for these particle fluxes requires at least an inventory of (horizontally immobile) B and M particles everywhere, and the particle transitions between them.

Jumars et al. [218] present a Markovian model of flow–sediment–organism interaction. They assume that once buried, a sediment particle can never go back to the free (mixing) state; and assume that the rate of particle burial μ is constant across all particles and independent of t. If there are N particles in M, then the burial rate would be $\mu = \frac{1}{N}\frac{dN}{dt}$. For an individual particle, the corresponding probability of burial during a time step is $\mu \Delta t$. The single free parameter is the particle burial rate μ , a constant.

Sanford [401], in his probabilistic approach to the burial problem, gives the *cumulative* probability of burial since deposition in M as:

$$P_B(t) = 1 - e^{-t/T_b} \tag{9.43}$$

with t the time elapsed since deposition. This is the exponential or waiting time distribution. For a particle in M, the probability of burial in the coming time step would be $\frac{\Delta t}{T_b}$, independent of t. This is the same as Jumars et al. [218] above, with

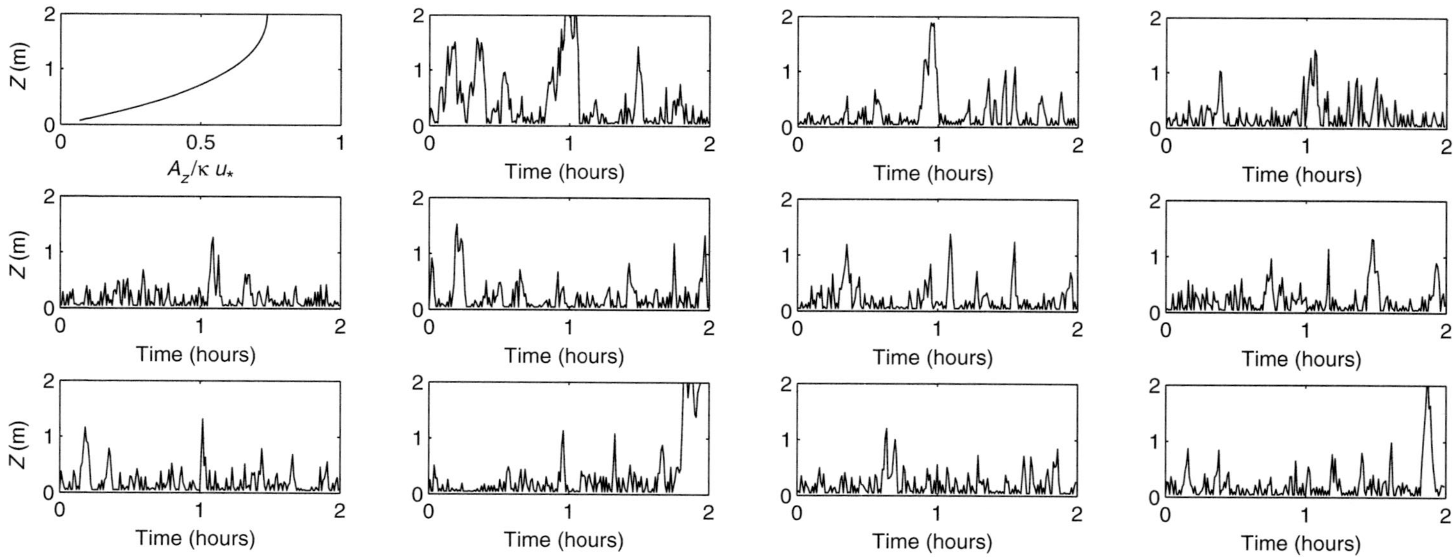

Figure 9.15. Saltating random walks. Case 1 with $P = 1.5$.

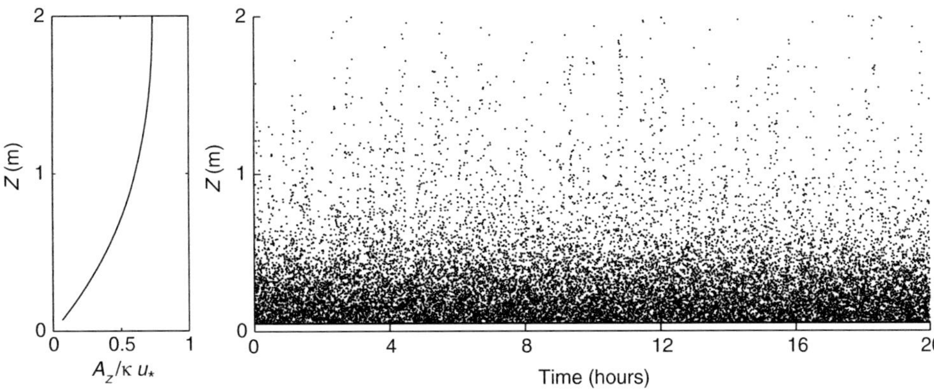

Figure 9.16. Case 1 with $P = 1.5$: Cloud depiction of saltating random walks. The walks are statistically the same as in Figure 9.15.

$\mu = \frac{1}{T_b}$. The parameter T_b is the natural time scale for burial, or the characteristic burial time; it is approximated as

$$T_b = nT_r \left(1 - \frac{0.25}{n^2}\right) \tag{9.44}$$

where n is an empirical mixing ratio and T_r is the duration of one resuspension cycle. It is obvious that the choice of n, T_r, and thus T_b is not straightforward. Sanford [401] suggests taking T_r as half the tidal cycle, that is, 0.259 days if fine sediments and only tidal resuspension are considered. He also gives values for n in the interval [26, 114], calculated from tidal resuspension and geochemical tracer data for Chesapeake Bay. Higher n values represent high mixing regions. We thus have two free parameters.

MacDonald et al. [291] use a probabilistic approach that accounts for the time-dependent depth of sediment mixing (based on Θ) and the particle burial depth. Both evolve with the sedimentary regime. This brings a great deal of complexity to bear, as one is well beyond particle simulation and requires potentially detailed (z,t) calculations of sediment evolution (deposition, scour, compaction, diagenesis) at all bottom locations, exchanging with the overlying water column. The formulations in this model represent a coupling of burial transitions with a vertical depiction of the buried particle transport in the sediment column, itself responsive to the bed stress, and the aggregate sediment transport rate.

Soulsby et al. [429] similarly address bedform evolution. Among other things, there is a special focus on tracked particles and the indigenous or background particles, the former serving like tracers without being required to do all the work of bedform evolution.

9.12 Theoretical Extensions

There are important theoretical extensions that examine the near-bed transport processes.

Wu and Chou [488] divide entrainment into "lifting" and "rolling" modes of initiation, and relate these theoretically to grain geometry plus turbulent velocity fluctuations near-bed. The focus is on separating the two forms of entrainment at low Θ. The theoretical results show lifting dominant for $\Theta > 0.6$, accounting for $\geq 90\%$ of entrainment; and rolling dominant for $\Theta < 0.05$.

Zhong et al. [503] developed an application of kinetic theory, wherein the Boltzmann equation is used to describe the distribution of particle velocities. The results recover the data supporting the Van Rijn pickup functions [459].

Several report advanced numerical simulations of two-phase turbulence near the bed, including motion and collisions among particles. Chan-Braun et al. [74] report the replication of Rouse profiles.

Duran et al. [122] examine particle entrainment and transport near the bed. Ji et al. [209] examine the interplay of entrainment and turbulence substructures for coarse particles near the bed.

At this high level of resolution, two-way coupling between sediment particles and water is necessary, in the context of a general theory of multiphase turbulence. Advanced computation will be instrumental.

9.13 A Simple Particle Model

We have arrived at a set of 3-D relations adequate for a simple 3-D model, Figure 9.17. A particle may be in only one state; the transition probabilities among them are given in terms of $P = r\Delta t$ where r is the appropriate rate as illustrated. Particles already entrained, undergo 3-D random walk within the hydrodynamic grid plus the log layer below it. If such a particle "walks" into the bottom, it enters M_F, ceases motion, and awaits transition to any of the other three states.

The Coastal Ocean Model provides gridded fields $[\mathbf{v}, A_h, A_z, \tau]$. Each particle k has properties $[D, s, w_s, \tau_{cr}]_k$. Variables for each particle include a logical location in one of the four boxes; and position and velocity $[\mathbf{X}, \mathbf{V}]_k$ appropriate to its logical location.

The intersection of fluid stress with a particle's parameters is captured in the dimensionless stress Θ, different for each particle. All the transition rates among the boxes depend on $\Theta > \Theta_{cr}$ – essentially, bottom stress must energize bottom particles before transitions are possible. The random walk embeds the hydrodynamic eddy dispersion coefficients, plus the particle's sinking velocity. The Rouse number P expresses the ratio of sinking velocity and fluid stress at the bottom, and conditions the vertical part of the random walk. Particles in either of the M boxes are susceptible to burying.

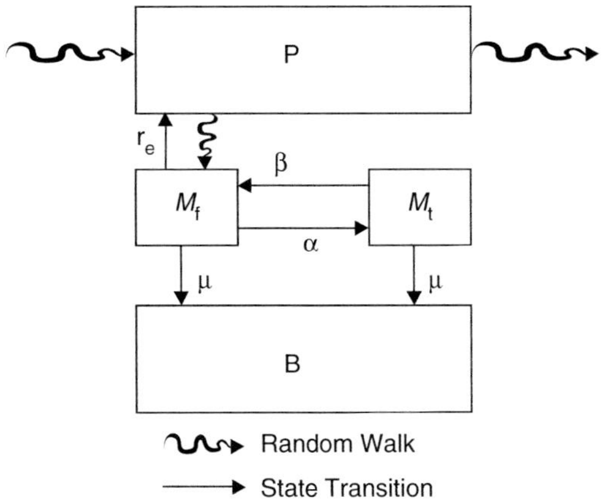

Figure 9.17. Four-state, 3-D particle model for noncohesive sediments.

In this framework, the distinction between "bed-load" and "suspended load" are not needed – these are modes of transport, outcomes that emerge from a simulation. Similarly, "saltation" as a mode of particle motion is a simulation outcome that emerges. No special relations are injected to represent these.

As in Figure 9.1, particles in this representation are capable of moving between a Coastal Ocean Model where they execute 3-D random walking and acquire horizontal displacement, and a vertical Sediment Model responsible for sediment biochemistry, bioturbation, compaction, formation of bedforms, and so forth with negligible horizontal movement.

An example is shown by Aretxabaleta et al. [10]. The concern was winter redistribution of bottom-dwelling cysts in deep locations in the Gulf of Maine. Particles were initialized in the benthos and entrained into a 3-D ocean model when critical stress was reached, with vertical position given by an erosion model. Storm events were capable of generating important horizontal displacement for some particles, before the sinking returned them to the benthos.

There are several additional requirements in 2-D; an example is described in Section 9.14.

9.14 2-D

Sediment particle motion is fundamentally 3-D. However there is a significant need to deal with 2-D transport data and model products. In those cases one has from the Coastal Ocean Model, only vertically averaged velocity and bottom stress for the hydraulic inputs. These have to produce horizontal sediment motion, which will be heavily dependent on the vertical position of a particle. In 3-D this is not a problem since individual particles will be tracked in 3-D and both velocity and A_z profiles are continuously available for the purpose. In 2-D the vertical needs to be parameterized before particle simulations begin.

There is a large amount of information available in this context. Among the useful 2-D concepts are (a) the distinction between bed-load and suspended load modes of transport, the former for particles close to the bottom; and (b) the use of Rouse profiles (an analytic result) for the vertical positioning of suspended particles, so that horizontal velocity can be obtained.

9.14.1 Bed-Load – Suspended Load Transport

Once a particle is entrained into the flow field (i.e., makes the transfer from the mixing phase M to the pelagic phase P), it may move as bed-load or suspended load. Whether a sediment particle will be rolling and hopping along the bed as bed-load or will be transported farther up in the fluid column as a suspended load is a complex function of sediment characteristics and turbulence properties of the flow. It is usually assumed that any form of transport in which particles are supported by momentum transfer from other particles more or less in contact with the bed can be called bed-load (Soulsby [427], Fredsoe and Deigaard [161]). At small transport rates, this transport occurs as particles rolling, sliding, or jumping along the bed in what is called the saltating regime, the top of which defines the thickness of the bed-load layer, δ. The idea here is that the turbulent mixing in this layer is so small that the particles cannot be entrained into suspension. δ has been the subject of several studies owing to its importance in transport rate calculations. Einstein [131] takes $\delta = 2D$. Van Rijn [460] relates δ to D_*, Θ_{cr}, and Θ in the following manner:

$$\delta = D\left[0.3D_*^{0.7}\left(\frac{\Theta - \Theta_{cr}}{\Theta_{cr}}\right)^{0.5}\right] \tag{9.45}$$

with an inaccuracy of approximately 10%. Davies [102] states the bed-load is assumed to be transported at a height of half the roughness height above the bed, that is, $k_s/2$, where k_s is the Nikuradse roughness length ($\sim 30z_r$). There are several other formulae derived for δ, a good summary of which can be found in Ghoshal and Pal [171]. At larger transport rates close to the limit of suspension, however, particles may move just above the bed in several layers, creating what is called a sheet flow. Although bed-load is usually considered to be a form of transport for coarse grains (i.e., coarse sand and gravel), it is intuitive that the limits are not clearly defined as finer sediments may also move as bed-load, especially as sheet flow on plane beds. As bed shear increases more, the limit of suspension is reached and the turbulent lift forces on the sediment particles become large enough to keep them in suspension in the water column. This situation lasts until the settling (or fall) velocities become large enough to deposit sediment particles back on the bed. As particles move at different speeds in these regimes, the flow velocities should be adjusted accordingly in the Lagrangian sediment model for each particle.

The traditional approach (e.g., Soulsby et al. [429]) is to use the suspension criterion $u_* > w_s$. In terms of the Rouse number, $P < \frac{1}{\kappa}$ gives suspension, with $\kappa = 0.41$ being the von Karman constant. So, once the entrainment is known, a sediment particle is considered suspended when $P < 2.43$. Above this, the transport mode is bed-load.

Alternatively, Davies [102] states that for $1 < \frac{\tau}{\tau_{cr}} \leq 8$, the bed-load transport occurs in the form of saltation and for $8 < \frac{\tau}{\tau_{cr}} \leq 14$, it occurs in the form of sheet flow. For values larger than 14, all particles move as suspended load, with a practical upper limit of 65. This classification is based on Θ alone, independent of w_s.

9.14.2 Bed-Load Particle Velocity

There are several methods to estimate the velocity of a bed-load particle. We will stick to the upper case/lower case convention: U is a Lagrangian particle velocity; u is the gridded fluid velocity. In 2-D, we have only $\overline{u}$ and u_*.

For particles undergoing bed-load transport, the simplest method would be to take an approach similar to Bogardi [42] and assume that the velocity of a bed-load particle, U_B, is on average a fraction of the vertically averaged fluid velocity $\overline{u}$. Bogardi suggests values in the range $\frac{1}{4}\overline{u} \geq U_B \geq \frac{1}{7}\overline{u}$. Following this method, Bilgili et al. [33] used a ratio of $\frac{1}{6}$ and obtained Lagrangian bed-load transport residuals consistent with field observations of shoal build-up. Soulsby [429] states that for bed-load particles traveling just above the threshold, typical near bed speeds are on the order of 15% of the current speed. These increase up to 50% of the near-bed current speeds as shear stresses increase toward sheet flow conditions.

Hunt et al. [201] choose Nino and Garcia's [331] dimensionless particle velocity formula in their model of saltating bed-load particles of juvenile bivalves:

$$U_B = u_* \cdot \left[a - b\,\Theta^{-0.5}\right] \tag{9.46}$$

Here a and b are coefficients ranging between $[11.8, 12.3]$ and $[1.4, 1.8]$ respectively for various grain sizes, with averages of 12 and 1.6.

Soulsby et al. [429], Davies [102], and MacDonald et al. [291] all adopt Fredsoe and Deigaard's [161] formula for the mean bed-load particle transport velocity, based on the bed-load transport method already developed and published by Fernandez Luque and Van Beek [145] and Engelund and Fredsoe [136]:

$$U_B = u_* \cdot \left[10 - 7\sqrt{\Theta_{cr}/\Theta}\right] \tag{9.47}$$

Here, u_* and Θ should ideally be the skin friction velocity and the skin friction Shields parameter if they are known.[10]

Abrahams and Gao [1], however, state that Equation 9.47 overestimates U_B when $\Theta \geq 1$ since it was developed for flows with low bed shear stresses. For higher values of Θ, they suggest using:

$$U_B = \overline{u} \cdot \left[0.6 \left(1 - \frac{\Theta_{cr}}{\Theta} \right)^{0.4} \right] \tag{9.48}$$

with the warning that further testing with more datasets is needed to establish its general applicability.

Van Rijn [460] gives a similar expression:

$$U_B = \overline{u} \cdot \left[9 + 2.6 \log D_* - 8 \left(\frac{\Theta_{cr}}{\Theta} \right)^{0.5} \right] \tag{9.49}$$

where D_* is the dimensionless particle size based on D_{50}. Van Rijn reports that Equation 9.49's inaccuracy is on the order of 10% when fitted to measured sand data by Fernandez Luque [144] and Francis [158].

Finally, Julien and Bounvilay [217] propose an equation for bed-load particles rolling on smooth and rough plane beds. Their experimental results lead to:

$$U_B = u_* \cdot [3.3 \ln \Theta_{br} + 17.7] \tag{9.50}$$

where Θ_{br} is the boundary roughness Shields parameter using k_s (the Nikuradse roughness length, $\sim 30 z_r$) rather than D.

Each of these relations is accompanied by important provisos reflecting the experimental context.

9.14.3 Suspended Load Particle Velocity

Determining the appropriate speed at which suspended sediment particles are advected horizontally is also of importance. The simplest approach is to assume that all suspended particles travel at the same speed as the water particles around them. For the depth averaged case, this would simply be $\overline{u}$. For a more realistic approach, however, one needs to predict at which depth an active particle is located in the water column at a specific time to calculate the advection velocity U_S. Davies [102] and McDonald et al. [291] both adopt the concentration curves of Rouse [392] to calculate

[10] Soulsby et al. [429] report that Θ and u_* should be replaced by Θ_{max} and the mean friction velocity ($u_{*,m} = \sqrt{\tau_m/\rho_w}$) over a wave cycle for the wave plus current case.

the centroid height, z_S:

$$\frac{z_S}{H} = 0.0398 \cdot 10^{-1.8 \cdot \tanh[1.2(\ln P) - 0.4]} \tag{9.51}$$

For the 2-D depth averaged case, the corresponding particle advection velocity, U_S, can be extracted using the law of the wall logarithmic profile once z_s is known:

$$U_S = \frac{u_*}{\kappa} \ln \frac{z_S}{z_r} \tag{9.52}$$

z_r is the hydraulic roughness length as used throughout this chapter (distance from the bed at which the law of the wall velocity goes to zero).[11]

9.14.4 Sediment Dispersion Coefficients

Particles in motion will have both advective velocity plus a random walk. The latter requires that the random part have variance that scales with a dispersion coefficient. Dispersion coefficients are part of the gridded output of many 3-D coastal ocean models; in 2-D situations, one often needs to add relations for these.

Horizontal Dispersion. Soulsby et al. [429] report that a constant horizontal dispersivity of $A_{ho} = 0.2 \ \mathrm{m^2 s^{-1}}$ is often used in coastal sediment transport applications. Chapra [78], Rathbun and Kennedy [368] and Fischer et al. [153] relate the horizontal dispersivity to the flow depth, H, and the intensity of turbulence, the latter being quantified by the shear velocity, u_*. The resulting horizontal dispersion coefficient, is given as:

$$A_h = A_{ho} \cdot H u_* \tag{9.53}$$

Rathbun and Kennedy [368] report that for suspended sands, the value of A_{ho} should be on the order of 0.22. Fischer et al. (1979) [153] further speculate that although the values of A_{ho} are within the $[0.1 - 0.2]$ range for straight uniform channels, they are usually in the $[0.4 - 0.8]$ range for natural channels, with increasing values corresponding to increasing channel irregularities and complexities. MacDonald et al. [291] also use this approach to quantify dispersion with a A_{ho} range of [0.1–0.6]. For larger particles moving as bed-load, A_{ho} may be more than two orders of magnitude smaller, as reported by Rathbun and Kennedy [368]. This suggests that the importance of the dispersive processes diminishes with increasing sediment grain diameter, an expected result.

[11] The Nikuradse roughness length $k_s \simeq 30 z_r$ for hydraulically rough turbulent flow (boundary Reynolds numbers $R_b = u_* k_s / \nu$ larger than approximately 70–100).

Vertical Dispersion. Considering only the effect of bottom turbulence and in the absence of stratification, Jobson and Sayre [211] and Fischer et al. [152, 153] suggest an approach similar to Equation 9.53, taking $A_{zo} = 0.067$. If there is stratification, the formula by Pritchard [364] can be used:

$$A_z = 8.5910^{-3}\left[\frac{H\bar{u}}{(1+0.276Ri)^2}\right]\left[\frac{z}{H}\frac{(H-z)}{H}\right]^2 \tag{9.54}$$

Here, Ri is the gradient Richardson number and z is the particle height above bottom. This is an example of a "mixing length" type of closure reviewed in Chapter 6. There are others, all of which seek the same effect – the reduction of vertical mixing under stratification.

9.14.5 A Generic Scheme

Figure 9.18 shows a generic flowchart for noncohesive 2-D sediment applications. Although the input velocity field is vertically averaged, the velocity gradient in the vertical is roughly represented by the incorporation of the Rouse centroid height concept. This allows particles to be advected by the horizontal velocities at this height, approximated using the law of the wall.

Within this generic framework, the possibilities are many, and new knowledge can be incorporated as it becomes available. For instance, another bed-load model could be incorporated to generate velocity. Or, since the mobility of a particle does not necessarily imply entrainment, a probabilistic approach could be built into the scheme once $\Theta \geq \Theta_{cr}$, as is developed in the 3-D section earlier.

Another interesting option is to incorporate the bedform predictors. These calculate the height, length, and steepness of bedforms as a function of mobility, depth, and grain diameter. The bed friction can then be updated at each time step via the roughness length k_s. A good summary is in Soulsby [427].

9.14.6 Results

As an example, the Great Bay Estuarine System in New Hampshire was simulated using previously studied hydrodynamics [31, 33, 307].

Figure 9.19 shows the vertical position of six fine sand particles ($D = 0.1$ mm) during approximately four tidal cycles. z's are calculated using P. While some particles find time to settle during tidal cycles (upper right and middle left), others in more dynamic and deeper regimes stay suspended for longer periods of time (middle right). The particle in the lower left corner gets stranded on a drying tidal flat around time step 2230. It should be noted that $z = 0$ does not imply a static particle because

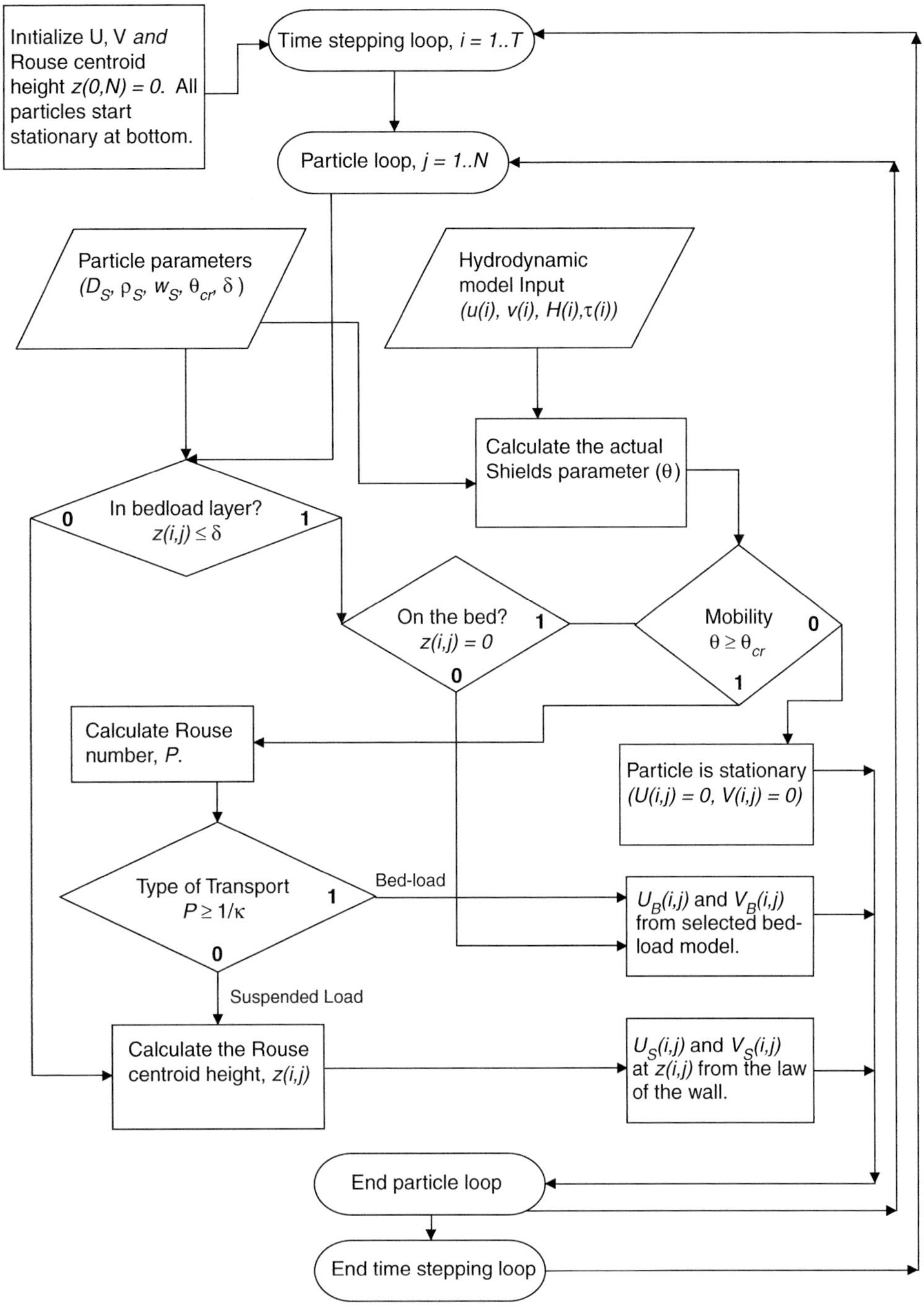

Figure 9.18. 2-D flowchart for noncohesive sediment particle simulation.

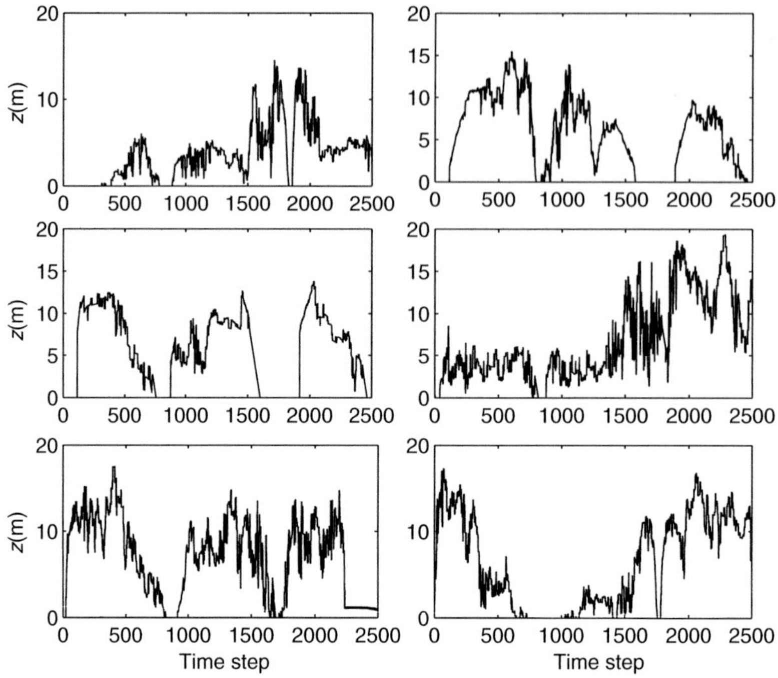

Figure 9.19. Vertical excursions for six individual fine sand particles, $D = 0.1$ mm over approximately four tidal cycles in Great Bay.

it may be moving as bed-load during that time. It should also be noted that $z = 20$ m does not indicate the water surface.[12]

Figure 9.20 shows the trajectories of 6 randomly chosen fine sand ($D = 0.1$ mm) particles. The top panel shows the total distance traveled; the bottom panel shows the net (subtidal) displacements over one tidal cycle. The maximum total distance covered is on the order of 15 km during flood and ebb. The net displacements are smaller (on the order of 1.5 km) and show a trend of upriver transport except the northeastern side of the channel, where a reversal is observed. The magenta particle is stranded on a tidal flat and never goes back with ebb. Figure 9.21 is similar but for coarse sand particles ($D = 1$ mm). The total and net distances traveled by each particle are shorter as expected. Net displacements are predicted to be toward the ocean along the coasts and upriver in the main channel, a result consistent with coarse sand shoal formation in the main channel in this area (Bilgili et al. [33, 34]).

Figure 9.22 shows movement of fine (0.1 mm – blue) and coarse (1 mm – red) sand particles during approximately 2.5 tidal cycles after they're released at low tide in the central channel (upper left frame). There are approximately 400,000 particles in each sediment class. The panels show the large displacements that sediment particles

[12] Note that the generic 3-D equivalent of Figure 9.19 is Figure 9.12.

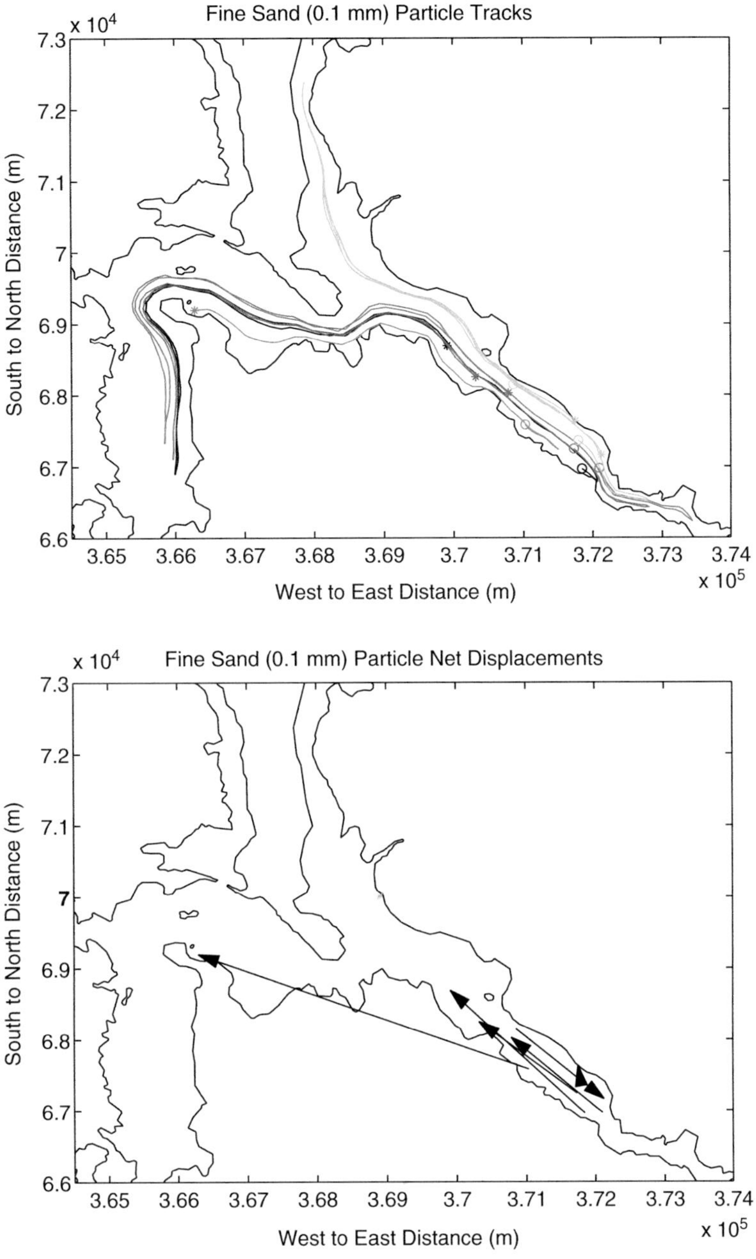

Figure 9.20. Tidal displacement (top) and net displacement (bottom) for fine sand ($D = 0.1$ mm). "o"s mark the starting point and "$*$"s mark the final location. Net displacement arrows are drawn from "o" to "x" for each particle. For a color version of this figure, please see the color plate section.

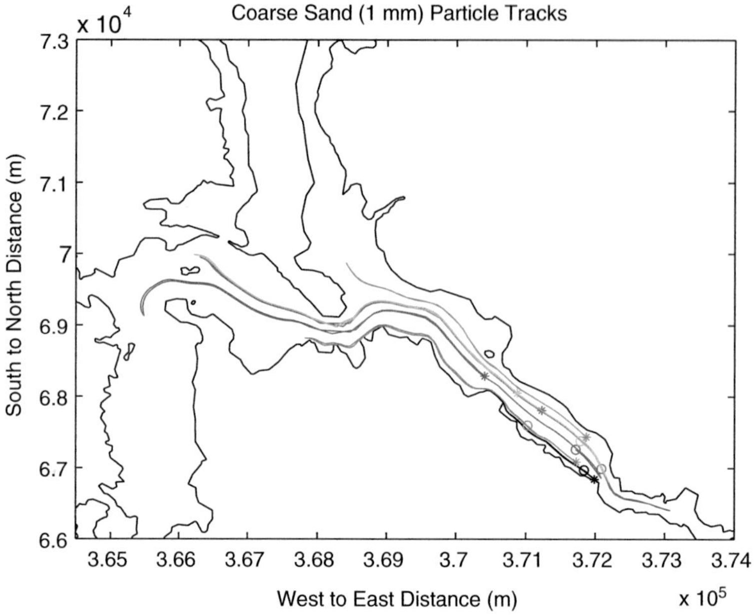

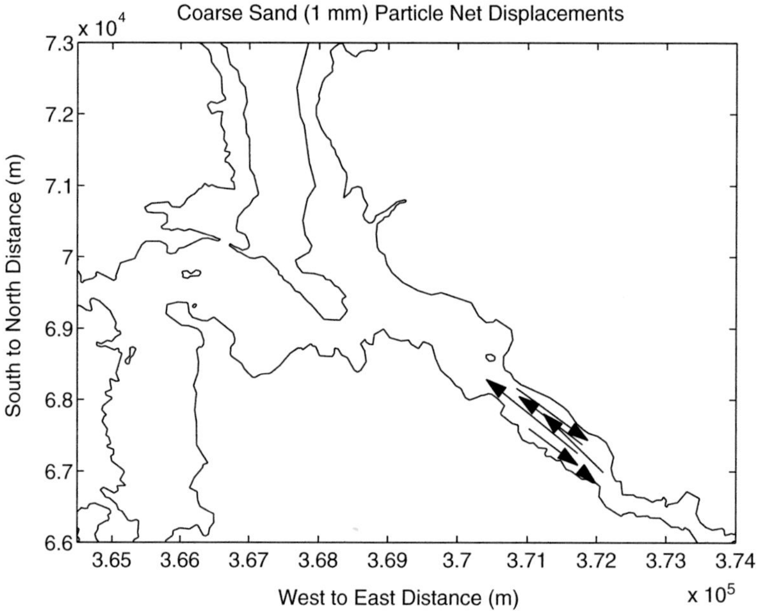

Figure 9.21. Tidal displacement (top) and net displacement (bottom) for coarse sand ($D = 1$ mm). "o"s mark the starting point and "$*$"s mark the final location. Net displacement arrows are drawn from "o" to "x" for each particle. For a color version of this figure, please see the color plate section.

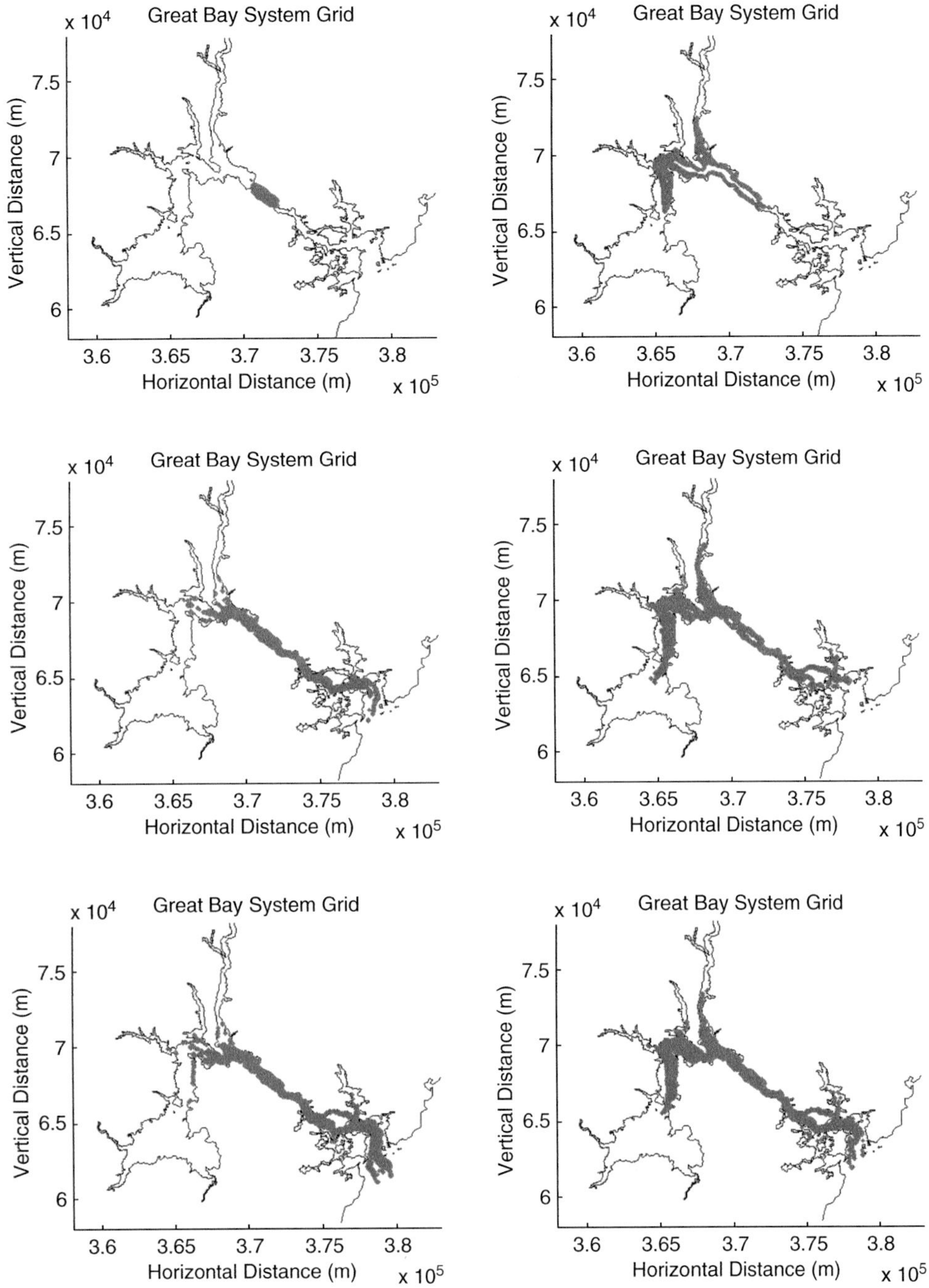

Figure 9.22. Displacement of particles over 2.5 tidal cycles. Fine: blue; coarse: red. Release at low tide (upper left panel). The panels depict subsequent high and low tide points in sequence. For a color version of this figure, please see the color plate section.

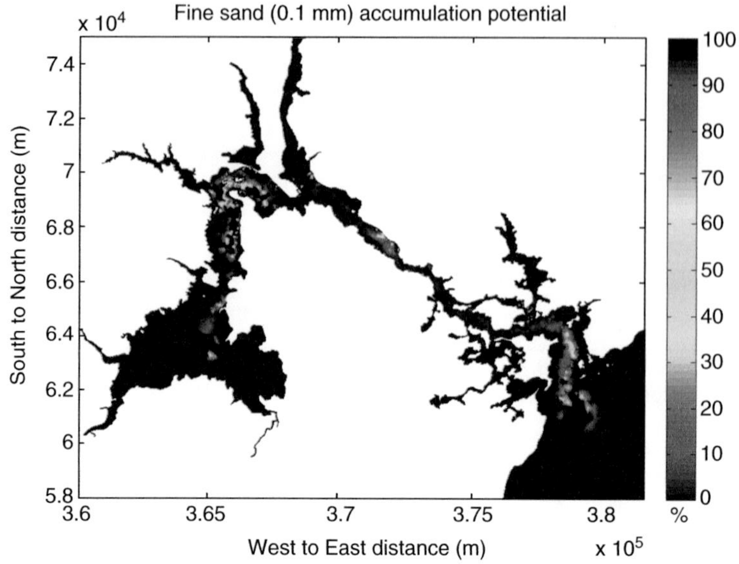

Figure 9.23. Accumulation potential for fine ($D = 0.1$ mm) sediment particles. The color bar gives accumulation potential in percentage terms. For a color version of this figure, please see the color plate section.

undergo during each tidal cycle. It is observed that a large population of both types of sands are transported back and forth confined in the tidal channels between the estuarine mouth and the upper reaches during each tidal cycle, a result consistent with Armstrong [12]. Note that the blue particles are plotted below the red ones and they may be hidden by the latter if there's overlap.

Figures 9.23 and 9.24 show the steady-state sediment accumulation potential respectively for coarse and fine sands throughout the estuary, given availability. It is observed that there are two major accumulation zones for finer sands, the one at the mouth (lower right) being larger than the one in the upper reaches (lower left). The coarse sands are mostly accumulated at the mouth and in the middle sections of the estuary where higher velocities are experienced due to channel geometry. The results indicate a sorting mechanism which transports finer particles to low energy zones. These results are consistent with the findings of Bilgili et al. [33], Ward [468], Armstrong [12], and Armstrong et al. [13].

Problem 9.12 Track a statistically significant number of very fine sand particles in a virtual 2-D steady flow field over a flat bottom. Use the framework of Figure 9.18. Run your simulations without and with horizontal dispersion, Equation 9.53, and show the effect of the latter in sediment accumulation potential.

Problem 9.13 Simulate the same problem above with very coarse sediment particles. Comment on the sensitivity of the results to the existence of dispersion and to the choice of the dispersion coefficient.

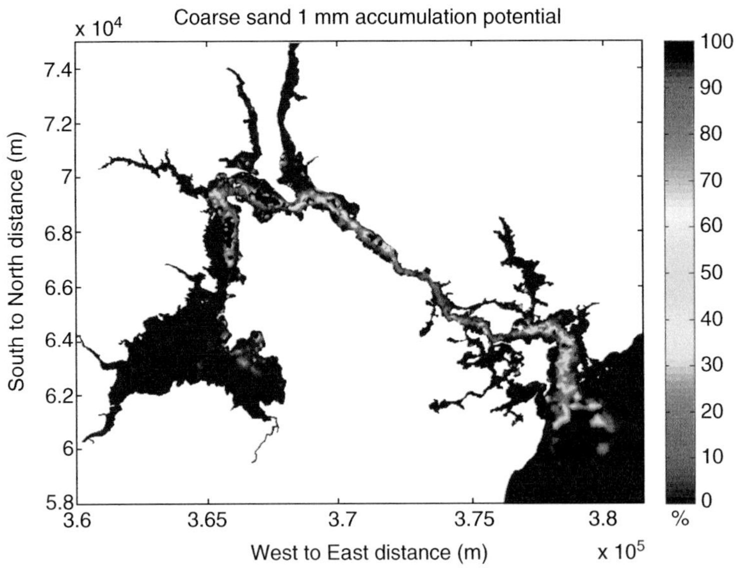

Figure 9.24. Accumulation potential for coarse ($D = 1$ mm) sediment particles. The color bar gives accumulation potential in percentage terms. For a color version of this figure, please see the color plate section.

Problem 9.14 The methodology given in Figure 9.18 is deterministic on particle mobility. This means that once a sediment particle is mobile, it is always transported. Formulate a probabilistic approach for sediment entrainment into this framework.

Problem 9.15 Using your probabilistic formulation, simulate the problems given in Problems 9.12 and 9.13. Do your results make sense? What modifications to your formulation are reasonable?

9.15 Summary of Notation

D is the grain size.
D_r is a reference grain size.
D_n is a nominal diameter for an irregular particle...its volume-equivalent spherical D.
D_{50} is the median grain size.
ρ_s is the particle density.
s is the dimensionless density normalized by that of water, $\frac{\rho_s}{\rho_w}$.
w_s is the settling velocity.

Notation Summary Continued

z_0 is the distance between the computational and the true bottom.
z_r is the bed roughness height.
k_s is the Nikuradse roughness length.
δ is the bed-load thickness.

$\Phi = -\log_2 \frac{D}{D_r}$ is the Krumbein parameter, with reference size D_r.
$\Theta = \frac{u_*^2}{(s-1)gD}$ is the Shields parameter.
Θ_{cr} is the critical Shields parameter.
$\Theta_{br} = \frac{u_*^2}{(s-1)gk_s}$ is the boundary roughness Shields parameter.
$P = \frac{w_s}{\kappa u^*}$ is the Rouse number.
$R_* = \frac{u_* D}{\nu}$ is the bed Reynolds number.
$R_s = \frac{w_s D}{\nu}$ is the settling Reynolds number.

τ is the bottom stress.
$\kappa = 0.41$ is the von Karman constant.
$u_* = \sqrt{\frac{\tau}{\rho}}$ is the friction velocity.
c_d is the quadratic drag coefficient at the computational bottom.
q^2 is twice the turbulent energy.
l is the turbulent mixing length.
A_z is the vertical eddy viscosity.

τ_{cr} is the critical bottom stress.
$D_* = \left[\frac{(s-1)gD^3}{\nu^2}\right]^{1/3}$ is the normalized particle size.
$S_* = \frac{1}{4}\left[\frac{(s-1)gD_n^3}{\nu^2}\right]^{1/2}$ is a particle size parameter.
r_e is the particle entrainment rate at the bed.
E is the sediment pickup rate, mass/time/unit area.
$\phi_p = \frac{E}{\rho_s\sqrt{(s-1)gD}}$ is the dimensionless pickup rate.

ρ_w is the density of water.
ν is the kinematic viscosity of water.
g is gravity.

10

Oil – Chemically Active Particles

10.1 Introduction

Oil is a particular case of a chemically active substance. Of interest here is its oceanic transport, chemical transformations, and fate. In the present context (high abundance, small space/time scales) oil is not "natural" in the sea, but is introduced unintentionally by accidents – spills, exploration/production blowouts – or clandestinely by dumping of waste products.

Oil is not a pure substance, but a mixture of hundreds of specific compounds. For the commercially common oil products, there are large databases of chemical composition and properties. Some are maintained by government agencies, some by transboundary corporations or their associations; some are embedded in operational models. Crude oils are distinctive and discernible, by origin; refinery products are quite specific about chemical properties, and subject to commercial norms of identity and quality.

Generally, oil will be a buoyant liquid mixture, immiscible in water, with density, viscosity, and other physical properties sensitive to its weathering history. In the sea it will occur either as a surface slick or as subsurface droplets. It will be transported by oceanic motion while weathering. Evaporation tends to account for the lion's share of most surface spills, unless a slick makes landfall early or emulsifies. Exposure in the environment will alter the chemical mix and its bulk properties. Destinations for present purposes include the sea floor, mingled with the benthos; the shoreline; in the biota; in the atmosphere as evaporation products. Some subsurface oil will dissolve or biodegrade in the sea.

We use *parcels* to represent discrete quantities of oil in the environment. A parcel may be floating at the surface as a slick, or submerged as an aggregation of identical small droplets. Each parcel moves as a Lagrangian particle; its content is subject to the operative processes summarized in Figure 10.1. A surface parcel could be thought of as a "spillet" – a physically well-defined subdivision of a larger spill. A subsurface

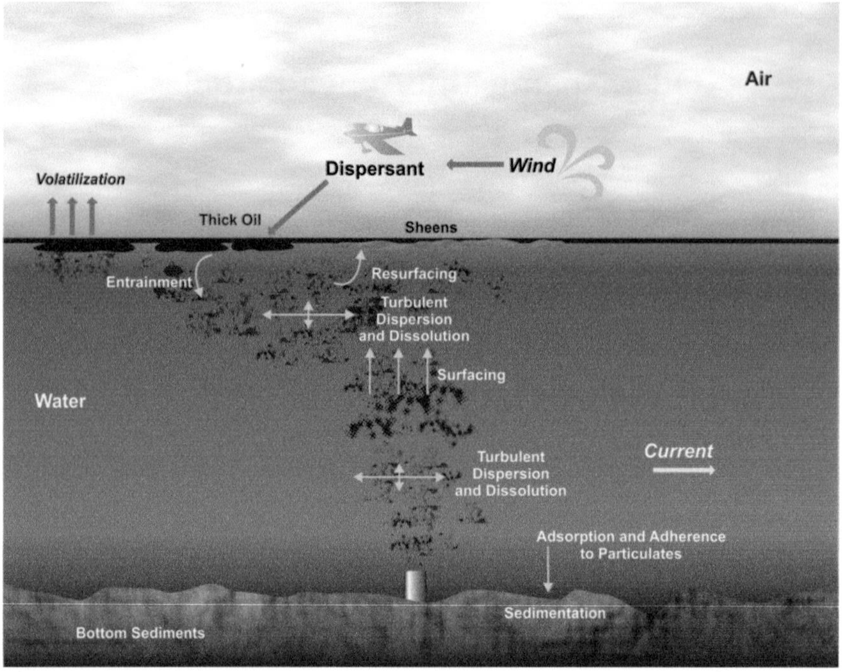

Figure 10.1. Occurrence and processes for marine oil. (Image provided by ASA-RPS, courtesy M. Spaulding.) For a color version of this figure, please see the color plate section.

parcel could be thought of as a "group of droplets," an aggregation of many very small quantities.

10.1.1 Composition

Oil is a complex mixture of hydrocarbons, organic material, and metals. It is crucial for oil models to access information about chemical content, and several very good databases are provided online ([119, 125, 129, 332]), with the Environment Canada [125] and NOAA [332] databases being exemplary at the moment.

The major players in the oil mix, the hydrocarbons, are generally classified according to their molecular structures using what is commonly referred to as SARA: Saturates, Aromatics, Resins, and Asphaltenes. The amount and composition of these structures in the mix is the determining factor in oil properties and thus in how oil behaves when exposed to environmental processes.

For example:

The aromatics and the light saturates are components that are subject to high rates of entrainment, evaporation, and dissolution. Both evaporation and dissolution separate these components from the rest, which remain concentrated in residual products of weathering. Entrainment indicates the creation of oil droplets beneath the sea surface by physical surface processes (turbulence and mixing) operating on floating oil; unless chemically induced, it affects all components.

The asphaltenes are heavy, nonvolatile components that can produce stable water-in-oil emulsions. These emulsions are very dense and viscous (orders of magnitude increase) end-products of weathering that rapidly develop from surface oil subject to wave action, under the right conditions.

The light aromatics such as benzene and toluene contain material toxic to marine life and even small dissolved quantities can be important.

A crude with high resin and wax content is more susceptible to sedimentation due to increased adhesion properties. Waxes are heavier saturates.

The primary physical properties are the viscosity, density, volatility, and pour point. All can be discussed generally in terms of the SARA configuration.

Generally: As the percentage of saturates and aromatics decreases and the percentage of resins and asphaltenes goes up, the viscosity increases. This means that after an oil spill, the viscosity has no other choice than to go up as lighter components evaporate and dissolve. Viscosity is a primary player in droplet formation, thus in entrainment of surface oil into the water column.

The density mainly controls the buoyancy. A large majority of oils have density smaller than water and they float unless they acquire dense sediment particles through adhesion. In that case, sedimentation becomes important in the delivery of oil to the benthos. Oil droplets will generally be buoyant, although very small subsurface droplets will rise very slowly, and the tendency to float can be overcome by other transport processes.

Emulsification – the formation of water-in-oil emulsions – involves the incorporation of water into the floating oil. As the higher density water is added to oil, the composite density increases. Stable emulsions can be very viscous, referred to as mousse; stability depends on the asphaltene content and other factors. The pour point, defined as the temperature a liquid stops flowing, depends on the asphaltene and resin content. It becomes an important parameter if it is close to water temperature for a specific oil type.

Table 10.1 shows some typical parameters.

10.1.2 Motion

The motion of an oil parcel must include surface spreading, advection, buoyancy, and nonadvective transport (eddy diffusion). For three decades, oil modelers have been treating eddy diffusion as a random walk process. During that time, the workability and availability of 3-D circulation models has greatly improved, a premise of this book. Theoretically one can obtain and use high-quality fields for the major variables

Table 10.1. *Some Typical Oil Properties at 15°C*

Oil Type	Density ($kg\cdot m^{-3}$)	Dynamic Viscosity ($N.sec.m^{-2}$)	Emulsion Stability	SARA- Wax%
Gasoline (premium)	750	0.0006	No emulsification	37-64-0-0-0
Jet Fuel (Anchorage)	811	0.002	No emulsification	81-19-0-0-0
Ekofisk	828	0.005	Stable	68-25-7-1-5
Diesel	830	0.002	No emulsification	74-24-1-0-0
Alberta	840	0.006	n/a	78-18-3-1-7
West Texas (2000)	847	0.0086	Unstable	78.5-14.8-6-0.7-2.8
Hibernia (1999)	850	0.013	n/a	62-31-6-2-5
Arabian Light	866	0.014	Stable	51-39-6-3-5
Alaska North Slope Northern Pipeline	872	0.014	Stable	51-34-9-5-3
Fuel Oil No: 2	874	0.005	No	57-42-0-0-0
Arabian Medium	878	0.029	Stable	54-32-7-6-6
Prudhoe Bay (1995)	905	0.068	Mesostable	53-34-10-4-4
Belridge Heavy	975	12.6	Mesostable	28-39-30-3-1
Bunker C	983	45	Entrained	24-55-15-7-12

governing motion, including advective velocity, density, and turbulence, sufficient to support random-walk trajectories.

Important physical areas still requiring focus are coastal resolution and the description of horizontal boundary interaction with oil; the free surface, where subgrid wind and wave effects are critical to motion and weathering; and the near-field of subsurface releases (discussed later). The description of short-time oil slick spreading was among the earliest advances in spill modeling; but there remain imperfections in that description, and improvements will surely be important in establishing the extent of coverage and thickness of an oil spill. Finally, the interaction among surface parcels – the "crowding" effect – continues to be problematic when surface oil is concentrated by wind or convergence.

10.1.3 Weathering

Most of the important weathering processes are illustrated in Figure 10.1. Included are evaporation, entrainment, emulsification, dissolution, biodegradation, oxidation, and sedimentation. A full weathering model would include descriptions of all these processes; practical descriptions will concentrate on the major ones.

NOTE: The industry identifies "dispersion" as the process of forming surface oil into droplets and injecting them into the underlying water column. We are using the more conventional term "entrainment" for this process, reserving "dispersion" for its wider, more comprehensive set of environmental transport processes.

In the period 1980–2000, a general pattern emerged for process description for surface spills. Good reviews are given by Spaulding [432], Cekirge et al. [72], ASCE [14], Reed et al. [371], Lehr [243].

Many early process descriptions are based on the work of Mackay et al. [293, 294, 295]. Variations of these appear to be encoded in many extant models that have concentrated on surface oil, and in use by the major operating agencies and corporations. Generally these are the organizations that maintain databases for crudes and refinery products; a set of weathering relations is a natural complement.

A typical, simple description for a surface spill accounts for the volume fraction of oil lost to weathering, and the volume fraction of water in emulsification; with bulk properties (density, viscosity) and weathering rates given in terms of these. This would all be specific to a particular oil's pre-release, initial composition.[1] An equivalent description would make explicit the nonvolatile portion of the mix. The extension to more detailed chemical representation of the mixture is clear: more functional chemical groups, a mass basis of accounting (instead of volume), and greater precision in describing transport processes and interfaces.

Among the surface processes, entrainment is special in that it moves material from the surface and injects it below in the form of small droplets. The subsequent 3-D transport and evolution of the entrained oil droplets would have to be followed subsurface.

10.1.4 Subsurface Releases

Much recent activity has focused on subsurface releases, typically at the sea floor. These require a near-field description of physical and chemical interactions within high-pressure mixtures of oil, gas, and entrained seawater. Especially for deep releases, temperature can be important. These processes are neither described nor resolved in the type of Coastal Ocean Models we envision being available. It is necessary to rely on a separate near-field description to initiate droplets – location, flow rate, droplet size, gas-to-oil ratio, chemical makeup. Once droplets are formed and clear of the near-field, the far-field motion takes over, and dissolution, biodegradation, and buoyant rising become very important.

Near-field models are beyond the current scope; but far-field model success depends on initialization by such a model. Current reviews are available (e.g., Dasanayaka and Yapa [97], Socolofsky et al. [426], Johansen [213]).

[1] A clear example is reported by Buchanan and Hurford [58], reviewed later.

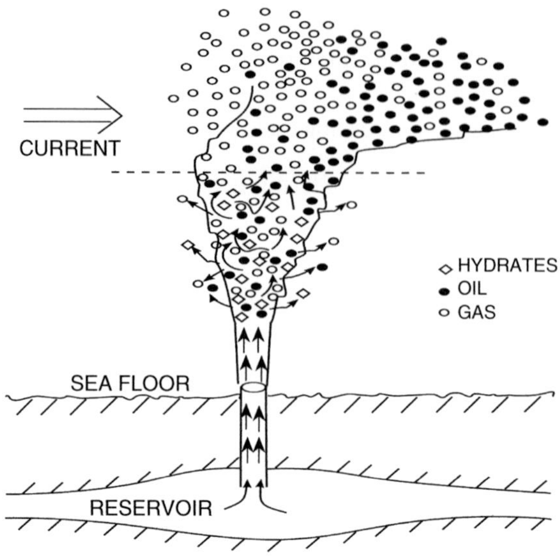

Figure 10.2. A subsurface release. A near-field model forms droplets (size, rate, and composition) and injects them into the far-field. A Coastal Ocean Model takes over from this initialization.

10.2 Oil as Parcels

Oil will occur in two idealized forms: floating slicks and subsurface droplets. In either case, we have *parcels* of oil – discrete amounts, treated mathematically as "particles."

Surface oil is typically very thin. Early oil slick thickness is typically of order 1–2 cm, ultimately reaching a sheen thickness < 0.1 mm. It is exposed to the atmosphere above and the ocean below.

Subsurface oil will occur as small buoyant droplets, in the size range 0.01–1 mm. The spectrum of sizes will depend upon the formational processes. Typically, droplets will be buoyant (there are exceptions) and the larger ones rise faster. Droplets are subject to oceanic processes, but of course, not to atmospheric ones.

A late-stage slick can form into tar balls. Weathering is effectively over for these; processes of biological and chemical degradation accompany potentially long distance transport. These are a relatively persistent aquatic pollutant.[2]

Parcel initiation can occur in three ways:

Spills will originate as surface deposits of slick parcels of material with known composition (crude oil from a commercial supplier, or a refinery product.)

[2] Tarballs are relatively crusty on the outside; their density is usually lower than seawater. However, if weathering has occurred in a region of high suspended sediment activity, a tarball may incorporate sediment particles and be denser than seawater. These are often treated as passive particles since tar ball degradation in the environment is slow.

Table 10.2. *Some Typical Dimensions of Oil*

Sheen thickness	0.0001–0.1 mm
Nonemulsified slick thickness	0.1–1 mm
Mousse (emulsified slick) thickness	1–20 mm
Subsurface droplet diameter	0.01–1 mm
Tarball diameter	1–300 mm

Subsea releases will originate as parcels of droplets escaping from blowout jets and plumes, containing gas, water, and oil of potentially uncertain composition.

Transformations: Surface parcels can be broken up by breaking waves and droplets injected below – the process of entrainment. Subsurface parcels of droplets can reach the surface by buoyant 3-D transport; they become surface parcels at that point.

Spills and blowouts create new parcels. Entrainment creates new subsurface parcels and removes matter from extant surface parcels. Surfacing parcels simply change status; no new parcels are made.

10.2.1 Surface and Subsurface Parcels

A surface parcel represents a quantum of oil deposited at a point on the surface. An isolated parcel will spread early and fast, slowing to a terminal thickness. Its surface area will regulate several weathering processes.

Neighboring surface parcels have the potential for overlap – a sea surface crowding effect. In that case they will share air/sea surface area, limiting the surface exposure and transport rates that depend on it. Parcels must be aggregated properly to obtain a composite exposure in that case. Aggregation will be a necessity whenever physical processes (point sources, downwelling, or other concentrating mechanisms) are operative. This is a generic problem, a "parcel density" effect. An efficient simulation approach involves "projection" of parcel quantities into a jointly defined field, and using the field quantities to compute the joint effect on each parcel.

Subsurface parcels comprise groups of droplets, assumed spherical and isolated. They do not interact with each other. They have high surface-to-volume ratios, and non-atmospheric processes including dissolution and biodegradation can be important. These parcels can persist for a long time in deep water. Generally, larger droplets of the same material will rise faster; the rise velocity depends on buoyancy, size, and ambient density. Droplet size reflects the formational processes and subsequent evolution.

Because of their small size, many subsurface droplets must be aggregated into a single subsurface parcel, retaining individual properties dependent on size – notably buoyant velocity and surface-to-volume ratio.

10.2.2 Generic Model Needs

Every parcel will be treated as a particle in motion, integrating advection and random walk in each time step. For each parcel, there needs to be:

- A logical marker indicating whether this is a slick or droplet
- An initialization procedure
- An equation of motion
- A mass balance for each substance in the mixture
- A description for each process rate involved
- Equations of state for density and viscosity
- A criterion for determining if a parcel must change from slick to droplet, or vice versa

For processes limited by exposure at the oil–air and oil–water interfaces, the surface-to-volume ratio is required. For floating parcels, this is $\frac{1}{h}$; for spherical droplets, $\frac{6}{D}$.[3] For an ideal[4] droplet, D is constant; the statistical description of D reflects the formational process. For surface slicks, spreading and mass loss will cause h to decline, resulting in greater exposure; so a relation for the evolution of h, the effective "thickness," is needed.[5]

10.3 Surface Parcels

Figure 10.3 shows a surface parcel with mass M, density ρ.

If F is the rate of mass transfer into the air (units of mass/time), the balance is

$$\frac{dM}{dt} = -F = -\rho \cdot A \cdot K(T, U, E, \dots) \tag{10.1}$$

with K the mass transfer coefficient, dependent on temperature T, windspeed U, turbulent energy E, and so on, Since $M = \rho A h$,

$$\frac{dM}{dt} = -M\frac{K}{h} \iff \frac{1}{M}\frac{dM}{dt} = -\frac{K}{h} \tag{10.2}$$

[3] h is slick thickness; D is droplet diameter.

[4] ideal = isolated, incompressible, inert

[5] Equivalently, with $V = hA = M/\rho$, the evolution of the surface area A is needed.

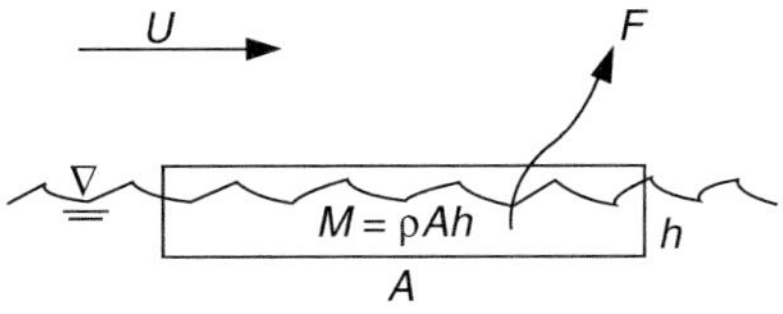

Figure 10.3. A surface parcel model: idealized geometry.

The decay rate $\frac{1}{M}\frac{dM}{dt}$ is the ratio of two quantities. K is an intrinsic property of the mass transfer kinetics. h represents the geometry realized in the field, the joint effects of spreading and sea surface convergence. We expect to get K from controlled experiments; h requires a simulation of field conditions.[6]

- For a limited time following parcel initialization, the fluid mechanics of oil will control its shape adjustment into a slick. This is "Fay spreading."
- Hydrodynamic eddy diffusivity will take over once the fluid mechanics of oil per se are no longer limiting spreading.
- Hydrodynamics may re-concentrate parcels over convergence zones or down-welling zones near boundaries. The oil will increase in thickness, then relax via Fay spreading again.
- There will be a minimum thickness h_m. Further removal of oil, for example by evaporation, will occur by reduction in A, by the retreat of the slick or perhaps by the exposure of open water within the slick.
- Both h and A represent idealized geometry; their product is $V = \frac{M}{\rho}$, the ratio of two primary properties of a parcel. h need only be understood as a form of average, to parameterize the rate F (Equation 10.2) with a single shape parameter.

Oil will pile up where surface motion converges or where spills are being introduced. In simulation h needs to be obtained from the distribution of parcels, each with its own $[\rho, M, V, A]$. For a single parcel, $h = V/A$. If there are two parcels, they will be independent unless $A_1 + A_2$ exceeds the available local area; in that case the parcels share a composite h. Exposure is limited by parcel proximity ("particle density").[7] This is addressed below.

Generalizations are natural for a mixture of several components, M_i.

10.3.1 Spreading

Mechanical spreading is a sea surface phenomenon. For a concentrated (point) release, the immediate outcome is an increase in the spill area over time and the corresponding decrease in the slick thickness. The gravitational, viscous, inertial,

[6] Units: $[K] = [l]/[t]$; $[h] = [l]$

[7] A similar issue concerns surface divergence.

and surface tension forces are in effect for calm conditions, with the additional contribution from wind, currents, and waves in dynamic systems. The area increase is a relatively rapid process and has important ramifications as far as the weathering processes are concerned. One of the earliest algorithms comes from Blokker [36], but it only considers the gravity in calm water and consequently it is applicable only during the very early stages of a spill. Fay's algorithm [137, 138] adds viscous forces and surface tension and is still one of the most commonly used spreading models today. It is usually referred to as Fay–Hoult type spreading, owing to the companion work by Hoult [200]. The Fay–Hoult spreading analysis indicates three sequential stages: (1) an initial gravity-inertial phase where the spill is thick and the driving force is gravity; (2) an intermediate gravity-viscous phase where the viscous forces from the water start to act against gravity; and (3) a final surface tension-viscous phase where the slick is thin and the major driving force becomes the interfacial tension between the oil and the water.

The initial gravity-inertial phase is rapid and occurs in time scales much shorter than the other processes. Cekirge and Palmer [71] indicate that this phase is important only in the first hour of the spill; others suggest even shorter times.

Experience does not support the Fay description of the terminal phase. Among other things, slicks appear to reach a finite, terminal thickness and stop spreading; weathering, hydrodynamic transport, and emulsifying processes at this point cannot be ignored.

Accordingly, the following pattern of use has evolved. The first phase is used to introduce spills into simulation models; this describes the initial near-field of the spill during the first hour or so. Transport and weathering are simulated from there, with the Phase 2 Fay relations used to describe the evolution of slick area or thickness. As thickness approaches a small finite minimum, a simulation will continue but the Fay analysis loses relevance.

The governing equations for all phases, as well as the spreading coefficients and transition times are given in Table 10.3 for radial spreading of an instantaneous spill. The conditions at the end of the initial inertial phase t_{iv} are plotted in Figure 10.4, versus volume of spill, for a representative oil. This phase is over fast (of order an hour for a spill $\mathcal{O}(10^4\ \mathrm{m}^3)$) and provides a useful point of initialization for a spill parcel.

For perspective: some volume estimates are:

Tanker cargo volume: 80,000–640,000 m^3
Exxon Valdez spill estimate: 41,000–120,000 m^3
Deepwater Horizon blowout estimate: 651,490–778,610 m^3

Phase 2: Spreading with Loss

Beyond Phase 1 initialization, simulation will require accommodating evaporative and other losses. We can accommodate this, following Dodge et al. [118].

Table 10.3. *Fay's Circular Spreading Laws for Instantaneous Spills*

Phase	Slick Radius	Coefficient	Termination Time
1. Inertial-gravity	$r_i = K_i\,(\beta g V)^{1/4} t^{1/2}$	$K_i = 1.14, 1.14, 1.53$	$t_{iv} = \left(\frac{K_v}{K_i}\right)^4 \left(\frac{V}{\beta g \nu_w}\right)^{1/3}$
2. Viscous-gravity	$r_v = K_v \left(\frac{\beta g V^2}{\nu_w^{1/2}}\right)^{1/6} t^{1/4}$	$K_v = 1.45, 0.98, 1.21$	$t_{vs} = \left(\frac{K_v}{K_s}\right)^2 \left(\frac{\rho_w}{\sigma}\right) \left(V^2 \beta g \nu_w\right)^{1/3}$
3. Surface tension-viscous	$r_s = K_s \left(\frac{\sigma^2}{\rho_w^2 \nu_w}\right)^{1/4} t^{3/4}$	$K_s = 2.30, 1.60$	–

Note: The proportionality coefficients K given are from Fay [138], Shen et al. [410] and Dodge et al. [118] respectively. r is the radius of the spill, t is time, and V is the volume of the spill. $\beta = (\rho_w - \rho_o)/\rho_w$.

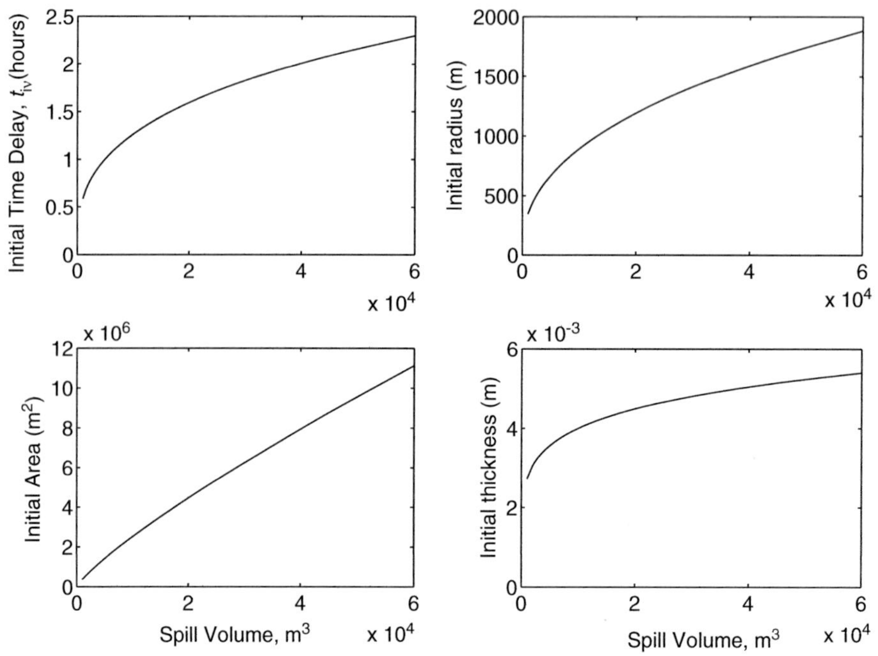

Figure 10.4. Initial conditions for a spill, volume V, at the transition time t_{iv} between the initial and second phases of Fay spreading. These are used to initiate spilled parcels of oil for simulation. The slick area at this point is very nearly linear in the spill volume. $\rho_o = 840$ kg/m^3.

The time rate of change of the spill area can be obtained by taking the derivative of $A = \pi r^2$ with respect to t and V:

$$\frac{dA}{dt} = \frac{K_v^2 \pi}{2}\left(\frac{\beta g A^2 h^2}{\nu_w^{1/2}}\right)^{1/3} \cdot t^{-1/2} + \frac{2K_v^2 \pi}{3}\left(\frac{\beta g A^{-1} h^{-1}}{\nu_w^{1/2}}\right)^{1/3} \cdot t^{1/2} \frac{1}{\rho_o}\frac{dM}{dt} \tag{10.3}$$

with the initial conditions $A_i = K_v^2\left(\frac{K_v}{K_i}\right)^2 \cdot \pi\left(\frac{V^5 \beta g}{\nu_w^2}\right)^{1/6}$ and $h_i = V_i/A_i$. Here, V_i is the initial spill volume and A_i is the area of the slick in the beginning of the simulation, $A_i = \pi r_i^2$ evaluated at t_{iv}. During Phase 2, V evolves as a function of time.

The Phase 2 dynamics give $r(t)$ and therefore $A(t)$ (Table 10.3):

$$A(t) = \pi r^2 = \pi\left[K_v\left(\frac{\beta g V^2}{\nu_w^{1/2}}\right)^{1/6}\right]^2 t^{1/2} \tag{10.4}$$

Solving Equation 10.4 for t and substituting in Equation 10.3 yields:

$$\frac{dA}{dt} = \left(\frac{(K_v^2\pi)^2}{2}\right)\left(\frac{\beta g}{\nu_w^{1/2}}\right)^{2/3} h^{4/3} A^{1/3} + \frac{A}{3}\left[\frac{2}{M}\frac{dM}{dt}\right] \tag{10.5}$$

A relation for the time rate of change of thickness h can now be written compactly, recognizing that the total mass of the spill $M = \rho_o A h$:

$$\frac{dh}{dt} = \frac{\frac{dM}{dt} - \rho_o h \frac{dA}{dt}}{\rho_o A} \tag{10.6}$$

The $\frac{dM}{dt}$ term here incorporates any mass gain (continuous spill) or loss (evaporation, dissolution, etc.). The dynamics of h and A would evolve according to Equations (10.5, 10.6) until a minimum thickness or a maximum area is reached, or the oil is completely gone. Fay [138], Lehr et al. [244], and Dodge et al. [118] report that the minimum thickness is on the order of 0.01 to 0.1 mm going from lighter oils to crude. The maximum slick area is given by Fay as $A_{\max} = 10^5 V^{3/4}$. (This is dimensional; V is in m^3.)

The case of no loss: A spreading relation of the form 10.5 is in common use, without the $\frac{dM}{dt}$ term.

$$\frac{dA}{dt} = K h^{4/3} A^{1/3} \tag{10.7}$$

Without the dM/dt term, Equation 10.6 reverts to

$$\frac{1}{h}\frac{dh}{dt} + \frac{1}{A}\frac{dA}{dt} = 0 \tag{10.8}$$

In simpler form: The product hA is constant; algebra suffices for h.

As an illustration, suppose the loss rate were given as

$$\frac{dV}{dt} = -\left[c\frac{1}{\tau}\right]\left[\frac{h_{\min}}{h}\right] V \tag{10.9}$$

When $h_{\min}$ is reached, this is a first-order decay with time constant $\frac{\tau}{c}$. Prior to that, $h > h_{\min}$ and thickness limits – effectively the process is slowed down. The $\frac{1}{h}$ effect is the generic surface-to-volume effect for slick geometry. c is a dimensionless parameter; increasing c makes the loss go faster.

Figure 10.5 illustrates this lossy phase spreading, according to Equations (10.5, 10.6, 10.9), initialized following Phase 1 completion. The time constant τ is 1/2 day; the parameter c ranges from 0 (no loss) to 2. When $h_{\min}$ is reached, the system reverts to simple decay of V with constant rate $\frac{c}{\tau}$. A change in slope is evident at that point; subsequently, all the loss goes into shrinking A. As the process goes faster, the peaks are lower and sooner.

Problem 10.1 The results in Figure 10.5 show a kink as the minimum thickness is reached. At that point, the dynamics change abruptly as A is forced to absorb all

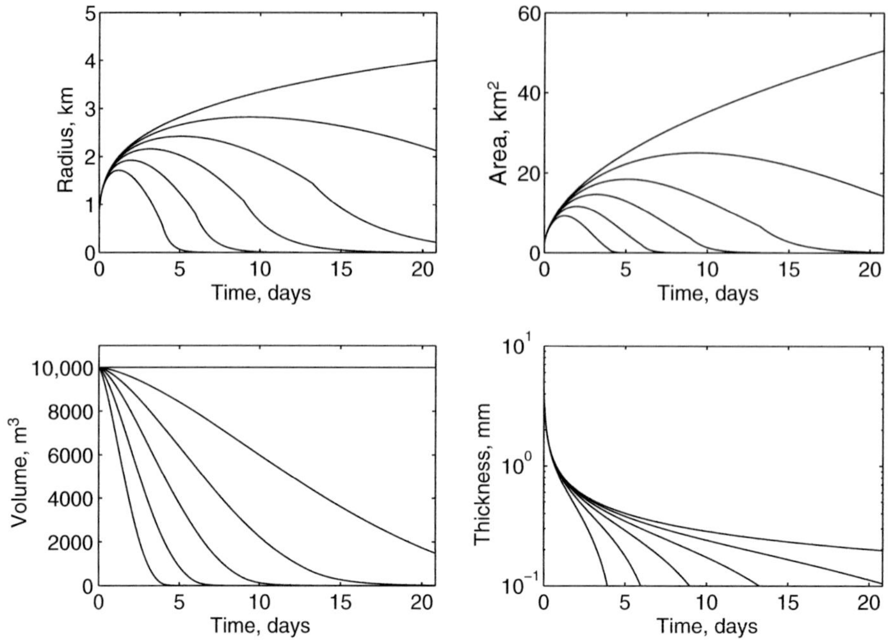

Figure 10.5. Phase 2 spreading with loss. $\tau = 1/2$ day; $h_{\min} = 0.0001$ m. The six cases represent increasingly fast loss: $c = [0, 0.10, 0.25, 0.50, 1.0, 2.0]$. Note that the h axis is logarithmic.

losses. Propose a less abrupt transition by altering the rate expressions as h gets close to $h_{\min}$.

Problem 10.2 Evaluate your solution for Problem 10.1, by simulation. Interpret the changes. Is your proposed change effective?

Density Changes

ρ_o is constant in the equations above. However, weathering will generally increase the slick density over time. Going over the development leading to Equation 10.5, and treating β as a variable, adds a simple extra term

$$\frac{dA}{dt} = \left(\frac{(K_v^2\pi)^2}{2}\right)\left(\frac{\beta g}{\nu_W^{1/2}}\right)^{2/3} h^{4/3}A^{1/3} + \frac{A}{3}\left[\frac{2}{M}\frac{dM}{dt} + \frac{1}{\beta}\frac{d\beta}{dt}\right] \tag{10.10}$$

Increasing ρ means decreasing β; a shrinking influence on A. From the constraint $M = \rho h A$ we get the generalization of Equation 10.6 for h:

$$\frac{d}{dt}(\rho_o h) = \frac{\frac{dM}{dt} - \rho_o h \frac{dA}{dt}}{A} \tag{10.11}$$

and there are other rearrangements of $\frac{dM}{M} = \frac{d\rho_o}{\rho_o} + \frac{dh}{h} + \frac{dA}{A}$.

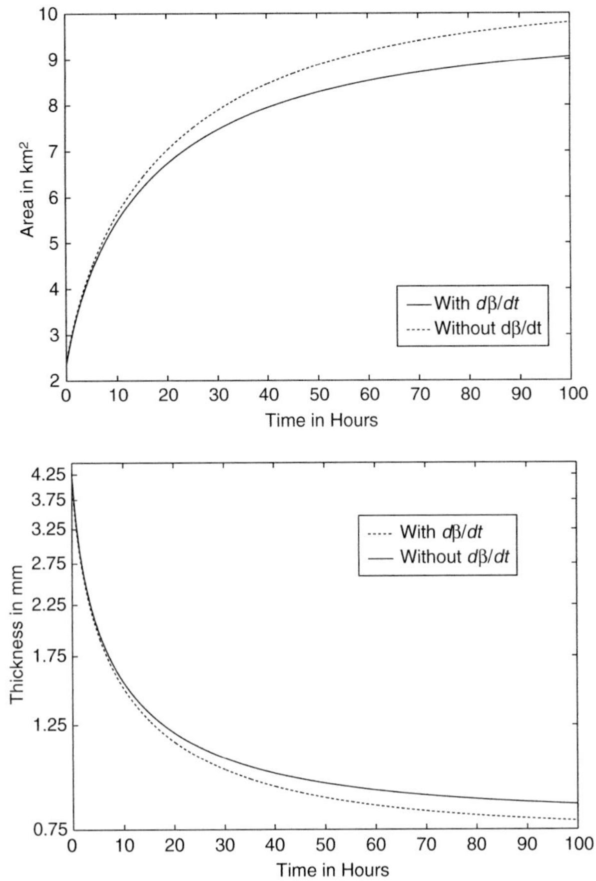

Figure 10.6. Effect of including the $\frac{d\beta}{dt}$ term in Equation 10.10.

Density is constrained by the equation of state for a particular oil. An example density function is given in Section 10.4.3. Assuming evaporative loss only, the form is

$$\rho_o = \rho_{\text{ref}}\,[1 + cF_v] \tag{10.12}$$

with F_v the fraction of the oil evaporated. Density increases as evaporation proceeds. Figure 10.6 shows the effect of the $\frac{d\beta}{dt}$ term in Equation 10.10.

Problem 10.3 Suppose evaporation were the only loss for a surface parcel, and the equation of state is 10.12. Using the chain rule, find $\frac{d\rho}{dt}$ in terms of ρ, M, and $\frac{dM}{dt}$.

Problem 10.4 Using the results from Problem 10.3: reexpress the relations for $\frac{dA}{dt}$ (10.10) and $\frac{dh}{dt}$ (10.11) by eliminating $\frac{d\rho}{dt}$. Does this make sense? Why or why not?

Elliptical Parcels

Throughout we have used a two-parameter description of horizontal spreading (r, h) based on analogy with circular geometry. Lehr et al. [242] added a third parameter, to allow elliptical geometry with the major axis aligned with the wind. Berry et al. [26] provide a recent summary.

Initialization

The strategy outlined above has a new surface parcel at the completion of Phase 1 (inertial phase), with M, A, h, and elapsed time t_{iv} since spill. This parcel has been in motion (advection + random walk) throughout Phase I, and hence has accumulated distance (advection + diffusion) from the spill point. So a simulation strategy would be: at spill time, initiate a parcel with mass M and assign it a Phase 1 duration t_{iv}. Motion begins instantly, and when the elapsed time has reached t_{iv}, begin Phase 2 (viscous) spreading according to the relations above for A and h; begin weathering; and continue the advection + random walk.

For a continuous spill, rate $\dot{M}$: sequential parcels of mass $\dot{M}\Delta t$ will be released at the spill point, each accumulating a distance $\Delta\mathbf{x}$ from the spill, before beginning weathering:

$$\Delta\mathbf{x} = \int^{t_{iv}} \mathbf{v}\, dt + \mathcal{R} \qquad \mathcal{R}_x \sim \sqrt{2A_h t_{iv}} \cdot \mathcal{G}(0,1) \tag{10.13}$$

$$\mathcal{R}_y \sim \sqrt{2A_h t_{iv}} \cdot \mathcal{G}(0,1) \tag{10.14}$$

Figure 10.7 illustrates this. The relations are:

$$r_{iv} = \frac{K_v^2}{K_i}\left[\frac{\beta g}{\nu_w^2}\right]^{1/12} [\Delta V]^{5/12} \qquad t_{iv} = \left(\frac{K_v}{K_i}\right)^4 \left[\frac{\Delta V}{\beta g \nu_w}\right]^{1/3} \tag{10.15}$$

$$\sigma_a = \sqrt{4A_h t_{iv}} \qquad \sigma_b = \sqrt{4A_h \Delta t} \qquad \Delta V = \frac{\dot{M}\Delta t}{\rho} \tag{10.16}$$

Figure 10.8 shows the first six parcels launched from a continuous spill of Alberta Crude, at 5000 m^3 per hour. The time step is 1 h; the release is at the origin. The ambient velocity is 0.5 m/s in both directions; eddy viscosity $A_h = 100$ m^2/s. The area of each parcel is plotted; it continues growing after initial release; there is no loss.

Problem 10.5 A continuous release of oil, $\rho_0 = 0.85\rho_w$, occurs at the rate 2000 m^3/h. There is no weathering. The ambient tidal current is 0.5 m/s; $A_h = 10$ m^2/s. Compute the displacement parameters of Figure 10.7 for $\Delta t = 0.5$ h. Compare the relative sizes; comment on whether these parcels overlap.

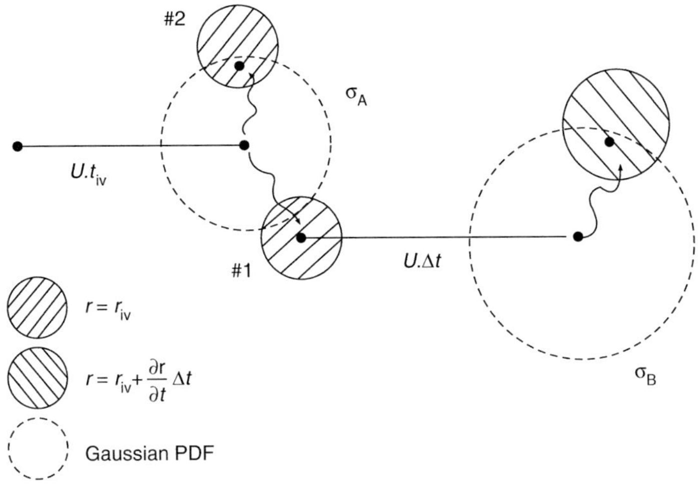

Figure 10.7. Release of two equal parcels in sequence, with subsequent displacement and spreading. Parcel 1 undergoes two sequential steps; Parcel 2 is newer, undergoes only one. Advective displacements are the straight vectors. Random walks are squiggled. The first parcel will have advected an added distance $U\Delta t$ and will have larger overall variance due to the random walks.

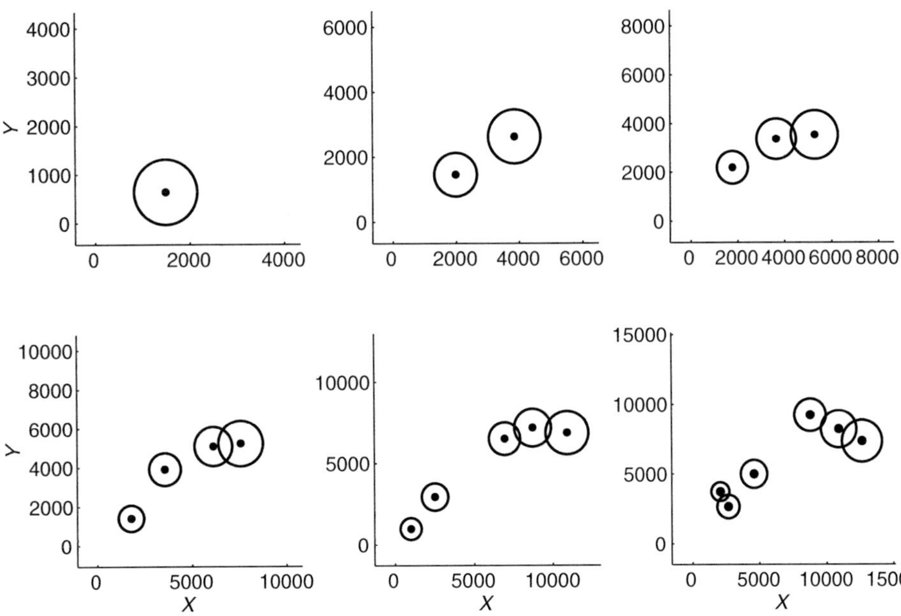

Figure 10.8. Continuous release of 5000 m^3/h at the origin in 1-hour time steps. One parcel is released at the origin in each time step. The displacement by advection and random walk is shown over t_{iv} and six subsequent timesteps. MKS units.

Problem 10.6 For the spill in Problem 10.5: Simulate the first five time steps and show the parcel areas as circles at each time. Is this compatible with your analysis in Problem 10.5 above? Explain.

Problem 10.7 For the spill in Problem 10.5: Add rotary tidal movement, $|v| = 0.5$ m/s, period $P = 12$ h. Simulate 12 time steps and show the parcel areas as circles at each time.

10.4 Weathering Processes

There are three major processes for surface oil:

Evaporation is generally the biggest flux. The lighter portions of the oil mixture are permanently removed into the atmosphere.

Emulsification is water uptake into surface oil. The process occurs under wind/wave action and can rapidly produce emulsions with very high viscosity. There are four classes of stability, discussed below. The more unstable emulsions can be essentially reversed when the surface forcing subsides.

Entrainment ("dispersion") is the conversion (normally associated with wave breaking) of surface oil into oil droplets into the water column below, entraining them in the 3-D flow. There is the mass entrainment rate; the size of the new droplets being formed; and the entrainment depth for their initial placement. Entrained droplets will generally be buoyant, and may regain the surface.

Other surface processes include dissolution, biodegradation, sedimentation, and stranding. Each can be very important environmentally, possibly among the critical concerns. Stranding is potentially important from an overall mass balance point of view.

10.4.1 Evaporation

Stiver and Mackay [439] examined this process for laboratory mixtures and related field occurrences of floating mixtures. Relations are derived from basic mass transfer theory, and procedures given for estimating the empirical constants for particular oil mixtures.

For a given spill of initial volume V_o: let F_v be the evaporated volume fraction. In dimensionless form:

$$F_v = H\theta \qquad \Longleftrightarrow \qquad dF_v = Hd\theta \tag{10.17}$$

There are two dimensionless groups:

- θ is the accumulated evaporative exposure; $d\theta$ is its increment:

$$\theta = \frac{kAt}{V_o} \qquad \Longleftrightarrow \qquad d\theta = \frac{kAdt}{V_o} \tag{10.18}$$

  k is the mass transfer coefficient (m/s) from liquid into the air, a function of the ambient wind speed; A is the area of the slick; V_o is the initial slick volume.
- H is the dimensionless Henry's law constant (ratio of equilibrium concentration of the substance in the vapor phase, to that in the liquid):

$$\ln(H) = a - b\frac{T_B}{T} \tag{10.19}$$

  T is the absolute temperature (°K); T_B is the boiling point; a and b are empirical constants that characterize the oil mixture.

The rate of loss of volume is $V_0\frac{dF_v}{dt}$:

$$\frac{dF_v}{dt} = \frac{kAH}{V_o} \Longleftrightarrow \frac{dV}{dt} = -kAH \tag{10.20}$$

The ratio V/A is the film thickness h; and $\frac{V}{V_0} = 1 - F_v$, so

$$\frac{dF_v}{dt} = \left(\frac{k}{h}H\right)[1 - F_v] \Longleftrightarrow \frac{dV}{dt} = -\left(\frac{k}{h}H\right)V \tag{10.21}$$

The rate of change for V is locally first-order, with rate constant $\frac{k}{h}H$. k depends on the wind speed; H depends on the ambient temperature; h is the thickness of the slick.

An alternate form uses the linearization $T_B = T_0 + gF_v$, with T_0 the boiling point before evaporation begins, and $g = dT_B/dF_v$

$$\ln(H) = a - b\frac{(T_0 + gF_v)}{T} \tag{10.22}$$

This is the mathematical form reported by Buchanan and Hurford [58].

Buchanan and Hurford [58] give the typical form in early use. For F_v, the integrated form used is

$$F_v(t) = K_1 \ln\left(K_2\frac{U^{0.78}}{h}t + 1\right) \tag{10.23}$$

U is the wind speed; temperature dependence is embedded in the coefficients. This integral is the result of the process represented; a little rearrangement yields the equivalent marginal description:

$$\frac{dF_v}{dt} = \frac{K_1 K_2}{h} e^{-F_v/K_1} U^{0.78} \tag{10.24}$$

This reduces to the form of Stiver and MacKay above, if the linearization 10.22 for the boiling point is used. ($H \sim e^{-cF_v}$).

Problem 10.8 Demonstrate the relationship between the Stiver/MacKay and Buchanan/Hurford formulations for evaporation, provided the linearization 10.22 for the boiling point is used in the former. Express the relations among the coefficients of the two formulations; note any discrepancies.

A generalization, the pseudo-component method, is widely used (Berry et al. [26]; Chao et al. [76].) This consists of separating the oil into independent components based on volatility and tracking the evaporation mass flux f_v (units: moles/s) of each component j separately:

$$[f_v]_j = \frac{kA}{RT} [\mathcal{X} P_v]_j \tag{10.25}$$

Here, $\mathcal{X}$ (function of time) and P_v are the mole fraction and the vapor pressure of component j respectively. R is the universal gas constant and T is the water temperature. The mass transfer coefficient k is usually taken as:

$$k = 0.029 U^{0.78} L^{-0.11} Sc^{-0.67} \tag{10.26}$$

Here Sc is the Schmidt number and L is a length scale, usually taken as the slick diameter (Betancourt et al. [27]) or the slick length in the wind direction (Berry et al. [26]). The total mass lost to evaporation would then be the summation over j with each component weighted by its molecular weight.

Fingas [149] reports experiments that result in a simpler formulation for several oils. In these, the evaporation rate depends on exposure time and temperature only. *There is no dependence on wind or thickness.* The formulations, applicable for thickness $h > 1.5$ mm, are reported in integrated form F_v, the total evaporated volume as a fraction of the initial volume.

$$F_v^1 = g(T) \cdot \ln(t) \qquad F_v^2 = f(T) \cdot \sqrt{t} \tag{10.27}$$

Form 1 applies to most oils; form 2 applies to oils like diesel fuel.[8] The differential equivalents are:

$$\frac{dF_v^1}{dt} = g(T) \cdot e^{-F_v^1/g(T)} \qquad \frac{dF_v^2}{dt} = \frac{f^2(T)}{2}\frac{1}{F_v^2} \tag{10.28}$$

The temperature variations are

$$g(T) = [0.165\alpha + 0.045(T-15)] \qquad f(T) = [0.0254\alpha + 0.01(T-15)] \tag{10.29}$$

α is a refinery constant (mass fraction distilled at 180°C).

If thickness does not matter, then the modeling is greatly simplified. These formulations are unique in this feature, as well as in having no dependence on wind.

10.4.2 Emulsification

In this process, surface oil takes up water, forming a water-in-oil emulsion. Surface energy enables this process; the emulsions can become very viscous.

Buchanan and Hurford [58] report a typical early formulation. For a surface slick, let Y be the fractional water content by volume. The process is expressed in integral form

$$Y = Y_s\left[1 - e^{\left(-\frac{Kt}{Y_s}\right)}\right] \tag{10.30}$$

where Y_s is the ultimate, steady Y limiting the process – its endpoint. This is equivalent to the general differential form:

$$\frac{dY}{dt} = K\left[1 - \frac{Y}{Y_s}\right] \tag{10.31}$$

Fingas et al. [148] report a similar form with K dependent on wind speed U:

$$\frac{dY}{dt} = K_a(1+U)^2\left[1 - \frac{Y}{Y_s}\right] \tag{10.32}$$

For some oils, emulsions are unstable and the emulsion falls apart when the surface energy driving the process falls off. Very low and very high water contents, $Y < 0.3$ or $Y > 0.9$, do not yield stable emulsions.

Stable emulsions require asphaltene content $> 7\%$, and $0.55 < Y_s < 0.85$. For stable emulsions, the process runs to completion in 0.1 – 0.3 hours for moderate surface energy. The transition to the emulsified state is very rapid, nearly instantaneous when

[8] There is an obvious problem at $t = 0$ for both forms.

$\Delta t \sim 0.5$ h. The emulsion can be very viscous – μ increases by 10^3, a semi-solid. $Y_s \simeq \frac{3}{4}$ is reported as a reasonable choice for modeling stable emulsions. [148].

Fingas [150] provides a good summary of stability conditions.

Xie et al. [490] provide a comprehensive summary. Their conceptual model includes the formation, stability, water uptake, and effect on evaporation. Whether the emulsification process begins or not is treated using a minimum turbulent energy approach. An example problem is solved to illustrate the use of the equations.

10.4.3 Density and Viscosity

Weathering changes the density and viscosity of the oil as a function of the emulsification and evaporation rates and the temperature of the slick. An example density function comes from Lehr et al. [242]

$$\rho_o = \rho_{\text{ref}}(1 + c_e F_v)[1 - c_T(T - T_{\text{ref}})] \tag{10.33}$$

$$\rho_e = \rho_w Y + \rho_o(1 - Y) \tag{10.34}$$

Here, (ρ_w, ρ_o, ρ_e) are the density of water, oil, and the water-in-oil emulsion respectively. T is the temperature of the oil (°K), usually taken as the surrounding water temperature, and ρ_{ref} is the initial density of oil at T_{ref}. F_v and Y are the fraction of the oil evaporated and the water content of the water-in-oil emulsion respectively, both explained elsewhere. The corrected values for the empirical constants c_e and c_T are given as 1.8×10^{-2} and 8.0×10^{-4} respectively by Berry et al. [26].

Similarly, a convenient formula for viscosity change is given by [242]:

$$\mu_o = \mu_{\text{ref}} \cdot \exp\left[c_{ve} F_v + c_{vT}\left(\frac{1}{T} - \frac{1}{T_{\text{ref}}}\right)\right] \tag{10.35}$$

$$\mu_e = \mu_{\text{ref}} \cdot \exp\left[\frac{aY}{1 - bY}\right] \tag{10.36}$$

Here, (μ_o, μ_e) are the dynamic viscosity of oil and the water-in-oil emulsion. μ_{ref} is the initial viscosity of oil at T_{ref}. The empirical constants: c_{ve} ranges from 1 to 10 going from light to crude oils (Reed [370], Fingas [150]); c_{vT} is equal to 5×10^3, but larger values are reported [242]; a and b are constants whose values are usually taken as 2.5 and 0.65. Lehr ([243]) reports that values within the range $[0,5]$ and $[-0.9, 0.9]$ are possible. Buchanan and Hurford [58] make explicit the dependence on the asphaltene fraction P_A: $\mu_{\text{ref}} = 224\sqrt{P_A}$.

The increase in μ_e is especially important and fast, on the order of hours, for crudes that are subject to emulsification in energetic environments.

10.4.4 A Simple Weathering Model

A simple weathering model can be assembled, generally following Stiver and MacKay [439] and Buchanan and Hurford [58], and using the nomenclature already established. The Fay Phase 1 termination can be used for initialization, followed by Equations 10.10 and 10.11 for spreading. The parcel will have two state variables: F_v, the volume fraction oil evaporated relative to the initial volume V_o; and Y, the volume fraction of water present in the emulsion. There are two differential equations: 10.20 and 10.22 for evaporation; 10.32 for emulsification. The oil density will increase with evaporative loss. For the water-in-oil emulsion, both density and viscosity increase with water uptake. See Section 10.4.3 for typical functions.

Two limiters are added: the rate of spreading, and the asymptotic thickness.

- **Spreading** ceases as before, when $h = h_{\min}$, the limiting thickness. To avoid an abrupt transition, we use the rate limiter

$$\frac{d(\rho_o h)}{dt} = \left[\frac{\frac{dM}{dt} - \rho_o h \frac{dA}{dt}}{A}\right] \cdot \left[1 - \left(\frac{h_{\min}}{h}\right)^2\right] \tag{10.37}$$

  which reduces the spreading rate as $h_{\min}$ is approached, ultimately shutting it down.
- **With emulsification**: $h_{\min}$ is variable and rises with water content. At any time, its value is a weighted average of h and its base value h_{min0}.

$$h_{\min}(t) = f \cdot h + (1 - f) \cdot h_{\min 0} \tag{10.38}$$

  The weight f is a linear metric of emulsification: $f = Y/Y_s$. As water content approaches its limit Y_s, $h_{\min}$ rises to the current h and spreading ceases.

Figures 10.9 and 10.10 show simulations for an isolated surface parcel, using an assembly of the above rates. The oil is the Alaska North Slope crude, whose properties can be found in Table 10.1. The initial spill is 10,000 m^3 in seawater. We investigate two steady scenarios: low wind (1 m/s, Figure 10.9) and high wind (10 m/s, Figure 10.10).

It is observed that the overall weathering process is approximately 20 times faster in higher wind. The dip in emulsion volume at the start corresponds to evaporation before emulsification kicks in. Faster emulsification due to higher wind masks this effect. Note that the water uptake increases the slick volume by a factor of 2. The same effects are also observed in emulsion mass calculations. The maximum mass reached in time is the combined masses of seawater and oil. The decreasing trend after the maximum indicates that evaporation continues after the parcel is completely emulsified, an illustration of the idea in the previous paragraph. The density and viscosity both increase toward limiting values of the functions in Section

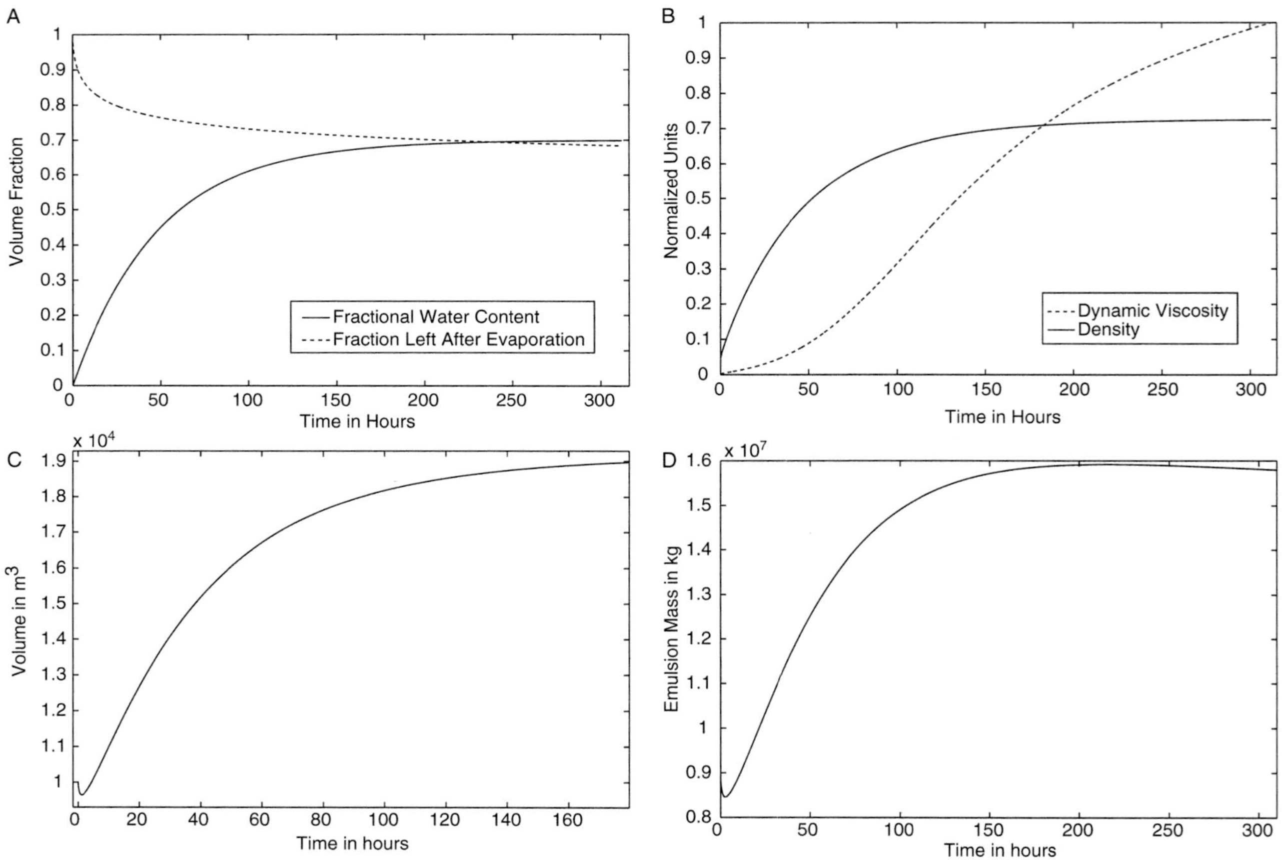

Figure 10.9. Simple weathering with **low wind**, 1 m/s. (A) Fractional water content (Y) and oil fraction remaining ($1 - F_v$) (B) Viscosity and density. Viscosity is normalized by maximum value of 25.41 N s/m^2. Density is normalized as $(\rho_o - \rho_i)/(\rho_w - \rho_i)$; $\rho_i = 890$ kg/m^3, $\rho_w = 1030$ kg/m^3 (C) Volume of the emulsion. The dip in the beginning corresponds to evaporation before emulsification kicks in. (D) Mass of the emulsion. The dip in the beginning corresponds to mass loss due to evaporation before emulsification kicks in. After emulsification completes ($Y = 0.7$), there is still mass lost due to evaporation.

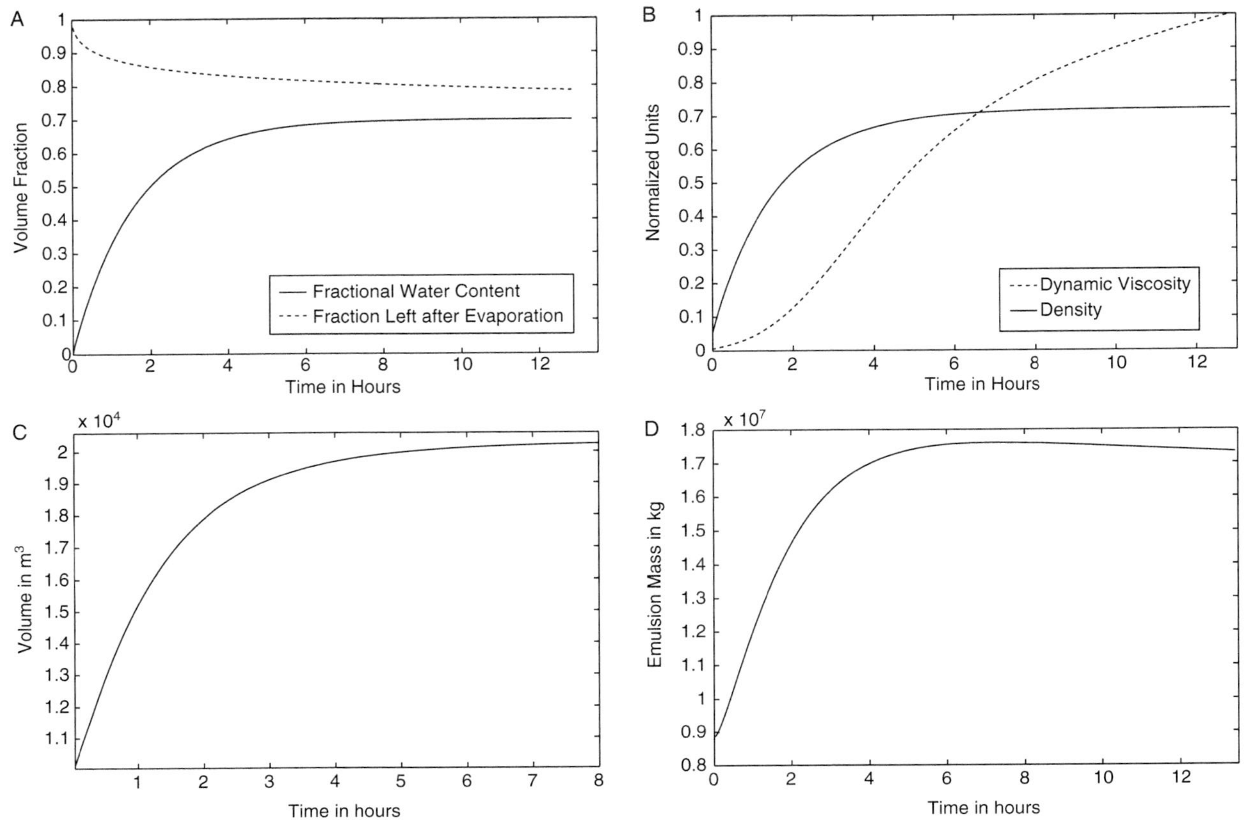

Figure 10.10. Simple weathering with **high wind**, 10 m/s. (A) Fractional water content (Y) and oil fraction remaining ($1-F_v$) (B) Viscosity and density. Viscosity is normalized by maximum value of 9.2 N s/m^2. Density is normalized as $(\rho_o-\rho_i)/(\rho_w-\rho_i)$; $\rho_i=890$ kg/m^3, $\rho_w=1030$ kg/m^3 (C) Faster emulsification due to higher wind hides the effect of evaporation. (D) Mass of the emulsion. Faster emulsification due to higher wind hides the effect of evaporation in the beginning. After emulsification completes ($Y=0.7$), there is still mass lost due to evaporation.

10.4.3 (note normalized coordinates). The viscosity increase is especially noticeable, by a factor of 290 and 800 for high and low wind cases, respectively. It should be noted that longer weathering causes higher viscosity, as oil has more time to experience environmental conditions. The fractional water content goes up to the user-set limiting value of 70% as a result of emulsification. It is also observed that the volume fraction evaporated is 10% more in low wind due to longer exposure.

One can elaborate on this basic description. For example, as stated, all the oil will ultimately evaporate, including that portion participating in emulsion. A refinement could introduce a limiting value $F_{vs} < 1$ (or equivalently $V_s > 0$), by analogy with the Y_s implementation. Emulsification has been described in terms of the water droplet spectrum, and its dynamic (Eley et al. [132]). The formational process for the water droplets would need to be embedded; at least one parameter of size distribution (e.g., the total surface area) would have to be accounted for, in addition to the total water in emulsion. Another improvement would be the addition of emulsion stability, thus de-emulsification (Xie et al. [490].) At this level it could be useful to separate the oil fractions into volatile and nonvolatile parts, and account for them separately throughout. A generalization would be to account for oil constituents separately. These types of refinements come with the burden of describing all rates for constituents and their synergy. There is a great deal of flexibility in formulation at this point, and all of the forms of rate expressions and rate limiters (Appendix D) are potentially useful. An accounting based on constituent mass, rather than volume, would be the most natural in all cases.

Problem 10.9 Change the limiters in Equations 10.37 and 10.38 to another form (Appendix D) in order to eliminate slick spreading at high Y. Compare properties and computed results with that shown in the text.

Problem 10.10 In the simple model above, wind drives irreversible emulsification. What form of modifications are required to achieve de-emulsification when wind energy subsides? Show that your modifications achieve the desired end without unintended side effects.

Problem 10.11 In the simple model above, evaporation progresses independently from emulsification. A direct effect is that it continues after emulsification is complete (Figure 10.10). What can be done to eliminate this effect? Test your idea by simulation; explain the results.

10.4.5 Entrainment

Entrainment is the conversion of surface oil into droplets, and their injection subsurface. There they undergo 3-D motion without exposure to the atmosphere. The process is driven by surface turbulence; typically breaking surface waves. The process is referred to in the oil community as "natural dispersion."[9] From a modeling perspective, this process involves a *change of state*[10]: matter is removed from the surface, and surface parcels are either decremented or vanish altogether; subsurface parcels are introduced, conserving mass. A threefold description is needed: the rate of oil entrainment; the size of the droplets formed; and the depth to which droplets are injected.

Delvigne and Sweeney [112] report fundamental experiments for this process; Delvigne [113] summarizes the findings.

- *Droplet size distribution:* For droplet sizes $D \pm \frac{dD}{2}$: the number of droplets formed dN is dependent on D:

$$dN = B \cdot D^{-2.3} dD \tag{10.39}$$

  with B an experimental constant. The volume of each droplet is $\frac{\pi}{6}D^3$, so the volume of oil entrained in this size class $dV \sim D^{0.7} dD$.
- *Intrusion Depth:* In terms of H_b, the breaking wave height, a maximum intrusion depth z_i is given for all droplets:

$$z_i = 1.5H_b \tag{10.40}$$

  Small droplets are uniformly distributed to this depth; large droplets concentrate higher.
- *Entrainment Rate*, [kg/m^2/s], for a single size class:

$$dQ = C_o E^{0.57} D^{0.7} S_{\text{cov}} F_{wc}\, dD \tag{10.41}$$

  - dQ is the mass entrainment rate for droplets in the size class $D \pm dD$, per unit surface area: [kg/m^2/s].
  - E is the wave energy dissipated per unit surface area, for a single breaking wave, J/m^2.
  - F_{wc} is the fraction of sea surface hit with waves per unit time, s^{-1}.
  - S_{cov} is the fraction of sea surface covered with oil.
  - C_o is an empirical constant for a particular oil, $\sim \frac{1}{\nu(T)}$.

[9] "Chemical dispersion" would be inducing the same changes, by addition of chemical agents.

[10] Here we refer to the logical state of a parcel: surface or subsurface. A chemical change is not implied.

C_0 for a few oils ranged from 510 to 1800. Empirical relationships are offered for F_{wc} and E:

$$F_{wc} = c_b(U - U_i)/P_w \tag{10.42}$$

$$E = 0.0034\rho_w g H_{rms}^2 \tag{10.43}$$

with

- H_{rms} the rms wave height (m)
- P_w the wave period (s)
- U the wind speed (m/s)
- U_i the wind speed at initiation of breaking ($\simeq$ 5 m/s)
- $c_b = 0.032\,\mathrm{s/m}$

This characterization lacks a maximum droplet size; and the entrainment rate grows as droplet size D increases. The accumulated entrainment rate for all droplets smaller than $D = \mathcal{D}$ is gotten by integrating Equation 10.41:

$$Q(\mathcal{D}) = \int_0^{\mathcal{D}} dQ = K \cdot \mathcal{D}^{1.7} \tag{10.44}$$

with $K \equiv \left[C_o E^{0.57} S_{\mathrm{cov}} F_{wc}/1.7\right]$. $Q(\mathcal{D})$ grows without bound, as $\mathcal{D}^{1.7}$. A criterion for limiting the droplet size must be added.[11]

Assuming a finite entrainment rate Q, (by setting $D_{\max}$ or otherwise altering the droplet distribution): the mass balance on the donor slick parcel is

$$\frac{dM}{dt} = -AQ = -\frac{Q}{\rho h}M \tag{10.45}$$

and Q is the rate of initiation of new subsurface parcels.

Droplet Initiation

Surface oil abundance h is finite and must be apportioned among the droplet spectrum.

The smallest droplets have the slowest rise velocity and therefore are the most persistent sub-surface; the largest return to the surface quickly. So the recommended procedure is: entrain all the available surface oil by filling up the lower end of the droplet spectrum [113]. From Equation 10.44, increase $\mathcal{D}$ until there is no more

[11] This is a consequence of the basic droplet distribution, Equation 10.39. A distribution function for droplet size cannot have this shape at the top end – it implies unbounded cumulative droplet volume. In fact, the high end of the D spectrum becomes harder to observe experimentally (large droplets return to the surface quickly) so enforcing theoretical distributional properties there is a necessity.

surface oil:

$$h = Q\Delta t = \Delta t K \cdot \mathcal{D}^{1.7} \tag{10.46}$$

This provides an operational form for $D_{\max}$:

$$D_{\max} = \left[\frac{h}{K\Delta t} \right]^{1/1.7} \tag{10.47}$$

Procedurally, start with the smallest droplets; and accumulate $\Delta t \cdot dQ(D)$ in each size class, until the total available is exhausted. All surface oil will be entrained in this fashion; the bigger droplets will surface quickly and offset this over-entrainment.

To take this procedure one step further, when Δt is finite: if the rise velocity of a droplet exceeds $z_i/\Delta t$, then those droplets need not be entrained; they will resurface in the coming Δt. This can be used as a criterion for the maximum droplet size. An estimate of rise velocity is $w = \frac{1}{18} \frac{\beta g}{\nu_w} D^2$ for $D < 0.2$ mm (Section 10.5.3).[12] Balancing these, the biggest practical droplet would be

$$D_{\max} = \left[\frac{18 z_i}{\Delta t} \frac{\nu_w}{\beta g} \right]^{1/2} \tag{10.48}$$

If the surface oil is very abundant, this would limit the entrainment. The time scale Δt could be replaced with a physically meaningful time for example, the wave period P_w.[13] (Very small droplets are often treated as an instant loss to dissolution and biodegradation, so the low end of the entrained D spectrum might be truncated as well.)

These are operational strategies, making up for the inadequate top end to the D spectrum in use. Without that, we are extrapolating beyond the data support.

Tkalich and Chan [450] discuss generalizations of the Delvigne formulation, including relations for the $D_{\max}$ to be used with this particular distribution. Vos [465] found that four different weathering models in common use employed the Delvigne and Sweeney [112] formulation. Li and Garrett [250] examined the formational hypotheses for entrainment theoretically, and provide a foundation and context for this process with some directions for improvement.

Problem 10.12 Suppose there is a limited quantity of oil at the surface, density ρ_o, thickness h, and that entrainment is proceeding over a period Δt. (1) What is the maximum droplet size that will be entrained? (2) what is the distribution of D, per unit of surface area?

[12] The data in these 1988 experiments [112] are reported to a maximum $D \simeq 1.0$ mm, just beyond the range of this approximation.

[13] One could add vertical diffusion to this method.

Dunnewind et al. [120] advocate a new approach based more fundamentally on the physics of the process. Factors not included into the empirical model of Delvigne and Sweeney [112] are shown to be important; and some mathematical forms (including the relevant probability density functions) are examined. One suggested formulation that results is a very simple statement of entrainment rate Q, m^3/s:

$$Q = \phi_{100} \frac{V}{T} \tag{10.49}$$

where ϕ_{100} is the fraction of oil droplets smaller than 100 microns; V is the volume of surface oil; and T is the wave period. Per unit of surface area, this is $q = Q/A$:

$$q = \phi_{100} \frac{h}{T} \tag{10.50}$$

This is the rate at which oil leaves the surface. The details of the droplet formational process are embedded in the relations for the droplet size spectrum. For the oil entering the subsurface, one needs the full spectrum description. This more general approach would be capable of handling both chemical and natural emulsification, and is recommended for future theoretical and experimental work.

Current research on droplet formation in plumes from deep blowouts uses the log-normal, Rosin-Rammler (Weibull) and related distributions for droplet size (Chen and Yapa [80], Johansen et al. [214]). These have a sigmoid cumulative distribution function, approaching a finite asymptote at high D. It is likely that these distributions will be applicable in formulations of surface oil entrainment.

10.5 Motion

3-D gridded fields are assumed available for $(\mathbf{v}, T, S, q^2, l)$. The generic problem of computing Lagrangian motion in these fields is treated throughout this book. There are a few things that deserve special attention in the oil context.

10.5.1 Stokes Drift and Surface Velocity

The motion of particles is a combined function of currents (tidal, density, geostrophic, wind, etc.) and waves. The current field is the expected product of a hydrodynamic model. This may or may not come with wind contribution. However, unless the surface layer velocities (especially the top few centimeters) in the input flow field are adequately resolved by the hydrodynamic model, a wind correction is needed. For floating oil, this will be important. During a spill event, the ability to incorporate available real-time wind data would also require a correction for improved spill response. The conventional rule of thumb is to add 2.5 to 4% of the wind speed at 10 m above sea level (W_{10}) to the current input data. The directional corrections occur as Ekman type deflection angles between 0 and 25 degrees to the right of the wind direction in the Northern Hemisphere (Berry et al. [26], Perianez and Caravaca [353]). The mean values are good candidates. Samuels et al. [400] suggest an empirically derived deflection angle that is a function of wind speed:

$$\theta = 25^\circ e^{-\left[10^{-8}\frac{W_{10}}{g\nu}\right]} \tag{10.51}$$

Here θ is the deflection angle and ν is the kinematic viscosity of seawater. The formula allows the reduction of θ in high wind situations. A good summary, which also includes the conditions under which the correction factors are derived, is given in [400].

The downward progression of the wind-induced current in the water column can be modeled according to Pugh [354, 367]:

$$U_w(z) = U_{w0} - \frac{U_*}{\kappa}\ln\left|\frac{z}{z_0}\right| \tag{10.52}$$

where $U_w(z)$ is the current speed due to wind at depth z, U_{w0} is its value at the surface (typically 3% of W_{10}), κ is the von Karman constant, z is the distance below the free surface, and z_0 is the roughness length (0.5 mm $\leq z_0 \leq$ 1.5 mm). U_* is the frictional velocity and can be approximated by $0.0012W_{10}$.

The wave contribution to the advective transport ideally comes from a wave model in the direction of wave propagation as Stokes drift, given by:

$$U_S = \left(\frac{\pi H}{L}\right)^2 Ce^{-2kz} \tag{10.53}$$

for deep water ($h > L/2$), and

$$U_S = \left(\frac{\pi H}{L}\right)^2 C\cdot\left[\frac{\cosh(2k(h-z))}{\sinh^2 kh}\right] \tag{10.54}$$

for shallow water ($h < L/20$). Here H is the wave height, L is the wavelength, k is the wave number ($2\pi/L$), and h is the water depth below the surface. C is the celerity of the wave, that is, $\sqrt{gL/2\pi}$ for deep water waves, $\sqrt{gh}$ for shallow water waves.[14] z is the vertical distance from the mean sea surface down, $z = 0$ at the surface. The significant wave height can be used for H if the wave field is assumed to be steady (Wang et al. [467]). This mechanism is especially important in coastal areas due to wave refraction, causing a transport biased toward shore. It also decouples wind and waves, allowing the transport from ocean swells to be incorporated in the absence of

[14] For intermediate waves, celerity is governed by the important dispersion relation, see for example, Dean and Dalrymple [106].

wind. It should be noted Equations 10.53 and 10.54 give the transport velocities for a single wave [467].

A better and more natural approach would couple the wave and shear effects using a realistic vertical eddy viscosity profile to resolve subsurface transport. Youssef and Spaulding [431, 499] present a model for combined wind and Stokes drift currents, acknowledging that the vertical eddy viscosity profile is a property of the flow itself and that it should be calculated from the turbulent kinetic energy. The model assumes deep water (Equation 10.53) with a no-slip condition at the bottom to incorporate dissipation due to friction. They test their models with winds ranging between 1 and 30 m/s. It is observed that in low wind conditions, the wind energy is confined at the surface due to low turbulence levels and thus low eddy viscosity, causing larger drift factors on the order of 4%. As winds get higher and waves break more frequently, the higher eddy viscosity increases the energy transfer towards the bottom, causing lower drift factors, on the order of 3%. These results are in good agreement with observations, especially when depth averaged over a depth range of $2.5H - 5H$ where H is the significant wave height (Reed et al. [369, 371]). At the surface, a curve fit gives:

$$F = 3.91 - 0.0318 \cdot W_{10} \tag{10.55}$$

$$\theta = 23.627 - 7.97 \cdot \log(W_{10}) \tag{10.56}$$

for the drift factor and deflection angle respectively, under steady-state deep water conditions. The model can be used to predict time dependent F and θ at various depths. It is only natural that similar models that resolve underlying processes would take the place of the empirical approach in the future, especially considering that parameters such as the turbulent kinetic energy are readily computed in hydrodynamic models.

10.5.2 Random Walk Models

For more than three decades oil motion has been simulated with model-based advection, with random walk added, based on an eddy diffusion coefficient (A_h,A_z). Most contemporary models have turbulence closure, in which case the diffusion coefficient is a hydrodynamic output.

Chapter 4 develops a family of such models and their properties. Efforts have largely centered on the simplest AR0 models; autoregressive models are being used by some, with increasing frequency. Lonin [258]) reports the use of AR1 or "random flight" models. Isaji and Spaulding [206] and Spaulding et al. [433] report use of AR1 and AR2 ("random push") models. These authors are finding that good results and general improvement over AR0.

Kitanidis [228] develops and uses the AR0-type model in the groundwater arena. The pseudo-velocity ∇A is explicitly studied and found to be important, corroborating experience in many fields, notably the Visser formulation, which is widely followed. There is some theoretical generalization that supports more advanced time-stepping schemes. North [334] uses the pseudo-velocity as in Visser [464], and it is built into the LTRANS coding. Berry et al. [26] adapted the LTRANS implementation to the oil context; OILTRANS is the product.

10.5.3 Droplet Rise Velocity

Subsurface oil will occur as buoyant droplets. The rise velocity w_o depends on droplet size D and density contrast $\Delta\rho = \rho_w - \rho_o$ (in normalized form, the buoyancy $\beta = \frac{\Delta\rho}{\rho_w}$.) Generally, larger droplets rise faster and present a smaller surface area per unit of oil.

Stokes' law governs in the laminar range, applicable to very small spheres:

$$w_o = \frac{1}{18}\frac{\beta}{\nu_w} g D^2 \qquad D < 0.0002\,\mathrm{m} \tag{10.57}$$

The smallest spheres will rise very slowly; and approach the case of a passive tracer. These droplets can be expected to be vulnerable to dissolution in the surrounding water, due to their longer exposure time and large surface-to-volume ratio. The rise velocity increases quadratically with D.

For large spherical droplets, form drag (Newtonian drag) will dominate the vertical velocity,

$$w_o = \sqrt{\frac{8}{3}\beta g D} \qquad D > 0.01\ \mathrm{m} \tag{10.58}$$

Intermediate size droplets will be governed by both mechanisms – viscous drag and form drag. The general force balance between drag and buoyancy still governs; for a sphere,

$$w_o^2 = \frac{4\beta g}{3}\frac{D}{C_d} \tag{10.59}$$

with C_d a function of Reynolds number $R_e = w_o D/\nu$ (White [482]):

$$C_d = \left[\frac{24}{R_e} + \frac{6}{1+\sqrt{R_e}} + 0.4\right]\left[1 + 1.9 R_i^{1/2}\right] \qquad 0 \leq R_e \leq 2\times 10^5 \tag{10.60}$$

The stratification multiplier in Equation 10.60 was added by Yick et al. [498].[15] Iteration is required to find w_o for a given D; the result matches both high and low D limits.

Of course fluid droplet shape is not so ideal. Approximately, small droplets are spherical; intermediate ones are ellipsoids; and large ones are spherical-cap shaped (acorn tops). Zheng and Yapa [501] show an integrated formulation that relaxes the assumption of spherical droplets, and has good experimental support. At the high D end, the shape distortion is serious and rise velocities can be of order 50% less than for rigid spheres [501].

10.5.4 Droplet Size Distribution

Delvigne and Sweeney [112] found an orderly droplet size distribution dependent on turbulent energy and oil viscosity, applicable to oil introduced in initially large droplets subsurface, and distributed by the ambient turbulence. The same spectrum was created at the surface by waves breaking over floating parcels. The larger droplets surfaced quickly, placing a limit on their observability. This is discussed in the entrainment section, 10.4.5.

The size of the droplets observed is 10–1000 microns. In dealing with thousands of cubic meters of oil, it is apparent that it is not practical to simulate every droplet as a particle, certainly in the operational context.

A more practical approach would recognize that both ends of the spectrum are short-lived. But even with this truncated spectrum, one will need to associate many droplets with a single Lagrangian parcel. Because the rise velocity depends on droplet buoyancy and size, it will be important to make all droplets within a single parcel have the same size.

10.6 Density and Crowding

An isolated parcel evolves, weathering and spreading, with rates updated along the way. Thickness h is a surrogate for the volume/surface ratio for the parcel. It will reach h_m in finite time; thereafter, continued weathering will cause A to shrink, at constant h_m, presumably exposing open water subareas within the slick parcel. The spreading dynamic is based on very simplistic geometry, and the nomenclature reflects that. Simultaneously of course, the parcel is moving following hydrodynamic advection and dispersion by random walk.

15 The "viscous Richardson number" is $R_i = \frac{r^3 N^2}{\nu_w U}$; the buoyancy frequency $N = \left[-\frac{g}{\rho_w}\frac{d\rho}{dz}\right]^{1/2}$

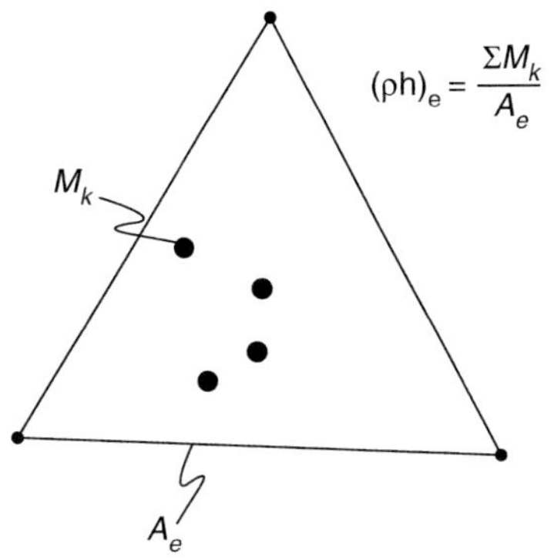

Figure 10.11. The simplest piecewise constant projection. Chapter 8 develops the projection idea more fully.

10.6.1 Mass Field

When many parcels are involved, it is useful to define an equivalent density field, a continuous representation of many point masses. This is the problem of projection, discussed in Chapter 8. It is readily achieved as needed, without change to the particles. The operative sense of "density" is the horizontal distribution $h\rho$, the mass per unit of horizontal area. The projection sought is

$$\sum_k M_k \delta_k \Longrightarrow \sum_i (h\rho)_i \phi_i(\mathbf{x}) \tag{10.61}$$

with $\phi_i(\mathbf{x})$ a continuous basis, and parameters $(h\rho)_i$.

There are some simple projections in use. The common default is to gather all parcels within a computational cell or element, and compute an element average $h\rho$:

$$(h\rho)_e = \frac{1}{A_e} \sum_e M_k \tag{10.62}$$

This would be the "element" projection, illustrated in Figure 10.11. It is the simplest and most intuitive, resulting in a discontinuous (piecewise constant) $h\rho$ field. Its generalization to a matrix form describes a large class of projections, many of which provide differentiability. Galt [167] uses a Thiessen polygon (Voronoi) approach; Zhang et al. [500] use a Gaussian kernel method.

10.6.2 Crowding

Many parcels have the potential to overlap, pile up. This can happen initially at the point of a continuous release; or at fronts, downwelling boundaries, and so forth where parcels are concentrated by hydrodynamic transport. In these cases

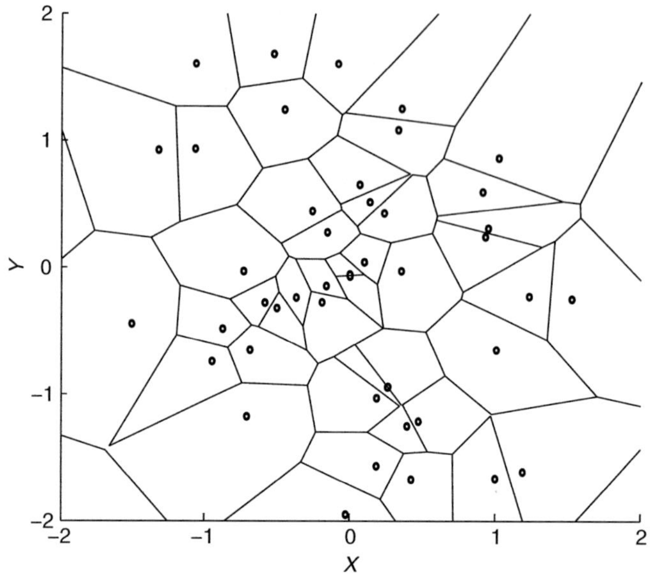

Figure 10.12. Voronoi tesselation of particles placed at random, $(X,Y) \sim \mathcal{G}(0,1)$.

the decay of h is interrupted as parcels must share the available sea surface. (Recall that enlarging h will decrease weathering rates.)

The Voronoi tesselation of parcel locations assigns a unique area surrounding each parcel, illustrated in Figure 10.12. This is commonly referred to as "tiling." The tiles fill the plane contiguously; each is convex and contains exactly one parcel. All points in a tile are closer to the parcel it contains, than to any other parcel. And so, the area of tile k, A_k^*, is a metric of the area available to parcel k. When $A_k < A_k^*$, there is spare area and that parcel is not crowded; it has room to grow without interference.

If $A_k > A_k^*$, then an adjustment must be made – keeping ρ_k and M_k unchanged, shrink A_k to A_k^* and increase h_k, conserving mass:

$$h_k = \frac{M_k}{\rho_k} \frac{1}{A_k^*} \tag{10.63}$$

By doing this, we are assuming that the intrinsic spreading in each parcel alone, will fill the available nearby space if needed; and once that is filled, parcels will be squeezed together and must grow vertically.

This uneven free surface results of course in a repellant force between contiguous parcels with different $h\rho$ and hence an extra motion is needed. Intuitively, the horizontal pressure gradient in the oil will have to balance the viscous shear force at the underlying oil-water interface. (This is generally the Phase 2 assumption already in use.)

A continuous $h\rho$ is necessary in order to get its gradient. Assume this is done. The balance of pressure gradient and viscous shear will have to be of the general form

$$-gh\nabla(h\rho_o) = \rho_w \nu_w \frac{\dot{\mathbf{X}}}{L} \tag{10.64}$$

with $\dot{\mathbf{X}}$ the velocity of the oil parcel relative to the water, and L a vertical shear length in the water. So we end up with a repellant velocity $\dot{\mathbf{X}}$:

$$\dot{\mathbf{X}} = -L\frac{gh}{\rho_w \nu_w}\nabla(h\rho_o) \tag{10.65}$$

$\dot{\mathbf{X}}$ is a velocity of the oil parcel due to crowding; it is a sideways repellant response to neighbors, directed toward decreasing oil abundance. It represents an extra motion for the parcel. There is a single parameter L.

There are therefore three spreading effects:

- the basic oceanic transport of the parcel, advection and random walk, affecting the parcel center
- the intrinsic spreading of an individual parcel, represented in $\frac{dh}{dt}$ and $\frac{dA}{dt}$ – spreading within a parcel, affecting its shape
- interparcel spreading – the repulsive effect of all neighbors, expressed in the gradient of the projected $h\rho$ field and adding to the motion of the center of mass of the parcel

The third of these is the crowding effect.

10.7 Submerged Parcels

Subsurface oil is idealized as occurring as small buoyant spherical droplets. Typical diameter is $10 - 10^4$ microns. Individual droplets will have kinetics representing dissolution and biodegradation. Droplets are not exposed to the atmosphere, so evaporation is not relevant. There are potentially huge numbers of droplets, so aggregation of droplets into parcels is natural.

A single parcel will contain many droplets of identical size D. Each has the same density ρ, viscosity ν, buoyant velocity $w_b(D,\rho)$, and surface area πD^2. The parcel will have the same properties (ρ, ν, w_b); the aggregate amount of oil (mass M, volume V, surface area A), will be the summation of many droplets. The parcel will have biochemical kinetics representing the aggregate of all its droplets; within a parcel, each droplet is assumed independent.

10.7.1 Surface Source – Entrainment

Suppose a quantity of oil M with properties (ρ,ν) is entrained from a surface slick by breaking waves. The mass is assigned to a new submerged parcel of N droplets; the probability density function (PDF) of droplet diameter D is associated with the turbulent dispersal event.

There are several options for establishing a proper particle size distribution. Among them:

- Assign the median D_{50}.
- Draw D at random from the known PDF.
- Partition D into a few bins, according to the known PDF.

All of these pertain to a single event, one Δt duration. Over time, there will be many, and the PDF imperfections need to cancel on average.

In all cases, the droplets retain the properties (ρ,ν) of the slick parcel whence they came. The mass balance allows the computation of the combined surface area exposed.

$$A = \frac{M}{\rho}\frac{6}{D} \tag{10.66}$$

So the newly submerged parcel has M,ρ,ν; uniform droplet size $D(\varepsilon)$ drawn from the spectrum associated with the turbulence ε; buoyant velocity $W(D,\rho)$; and total oil–water interfacial area A for the purposes of surficial transfer processes. 3-D transport by advection, buoyancy, and dispersion (random walk) begins.

If the parcel resurfaces, it becomes a new slick parcel. During the time spent submerged, chemical transformations are different, as there is no air interface. The newly emerged slick parcel will differ in location and chemistry from its "parent" slick parcel.

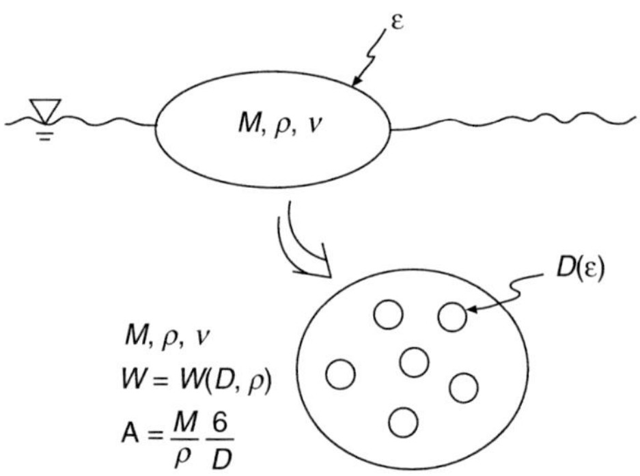

Figure 10.13. Breakup of a slick parcel into a submerged parcel of small droplets, size D. The turbulent energy driving the process is ε.

10.7.2 Shear Dispersion

Surface oil is the source of parcels of droplets entrained into the 3-D flow. The formational process is described previously. Briefly, breaking of surface waves is the primary mechanism; there will be a spectrum of droplet sizes; the droplets are tiny so large numbers are required in any single parcel; and the entrainment depth is comparable to the surface wave height. 3-D advection plus buoyant random walk takes over at that point.

It is evident that entrained oil will be "protected" from the surface motion; and have different transport speed and direction, depending on the aggregate of oceanographic processes. One can expect entrained oil to depart from its surface source. And being buoyant, it will resurface separated from the source oil. During high wind events, we can expect the subsurface flow to be slower and deflected by Coriolis force (to the right in Northern Hemisphere); and there will be a Stokes drift effect. The result is shear dispersion – and the smaller, slower-rising particles will be more susceptible.

Field observations are consistent with this description (Elliott et al. [133]; Elliott [134]; Elliott and Wallace [135]; Reed et al. [372]) and it is reproduced in modeling outcomes. The mechanism is observed in many contexts.

Example 10.1 Consider the case of uniform gravity forcing toward the west, with surface wind opposing it, and uniform Eddy viscosity, isotropic in the horizontal. In the absence of rotation, the velocity profile is quadratic, illustrated in Figure 10.16. An ensemble of identical spherical, buoyant particles was released at the surface, and motion ensued. Very elementary BCs were enforced for simplicity.[16] The random outcomes are depicted at four points in time in Figure 10.14. The expected deepening of the ensemble is evident, as is the (vertically) sheared horizontal advection; the lower particles are flowing westward, the near-surface ones, eastward. The horizontal consequences are shown in Figure 10.15. The ensemble is elongated, and due to the particle buoyancy, favors the surface eastward displacement with the wind. Figure 10.16 shows ensemble growth in variance in the two horizontal directions. As expected in the cross-wind direction, $\sigma_y \sim \sqrt{t}$; while σ_x is faster.

Example 10.2 For this example, rotation was added; wind alone is the forcing. In MKS units: $H = 20; A_z = 0.01; f = 0.0001; \beta = 0.85; A_h = 1$. The velocity profile is the classic Ekman spiral, in shallow water (Figure 10.17).

Four particle diameters were used: $D = 0.001 * [0.005\ 0.01\ 0.02\ 0.04]$. Their abundance was scaled according to Delvigne and Sweeney [112]: $N = [500\ 308\ 190\ 117]$. Rise velocity was according to Stokes' law, Equation 10.57. Particles reaching the

[16] Random walk was constrained to remain in the water column: if a particle escaped through the bottom, it would simply be restored to the bottom, likewise at the top.

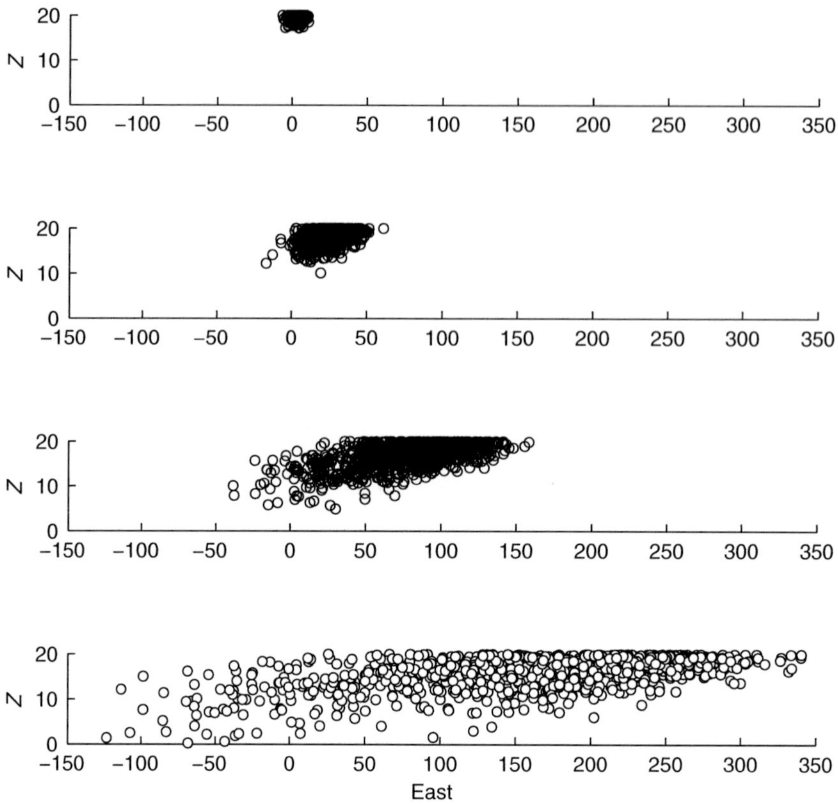

Figure 10.14. Example 1. Profile display of particle locations at four sequential points in time. The wind is eastward.

surface were re-entrained with probability $P_e = \Delta t/T_e$, $T_e = 1/2$ hour being the expected waiting time.

Particle locations are shown in Figure 10.18 at four sequential times, the first shortly after release. Particle size scales with dot size. All particles are plotted, surface and subsurface. The horizontal distribution resembles a curving comet. At the front of the distribution (the plus sign) is the displaced launch position, with surface velocity and no random walk.

Figure 10.19 is the same plots rendered on a single axis. Figure 10.20 is a 3-D rendering, which illustrates the development of the vertical distribution and the tendency of small particles to linger at depth. The combined message of this example: small particles spend more time at depth; as a result they lag behind the larger more buoyant particles that dominate the surface; and the rotational veering with depth is evident in the rightward-leaning distributions.

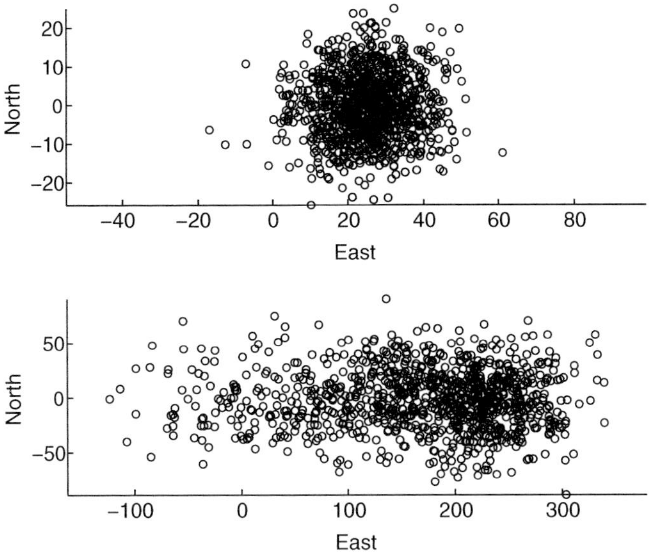

Figure 10.15. Example 1. Horizontal plots of particle locations at time points 2 and 4. All particles are displayed, regardless of depth.

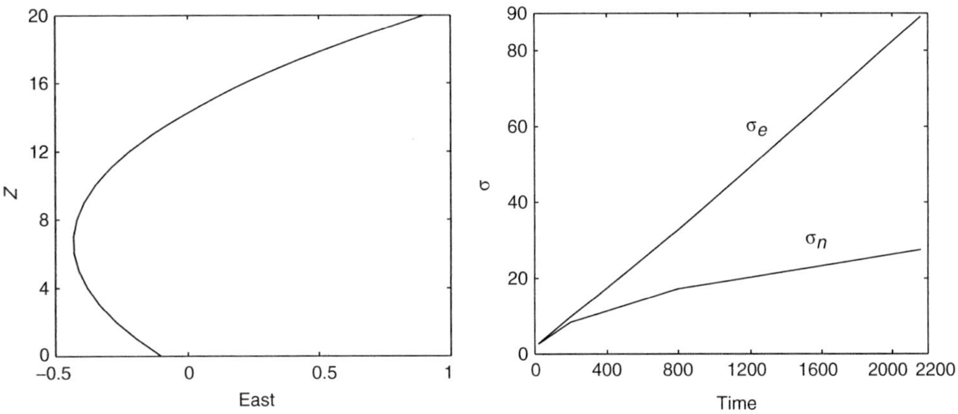

Figure 10.16. Example 1. Left: Eastward velocity profile. Right: Temporal growth of particle standard deviations σ_e (along flow) and σ_n (crossflow), plotted at the four points in time. Horizontal eddy viscosity is isotropic.

Overall: The shear dispersion process is embedded in an elementary way in the random walk. It will occur in simulations with no special inputs. Given realistic circulation fields and particle distributions, this is a result consistent with basic theory. The field comparisons cited above indicate that it is of practical importance.

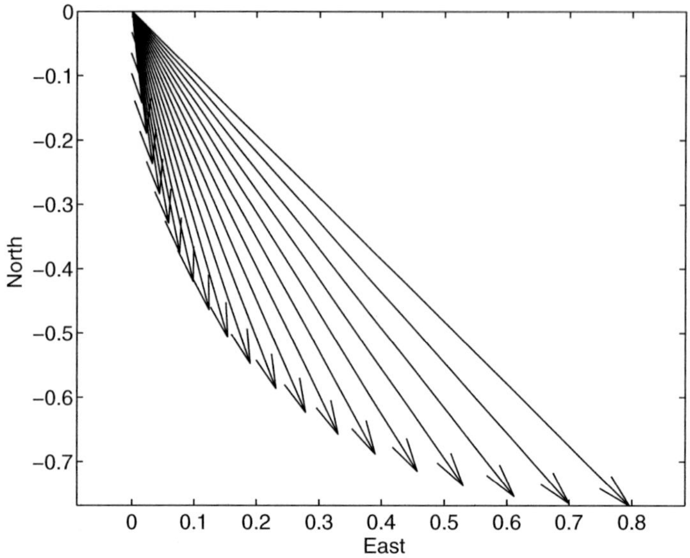

Figure 10.17. Example 2: Velocity profile. The wind is toward the east.

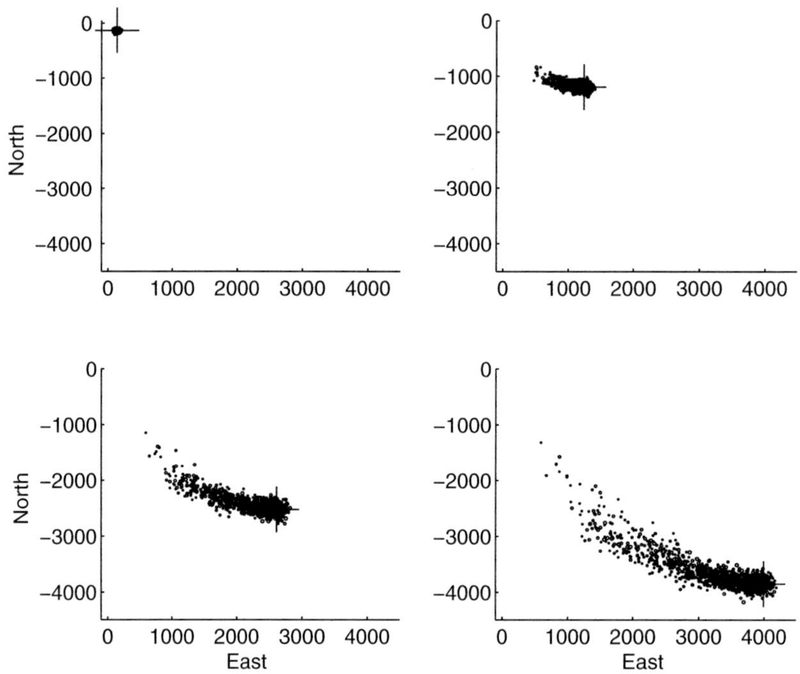

Figure 10.18. Example 2. Horizontal distribution at four sequential times. Dot size scales with particle size. All particles are plotted, regardless of depth. The plus sign shows the purely advective displacement of a hypothetical surface particle.

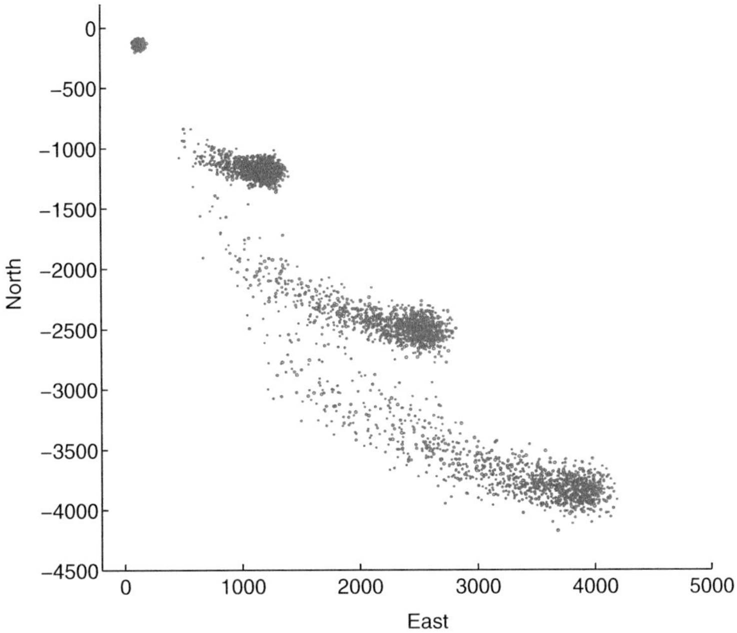

Figure 10.19. As in Figure 10.18: The particle locations at four times, on the same axis. For a color version of this figure, please see the color plate section.

10.7.3 Subsurface Source – Blowout

Subsurface parcels can originate at the sea floor, the result of an exploration or production accident. Oil will occur as droplets, aggregated into parcels as in the surface entrainment case. But the operative processes near the point of introduction require special attention. These processes need to account for droplet formation and injection into a Coastal Ocean Model (Figure 10.2).

There is a long history of plume studies focused on shallower releases of wastewater and cooling water, and in the case of aeration, gas. These support a near-field analysis that includes small-scale hydrodynamics of buoyancy, jets, entrainment, and droplet formation, and ultimate placement of discharged fluid in the ambient fluid where far-field relations govern.

There are many factors that can complicate things in deep water: a potentially uncertain mix of oil, gas, and water under high pressure and low temperature; formations of gas bubbles and oil droplets; the formation of gas hydrates and their transport/fate; and the presence of deep oceanic stratification. A further complication would be intentional injection of chemical agents into a blowout plume or discharge pipe, and the resultant effect on the dynamics. The problem involves a thermodynamic description of multiphase turbulent flow and the injection of heat, mass, and momentum into the ocean.

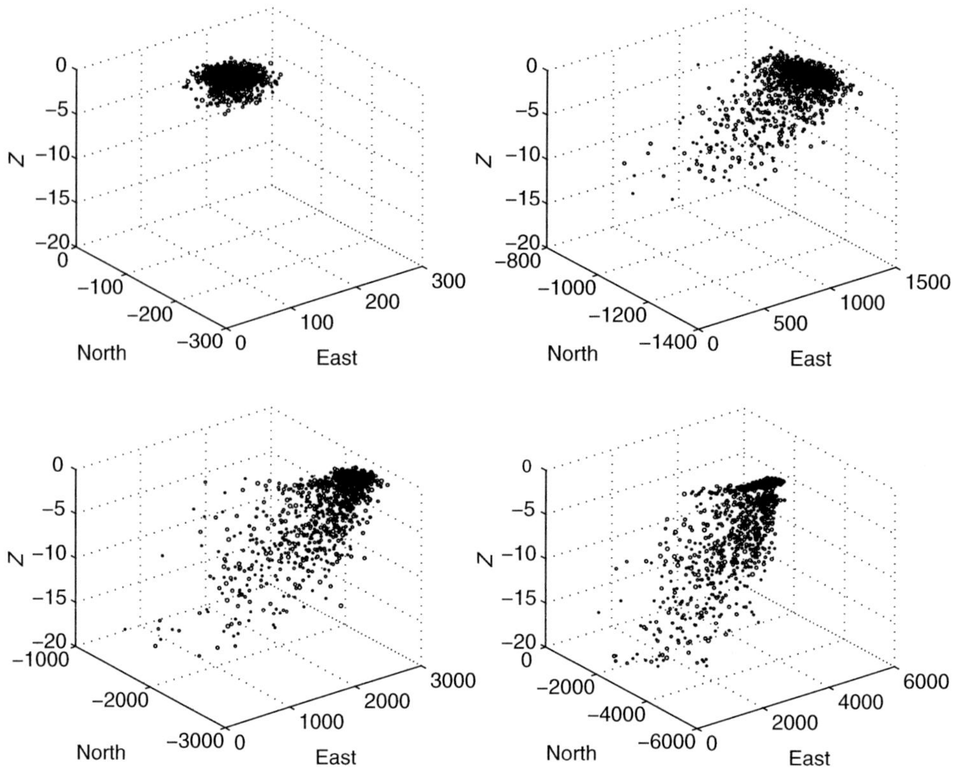

Figure 10.20. Example 2: 3-D rendering at four times, illustrating the vertical distribution of particles.

A particle model requires initialization by such a model: droplet size and abundance, composition, position, buoyancy, and sufficient information to allow a particle to interact with oceanic conditions over potentially long periods of transport. Beyond this initialization, particle dynamics are relatively straightforward: random walk of buoyant droplets in a 3-D flow field. Some blowout parcels will eventually reach the surface, where a change of parcel identity occurs.

The vertical injection height can be of paramount importance. For example, in the Macondo blowout incident (Section 10.8.3) the injection height was 350 m above the seabed where the blowout occurred.

Currently there is significant effort directed at this problem. Reviews and general model development are reported by Spaulding et al. [434]; Johansen [212, 213]; Socolofsky, Adams et al. [424, 425]; and Yapa et al. [494, 497]. These represent a blend of experiment, analysis, and simulation; and field-scale comparisons are possible with the DeepSpill experiment and the Deepwater Horizon accident (Johansen et al. [214], Socolofsky et al. [426]). (See Section 10.8.) All forms of scientific investigation are needed to attack this significant problem. Yapa, Dasanayaka et al. [97, 496] provide recent overviews.

It is hoped that in the long run, a single particle model will serve both near-and far-field needs. The foundations for such a model are beyond the present exposition; for present purposes, we look to a near-field model for initialization.

10.7.4 Dissolution

Light aromatic constituents of oil such as benzene and toluene start to dissolve into the surrounding water as early as 5 minutes after a surface spill, and maximum concentrations are reached on the order of hours (NRC, 2005) [338]. This process, however, is a losing competitor to evaporation since the same constituents are also taken away from the environment much more rapidly by the latter. References state that dissolution is two or more orders of magnitude smaller than evaporation (Cekirge and Palmer [71], Fingas [150]. Although this may seem to justify the frequent exclusion of dissolution in mass balance calculations, one should keep in mind that these exact same constituents are the very components that can be toxic to marine life and they need to be incorporated in cases where toxicity is important (i.e., gasoline, diesel or light crude spills in areas with high biological activity). Dissolution also gains in importance in deep water spills due to larger underwater residence times and larger exposure area (droplets).

The dissolution mass flux f_d (units: moles/s) can be calculated for each petroleum hydrocarbon component separately. Following Mackay and Leinonen [294], the mass flux for each component j, is given by:

$$\left[f_d\right]_j = A\,[K_d(S_e - C_w)]_j \tag{10.67}$$

Here, A is an area across which the solute is moving, and K_d is a mass transfer coefficient (units: cm/s). C_w is the bulk water phase concentration, usually taken as zero in practice since it is much smaller than solubility S [294, 495]. S_e is the effective solubility in moles/cm^3, given by:

$$S_e = S\mathcal{X}E \tag{10.68}$$

$\mathcal{X}$ is the oil phase mole fraction of the component j in the simulated oil; E is the component-dependent solubility enhancement factor summarized in Table 10.4. S is the pure component solubility, its values are known for specific hydrocarbons [85, 337, 415]. The total mass flux lost from the slick to dissolution would then be the sum over j. Conversion from moles to mass is straightforward using the molar mass of a component. For a less specific approach, Equation 10.67 can be applied to the oil as a whole, for which example bulk solubilities are provided in Table 10.5.

Table 10.4. *Solubility Enhancement Factor E, after [294]*

Component Type	Solubility Enhancement Factor E
Alkanes (pentane, octane)	1.4
Cycloalkanes (cycloheptane, cyclopentane)	1.4
Aromatics (benzene, toluene)	2.2
Olefins (pentene, decene)	1.8

Note: Example hydrocarbons are given for each component type in parentheses.

Table 10.5. *Bulk Solubilities S_b for Some Oils of Interest around Room Temperature and in Distilled Water*

Oil Type	Solubility S_b (kg· m^{-3})
Gasoline (Premium)	0.112
Diesel	0.003
Alberta	0.025
Hibernia B	0.024
Arabian Light	0.019
Fuel Oil No:2	0.003
Arabian Medium	0.018
Prudhoe Bay (1995)	0.029
Bunker C	0.006

Note: In salt water these values should be reduced by 25–30% due to the salting-out effect [21, 415].

Equation 10.67 can be applied to floating slick parcels or to entrained droplets in the water column [71]. A needs to be the exposed area across which the mass flux is occurring: the horizontal area of a parcel, or the surface area of a droplet. The relevant dimensionless number is the Sherwood number S_n

$$S_n = \frac{K_d L}{N_m} \tag{10.69}$$

Here K_d and N_m are the mass transfer coefficient (m/s) and mass (molecular) diffusivity of oil in water (m^2/s) respectively. L is a length scale, that is, the thickness h of a floating parcel, or the diameter D for a submerged droplet. MacKay and Leinonen [294] suggest using $S_n = 2$ for droplets dissolving slowly in suspension and thus in

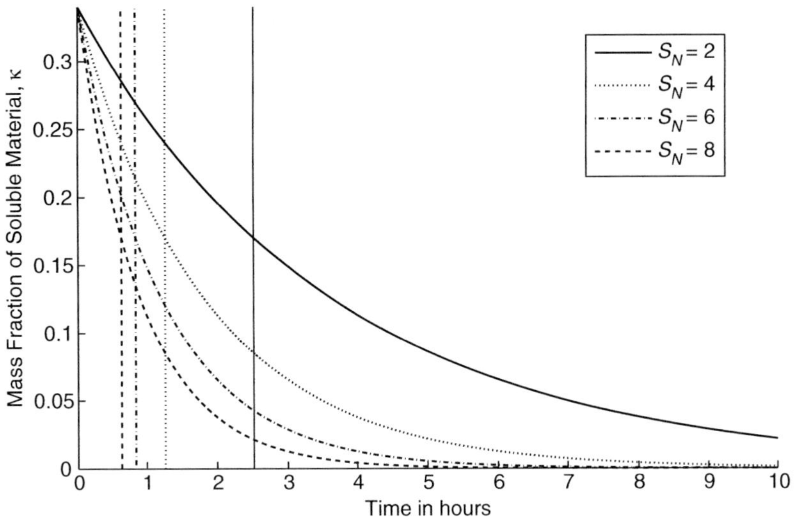

Figure 10.21. Mass fraction decay of soluble material from Prudhoe Bay crude oil droplets of 100 microns dissolving in seawater. Line types are for various values of the Sherwood number. Vertical lines are corresponding dissolution half lives.

relatively calm conditions (i.e. non-tidal lagoons and ponds). Using $S_n = 2$ and a benzene droplet of 100 microns would give a K_d of around 2.2×10^{-3} cm/s.[17]

Equations 10.67 and 10.69 can be combined to write the mass balance equations, assuming constant D or h:

$$\frac{1}{M\mathcal{X}}\frac{d(M\mathcal{X})}{dt} = -\frac{S_n N_m S_e}{\mathcal{X}\rho_o h^2} \quad \text{(surface parcel)} \tag{10.70}$$

$$\frac{1}{M\mathcal{X}}\frac{d(M\mathcal{X})}{dt} = -\frac{6 S_n N_m S_e}{\mathcal{X}\rho_o D^2} \quad \text{(subsurface droplet)} \tag{10.71}$$

M is the oil mass of a surface parcel ($\rho_o A h$) or of a droplet ($\rho_o \pi D^3/6$). $\mathcal{X}M$ and $\mathcal{X}\rho_o$ are the constituent mass and density.[18]

Figure 10.21 shows the decay of $\mathcal{X}$ for $D = 100$ microns and for increasing values of S_n. The oil is the Prudhoe Bay crude at 20° C. The initial soluble material content $\mathcal{X}_0$ is 34%; the saltwater bulk solubility S_b is 2.175×10^{-5} g/cm^3. The mass diffusivity N_m can be taken typically as 0.0324 cm^2/h for lighter components [44]. There is

[17] A bulk K_d value of 2.36×10^{-4} cm/s can also be used for the oil as a whole, given the uncertainty especially under dynamic environments [294].

[18] For submerged particles, it is possible to lump the dissolution process into the entrainment calculations – making no distinction between entrainment and dissolution. The idea behind this approach is that the entire soluble percentage of any "small" droplet entrained into the water column will dissolve instantaneously. The definition of "small" is related to the Sherwood number.

no evaporation. The effective solubility is assumed to decrease linearly according to

$$S_e = S_b \frac{\mathcal{X}}{\mathcal{X}_0} \tag{10.72}$$

It is observed that for calm conditions, the half life for the soluble component is approximately 2.5 hours. As the ratio S_n increases to 8, this value drops down to 35 minutes. Furthermore, in a real-world situation, a large percentage of the initial mass fraction $\mathcal{X}_0$ would be taken away by evaporation. Based on a similar analysis, Mackay and Leinonen [294] state that the dissolution process is instantaneous for droplets smaller than 100 microns due to reduced half-lives, especially for large S_n.[19]

10.8 Field Tests

10.8.1 Surface Releases

Lange [237] describes a surface-release experiment conducted in 1982, offshore of Norway, at Halten Bank. There are data from drifters and from airborne radar. 100 tons of crude were released over 2 hours, creating an initial slick 3.5 × 0.5 km. Over 7 days the slick drifted 90 km, achieved a maximum surface area of 86 km^2, and dispersed droplets below it. In addition to the physical features, biological and chemical properties were measured.

Reed et al. [372] report four experimental surface releases at two sites west of central Norway in 1989 and 1991. The monitoring was similar to the Halten Bank release, including drifters, airborne remote sensing, and oil biochemistry. Reed confirms the importance of shear dispersion in the field, as described in the Elliot and Delvigne and Sweeney papers.

Price et al. [363] report on "oilers" – drifters that imitate the motion of surface oil – and field experiments with same (20–30 drifters per experiment, five experiments) in the Gulf of Mexico, 1997–1999. The field trajectories were long (30 days) and wide (500 km).

10.8.2 Subsurface Releases

Deep experimental releases have accompanied the movement of exploration and production to greater depths. Rye et al. [394] report a 1996 experiment at 106 m depth, near the Frigg Field in the North Sea. A mixture of oil and air was discharged; the observations concentrate on the near-field of the discharge.

Johansen et al. [214] report on the June 2000 DeepSpill experiments. These were done near the Helland Hansen site, 25 miles west of central Norway, ocean depth 844 m. Gas and oil were discharged from the sea floor, over durations of 40–120 minutes. The greater depth allowed the plume to establish itself, and then evolve,

[19] This is based on fresh water solubility and $S_n = 2$.

with droplets exiting the near-field of plume formation and beginning their migration and evolution.

Generally, deep releases have focused on the near field of the release, plume formation, and its subsequent rise to the surface. The time frame is short.

The Deepwater Horizon blowout in the Gulf of Mexico was a full-scale accident that yielded a significant dataset with oil originating at ~1500 m depth and significant space and time scales. Its analysis, both observation and modeling, is documented by Liu et al. [254]. The time of observation is long enough for the oil to reach the surface, and be tracked remotely over long distances; and for small buoyant droplets to move significant distances with the deeper current. The model/field studies include the study of surface trajectories initialized from satellite observations (Liu et al. [255]; MacFadyen et al. [292]; Huntley et al. [203]); all used floating, inert particles; several different models provided the gridded circulation fields. There are also inert 3-D trajectories for (a) passive particles initiated at different depths (Weisberg et al. [471]); and b) inert, buoyant particles, initialized with a plume model of the blowout (North et al. [334]). This simulation is reviewed below.

10.8.3 The **Deepwater Horizon** *Incident*

Deepwater Horizon (DH) was a semi-submersible drilling rig in commercial use for deep exploration. In 2009, it drilled the deepest (~10 km) oil well in history, in the Keithley Canyon, Gulf of Mexico, in ~1,260 m of water, ~400 km southeast of Houston.

Subsequent drilling at the Macondo Prospect in the Mississippi Canyon led to the largest US oil spill in history, in 2010. Oil and gas was discharged under pressure at the sea floor, ~ 1500 m below the surface and ~ 65 km offshore.

North et al. [334] describe 3-D particle tracking for this incident. The particle tracking model LTRANS (Schlag and North [404]) was adapted, in concert with an empirical plume model for the near field, and a realistic circulation hindcast for the far field.

The plume model of Socolofsky et al. [426] provided particle initialization. Particles were apportioned among nine size classes: from 10 to 300 microns and assigned as oil or gas droplets. Abundance and timing of particle releases was based on estimates of discharge rate and chemistry.

The plume model allowed for the particle distribution, buoyancy and initial heights; particle simulation took over from this point. Particles were inert with constant (D, ρ, w); the ambient environment was served from a Regional Ocean Modeling System (ROMS) model of the Gulf. The random walk was AR0 (no memory), with the Visser temporal discretization including the pseudo-velocity ∇A_z.

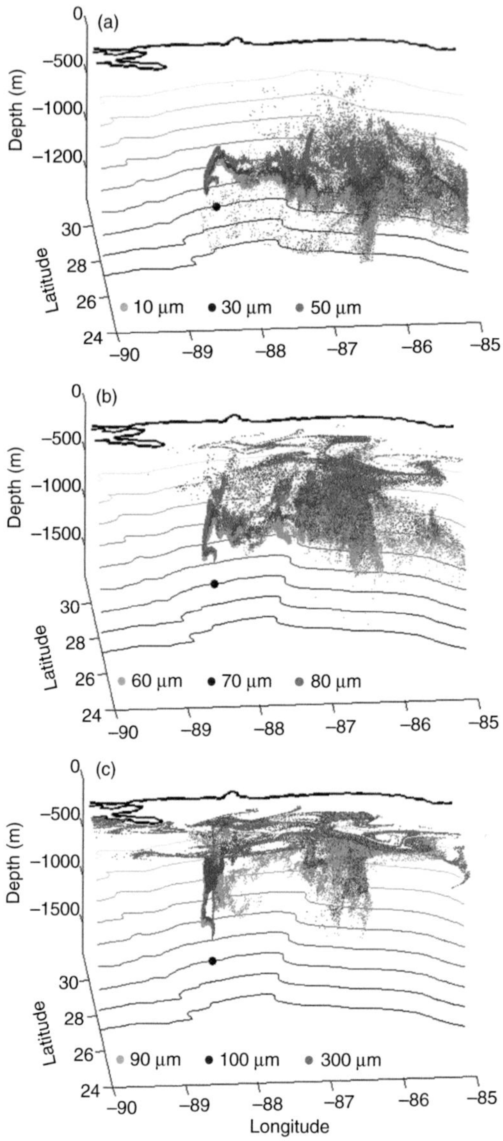

Figure 10.22. 3-D view of simulated particles originating from the Macondo Blowout on June 3, 42 days into the incident. Colors indicate particle diameter. Particles were released at the plume exit height above the well, indicated by the black dot. (From North et al. [334]. Source: E. North, pers. comm. Reproduced with permission.) For a color version of this figure, please see the color plate section.

It was found:

- Oil droplets for $D = [10, 30, 50]$ microns remained subsurface for over a month, at depths ~1000–1200 m. This is in accord with the observations of a subsurface plume in that vicinity (Camilli et al. 2010 [66]).

- Increasing size meant faster rise. For $D = [60, 80]$ microns, particles rose slowly, with some vertical dispersal throughout the water column.
- For $D = [90, 100]$ microns droplets rose faster to the surface, and the vertical dispersal confined them nearer the surface
- For $D = 300$ microns, droplets rose rapidly to the surface and remained there.

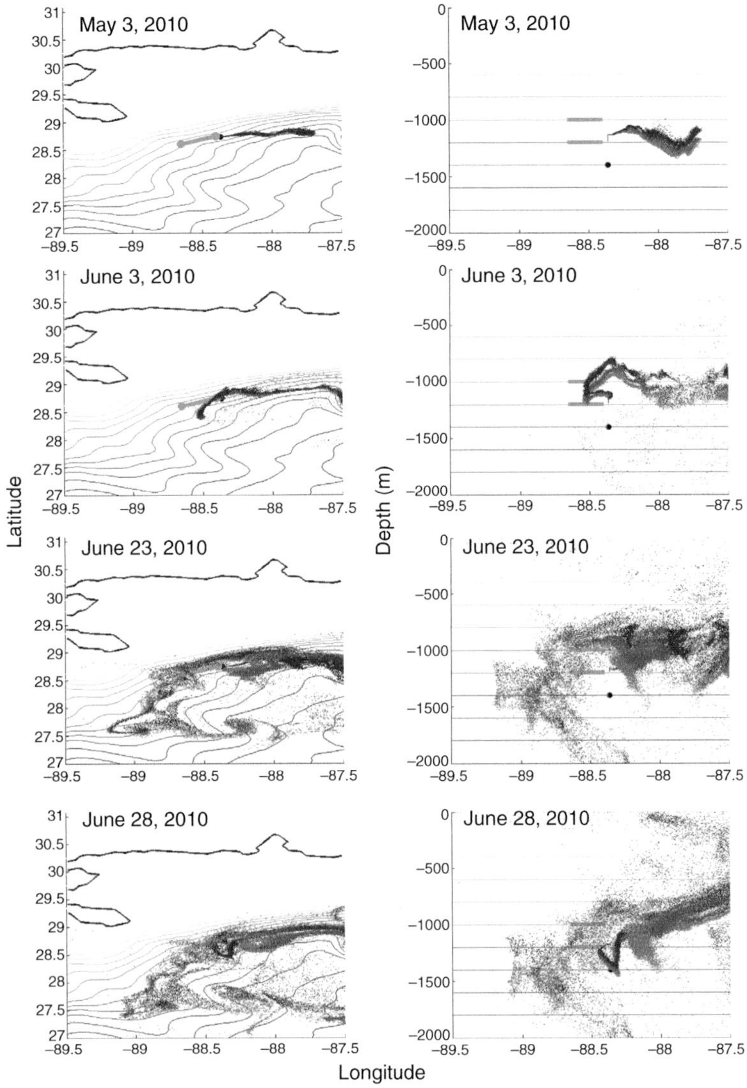

Figure 10.23. Temporal development of the simulated Macondo plume for the smaller particles, $D = [10, 30, 50]$ microns ([red, blue, dark blue]). Particles were released continuously from April 22 to July 15, 2010, at the estimated rates. Evidence of a subsurface plume was reported at the locations marked in orange. (From North et al. [334]. Source: E. North, pers. comm. Reproduced with permission.) For a color version of this figure, please see the color plate section.

These trends are evident in the detailed presentation by North et al. [334], a sample of which is reproduced here (Figures 10.22 and 10.23).

10.9 Impact Assessment

There are many ways that oil transported in the sea is finally assimilated into its surroundings. Evaporation, dissolution, biodegradation, sinking through sedimentation, and beaching are some of them. Each of these has its own complexities and ramifications, causing short- or long-term environmental effects. For instance, one might consider an oil parcel stranding at low tide under the sun on a porous sediment bed with biological activity; or simply, an oil particle dissolved in the water column causing short-term toxicity for the biota. Anthropogenic intervention, such as the use of dispersants or other cleanup efforts, makes the prediction of final fate more complicated and is accompanied by its own environmental consequences. The possibilities are endless and the ramifications are almost always negative, especially in the short term. It is intuitive that modeling of impact is an interdisciplinary task, requiring knowledge on chemical, biological, and physical sciences to simulate and evaluate the environmental interactions.

The overall methodology of starting with the simplest model applies. Good reviews can be found in Seymour and Geyer [407] and French-McCay [162, 163]. There is an enormous horizon for making comprehensive models, which embed particle methods in larger systems that include ultimate impact evaluation.

11
Individual-Based Models – Biotic Particles

11.1 Introduction

The individual-based model (IBM) is the logical progression of cohort-based fisheries models (e.g., Beverton-Holt, [28, 83]). In that case, the distinction is made between mortality of a cohort's members, and the growth of the members. Different cohorts are kept conceptually; they are recruited in different batches, tracked through time, identifiable in data. Individual members of a cohort are identical. Their vital rates and environment are the same; their number is subject to mortality. The IBM extends this idea to "the cohort of 1." Individuals bear the diversity of the population; each lives a Lagrangian life – in a combined physical and biotic space, from birth to death.

The essential features of the IBM are illustrated in Figure 11.1, emphasizing the comparison with a conventional Eulerian approach. The diffusive motion is represented as a random walk as described in Chapter 4; and lifecycle events are represented as state transition probabilities.[1] Continuous particle evolution is related to Eulerian rates as Lagrangian ordinary differential equations (ODEs) traveling with a particle, some forms of which are described in Appendix D.

11.2 Diversity in the Cohort

In the simplest cohort description (see, e.g., Lynch [280]) one describes a group of identical individuals in terms of two dynamic variables: the number of surviving individuals N alive at any one time, and the weight W of each individual.

$$\frac{dW}{dt} = G(W,t) \tag{11.1}$$

$$\frac{dN}{dt} = -\mu N \tag{11.2}$$

[1] **Note:** In this chapter, $Pr(X)$ indicates the probability of event X occurring.

Eulerian: $\frac{dN}{dt} = -kN$	Lagrangian: $\Pr(\text{event in } dt) = k\,dt$
Diffusion $Q = -A_h \frac{dN}{dx}$	Random Walk $dx \sim R\sqrt{(2A_h dt)}$
Advection $\frac{dx}{dt} = V$	Pseudo-velocity $\frac{dx}{dt} = V + \frac{dA_h}{dx}$

Figure 11.1. Essential relations in individual-based modeling, by comparison with conventional Eulerian modeling. The motion of an individual is represented as a random walk, as in Chapter 4; biotic lifecycle events are represented as state transitions with probabilities *Pr*. Continuous development is represented as an ODE traveling with a particle; some standard forms are reviewed in Appendix D.

G is the biological growth rate function for an individual; μ is the mortality rate acting on the cohort as a whole. Importantly, W does not include any mortality, whether natural or induced by harvesting or other anthropogenic influences.

If this idea is played out in a spatially explicit ecosystem model, the random walk will produce a different trajectory for each individual. If there is any reason to predicate variations in either G or μ as a function of position, for example as a result of varying temperature, nutrient availability, and so forth, then we have *individual variation* within the cohort – each individual will experience a unique life history, even if its response to environmental influences is identical among individuals. But, all individuals are not the same. Variability is observed across individuals in essentially all vital processes. If the vital rates G and μ are not identical among individuals, that is, there is intrinsic variation among individuals, then we again have individual variation. So we have two generic sources of variability – that intrinsic to the individual organism, and that due to its environment. Both accumulate as the individual lives and moves.

For the simplest of situations, one can handle either of these variabilities by carefully working the pdf of the various parameters through the basic model equations 11.1, 11.2. Of course, for practical situations this rapidly becomes onerous, nonlinear, and prone to error – all of which invite simplification. Parsimony in this arena is obviously valuable, a trademark of good scientific method.

A simulation approach to individual variation is quite practical with today's machines and tools, and with the availability of site-specific computational products. One *simulates* the motion, maturation and death of individuals representative of the cohort; *sums the results* when aggregate outcomes are desirable; and *retains the variation* across individuals.

This is the essence of the IBM. In the idealized examples that follow (Sections 11.3–11.10) we assume that the motion is a random walk in the environment; the details of that are covered in Chapter 4. Here we concentrate on the biotic rates – the vital rates describing growth, transformation and death; and the behavioral depth regulation that can develop with maturity. Realistic spatially explicit applications are described in Sections 11.11 and 11.12.

11.3 Individual-Based States

11.3.1 Mixing and Aggregation

Mixing is irreversible: things that are mixed cannot be unmixed. The information about the mixture is lost; what survives is the mean. Diversity in the mix is replaced with the mean of the mix. But – the mean of rates is not the mean rate of the mean individual. The distribution matters!

Typically, mixing is based on co-location in space-time. But, individuals do not yield their biotic condition to such an average; nor do they yield their vital rates to average rates. The fact that they might be co-located does not mean they lose their individual characteristics.

In contrast, things that are *aggregated* need not be mixed. By this we mean that mathematical aggregation averages over all the components of an ensemble, to find their collective properties – but *retains the components*, including higher moments of their distribution. The aggregation is an analytical device; properly done, it need not lose any information about the individuals themselves. There are certainly aggregate quantities of interest – amount, aggregate rates, and so forth.

Some individual rates depend on Eulerian location (space, time) – motion, light, bottom condition, benthic community, and so forth. Some are temperature- or salinity-dependent. These Eulerian influences affect the individual , and in various ways are integrated throughout an individual's development, as it moves through physical and biotic state-space.

To summarize, some quantities are properly mixed, in terms of mean, variance, higher moments – notably the environment that an individual encounters. Other quantities are embedded in individuals that move through the environment and bear the outcomes. The basic IBM strategy is to preserve the appropriate amount of detail about the biotic state of individuals, their vital rates, and their behavior. Ensemble or mean properties are derivable from the full variation among individuals.

11.3.2 Life Histories

Implied in the preceding discussion is the preservation of life history on an individual basis – the individual life history being the basis of diversity in biological states, rates, and behaviors. Life cycle transitions are a common characteristic in marine organisms. For example, many fish species have planktonic larvae that metamorphose into juveniles during settlement, later maturing into adults (Section 11.11.1). Mesozooplankton such as copepods have distinct life stages including naupliar, copepodite, and adult (Section 11.11.2). Even phytoplankton can have distinct life cycles, including a resting cyst stage (Section 11.11.3).

11.3.3 Eulerian and Lagrangian Quantities

We need to distinguish between two types of living quantities. Those that belong to place or time, and affect processes that occur there – these are effectively characterized by Eulerian space–time fields. An example is the benthic community. It comprises living entities with biomass metrics (abundance and distribution) and vital rates. The benthos is immobile (or moving slowly) and affects pelagic individuals passing through. For example: pelagic larvae experience mortality due to bottom-dwelling predators. Lagrangian qualities – those that accompany an individual – represent the biotic state of that individual and its particular vital rates. In that sense, the individual is unique, the only bearer of its particular history – both geographical and biotic – all of which contribute to individual identity.

11.3.4 Continuous and Logical States; State Transitions

Further distinction is useful between types of individual-based variables. Some biotic state variables are properly conceived as continuous variables that evolve with the individual over time – for example, individual weight or fitness. We imagine descriptions of these in terms of ODE's associated with the individual. These may be subjected to stochastic influences as well as parametric variation.

Other states are fundamentally binary or logical – dead or alive; reproductive stage, and so forth. In these cases, one must rely on a probabilistic description of state transitions for an individual.

There is a useful general relation between the state transition probabilities and the statistical hazard rate, developed in Chapter 3. It is applicable to several binary processes.

11.4 Vital Rates

To convert the cohort model above (Equations 11.1,11.2) into an IBM, we have to describe the two vital rates applicable to individual i. The weight gain requires no conceptual change, as it originally described an individual trajectory:

$$\frac{dW_i}{dt} = G_i(W_i, t) \tag{11.3}$$

Immediately, we have an opportunity to formulate variation among individuals. For continuous growth functions there are three generic ways:

1. By changing either the form of the function G or its parameters to achieve a distribution G_i across individuals
2. By making the parameters of G_i stochastic over time
3. By making the growth process itself subject to random forcing:

$$\frac{dW_i}{dt} = G_i(W_i,t) + \epsilon_i \tag{11.4}$$

The first two amount to distributing randomness among vital rates; the third amounts to distributing random forcing to a process. Many nonlinear forms of G that have been found useful. Several are summarized in Appendix D.

11.4.1 Growth Rate Distribution – Analytic Example

Consider the simple Bertalanffy growth function (Appendix D):

$$G(W) = g\,(K - W) \tag{11.5}$$

The two parameters g and K are the growth rate and mature weight for an individual. Integration for a single individual, starting at $W = 0$, gives the weight function

$$W(t) = K\left[1 - e^{-gt}\right] \tag{11.6}$$

Both parameters, g and K, can be expected to vary among otherwise identical individuals. Suppose that the constant g varies continuously across individuals, with a uniform distribution over the range $g_i = g_o \pm \Delta g$. Then we have a range of slow- and fast-growing individuals. Suppose further that the mature weight K is uniform across individuals. The mean over all individuals at a fixed time will be

$$\overline{W}(t) = \frac{\int W(t)dg}{\int dg} \tag{11.7}$$

with integration range from $(g_0 - \Delta g)$ to $(g_0 + \Delta g)$. The result is

$$\overline{W}(t) = K\left[1 - e^{-g_o t}\left(\frac{\sinh(t\Delta g)}{t\Delta g}\right)\right] \tag{11.8}$$

The cohort, on average, will mature more slowly than a single individual with the average $g = g_o$.[1] Growth essentially ceases as K is reached; the average individual reaches maturity while the slower individuals are still growing.

[1] sinh is the hyperbolic sine function; $\sinh(x)/x \geq 1$ for non-negative x.

This simple example opens many complexities – the need for a fuller description of g's variability in terms of its probability density function; the similar need for a statistical description of K's variability; how g and K covary; their autocorrelation, and so forth. Though fruitful, this rapidly becomes intractable analytically. One seeks an approach to variability that is simulation-oriented – put the variability into the parameters, across individuals; simulate each individual; then aggregate the individuals as needed. Such a simulation-based approach preserves the full variability of the ensemble across all individuals.

Problem 11.1 Confirm Equation 11.8 for the mean of $W(t)$.

Problem 11.2 Continuing Problem 11.1: What is the variance of $W(t)$ in this case?

Problem 11.3 Generate many particles that grow according to the Bertalanffy function (Equation 11.5) with variability in g as given in the example above. Plot the results; compare with the theory.

Problem 11.4 Continuation of Problem 11.3: Compute the ensemble mean and variance as functions of time; compare with theory.

11.4.2 Growth Rate Distribution: Simulation

Suppose, for example, G were to be described by a simple logistic model,

$$G_i(W) = g_i W \left(1 - \frac{W}{K_i}\right) \tag{11.9}$$

where the two parameters are $g = 1, 1.5, 2$ and $K = 10, 12, 14$, all equally likely; in addition the starting values are $W_0 = 0.10, 0.15, 0.20$. There are 27 individuals, assumed equally likely in this case.

The analytic solution is known,

$$\frac{W(t)}{K} = \frac{1}{1 + \left(\frac{K - W_0}{W_0}\right) e^{-gt}} \tag{11.10}$$

Individual growth results are indicated in Figure 11.2 as the ensemble of outcomes $W_i(t)$, plus a single outcome $\overline{W}(t)$ achieved with mean parameters and initial condition. Note that the latter is *not* the same as the ensemble mean of the outcomes $W(t)$.

The growth with mean properties provides a reasonable approximation of the ensemble mean of the outcomes $W_i(t)$. It is a little bit higher at any time, and of

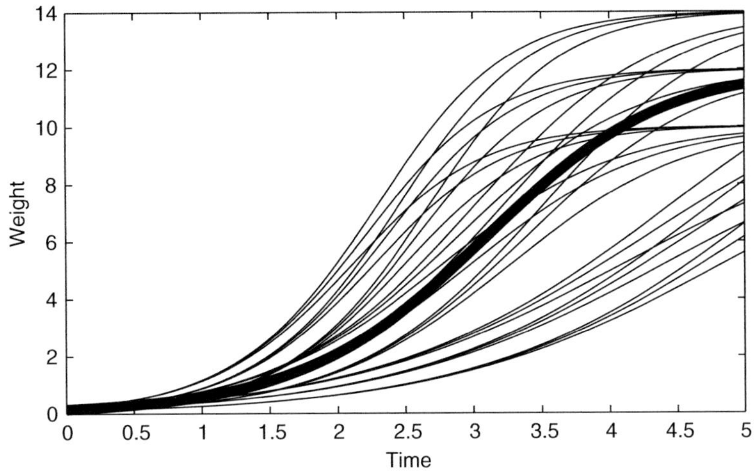

Figure 11.2. Logistic growth with individual variation in K, g, and W_0 for the 27 combinations as given in the text. Also shown (heavy line) is the growth of a particle with the mean of all properties.

course it is constrained at start ($W = \overline{W_0}$) and finish ($W = \overline{K}$) where the mean outcome is simply equal to the mean of the parameters.[2] At any time, the variance of the simulated ensemble is large. If one spreads these out in space via routine random walk, significant space–time variability would be generated.

Figure 11.3 shows the same simulation with an uncorrelated random component added in every timestep, $g^* = g \cdot (1 + \mathcal{G}(0,1))$. As expected, the variance is expanded. The details of the random formulations are substantive and invite detailed exploration. It is easy to imagine fairly complex variations in G_i and in the individual outcomes W_i, especially in cases where the nonlinear interaction of vital rate parameters is embedded in a growing individual moving by random walk in a heterogeneous space–time environment.

Problem 11.5 Replicate the results in Figure 11.3. Then, add autocorrelation (AR1) to the random influences on g, with autocorrelation coefficient $\rho = 1 - \frac{\Delta t}{T}$, with $T = 3$. Is the additional variation significant? What value of T is needed to make the variation significant?

Problem 11.6 Replicate the results in Figure 11.3. To that case, add an additional random variation in the parameter K at every time step. (1) How big do these additional variations have to be to be important at long time? (2) Same as (1) but for intermediate time, when growth is maximal.

[2] For low initial weight W_0, growth is very slow initially and its effect persists. Looking only at the $W(t)$ results in Figure 11.2, the variation due to W_0 is easily misinterpreted as the result of variation in g.

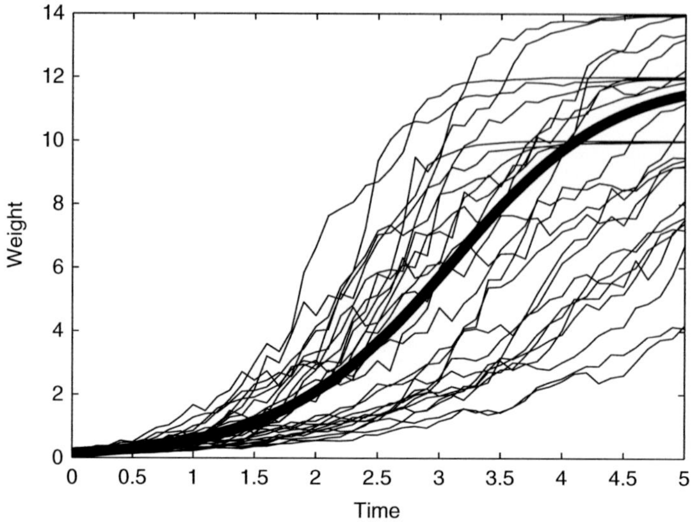

Figure 11.3. Same as Figure 11.2 but with a random influence on growth rate.

11.4.3 Multiple Equilibria

In the preceding examples we used the logistic growth function $G(W)$. This has the single stable equilibrium at $W = K$, which attracts all biological trajectories in (G, W) phase space (assuming $W > 0$). The origin ($W = 0$) is unstable and repels all trajectories. Negative W is meaningless here.

It is easy to simulate other nonlinear growth rates. Suppose for example we have growth rate cubic in W with three zeros as shown in Figure 11.4. The one at $W = 5$ is unstable and divides trajectories into those seeking one or the other of the two stable equilibria at $W = 2$ and $W = 9$. The approaches are sigmoid, accelerating or decelerating depending on the location relative the extrema of W. Such a population would be bimodal, reflecting the two mature states.

Needless to say, if one is near an unstable point, the trajectory is extremely sensitive to initial conditions. In practice, with many processes operative, including random influences, it is easy to flip attractions from one stable point to another. Figure 11.5 illustrates this. The starting point is slightly above the $W = 5$ divide; some trajectories are attracted to $W = 9$; some to $W = 2$; some remain ambivalent, with prolonged hovering around the initial condition.

11.4.4 Rate Estimation

The problem of deducing the operative vital rates is fundamental in ecology. Generally, there is a plethora of processes that are known to be relevant; identifying the key ones, their mathematical form, and their parameterization is critical. This problem

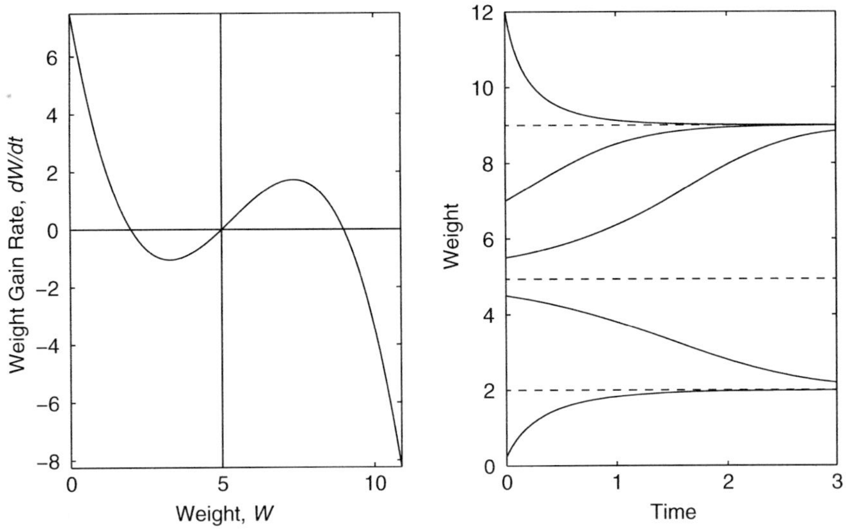

Figure 11.4. Left: Cubic Growth rate with two stable equilibria at $W = 2$ and 9. Right: Simulation from several different starting points. $W = 5$ is an unstable point; it repels trajectories, dividing them between the two stable equilibria.

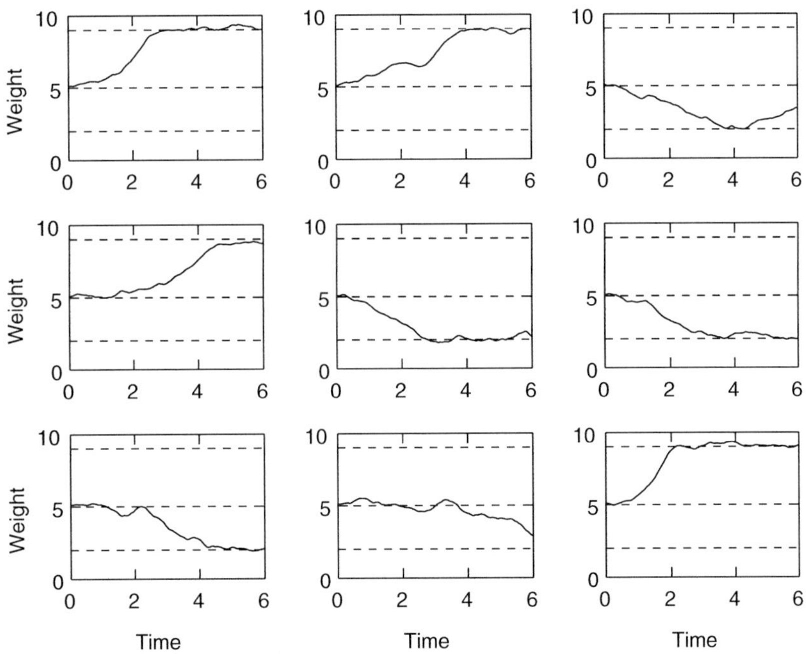

Figure 11.5. Nine trajectories of weight versus time as in Figure 11.4 (right), with AR1 growth added. All have the same initial weight $W = 5.1$. AR1 parameters: $\Delta t = 0.01$, $\rho = .95$, $\sigma_{\Delta W} = 0.01$. The AR1 memory time is $T = \frac{\Delta t}{1-\rho} = 0.2$.

is the focus of much laboratory work where basic biotic processes can be observed. Complexity is inevitable, and its reduction to tractable form is a basic scientific goal. *In the field*, this problem is exacerbated on several counts:

- Observations are necessarily limited and sparse in the field.
- Controlled "experiments" are nearly impossible in the field.
- Natural phenomena almost always involve a great number of organisms and processes, with geometric growth in the number of potential interactions.
- Intrinsic variability, which we seek to quantify, is always present but by definition, unknown a priori.
- The estimation problem involves
  - The form of rate expressions
  - Their parameters
  - Their forcing
- All observations are specific to place, time, very small-scale processes.

The IBM approach can make a special contribution by allowing simulation of many nonlinear physical/biological interactions among individuals and their environment, in a site-specific context, and embedding individual variation. Parsimony, a basic tenet in ecological modeling, is highly prized here as well as in conventional lab, field, and mathematical work. [3]

11.5 Mortality

We now turn to mortality, a special case of the "rates" described in Section 1.2.2. For the individual, mortality is binary – each either survives or dies. One needs the probability of death, P_d or its complement, the probability of survival $P_s = 1 - P_d$. Both finite probabilities concern the finite interval Δt. Needless to say, for small Δt we have $P_d \rightarrow 0$ and $P_s \rightarrow 1$; and for sufficiently small Δt it is reasonable to expect P_d to grow monotonically with Δt.

In the simplest cohort model, we have a first-order Eulerian process in N, the number of surviving individuals:

$$\frac{dN}{dt} = -\mu N \tag{11.11}$$

This Eulerian process is equivalent to a binary process for each individual particle:[4]

$$\Delta N = -\mu \Delta t N \iff P_d = \mu \Delta t \tag{11.12}$$

3 Fekedulegn et al. [142] discuss parameter estimation for several common processes.

4 The nomenclature: $\Delta p_d = \int dp_d = P_d$.

In the limit of small Δt,

$$\frac{1}{N}\frac{dN}{dt} = -\mu \iff dp_d = \mu\, dt \tag{11.13}$$

Suppose mortality were constant and acts simultaneously on a cohort of N individuals. Then the survival probability of each would aggregate to the number of surviving individuals:

$$N(t) = N_0 \left[1 - \frac{\mu t}{k}\right]^k \tag{11.14}$$

or its continuous limit. This is the essence of individual-based representation of mortality: assign a binary probability of survival as in Equation 11.12 for the small duration Δt; allow individual survival consistent with that probability, for all individuals separately. The aggregate effect on the population will achieve the desired mortality.

The uniform deviate $\mathcal{U}$ may be invoked to do the binary sorting. Suppose an individual has probability $\mu\Delta t$ of death. Then a random deviate from the uniform distribution can be used to assign survival for that individual. For example, if

$$\mathcal{U} > \mu\Delta t \tag{11.15}$$

then the individual survives; otherwise, not. This places a resolution requirement on the selection of Δt – clearly, $\mu\Delta t$ must be kept small (and certainly below 1) or results will be sensitive to Δt.

11.5.1 Simulations: Individual-Based Mortality

Here we provide examples comparing simulations of mortality with analytic results from the preceding section.

Figure 11.6 shows individual simulations for two different initial abundances. These demonstrate the equivalence of the mean abundance $N_0 e^{-\mu t}$ and the visual mean of the simulations. We note in passing that the higher initial abundance yields less relative variability in these outcomes; more on this point below (Section 11.5.4).

Next we examine the joint effects of mortality and logistic growth. Simulation of 27 individuals is shown in Figure 11.7. G is logistic as above (Figure 11.2). The logistic parameters are distributed in the same way; mortality is added at $\mu = 0.5$. The mortality in this case is sufficiently high that of 27 initial particles, only two survive. Repeated simulations recover the expected cohort mean, and provide an estimate of cohort variance.

Figure 11.8 shows the complete cohort evolution for this simulation. The survivors decline; the surviving cohort reflects the combined effects of growth in individual weight gain and mortality. At long time, the individual survivor weight reaches K,

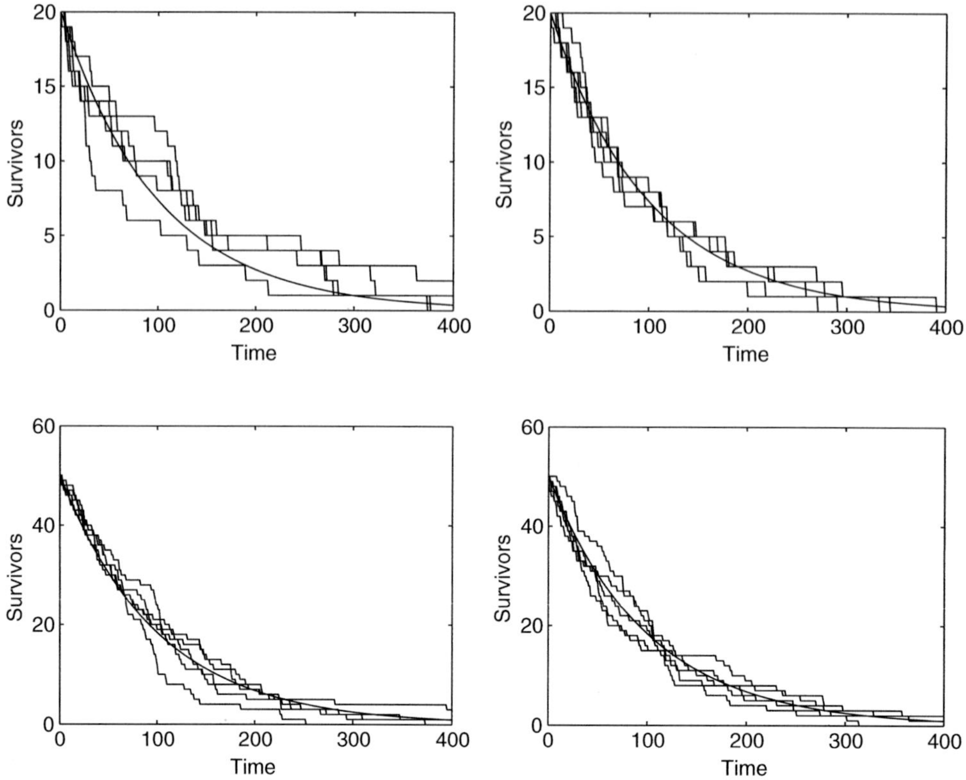

Figure 11.6. Simulations of binary individual mortality. Each panel shows number of survivors versus time, for five ensembles. In the top pair of panels, simulation is initialized with 20 individuals; in the bottom panels, 50. The smooth curve is the Eulerian equivalent for mean abundance, $N(t) = N_0 e^{-\mu t}$. $\mu = 0.01$; $\Delta t = 1$.

but the mortality steadily causes N to decrease. Hence the surviving cohort weight (combined weight of all survivors) peaks and then declines. The "deficit" shown is the difference between the mean cohort with zero mortality, and that actually realized.

The mortality rate μ is a blend of environmental and individual conditions. The former is external to the individual – Eulerian fields of nutrients, turbulence, heat, salt, chemicals, and so forth. The latter represents the individual's development over its own Lagrangian physical/biotic trajectory. The more suitable the environment, relative to the individual's age, fitness, life stage, and so forth, the less is the probability of mortality. We can conceive of μ as a function of individual accumulated fitness (where it has been, how well it has developed) and its current environment.

For the inevitable nonlinear or inhomogeneous situations, μ represents the linearization of mortality over a single small time step. Of course, the time step itself is specific to an individual, and should be adjusted to resolve individual dynamics appropriately.

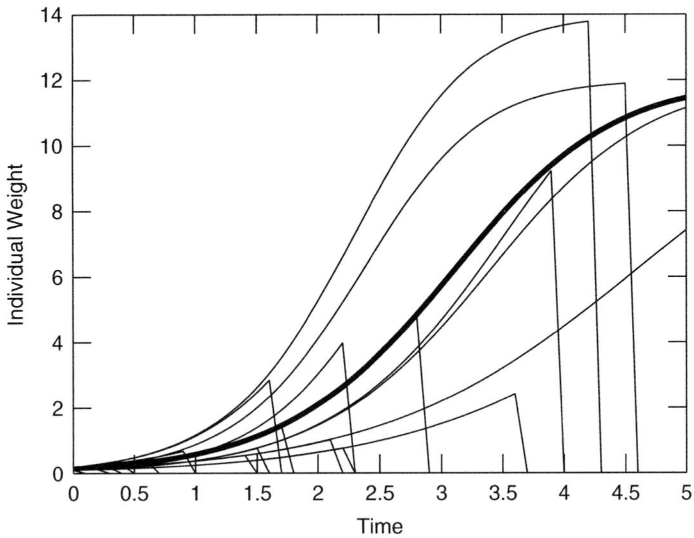

Figure 11.7. Logistic growth as in Figure 11.2, but with individual mortality $\mu = 0.5$ When a particle dies, its weight goes immediately to zero and it does not grow any more. Two of 27 survive.

11.5.2 Individual Survivorship Probability: Geometric Distribution

Recall from Section 11.5 that individual mortality in one timestep is $P_d = \mu\Delta t$ with μ the mortality rate. Individual survival probability over one timestep is $P_s = 1 - P_d = (1 - \mu\Delta t)$. For k steps the individual survival probability is $P_s(k) = (1 - \mu\Delta t)^k$. It declines geometrically with time. Survivorship to k does not indicate death immediately following; it includes all possibilities of mortality at times exceeding k.

The probability of mortality at exactly k time steps is the probability of survival to $k-1$, followed by mortality during the following step:

$$P_s(k) = (1 - \mu\Delta t)^k \tag{11.16}$$

$$P_d(k) = P_s^{k-1} P_d = (1 - \mu\Delta t)^{k-1} \mu\Delta t \tag{11.17}$$

P_d is the *geometric distribution* governing the probability of individual death in step k (an individual lifetime of exactly $k\Delta t$). The moments of k are

$$\bar{k} = \frac{1}{\mu\Delta t} \tag{11.18}$$

$$\sigma^2 = \bar{k}^2 (1 - \mu\Delta t) \tag{11.19}$$

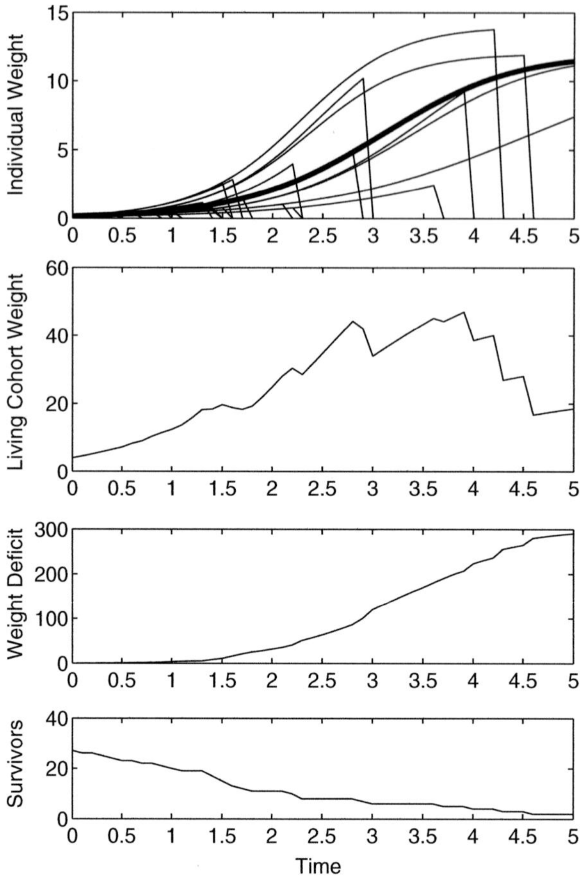

Figure 11.8. Cohort development with growth and mortality for the Figure 11.7 case. Initially, $N = 27$ individuals; individual mortality $\mu = 0.5$. The total cohort weight reflects the joint effects of individual growth and mortality; it peaks, then decays as $W \to K$ while mortality reduces the number of survivors.

The continuous limit is an exponential decline in survival probability. For $t = k\Delta t$, the limit for fixed t as $\Delta t \to 0$ is the *exponential distribution*:

$$P_s(t) = \lim_{k \to \infty} \left[1 - \frac{\mu t}{k}\right]^k = e^{-\mu t} \tag{11.20}$$

$$dP_d(t) = P_s(t)\mu \, dt \tag{11.21}$$

Naturally, $P_s(0) = 1$; $P_s(\infty) = 0$ in both continuous and discrete forms. Readers are encouraged to explore other properties of these distributions in Problems 11.7 through 11.9.

Problem 11.7 Compute the partial sum $\sum_{k=1}^{j} P_d(k)$. What do you expect and why?

Problem 11.8 Compute the sum $\sum_{k=1}^{\infty} P_d(k)$. What do you expect and why?

Problem 11.9 The same for the continuous limits: and $\int_{t=0}^{T} dP_d(t)$ and $\int_{t=0}^{\infty} dP_d(t)$.

Problem 11.10 Compare continuous and discrete mortality processes for $\mu = 0.1$, $\Delta t = 1$.

(1) Compute and plot discrete and continuous probabilities of survival as a function of time, for $0 \leq t \leq 40$.
(2) Compute and plot the difference between them as a function of time.
(3) Repeat (1) and (2) for $\mu = 0.2$ and for $\mu = 0.3$.
(4) Do the two different representations agree when $\mu \Delta t$ is small?

11.5.3 Lifetime Distribution

Discrete Form. The probability of a lifetime of exactly k time steps is then $P_s^{k-1} P_d$ (the probability of surviving $k-1$ steps, then failing in the coming step). Accordingly, the mean lifetime is then

$$\overline{k} = \frac{\sum k P_s^{k-1} P_d}{\sum P_s^{k-1} P_d} \tag{11.22}$$

Both sums run from $k = 1$ to ∞, all the possible values of k. Similarly, the second moment of the distribution, k^2, is

$$\overline{k^2} = \frac{\sum k^2 P_s^{k-1} P_d}{\sum P_s^{k-1} P_d} \tag{11.23}$$

The infinite series have simple expressions (see Appendix A):

$$\overline{k} = \frac{1}{1 - P_s} = \frac{1}{P_d} \tag{11.24}$$

$$\overline{k^2} = (1 + P_s)\overline{k}^2 \tag{11.25}$$

$$\sigma_k^2 = \overline{k^2} - \overline{k}^2 = P_s \overline{k}^2 \tag{11.26}$$

If for example, $P_d = \mu \Delta t$, then

$$\overline{k} = \frac{1}{\mu \Delta t} \tag{11.27}$$

$$\sigma_k^2 = (1 - \mu \Delta t)\overline{k}^2 \tag{11.28}$$

Problem 11.11 Continuous Form. For $\Delta t \to 0$, the relations in Section 3.1 converge to continuous forms. For a live individual at $t=0$, start with the discrete forms given, and develop the limiting continuous forms.

(1) What is $P_s(t)$, the probability of survival to t?
(2) Let t_d be the time at which death occurs. What is the mean and variance of t_d?
(3) What is the rate of change in P_s, that is, $\frac{dP_s}{dt}$?
(4) What is the incremental probability dP_d of mortality between time t and $t+dt$?
(5) Compare your answers with the discrete forms given. Do your results make sense?

11.5.4 Cohort Abundance: Binomial Distribution

For a cohort initialized with N individuals, and individual survival probability P_s: the *number of individuals i* surviving is governed by the *binomial distribution*[5]:

$$Pr(i) = \binom{N}{i} P_s^i (1-P_s)^{N-i} \tag{11.29}$$

where $\binom{N}{i} \equiv \frac{N}{i\,(N-i)}$ is the binomial coefficient. The number of survivors has mean and variance which vary in time:

$$\bar{i} = NP_s \qquad \sigma^2 = \bar{i}(1-P_s) \tag{11.30}$$

$$\frac{\sigma}{\bar{i}} = \frac{1}{\sqrt{N}} \left[\frac{1-P_s}{P_s} \right]^{\frac{1}{2}} \tag{11.31}$$

The analysis confirms the simulation results in Figure 11.6 for the mean. Figure 11.9 shows the complementary confirmation of the standard deviation.

This distribution concerns only integer values of i; this becomes important at low abundance or long time. The relative standard deviation is enlarged when N is small – the $1/\sqrt{N}$ factor in Equation 11.31. This is suggested in the simulation results reported here (Figure 11.9) and above (Figure 11.6).[6]

Problem 11.12 In the limit of large N and small P_s, the binomial distribution is approximated by another famous distribution.

(1) What is the limiting distribution?
(2) What are its parameters, in terms of N and P_s?
(3) What are the mean and variance of i?
(4) Make comparison plots. For the limit to be valid: how extreme need N and P_s be?

[5] For notational convenience here, P_s indicates the probability of survival to finite time t. The distributions developed above, $P_s(k)$ or its continuous limit $P_s(t)$, are implied for a given finite duration.

[6] The manner of plotting suggests (incorrectly) symmetry about the mean; but of course, for small i, $\bar{i}-\sigma$ may be negative while negative i is meaningless. This complexity is avoided in the plots.

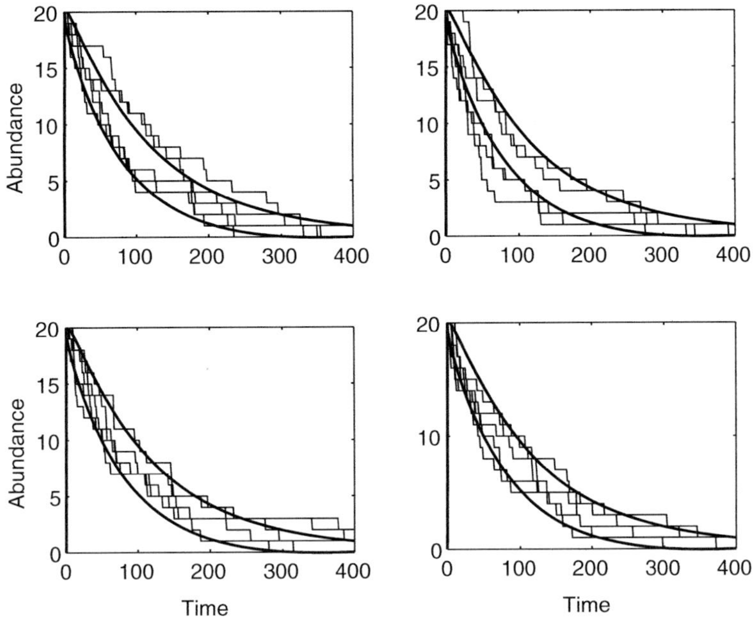

Figure 11.9. Survivors versus time: 20 Monte Carlo simulations with constant mortality. Each frame shows five different simulations; the smooth lines indicate the analytic envelope, mean $\pm\sigma$) (Equations 11.30–11.31). Parameters as in Figure 11.6; $\mu = 0.01$; $\Delta t = 1$; the analytic result is the small-time step limit.

11.5.5 Cohort Death Rate

Let the binary variable d_i indicate individual death during a given Δt, with $d_i = 1$ indicating death. All the d_i are independent random binaries. If the mortality rate is constant at P_d, then

$$\overline{d_i} = P_d \qquad Var(d_i) = P_d(1 - P_d) \tag{11.32}$$

For N individuals initially, the total number of deaths D is aggregated over the ensemble of individuals: $D = \sum_i^N d_i$. Assuming independent individuals, the mean and variance are

$$\overline{D} = NP_d \qquad Var(D) = NP_d(1 - P_d) \tag{11.33}$$

It is easy to add variability. For example, if P_d were distributed independently over the individuals with mean and variance $[\overline{P_d}, Var(P_d)]$ we obtain

$$\overline{D} = N\overline{P_d} \qquad Var(D) = N\overline{P_d}(1 - \overline{P_d}) + N(N-1)Var(P_d) \tag{11.34}$$

Intuitively, the mean is based on mean P_d, and the variance of P_d supplements the constant P_d case. This and several other forms of variability across an ensemble are explored in Pollard [358].

11.5.6 Example: Growth, Mortality, Motion

We placed N particles at the origin; and dispersed them with a random walk and zero mean motion (AR0, Chapter 4). Individual mortality rate μ and growth rate g depend on individual position and favor positive Y displacement:

$$g_i = g_0 + cY_i \tag{11.35}$$

$$\mu_i = \mu_0\, e^{-Y_i/L} \qquad Y_i > 0 \tag{11.36}$$

$$= \mu_0 \qquad Y_i \leq 0 \tag{11.37}$$

Growth of each particle is based on its current location,

$$\frac{dM_i}{dt} = g(Y_i). \tag{11.38}$$

Individual death is random with probability $\mu_i \Delta t$ in each time step. The individual mortality rate μ_i decays with increasing Y_i; for negative Y_i, it is constant at the high value μ_0. In addition, individual death becomes certain if M_i falls below $M_0/10$.

Results at $t = 1000$ are plotted in Figure 11.10, for an initial $N = 1500$. There are 143 surviving particles with mean position $[x,y] = [0.251,\ 154.5]$ marked by the $\times$. Dead particles are not plotted. The center of mass (marked by the +) advanced further north, $[-8.024,\ 201.1]$, reflecting the gradients in growth rate and mortality. The individual mass range was $[1.475 - 30.09] \times 10^4$ for the surviving particles; with total mass $1628. \times 10^4$. There is significant growth and significant mortality.

These results illustrate the nonlinear interplay of position, mortality and growth at the individual level. They are of course only one realization with one set of parameters and rate forms, and invite many experiments. For example, Figure 11.11 shows the effect of adding a sheared advective velocity $U_i(y) = aY_i$, with shear rate $a = 0.05$ for positive Y_i, otherwise $U = 0$; $V = 0$ everywhere; and all other parameters are unchanged. This outcome could easily be falsely interpreted – for example as a constant source of particles at the origin, advecting toward the northeast with constant vital rates.

11.6 Stage Progression

Individuals move through logical life stages, distinguished by logical or morphological qualities such as age, sexual maturity, size, vulnerability, behavior, and vital rates. An individual enters a stage, matures in-stage, and completes it, moving on to the

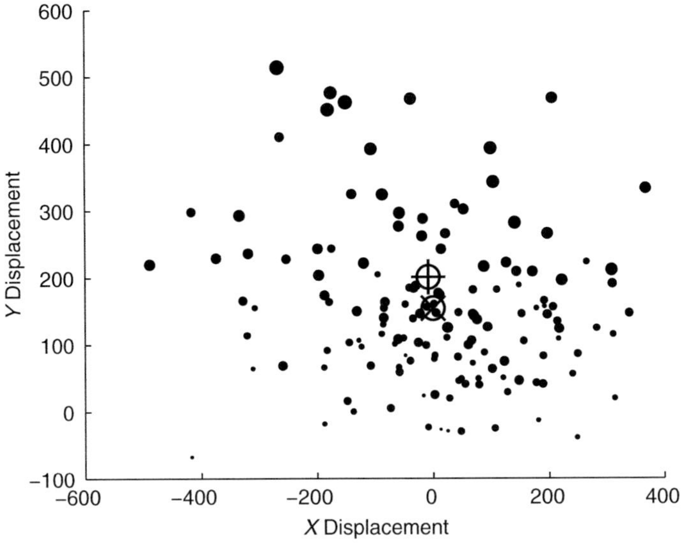

Figure 11.10. IBM outcomes, example in Section 11.5.6. Surviving particles are plotted, with area proportional to particle mass. The center of mass is marked +; the mean position is marked ×. Parameters: [$g_0 = 0.1$; $c = 1$; $L = 10$; $\mu_0 = 0.00015$; $\Delta t = 1$; $M_0 = 1$]. Initially: there are 1500 particles, each at the origin with the same $M = M_0$.

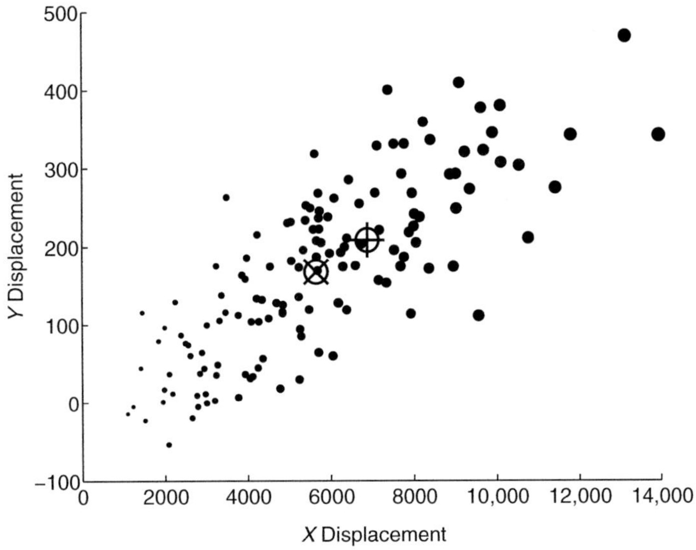

Figure 11.11. IBM outcomes, example in Section 11.5.6 but with velocity shear. Same conventions as in the no-motion case, Figure 11.10. This result invites misinterpretation, as described in the text.

next. There is a predefined ordering of stages, or maybe a set of alternatives, that defines pathways among stages.

Metrics of maturation in stage include weight, size, time-in-stage, and general biotic state of an individual. These are affected by the individual's environment, for example temperature, nutrient availability, or light. Individual vital rates, appropriate to each stage, integrate all these factors.

Stage completion is binary – yes or no – at every time step. Consistent with the IBM approach, the probability of stage completion for an individual is necessary, presumably formulated in terms of an individual's maturation or readiness, and its environment.

The state space for an individual includes at least the stage it is in, some information about stage entry, and metrics of progress toward stage completion. An individual "knows" its stage – it is a logical part of its state.

11.6.1 Example: Three Stages

We imagine an individual that progresses through three sequential stages. For example, a species of fish with the morphological stages juvenile (S1), adult (S2), and post-reproductive adult (S3). Stage 1 has first-order retention with mean residence T1; stage 2 has two substages: first, there is a mandatory residence time T2a, followed by a first-order promotion to the next stage, time scale T2b. Stage three individuals accumulate at the top end of development.

This is easy to achieve. A particle in S1 grows to S2 with per-step probability $\frac{\Delta t}{T_1}$. One records the time of entry t_e for an individual particle into S2; that particle remains in S2 until time advances to $t_e + T_{2a}$; thereafter, it matures to S3 at probability $\frac{\Delta t}{T_{2b}}$.

Figure 11.12 illustrates this for a small cohort of recruited to stage 1 at t_0. The top shows the pure delay achieved in S2. The lower panel shows the same simulation with S2a and S2b plotted explicitly. Parameters: $T_1 = 1$, $T_{2a} = 10$, $T_{2b} = 4$.

This is done with no mortality. It is easy to add mortality by stage. Suppose that the first two stages have finite mortality: $\mu_1 = 0.25$, $\mu_{2a} = 0.10$, and subsequent stages have $\mu = 0$. This appears in Figure 11.13. The effect of early mortality is evident.

It is straightforward to characterize stage progression (or any other binary event) with more general distributions. Here we have used three simple cases:

- The pure delay case for S2a
- The constant-in-time completion rate (the "first order" retention, exponential waiting time distribution) in S1 and S2b
- Constant mortality rate, by stage

More general distributions are available, as suggested in Figure 11.14. In the example case the Erlang family (Figure 3.6) could be quite useful. This provides for increasing completion rate as individual residence-in-stage increases. The hybrid stages S2a and

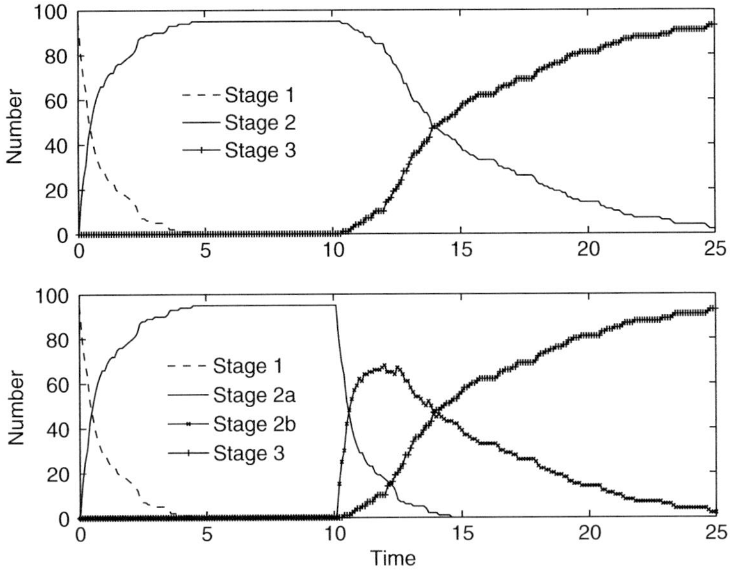

Figure 11.12. Cohort development with three stages. Stage 1 is loaded initially, and individuals make a first-order progression to stage 2. There is a pure lag in stage 2, followed by a first-order progression to stage 3. In the bottom panel, stage 2 is divided into its two substages, pure delay followed by first-order progression. There is no mortality.

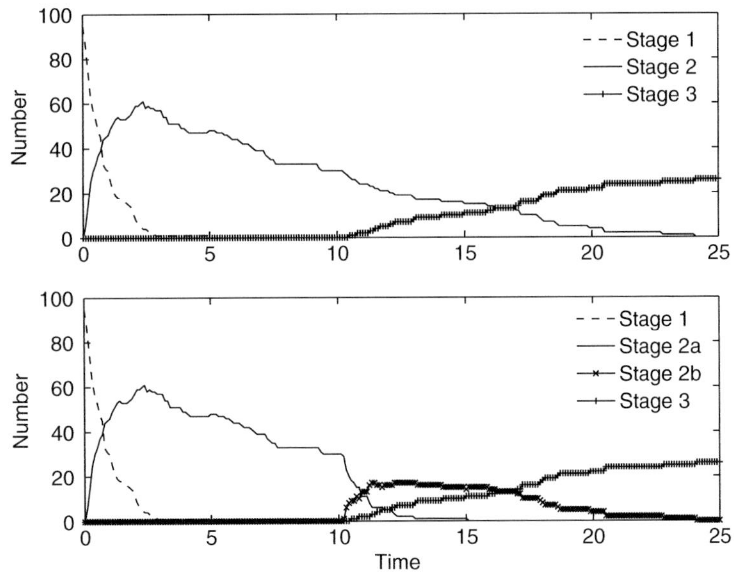

Figure 11.13. Same as Figure 11.12 except that mortality is active in the first two stages: $\mu_1 = 0.25$, $\mu_{2a} = 0.10$.

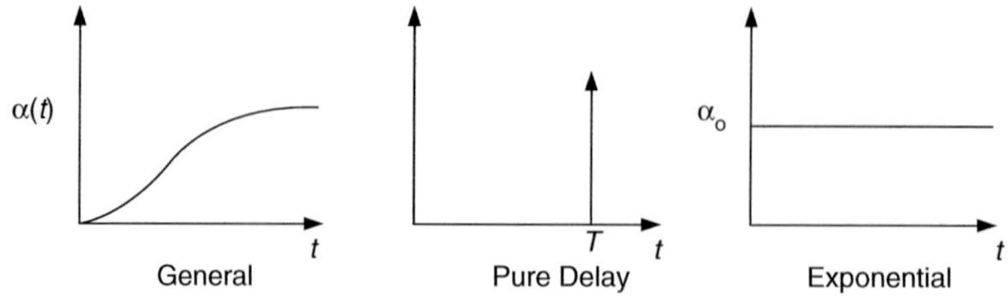

Figure 11.14. Examples illustrating the stage completion rate α versus time-in-stage. For the exponential, $\overline{\tau} = \frac{1}{\alpha_o}$ and $\mathrm{Var}(\tau) = \frac{1}{\alpha_o^2}$. For the pure delay, $\overline{\tau} = T$ with no variance.

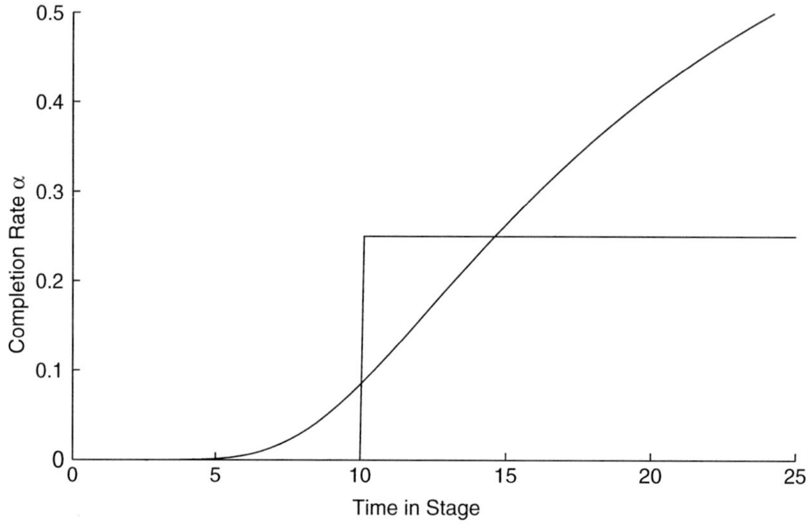

Figure 11.15. Example composite of stage S2a and S2b; completion rate versus time. The step function is the perfect 10-year delay followed by the exponential distribution, $[k,\alpha] = [1, 1/4]$. The smooth composite is Erlang with $[k,\alpha] = [14, 1]$. The Erlang provides a smooth transition in completion rate, initially rising slowly and then accelerating.

S2b represent the two Erlang extremes $k = \infty$ followed by $k = 1$; they could be combined into a single Erlang stage with high k, as suggested in Figure 11.15.

One of the great strengths of simulation is that individuals need only advance one time step at a time. So the problem of specifying a general stage progression is reduced to specifying a simple transition probability during the coming Δt. Ideally, Δt is small enough to support a constant probability over that interval; it could change in the coming timestep for many reasons including processes internal to the individual, or those imposed by location, environment, or other influences external to the individual.

11.6.2 Stage Completion Rate, Residence Time – Constant α

In general, the process of stage transition resembles mortality – a binary outcome with finite probability of occurring in a time interval.

Suppose there are N individuals in a stage; and that they complete at the rate αN:

$$\frac{dN}{dt} = -\alpha N \;\Rightarrow\; N(t) = N_0 e^{-\alpha t} \tag{11.39}$$

At any point in time t there are $\alpha N(t)\Delta t$ individuals completing; each has "age" t. The average age of maturing individuals over the interval $0 \le t \le \tau$ is

$$\overline{T} = \frac{1}{N_0 - N(\tau)} \int_{t=0}^{t=\tau} t \cdot \alpha N dt \tag{11.40}$$

We have $N(t)$ above; its substitution allows integration of (11.40). In the limit as $\tau \to \infty$, all individuals mature and their average residence time is:

$$\overline{T} \to \frac{1}{\alpha} \tag{11.41}$$

Other moments can be similarly evaluated. The complete distribution of time-in-stage turns out to be the exponential or waiting time distribution, or its discrete version, the geometric distribution. This distribution is skewed; there are no negative values; small values are the most common, but very high values are possible with diminishing probability. Accordingly, the median is lower than the mean.

The summary result: a cohort with first-order stage completion as in Equation (11.39) has in-stage residence times distributed according to the exponential distribution with parameter α. The properties are (a) **mean** $\overline{T} \equiv 1/\alpha$; (b) **variance** $\overline{T}^2$; (c) **median** $0.69\overline{T}$. It would be common to observe the outcome $\overline{T}$, and represent the process as a distribution of individual waiting times.

From an individual's point of view, the discrete probability P of exit from stage during the next Δt is

$$P \approx \alpha \Delta t = \frac{1}{\overline{T}} \Delta t \tag{11.42}$$

This probability of completion is independent of prior events. Specifically, it does not depend on the time already spent in residence. This is a remarkable feature of this distribution. It facilitates implementation of the waiting time distribution on a very local basis – at the level of the individual undergoing a single small time step Δt. The aggregate properties result.

In this simple example, α has a single value, constant over time and over individuals. Nevertheless, variability in individual residence time is achieved, simply by

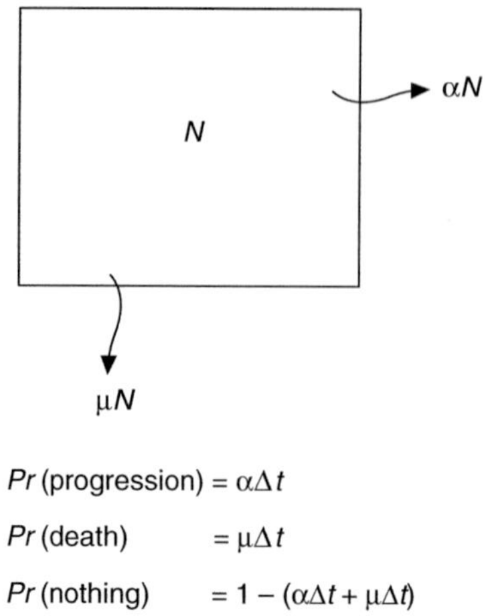

Figure 11.16. Residence in-stage with two outlets: stage progression and mortality

assigning the probability of completing the stage and making a binary statistical determination for each individual (as in Equation 11.15). The cohort will have the proper aggregate properties, with the desired distribution fully represented at the individual level.

Operationally, stage completion for an individual is the same as individual mortality. They both invoke the same discrete simulation apparatus even though the two processes correspond to very different realities. Figure 11.16 illustrates the combination of these two processes combined. As far as survivorship (retention) is concerned, the two probabilities (mortality and progression) add. But they represent different outcomes with their own separate probabilities.

11.6.3 The General Case –Variable α

Suppose there were a cohort of N identical individuals entering a stage at the same time. Over time, N will decrease as completions occur:

$$\frac{dN}{dt} = -R(N,t) \qquad \frac{1}{N}\frac{dN}{dt} = -\alpha(t) \tag{11.43}$$

where $\alpha(t) = \frac{R}{N}$ is the completion rate and t represents elapsed time-in-stage. Linearizing over a finite time step,

$$\frac{\Delta N}{N} = -\alpha_i \Delta t \tag{11.44}$$

For a cohort of one, P_i is the equivalent probability of stage completion

$$P_i = \alpha_i \Delta t \tag{11.45}$$

Generally, this probability is dependent on time-in-stage for the individual. Some examples appear in Figure 11.14.

The P_i are conditional probabilities – assuming one has been in-state for i time steps, P_i is the probability of completion in the coming time step; $(1 - P_i)$ is the probability of remaining in state.

Once the P_i or the rates α_i are formulated, all else is statistical outcome. Projecting forward we have P_k^r, the probability of retention-in-state after k steps:

$$P_k^r = (1 - P_k)P_{k-1}^r = \prod_{i=1}^{k}(1 - P_i) \tag{11.46}$$

and P_k^c, the probability of stage completion during step k:

$$P_k^c = P_k P_{k-1}^r = P_k \prod_{i=1}^{k-1}(1 - P_i) \tag{11.47}$$

Some additional consequences of the definitions:

$$\sum_{k=1}^{\infty} P_k^c = 1 \tag{11.48}$$

$$P_k^c = P_{k-1}^r - P_k^r \tag{11.49}$$

$$P_k^r = 1 - \sum_{i=1}^{k} P_i^c \tag{11.50}$$

The waiting time τ or completion time[7] (the time spent in-stage) is distributed as P_k^c:

$$Pr(\frac{\tau}{\Delta t} > k) = P_k^r \tag{11.51}$$

$$Pr(\frac{\tau}{\Delta t} = k) = P_k^c \tag{11.52}$$

For example: suppose P_i is given in Table 11.1. The values of the rest of the distribution are listed; readers should confirm that these are correct, and that the properties listed are correct. The mean stage residence time is $\overline{\tau} = 4.9\Delta t$.

[7] This is the discrete case; τ is an integer multiple of Δt.

Table 11.1. *Example of Probabilities for Completion and Retention, Based on the Contingent Probability of Completion*

i	P_i	P_i^r	P_i^c
1	0	1.0	0
2	0	1.0	0
3	0.1	0.9	0.1
4	0.2	0.72	0.18
5	0.6	0.288	0.432
6	1.0	0	0.288
7	0	0	0

In general, we are defining the discrete version of the general hazard rate as introduced in Chapter 3. The correspondences are:

$$\alpha_i \sim h(t) \tag{11.53}$$

$$P_i = \alpha_i \Delta t \sim h(t)\Delta t \tag{11.54}$$

$$P_k^r \sim S(t) = 1 - F(t) \tag{11.55}$$

$$P_k^c \sim f(t)\Delta t = -\Delta S(t) \tag{11.56}$$

The "hazard" here is stage completion. The algebra mimics that for mortality, though the process itself is quite different.

11.6.4 Recap: Stage Completion

Needless to say, the simulation approach provides enormous opportunity to have rates and transition probabilities depend on the current environment as well as the biotic state of the individual. And the IBM approach allows each individual to have its own memory of the past, its own timing and of course its own environment. It is not necessary to assume rates are constant over time, or over individuals. These assumptions are extremely useful analytically, allowing us to project forward over many time steps. The Erlang and gamma distributions are well-developed as two-parameter extensions of the popular exponential distribution, and provide considerable expansion of formulation possibilities – for example, the practical need for a variable hazard rate over time, initially low then peaking and then descending. (See, e.g., Figures 3.5, 3.6.)

In using these idealized solutions, we are essentially developing an analytic solution for the *outcome* of an idealized case resulting from simplified formulation. In practice, this approach can be limiting. Practical formulations need not be so constrained since simulation offers a straightforward way to compute the outcomes. It is the *formulation* that needs to be the focus: how does the probability of an event depend on individual state and environment?

For example: *a pure delay is entirely possible* at the individual level; each individual's probability of exit can be set to zero for a given delay time, and thereafter rise, for example according to an Erlang distribution.

Generally, one only needs the probability of stage completion $P_j \equiv \alpha_j \, \Delta t$ as a function of accumulated time-in-stage $j\Delta t$. This governs for only the coming Δt. Some basic constraints of probability are necessary; beyond those, the parameterization of P_j is quite flexible. It can be dependent on ambient temperature, nutrients, turbulence, or anything in the environment thought to affect maturation in stage; and of course, dependent on the biotic state of the individual.

These P_j are conditional transition probabilities – assuming an individual is in-stage, the outcomes are binary: complete the stage in the coming time step with probability P_j; or continue in stage with probability $1-P_j$. Notice that the time index is individual-based: each individual on entry to a stage acquires its own time-stamp indicating stage initiation.

As with the formulation of mortality described in Section 11.5, there is a correspondence between a density-based formulation and its statistical equivalence:

$$\frac{dN}{dt} = -\alpha N \;\Leftrightarrow\; \frac{\Delta N}{N} = -\alpha_j \Delta t = -P_j \tag{11.57}$$

with N the number of particles in stage; α_i the stage completion rate; and P_j the probability of stage completion in the coming timestep.

Figure 11.17 illustrates $\alpha(t)$. t is an individual's accumulated time-in-stage. Specification of α and its closure in terms of an individual's state and environment specifies the stage. Figure 11.18 depicts the analogous case for finite Δt. $P_j = \alpha_i \Delta t$ is the conditional probability of stage completion in the coming timestep, with α_j the mean of α over the timestep.

Specifying P_i is definitive. Outcomes include

- The probability of retention through k steps:
  $P^r_k = \prod_{i=1}^{k} (1-P_i)$
- The probability of completion during step k:
  $P^c_k = P_k \prod_{i=1}^{k-1} (1-P_i)$
- The waiting time distribution is the same as P^c_k.
- The mean waiting time is
  $\frac{\overline{\tau}}{\Delta t} = E(k) = \frac{\sum k P^c_k}{\sum P^c_k}$
- Its variance:
  $\mathrm{Var}(\frac{\tau}{\Delta t}) = E[(k-\overline{k})^2] = \frac{\sum (k-\overline{k})^2 P^c_k}{\sum P^c_k}$

These are all immediate consequences of P_i, the conditional probabilities of stage completion.

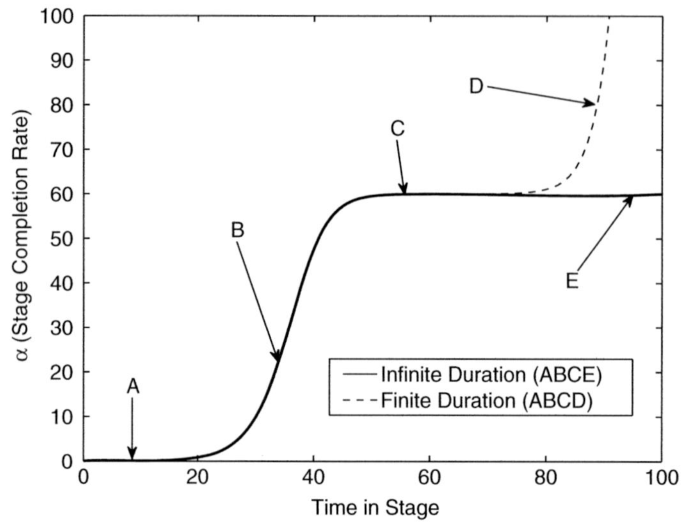

Figure 11.17. Illustration of stage completion rate $\alpha(t)$ and discrete completion probability $P_j = \alpha\Delta t$. The five phases are (A) pure delay; (B) Erlang; (C) exponential; and either (D) terminal or (E) continued exponential. Option (D) provides absolute stage termination at finite time T; alternatively, option (E) terminates asymptotically at large $t \to \infty$. The P_j are conditional probabilities – they assume an individual has survived to time j.

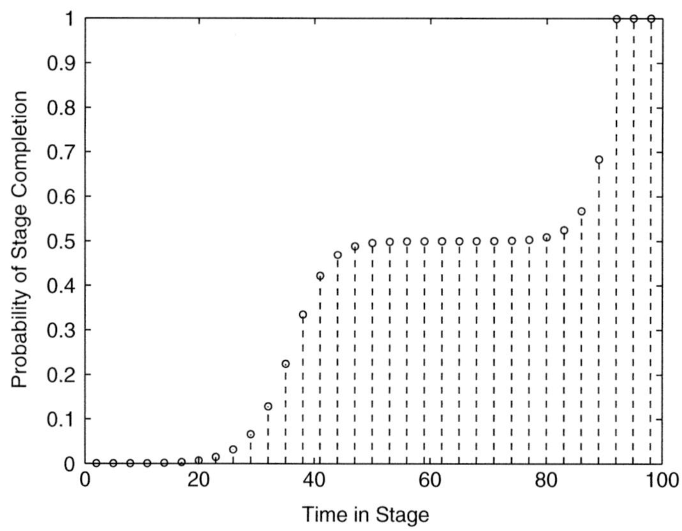

Figure 11.18. Discrete stage completion probabilities $P = \alpha\Delta t$ – discrete form of Figure 11.17. Shown is the finite stage duration, option D, for this Δt. The absolute completion of the stage occurs at $t = 93$; the transition probabilities saturate at $P = 1$ there and beyond.

Problem 11.13 A life stage has an absolute minimum duration of 3 years; beyond that, the median stage duration is one additional year and the waiting time (exponential) distribution is valid. There is no mortality.

(1) What is the mean overall stage duration?
(2) Construct a discrete model for individual stage duration.
(3) A group of 10 individuals enters the stage at $t = 0$. Simulate this stage; generate a completion time for each individual. Are the results in accord with theory?
(4) Repeat (3), N times. Compute the mean, median, and standard deviation of the stage durations as a function of N. Do you recover the result expected, and how big is N required to be to estimate these properties?

Problem 11.14 Repeat Problem 11.13. (2) through (4), with mortality as follows: $\mu = [0.1,\ 0.05,\ 0]$ in years 1, 2, and 3; and thereafter, $\mu = 0.05$.

Problem 11.15 A stage has conditional probability of completion during time step j as follows:
$P_j = [0,\ 0,\ 0,\ 0,\ 0.05,\ 0.10,\ 0.20,\ 0.25,\ 0.25,\ 0.25\ 0.30,\ 0.5,\ 0.5,\ 1.0,\ 0]$.

(1) Plot these as a function of discrete time j.
(2) Compute the probability of stage completion during time k, P^c_k.
(3) Compute and plot the probability of stage continuation beyond time k, P^r_k.
(4) What is the sum of P^e? Why?
(5) What is the mean waiting time to completion?
(6) What is the variance of the waiting time?

Problem 11.16 Continuation of Problem 11.15: There are $N = 1000$ individuals in stage initially. Derive and plot:

(1) The expected number of individuals N_i left versus time
(2) The variance of N_i
(3) The coefficient of variation of N_i
(4) The mean completion rate E_i as a function of time
(5) The variance of E_i as a function of time
(6) The coefficient of variation of E_i as a function of time

Problem 11.17 Continuation of Problems 11.15 and 11.16: Simulate the evolution of this system 100 times and compare with your derived answers. Do they agree? If not, describe why. Show that you can repair any error or ambiguity.

11.7 Reproduction

Reproduction is another binary process when considered from the individual point of view. Suppose the aggregate reproduction rate is

$$\frac{dN_0}{dt} = \rho N_a \tag{11.58}$$

with recruitment into stage 0, and N_0 the number of individuals in the earliest life stage, and N_a the number of reproductive-stage individuals. For an individual within the N_a cohort, the probability of reproduction in the period Δt would be

$$P = \rho \Delta t \tag{11.59}$$

Given the probability, reproduction is a binary matter: generate either 0 or 1 new recruits for a given adult, in a given Δt. The summation over all reproductive-stage individuals would lead to the correct number of recruits $\rho N_a \Delta t$. The same binary apparatus invoked in the other processes is relevant here. Initializing a new individual can be done in a way that inherits properties from the individual's parent. More generally, M offspring could be generated by one adult. M could be a constant, in which case the binary apparatus is the same; or M itself could be random, for example according to a Poisson process or other generator for nonnegative integers.

Figure 11.19 shows a simulation starting with a single mature individual with constant spawning probability undergoing simple random walk. The spawning season last 80 days. There are five reproductive events, each of which produces a single new individual. The new individuals are not mature so they do not themselves spawn; as a result there is one parent and five offspring trajectories marked.

For populations with sexual reproduction, male and female individuals would be differentiated randomly, with reproductive capability assigned only to female individuals. It is common to ignore the problem of sexual contact among individuals, assuming it does not limit the process.

Relative to Figure 11.16, the reproductive probability adds another destination, as in Figure 11.20. Reproduction is still binary and requires the presence of an individual capable of reproduction. It adds a new individual to the first life stage; but unlike death and progression, the reproducing individual remains in-stage. Reproduction is not a zero-sum event.

Of course it is easy to make each spawning event yield more than one new juvenile. One can use, for example, the binomial distribution that allows a variable number of new individuals per spawn, with a finite upper limit n and rate. The Poisson distribution is the limit of large n and small rate. The parameters of any of these processes would need to reflect the biology of the organism, and computationally, the duration of the time step size Δt, just as the simple exponential distribution does in Equation 11.59.

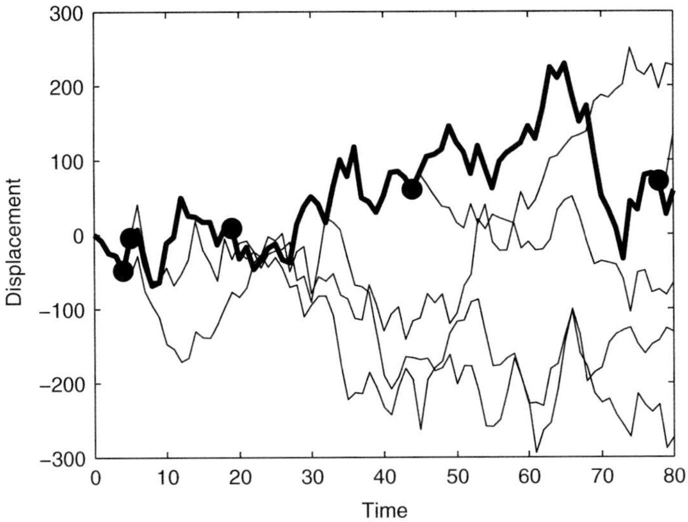

Figure 11.19. A single mature individual with constant probability of spawning. There are five spawning events, each producing a single new individual juvenile. The events are marked by the dots; each of the six trajectories is an individual random walk. The heavy line is the spawner; the five lighter lines are the five spawned individuals.

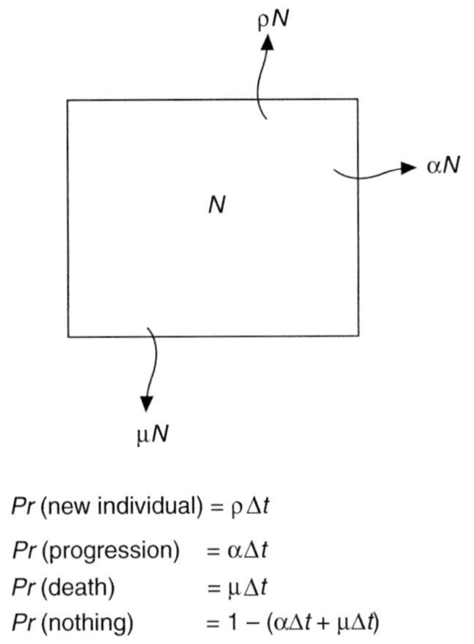

Figure 11.20. Residence in a box with three outlets: stage progression, mortality, and reproduction.

Some extensions are needed for broadcast spawners, where there can be millions of eggs per spawning event. This is common for marine fisheries; variations in M are substantive, and early mortality is high. Simulating millions of short-lived eggs is likely to be impractical in that case. An attractive alternative is to aggregate newly spawned eggs into a few parcels. A simulation could proceed by decrementing each parcel by an integer generator, until the early high-mortality period is over.[8] Following this early period, parcels could be unpacked and surviving larvae simulated individually.

11.8 Motion – Advection + Random Walk

Motion of passive particles is commonly represented by the sum of advection and nonadvective motions. The advective part is common to all individuals at a given location. The nonadvective part is typically represented as a diffusive process, accounting for unresolved motions typically below the grid scale. Each individual will experience a different random sample of the diffusive part. Both advective and diffusive motions are expected to be available as gridded, macroscale Eulerian fields.

$$\mathbf{V}_i = \mathbf{v}(\mathbf{X}_i) + \mathcal{R} \tag{11.60}$$

This is covered in some detail in Chapter 4; in particular the displacement due to $\mathcal{R}$ in a time step needs to be proportional to $\sqrt{D\Delta t}$; and the advective part needs to account for gradients in D.

For example, the simplest AR0 model of uncorrelated random walk in 1-D, individual displacement would be

$$X_i^{k+1} = X_i^k + \left[u(X_i) + \frac{dD}{dX_i}\right]\Delta t + \epsilon\sqrt{2D_i\Delta t} \tag{11.61}$$

with ϵ a zero-mean, unit-variance random variable and u the deterministic advective velocity. Here we indicate by $\frac{dD}{dX_i}$ the derivative of D along the trajectory; and have used an elementary Euler numerical integration. The random walk is configured so that the aggregate over many individuals is equivalent to diffusion; D is the diffusion coefficient. See, for example, [366].

These motions are physical in origin; there is additional motion due to *behavior*. Examples include light- or oxygen-seeking, predator avoidance, diel migration, etc. Behavioral capabilities will develop as does the organism; they will certainly be subject to physical constraints (effect of heat or turbulence) as well as biological ones – for example, constraints on speed. In addition, one can expect various sources

[8] For example, a Poisson generator could be used to represent mortality of a parcel of M individuals at rate μ: $\Delta M \sim \mathcal{P}ois(\mu\Delta t)$.

of randomness to be behavioral – hence an additional behavioral component to $\mathcal{R}$ beyond that due to hydrodynamic turbulence.

11.9 Motion – Behavior

Behavioral motion is a biotic trait that contributes to motion directly. The Eulerian (environmental) velocity $\mathbf{v}(\mathbf{x})$ is supplemented with a behavioral velocity or swimming speed $\mathbf{V}_{b_i}$:

$$\mathbf{V}_i = \left[\mathbf{v}(\mathbf{X}_i) + \mathbf{V}_{b_i}\right] + \mathcal{R} + \mathcal{R}_b \tag{11.62}$$

The random walk representation of hydrodynamic diffusion is $\mathcal{R}$ as in Equation 11.60; $\mathcal{R}_b$ is added as a potential behavioral contribution. A characteristic $\mathbf{V}_{b_i}$ is one body length per second, although much higher velocities are possible. We recall some approximate environmental velocities for the coastal ocean:

- 0.5–1.0 m/s tide
- 0.1 m/s subtidal (seasonal large-scale baroclinic pressure gradient, barotropic shelf setup, tidal rectification)
- 0.5 m/s wind events (surface layer)

Horizontal swimming is unimportant for small organisms, while vertical motion can have substantial consequences because of large gradients in that direction: temperature, light, velocity, and so forth. For larger animals, horizontal behavior may be strong enough to dominate the subtidal motion, and also be capable of hazard avoidance generally – for example, fast vertical motion to avoid wind-induced turbulence. Intratidal vertical migration may account for special subtidal horizontal displacements – for example, diving during flood, resurfacing during ebb.

To put these motions in context: persistent subtidal flows, 0.10 m/s translates to 8.64 km displacement in a day; a wind event sustained for 1/2 day results in a displacement of 21 km; a tidal displacement over 3 hours would be 10.8 km. These displacements are well above the scale resolved in typical Coastal Ocean Models where $\Delta x \sim 1$ km and certainly smaller near shore where topography influences trajectories greatly.

Some behavioral cues may be entirely inherited, for example, via a "biological clock." Others may be exogenous – turbulence, light, predators, or food supply. Generally, the behavior for a given individual needs to reflect its biotic identity and state; it can be sensitive to place, seasonality in the environment, and spawning, but must respect some basic physical laws of motion in a turbulent fluid. Life cycle closure is a result of coupled physical and biological processes, and the ability of simulations to build persistent, multigeneration populations in accord with observations, is a basic goal that requires careful behavioral closure.

As an example, one can easily imagine a bottom-spawning population whose icthyoplankton seek the near-surface where there is different advective motion – direction and magnitude – and extra dispersivity due to local wind-generated turbulence. In these conditions, older icthyoplankton would not occur above their bottom spawning location; their horizontal distribution would reflect a different mean motion and turbulent regime.

Virtually all studies of plankton reinforce the importance of vertical migration on larval fate and destination. For example, North et al. [336], in a comprehensive study of Chesapeake Bay oysters, found that vertical swimming details are more important that annual differences in hydrography/circulation. Etiene and Parisi [130] stress the point for zooplankton migration. Werner et al. [478] initiated a series of studies of larval fish that make complementary points about larval retention.

11.9.1 Example Formulations

Nonlinear Forms. There are many options for formulating the vertical behavioral velocity V_b. Suppose that an organism is currently located at Z and seeks the height Z_e due for example to a change in light, temperature, or other environmental cue. A simple two-parameter response would involve a reaction to the disequilibrium $\delta = Z_e - Z$, with rate constant r; limited by a maximum possible swimming *speed* V^* (based, e.g., on body length). The simplest would be the smaller of the two, either $r\delta$ or V_l; the latter being the limiting *velocity* V_l having speed V^* directed toward z_e:

$$V_l = sgn(\delta) \cdot V^* \tag{11.63}$$

Putting this together with attention to the directions, we have **form 1**:

$$V_b = V_l \left[\min \left(\frac{r\delta}{V_l}, 1 \right) \right] \tag{11.64}$$

The dimensionless variable θ emerges as the metric of saturation.

$$\theta = \frac{r\delta}{V_l} \tag{11.65}$$

θ is always positive. When $\theta < 1$ the simple first-order response occurs; when $\theta \geq 1$ the velocity is completely limited to V_l. In terms of θ, **form 1** has two useful expressions

$$V = V_l \left[\min(\theta, 1) \right] \tag{11.66}$$

$$= r\delta \left[\min\left(1, \frac{1}{\theta}\right) \right] \tag{11.67}$$

A smooth version, **form 2**, uses the exponential rate limiter (Appendix D), with the same asymptotes:

$$V_b = V_l \left[1 - e^{-\theta}\right] \tag{11.68}$$

$$= r\delta \left[\frac{1 - e^{-\theta}}{\theta} \right] \tag{11.69}$$

Form 3 achieves the same limits using the geometric limiter:

$$V_b = V_l \left[\frac{\theta}{1 + \theta} \right] \tag{11.70}$$

$$= r\delta \left[\frac{1}{1 + \theta} \right] \tag{11.71}$$

These three examples are shown in Figure 11.21. They all share the desired two-parameter formulation (r, V^*), with the same meanings. Progress toward saturation scales with the dimensionless $\frac{r\delta}{V_l}$, the ratio of the two extreme possibilities. All reduce to no motion when δ is achieved; they are stable about that point, with the same response time $1/r$ for small disequilibria δ.

An example simulation is reported in Figure 11.22 for an organism with swimming speed $V^* = 1$ cm/s; linear response time $\frac{1}{r} = 1$ min, at rest at $Z = -15$ m. At $t = 0$, Z_0 was raised instantly to -5 m and held there. The differences among the three responses are consistent with Figure 11.21, with form 1 responding fastest, and Form 3 the slowest. All three forms have the same two intuitive parameter settings, and lead to responses that are quite similar for this situation.

Periodic Behavior. Suppose that V^* is large relative to other factors – the response never saturates and each of the three formulations reduce to

$$\frac{dZ}{dt} + rZ = rZ_0 \tag{11.72}$$

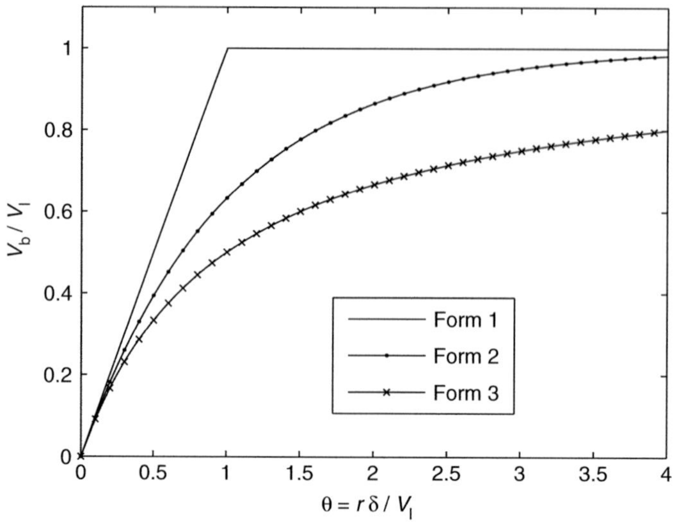

Figure 11.21. Three nonlinear vertical migration functions in dimensionless form. Form 1 is discontinuous and rises fastest. Forms 2 and 3 are both continuous. Form 3 (geometric saturation) rises slowest of the three. All three share the same two parameters $[r, V^*]$ and have the same slope at the origin and the same asymptote.

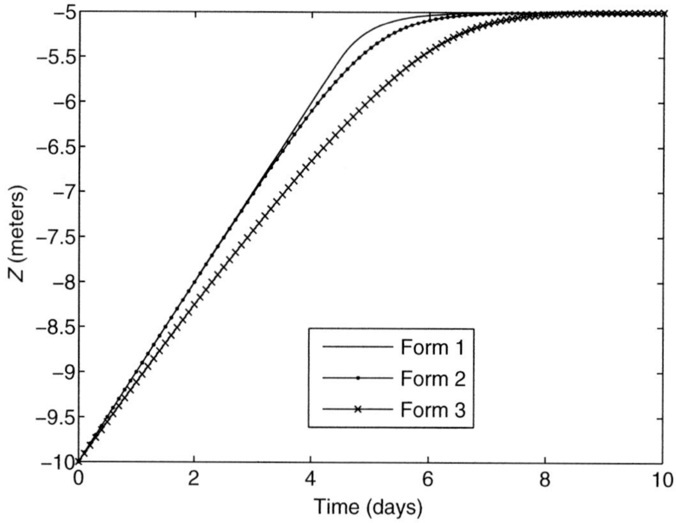

Figure 11.22. Simulated response to a $10m$ upward change in Z_0 for an organism with $V^* = 1$ cm/s, $r = 1$ min^{-1}. Form 1 is the fastest riser; form 3, the slowest.

Let the environment be oscillatory with real amplitude A; the response will be oscillatory too with complex amplitude B:

$$Z_0(t) = Ae^{j\omega t} \tag{11.73}$$

$$Z(t) = Be^{j\omega t} \tag{11.74}$$

$$\frac{B}{A} = \frac{r}{j\omega + r} \tag{11.75}$$

The amplitude and phase (radians) of the response B will be

$$\left|\frac{B}{A}\right| = \frac{1}{\sqrt{1 + \left[\frac{\omega}{r}\right]^2}} \tag{11.76}$$

$$\arg\left(\frac{B}{A}\right) = -\arctan\left[\frac{\omega}{r}\right] \tag{11.77}$$

Clearly the relevant measure is $\left[\frac{\omega}{r}\right]$. When small, the behavior is to follow Z_0; when it is $\mathcal{O}(1)$ the effect is to mute the response, and to create phase lag.

For example: consider a diurnal motion of thermocline; and an organism whose response time $1/r$ is 4 h. Then we have $\omega = \frac{2\pi}{24}$ h^{-1} and $r = \frac{1}{4}$; so $\frac{\omega}{r} = \frac{\pi}{3}$. The amplitude of the response is 0.69; the phase lag is 46° or about 3 h.

The situation is easy to saturate. For example, consider the case where vertical migration occurs in response to the daily light cycle, the effective penetration being surface to 40 m deep; and the individual motion is limited to 40 m/day. Clearly, the

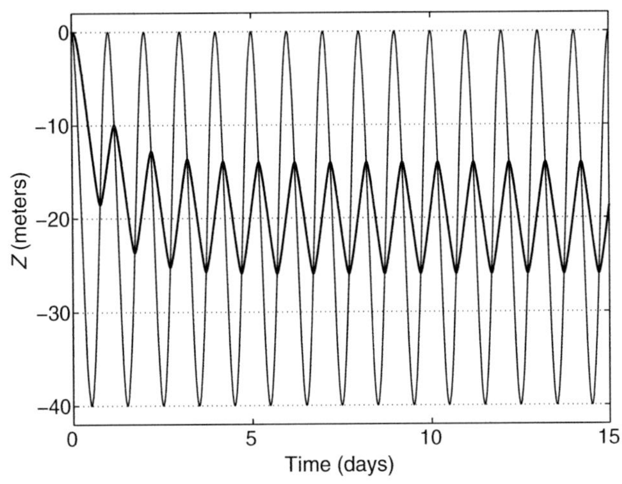

Figure 11.23. Simulated response to a large-amplitude variation in Z_0 (daily period, range ±20 m) and a relatively weak swimmer ($V^* = 40$ m/d; response time $\frac{1}{r} = 4$ h).

half-day displacement exceeds the organism's capability; there should be a phase lag and an amplitude and phase lag effect in the response. Simulation of this situation, with form 3, is illustrated in Figure 11.23. The resulting diel migration is roughly between 14 and 26 m deep (an amplitude reduction to 0.3 of the driving amplitude) and is phase-lagged by about a quarter of a day. From initiation at the surface, the cycle takes about 4 days to reach periodicity centered at mean depth 20 m.

11.10 Benthic Exchange

In this category we consider individuals entering the water column from the immobile bottom. In other chapters this was discussed for sediment particles elevated by bottom turbulence; or for biogeochemical fluxes between the water column and the benthos. For individuals, we have to generalize to include biotic processes in addition to physical and chemical ones. The process may involve resting or cyst phases of particular species that are activated by environmental or other signals; or bottom spawning, either by sedentary benthic organisms or by bottom-spawning adults, that release eggs or larvae.

Generally, there are three approaches:

1. Make the creation of new individuals a basic input.
2. Place adult spawners on the bottom at certain times/places, and allow them to respond to environmental and biotic cues.
3. Make benthic organisms emerge according to a Poisson process with spatially and temporally varying rate.

These are generally versions of the same idea and share some basic formulation challenges. The role of observation can influence the choice of strategy, as well as whether one is simulating sequential spawning by adults that were themselves spawned within a model. In some cases, the spawn themselves are in fact observed, and so strategy (1) would seem appropriate. One step removed would be to either observe the spawning stock itself, or to form hypotheses about the process and the behavior of the very earliest life stages. This might involve strategy (2). The third strategy could be relevant where the relative abundance of organism is known, but not in absolute terms. In this case, Poisson generation removes the bottom-dwelling adults and the burden of closing the life cycle within a simulation; of course it does not answer questions about the latter, only shifts the task from simulation to the specification of inputs.

The plankton examples below illustrate these. The Poisson distribution is discussed in Chapter 3, Section 1.2. Fundamentally, the Poisson rate λ needs to be proportional to the relative density of spawners (the Poisson intensity ν), the spatial subarea A_e generating the events, and the time step involved:

$$\lambda_e(\mathbf{x},t) = \nu(\mathbf{x},t) A_e \Delta t \tag{11.78}$$

11.11 Plankton IBMs

Planktonic organisms are uniquely dependent on hydrodynamics for their horizontal motion,[9] but the vertical is another matter. It is common to have planktonic organisms responding vertically to light, salt, nutrients, and so on and to regulate their depth with significant consequences to (a) horizontal displacement and (b) ecological function.

The general setting is illustrated in Figure 11.24. There is a box representing the model domain and of course the domain of the environmental data to be used. There are inputs and outputs – both vertical and horizontal. The horizontal reflects conventional hydrodynamic transport of individuals across all model boundaries – including the shore, rivers, the cross-shelf boundaries, and the along-shelf boundary. Individuals may have significant vertical behavior, depending on the individual state and numerous environmental influences. Representative processes include encystment in the benthos; its opposite, excystment to the water column; bottom-spawning of buoyant individuals; and so on. Within the box we have all the physical and biological process that matter – governing retention and export, mortality, growth/development, abundance, biotic state of individuals, reproduction – all contained within a "virtual environment."

In Figure 11.25 we illustrate some notions regarding the distribution of individuals within the box. If there are to be several aggregations of reproducing adults, based perhaps on habitat, fishing pressure, and so on, where do their spawn relocate to in the early life stages, while behavioral process have yet to mature? And, which areas are self-sustaining? Generally, what are the sizes of various aggregations; and what are the rates of exchange among them? Are some aggregations self-sustaining? Do others require restocking from boundaries or from internal exchanges? These are among the generic questions that we expect to address in any ecological study. The examples in this section illustrate some IBM formulations that have been found useful for specific problems. Many of these matters are relevant to population connectivity, which is addressed in Section 11.12.

11.11.1 Larval Fish

In this category, we are dealing with fish eggs and larvae, the very early life stages. These stages are typically the most vulnerable, and spent in the euphotic zone where there are many predators, highly variable wind-driven transport, and many competitors for food.

Spawning fish typically produce massive numbers of individuals, which as larvae seek the surface for feeding, are advected there, and are subject to substantial mortality. Survivors mature to be recruits in the population, quite often removed from the spawning ground by near-surface transport. The latter is highly variable; multiple

[9] A rule of thumb is that locomotion scales as one body length per second – trivial on horizontal scales.

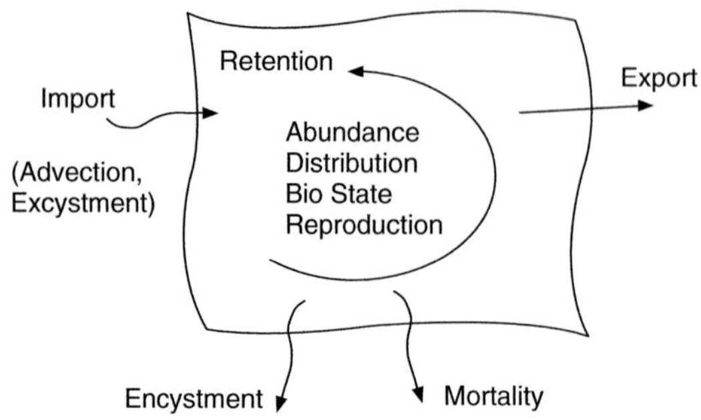

Figure 11.24. Individuals in a box. Abundance, distribution, vital rates.

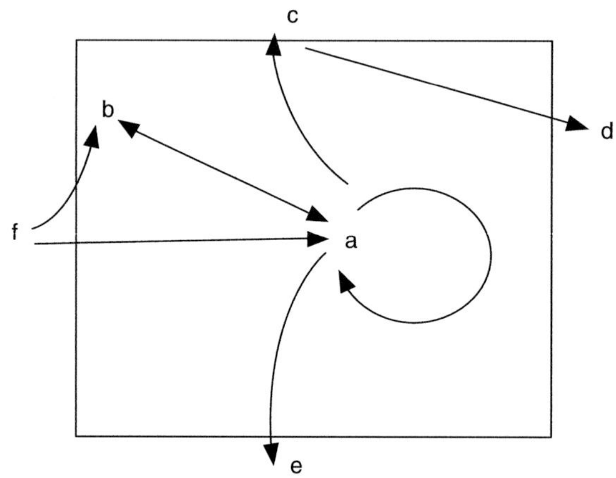

Figure 11.25. Network of locations (subdomains) within a box. Aggregations and transfers; pathways and exchange rates.

spawning locations/times can contribute to stock persistence. In some populations, depth-seeking develops during the larval period (Figure 11.26).

This is a broadcast spawning strategy. Most fish larvae die; a relative few can sustain a population if large numbers of eggs are spawned per adult, in variable locations and distributed through time.

This feature is thought in some populations to account for the absence of a stock-recruitment relationship, and in its place, a relation between spawning habitat and recruitment, with multiple spawning habitats/timing providing insurance against variability in hydrodynamic transport such that larvae are delivered to remote settlement areas. For a recent review of recruitment, see Lough and O'Brien [266].

11.11.1.1 Groundfish

Werner et al. [478] examined cod and haddock spawning with an early IBM. The spatial hypotheses are summarized in Figure 11.27. The principal spawning ground is location 1; larval fish are observed sequentially at locations 2 and 3; larval settlement

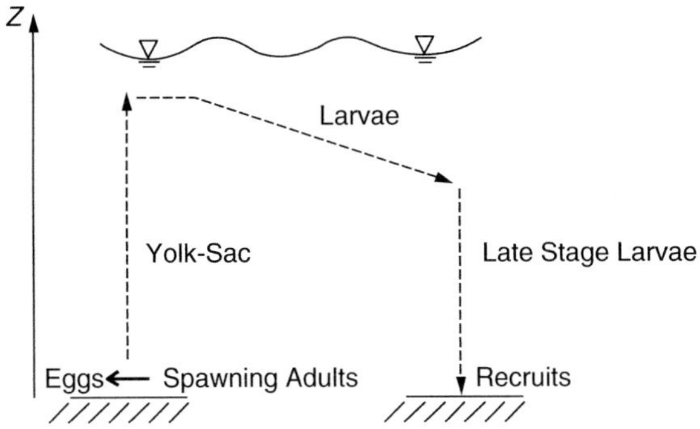

Figure 11.26. Larval fish and vertical positioning in the early life stages, leading to recruitment. In the planktonic phase shown, depth-seeking develops gradually.

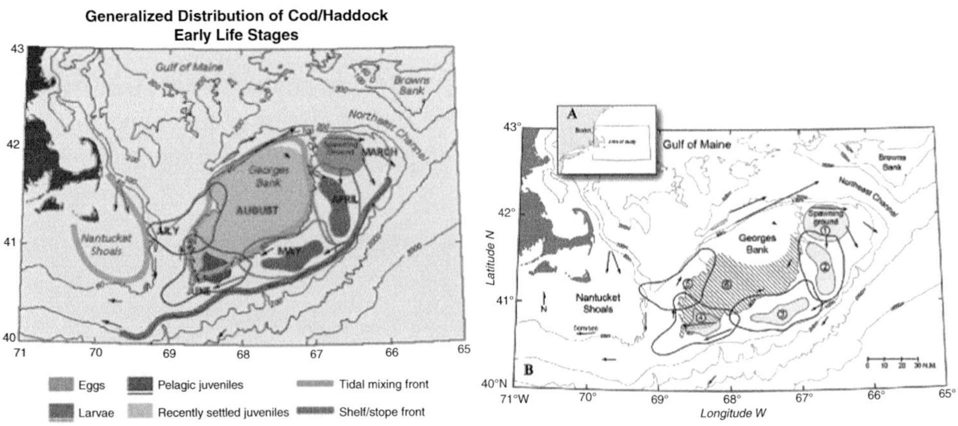

Figure 11.27. Spatial locations relevant to the Lough-Werner IBM for cod on Georges Bank. Version of this figure appear in several early documents. Left: As it appeared in Lough and Manning [264]. Right: As in Lough and O'Brien 2011[266]. (Original source courtesy of R. G. Lough, pers. comm. 2012.) For a color version of this figure, please see the color plate section.

at locations 4 and 5; and new recruits at location 6. This is an annual cycle, with spawning in March and juveniles reaching location 6 in summer. Many detailed hypotheses are possible to explain recruitment variability, concerning physical and biotic processes as occurring in a spatially explicit system. The available field data, though representing enormous effort, are necessarily sparse. In this scenario (supported by decades of observation and modeling), the population spawned on the bank is retained there by the hydrodynamics, despite the tendency to disperse larval fish by wind influence in the surface layer. The IBM for this case is as follows. Observed egg distributions in space/time were used to inject individuals at the bottom, location 1. Individuals grow through four egg stages during the first 20 days of life; and 7 larval stages during the next 100 days. Survivors at each stage have a preferred depth

(Lough Calderone [263]). The earliest stages stay deep, not needing external food source. Stage III reaches the surface layer where feeding occurs on planktonic organisms. Stage IV begins a descent to mid-depth. At stage VI, larvae began diel vertical migration, at about 60 days post-spawn. Settlement occurs at the end of roughly 120 days, the survivors recruited to the population as bottom-dwellers. This observed depth-age distribution is critical to enhancing the retentive nature of the system.

Environmental variability on several time scales influences recruitment variability, and its link to hypothesized stage-dependent behavior, growth rates, and mortality. Refinements of this basic idea account for wind variability (Lough et al. [260]), turbulence (Lough Mountain [265], Werner et al. [477], feeding (Werner et al. [476], Lewis et al. [284]), and light abundance (Lough et al. [262]). Werner and MacKenzie [475] examine recruitment data and interspecies differences between cod and haddock. Lough and Manning [264] used these methods for study of the southern flank front.

A more complex model is discussed in Lough, Broughton, Buckley et al. [261, 262] and Lough [261]. (See Figure 11.28.) Included are the effects of tide-generated turbulence and ambient light on feeding rates, temperature-dependent metabolic costs,

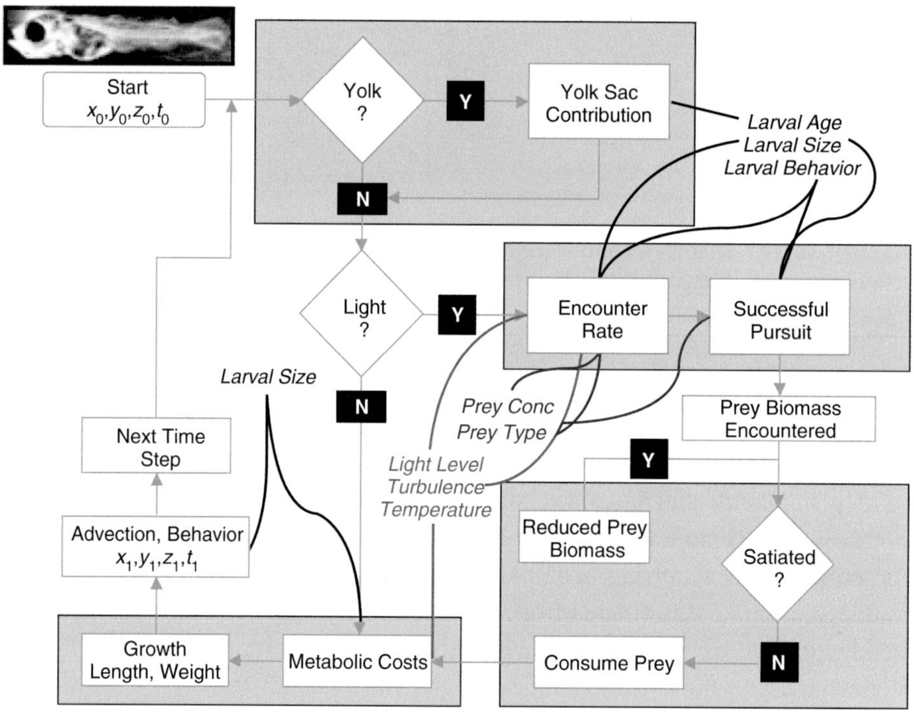

Figure 11.28. The larval fish IBM used by Lough, Broughton, Buckley et al. in [261, 262]. As it appears in Lough et al. [262]. (Original source courtesy R. G. Lough, pers. comm. 2012.) For a color version of this figure, please see the color plate section.

ingestion and digestion rate. Ambient prey are represented in four major species with prey abundance, prey selectivity, and maximum ingestion rates accounted for. Among the ecological issues considered was variation in recruitment between 1993 and 1994. The interannual variation was attributed to a combination of depth-dependent food limitation, faster growth in the warmer year, and mortality decreasing in the higher stages.

Stegman et al. [437] studied menhaden recruitment in the South Atlantic Bight; Edwards et al. [126] made similar studies of sea bass in the same area. These are representative of the practical, management-oriented studies enabled by a site-specific model of the organism. They build on basic laboratory research (biology), field sampling (oceanography), and two decades of steady progress in assembling these studies into site-specific models.

11.11.1.2 Shellfish

A related larval model was assembled by Tremblay et al. [454], and extended by Lewis et al. [249], to examine scallop recruitment in the same system (Georges Bank). Adults in this case are relatively sedentary and live many years. Reproduction is observed to occur during peak stratification, when eggs are generated in observed adult aggregations at the bottom. The buoyant eggs subsequently rise to the near-surface layer. The larval period lasts on the order of weeks-months, during which time larvae inhabit the euphotic zone and significant displacement occurs. Then there is period of diving/searching for habitat culminating in settlement to the benthos. Survivors recruit to the population and typically live many years, reproducing multiple times.

The life cycle of this organism is illustrated in Figure 11.29. It is possible to simulate the recruitment phase as in the larval fish studies above. After settlement, simulation can continue with bottom-settled juveniles maturing on the bottom and beginning reproduction in future years.

Tremblay et al. [454] simulated the larval phase over a single season. As in the cod studies above, this investigation was concerned with connecting observed spawning to recruitment at major observed aggregations of adults. The spawning is observed by presence of larvae during peak seasonal stratification when the system is most retentive. Hence, an issue is the extent to which the observed and fishable aggregations are self-seeded and thus sustainable, or whether they require an external source of larvae.

The larval period was modeled as follows. Eggs were placed at the bottom and began a roughly 2-day ascent to the surface layer. Once at the surface, larvae were transported for roughly 30 days, followed by a rapid descent to the bottom and a brief period of settlement. Sensitivity studies were made for the time of spawning, and for the duration of the surface phase (representing slow or fast growth.) The currents form a leaky gyre and the results indicate that retained larvae form a partly linked set of major aggregations of adults.

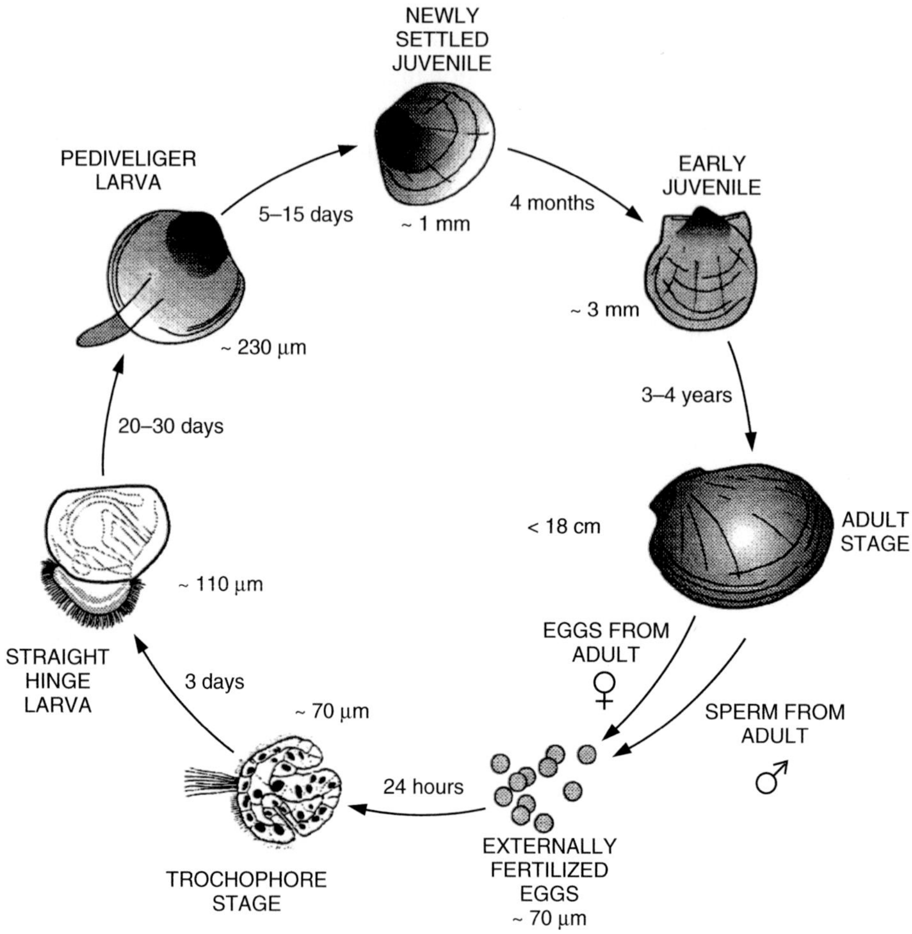

Figure 11.29. Scallop life cycle, from Stewart and Arnold [438]. (Source: Fisheries and Oceans Canada. Reproduced with permission.)

Because settled shellfish are largely sedentary, multiyear studies are possible, where idealized recruits do not move beyond settlement, and are responsible for future spawning. Lewis et al. [249] did such IBM-based simulations and addressed the management of fishing areas over many years. Settled larvae are held sedentary, subject to mortality and aging. Following a maturation delay, survivors begin generating offspring when the next year's spawning season comes around; and so the multiyear simulation continues. There is opportunity to exert fishing mortality in the settlement areas, and to examine the effects of managing the related effort by location, and examine the effects on the aggregate population, both its abundance and distribution (Figure 11.30). The management aspects of these scenarios are made more

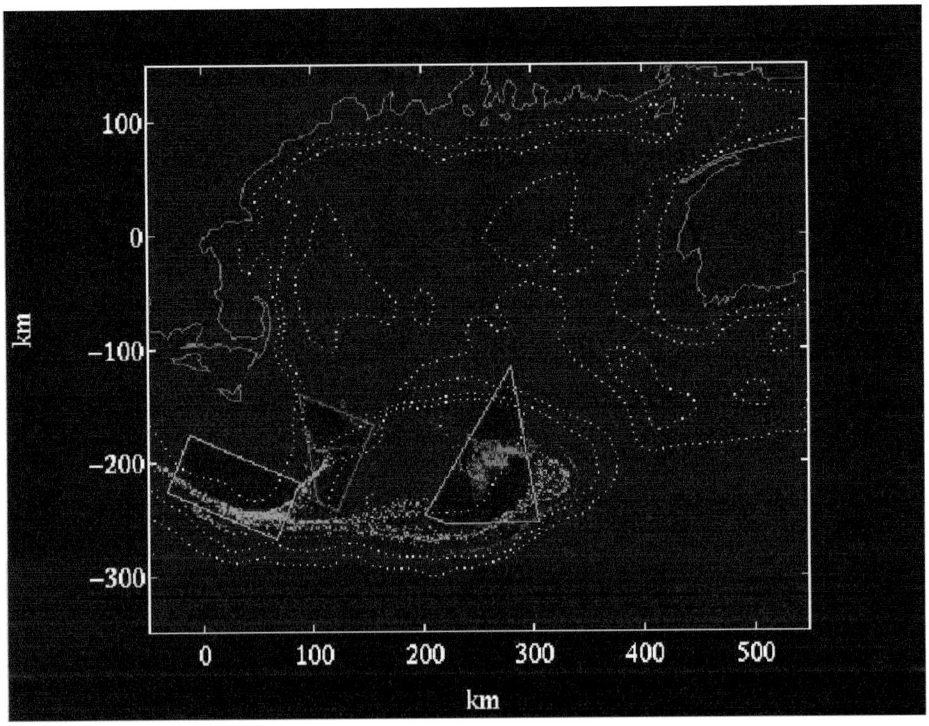

Figure 11.30. Results from the scallop IBM used by Lewis et al. [249]. The three polygons shown are observed adult aggregations. Numerical larvae were spawned there. The dot colors indicate spawning location; the dot locations indicate subsequent settlement. Reflecting management concerns, fishing mortality was reduced in the three adult aggregations. The system is partly self-sustaining, with the westernmost aggregation largely exporting larvae, and the other two, resupplying each other. Results reflect the reduction in fishing mortality in the three adult aggregations. For a color version of this figure, please see the color plate section.

complex as the population is distributed among the territorial seas of two different countries.

Incze et al. [205] utilized a very similar IBM with generic features similar to those of Tremblay et al. for the study of lobster populations. In this case, long-distance transport of larvae can occur in the surface-seeking postlarval stage. North et al. [336] report a comprehensive study of shellfish in the Chesapeake Bay, looking at recruitment variability and multiyear connections of various adult aggregations.

11.11.1.3 Recent Summaries – Larval Fish

Several recent collections and progress reports highlight the importance of IBM studies for fisheries research and management, including the Ortner ([346]), Peters [190, 356] and Gallego [164, 165] collections. North et al. [333] provide a manual of current recommended practices.

11.11.2 Zooplankton

The larval stage of fish populations separates adults and recruits via transit through the hazardous planktonic stage. Spawning and recruitment are observable; the IBM can be used to fill the gap in ways that can be very useful in understanding of recruitment variability. IBM formulation can be focused on the first few months of life; if needed and appropriate, a highly idealized IBM for a sedentary benthic phase can be added.

When we come to zooplankton, we deal with the opposite – short life, multiple generations per year, and in some cases a resting phase that provides overwintering of certain populations. There are thus two major states of a population, diapause (resting) and active (reproducing). Within the active phase, individuals pass through several morphological stages from egg to adult.

A typical life history, that of *Calanus finmarchicus*, involves six Naupliar stages and 6 copepodite stages, the last of which (C6) reaches reproductive maturity and produces eggs for the next generation (Figure 11.31). These are planktonic stages, seeking the surface. The higher copepodite stages are capable of entering an inactive phase ("diapause"), when environmental conditions are unfavorable. These sink into deeper waters where lateral currents are weaker and lower temperatures lead to reduced metabolic demands. Survivors reemerge the next year as reproducing C6s. The individuals are small, especially in the naupliar stages - making them both hard to observe and important sources of food for higher trophic levels (e.g., larval fish). Important questions concern the controls on population abundance and distribution; the oceanic location of diapause; the details of vital rates by stage; the cues for entry and exit from diapause; the number of generations that occur between diapause phases; and generally, the extent to which local populations are self-sustaining (retentive) versus being dependent on external stocking. Individual populations tend to be local, recurring in the same place and time, persistent with important variability superposed.

Simulations of the species *Metridia pacifica* were made by Batchelder and Miller [19] in terms of an IBM with 11 sequential stages (the first two were merged). Rate descriptions for reproduction, mortality, growth, and stage completion were employed in year-long simulations. Batchelder and Williams [20] adapted this to *Metridia lucens* in the North Atlantic, with suitable adaptations – vital rates generally, and adding diel behavior in the later copepodite stages. In that study, as many as four generations occur in a single year, before the population enters diapause.

Carlotti and Radach [68] describe a point IBM (z,t) for *Calanus finmarchicus* in the North Sea. The same general stage structure is used for *Calanus* individuals; they interact with nutrients, phytoplankton, and detritus modeled as Eulerian fields; details of rate relations are given. Miller and Tande [316] examined stage duration estimation for this copepod using this form of 12-stage model with North Sea observations.

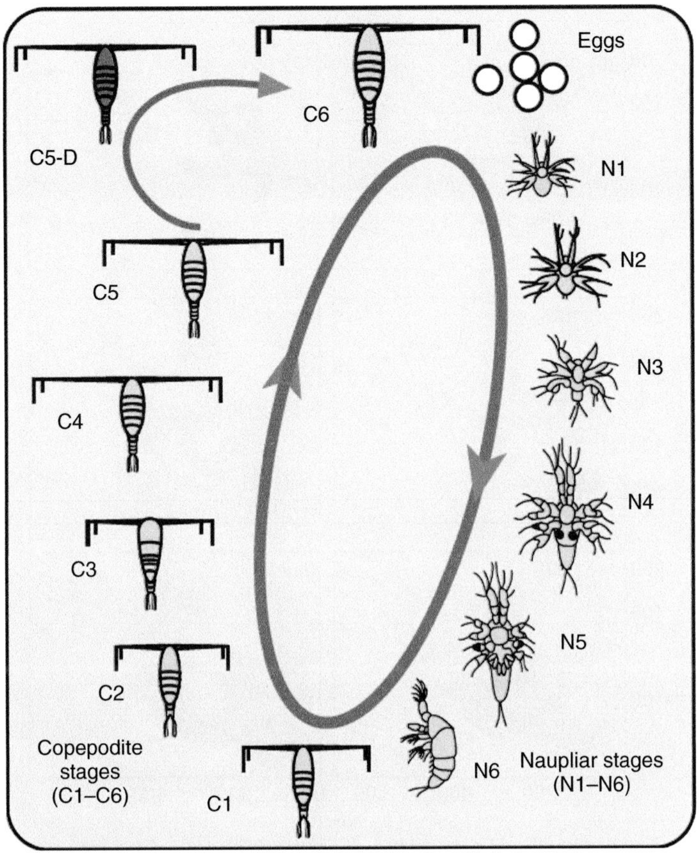

Figure 11.31. Generic stage structure for marine copepod populations. There are six naupliar stages (N_i) and six copepodite stages (C_i). (Image source: Fisheries Research Services [154]. Crown copyright, reproduced with permission, Marine Scotland.)

Meise et al. [310] describe a decadal survey of major copepod species in the Gulf of Maine; Meise and O'Reilly [309] report the abundance and distribution of *C. finmarchicus* by stage, for the larger stages C3 $\rightarrow$ C6.[10] There is evidence in the observed structure for significant yearly variation, but the absence observations of the early, smaller naupliar stages leaves that part of the population to inference – and hence, the generation questions. In addition, since the deep basins of the Gulf of Maine were not sampled in these data, the diapausing C5s were only partly observed. These are among the significant roles of the IBM here – filling in the population structure, constrained by (necessarily) partial observation.

Miller et al. [315] report an extension of the Batchelder/Miller and Carlotti/Radach models to a 3-D depiction of *C. finmarchicus* in the Gulf of Maine. The structure

[10] See Morse and Meise [320] for an accompanying animation.

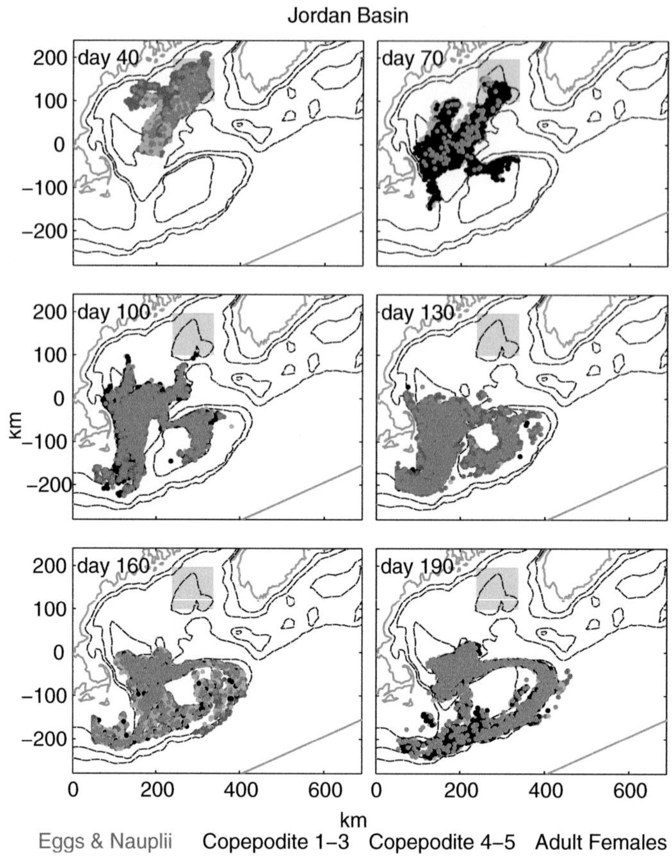

Figure 11.32. *Calanus finmarchicus* simulation results from Miller et al. [315]. Source: NML-98-7 [314]. Time evolution of population structure initiated from diapause in the shaded area on the winter solstice. For a color version of this figure, please see the color plate section.

of stage progression was monotonic. All individuals had stage-specific vital rates, sensitive to the environment; and carried the generic vector of information:

- Alive or dead
- Current stage
- Readiness to complete stage
- 3-D location

In addition, the individuals in the reproductive C6 stage carried age-in-stage, and a clutch readiness index to deal with the details of reproduction.

Example results appear in Figures 11.32, 11.33, and 11.34. In this series, a population of C5s in diapause were placed in the deep Jordan basin, and began to emerge and reproduce as C6s beginning at the winter solstice. Shown are the time evolution

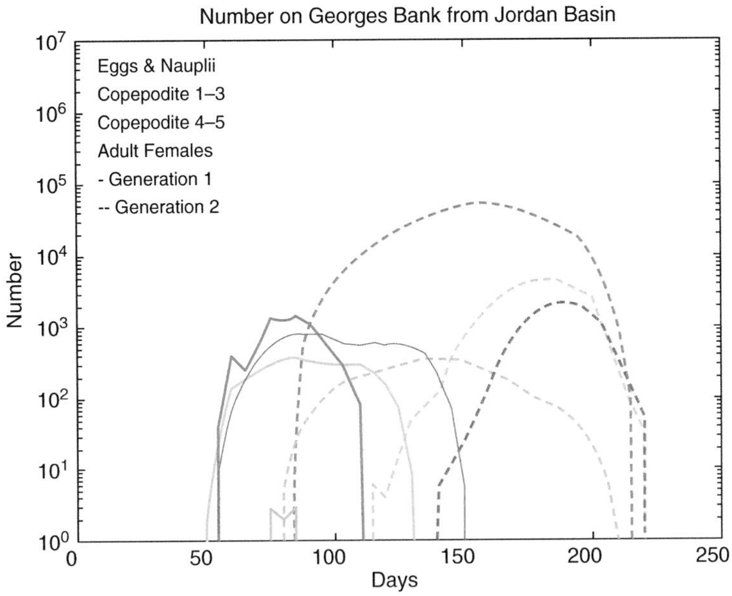

Figure 11.33. *C. finmarchicus* simulation results from Miller et al. [315]. Source: NML-98-7 [314]. Generational structure of population initiated from diapause on the winter solstice in the shaded area of Figures 11.32 and 11.34. For a color version of this figure, please see the color plate section.

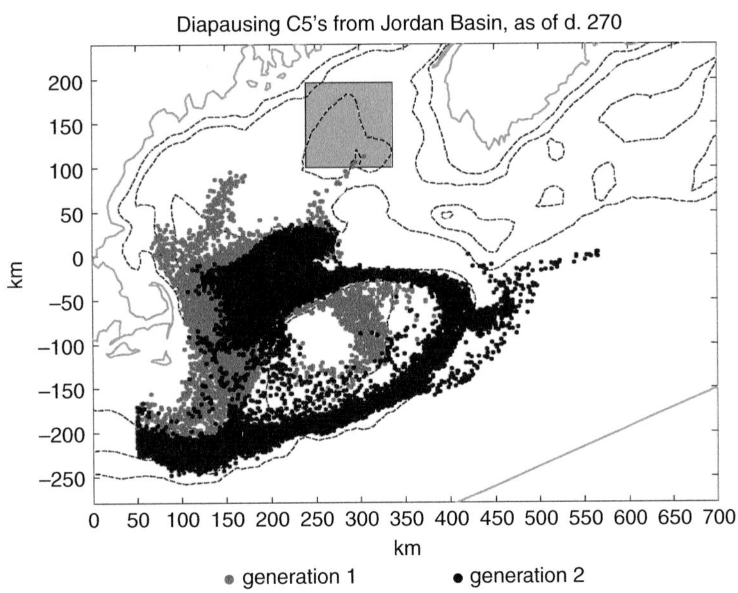

Figure 11.34. *C. finmarchicus* simulation results from Miller et al. [315]. Source: NML-98-7 [314]. Location of diapausing individuals, for a population initiated from diapause on the winter solstice, in the shaded area. For a color version of this figure, please see the color plate section.

of the population distribution (Figure 11.32); the stage structure indicating two simulated generations in one yearly cycle (Figure 11.33); and the location of the return to diapause when the growing season is finished (Figure 11.34). Animations from these runs are available at NML-98-7 [314].

Speirs et al. [435, 436] applied this type of model to the ocean basin-scale distribution of *C. finmarchicus*. This is the full scale of the population. Special attention is given to diapause and overwintering, to simulate multiyear population dynamics. Coverage of these large space and time scales, with full resolution of the stage structure, can be computationally intensive; these papers contain important methodological advances to achieve that.

11.11.3 Phytoplankton

The case reviewed here is that of the toxic dinoflagellate *Alexandrium fundyense* (Anderson [6]). It shares some lifecycle features with *Calanus* as reviewed previously – namely, a two-phase cycle where the population alternates between active and resting states. Blooms of this organism are in the "red tide" category. When the toxins accumulate in filter-feeding organisms like mussels and clams, they can threaten human health.

Figure 11.35 shows an example organism and its life stages from excystment to encystment. This organism rests in bottom sediments in the cyst stage over winter and potentially over many years; exits under environmental and/or endogenous cues, seeking the euphotic zone as a single-celled individual. Therein, individuals reproduce asexually, by cell division. When the growing season is complete, it begins formation of gametes that fuse to form zygotes that sink and become benthic cysts. The cycle repeats following excystment after one or more years in the benthos, with pelagic transport during the vegetative phases, alternating with benthic immobilization in the cyst phase.

A hypothetical system appears in Figure 11.36 where sources are linked by a spatial conveyor belt with partial retention. Critical to the self-sustaining nature of such a system is the balance of physical and biotic rates: hydrodynamic export versus the balance of growth, mortality, and encystment in the gyre. It is easy to imagine growth and subsequent encystment making up for hydrodynamic export.

11.11.3.1 The Gulf of Maine

A set of recent contributions relative to *Alexandrium* in the Gulf of Maine appears in the special issue of Deep-Sea Research [5]. Among the major concerns:

- When/how/at what rate do cells reach shore?
- When/how/at what rate do cysts form?

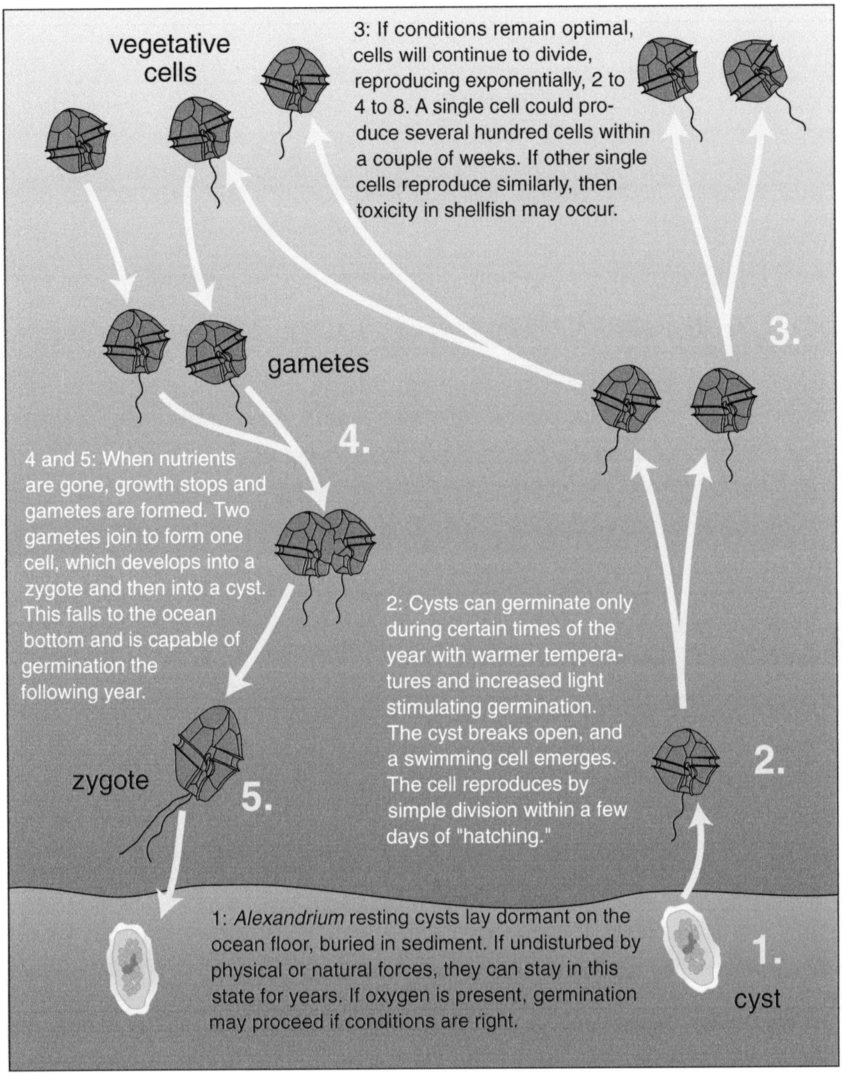

Figure 11.35. Life cycle for individual phytoplankton: *Alexandrium fundyense*. (Image courtesy of D. Anderson, B. Keafer and J. Kleindinst, pers. comm. 2014.) For a color version of this figure, please see the color plate section.

- What is the signal for cells to advance from vegetative growth to sexual differentiation?
- Does the population represent a closed (self-sustaining) or open (through-flow conveyor belt) system?
- What controls the population abundance and distribution?

Data and modeling suggest that the primary sources of resting cysts is offshore of mid-coast Maine and at the head of the Bay of Fundy. McGillicuddy et al. [305]

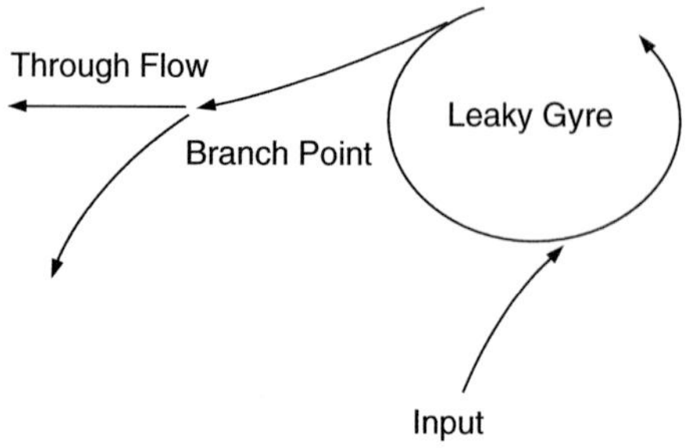

Figure 11.36. Hypothetical physical-biological system with through flow and a partially retentive gyre.

assembled a set of rate descriptions for excystment, reproduction by cell division, swimming speed, and gamete formation followed by encystment. Examination of the observed climatological cycle in the Gulf of Maine (McGillicuddy et al. [306]) revealed a credible cycle of partial persistence based on these rates, with two main features:

- A leaky gyre in the Bay of Fundy (Figure 11.37) providing a self-sustaining source of vegetative cells for westward export
- Transport of vegetative cells along shore in a persistent hydrodynamic coastal current
- Transition from vegetative phase toward encystment in the later summer, likely due to nutrient depletion
- Encystment occurring in the observed cyst beds, such that there is a westward movement of the population, and a persistent source in the east.
- The curious occurrence of cells in fresher water, formerly thought to indicate riverine sources, results from: shoreward near-surface transport of vegetative cells originating offshore, concentrated near the coast by downwelling-favorable wind.

All of these studies were completed using "conventional" Eulerian (concentration-based) modeling. An IBM representation is readily assembled. Major parts from the previous plankton studies are directly relevant, including the major stages, stage-based mortality, maturation, vegetative reproduction, encystment, and excystment. Assembling the relevant rates from [305] and inserting the IBM in the same simulated environment shows the replication of the basic behavior noted above, and the usual advantages of the IBM approach. An example outcome appears in Figure 11.38. Animations of these simulations are at NML-03-19 [273].

These studies generally focus on the settlement patterns of cysts following the growing season. The IBM approach has also been useful in examining resuspension and transport of settled cysts, providing quantitative estimates for the degree to which deposition patterns such as those shown in Figure 11.38 can be redistributed by bottom boundary layer dynamics [10].

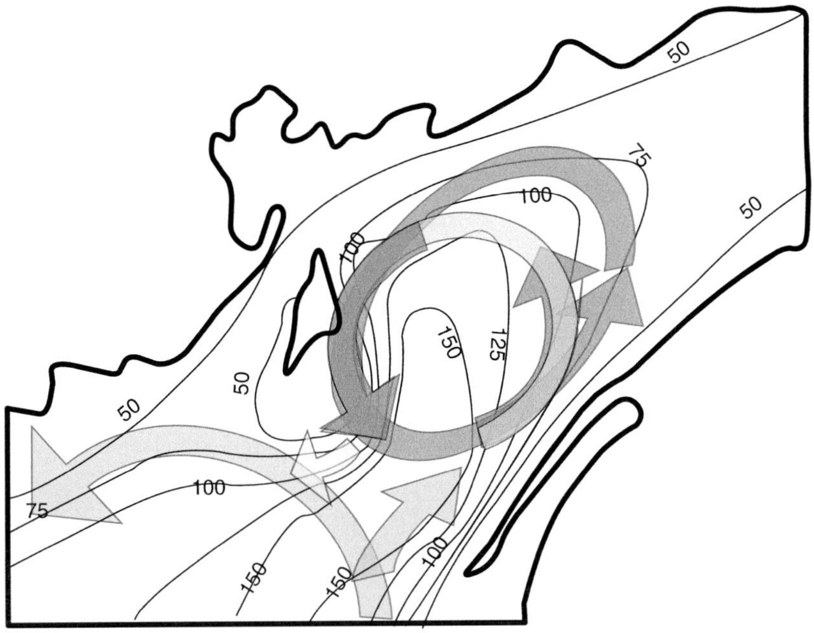

Figure 11.37. Schematic of leaky grye in the Bay of Fundy, as reported by Aretxabaleta et al. [9] and predicted by simulating passive Lagrangian drifters in the climatological flow fields (NML-05-2, [274]. Image from [9], Reproduced with permission.) For a color version of this figure, please see the color plate section.

11.11.3.2 Implementation

It is instructive to review the file structure supporting this IBM, as given in NML-05-6 [481].

- **General conventions** are simple and necessary to enable use by many people and institutions:
  - The MKS system is used.
  - Horizontal location is a Mercator projection centered at Boston.
  - Vertical location is positive upward; zero at mean sea level. Hence in the absence of tides, all particles have negative Z.
  - Time is registered in the DMY convention: day, month, year, elapsed time since the beginning of the day. Hence there are three integers and a real number of seconds.
  - Uniform *Dt* separates all data.
  - All files are generic ASCII format.
- **The biological states** for this organism are numbered as follows:
  - 0 = Dead
  - 1 = Starting cyst

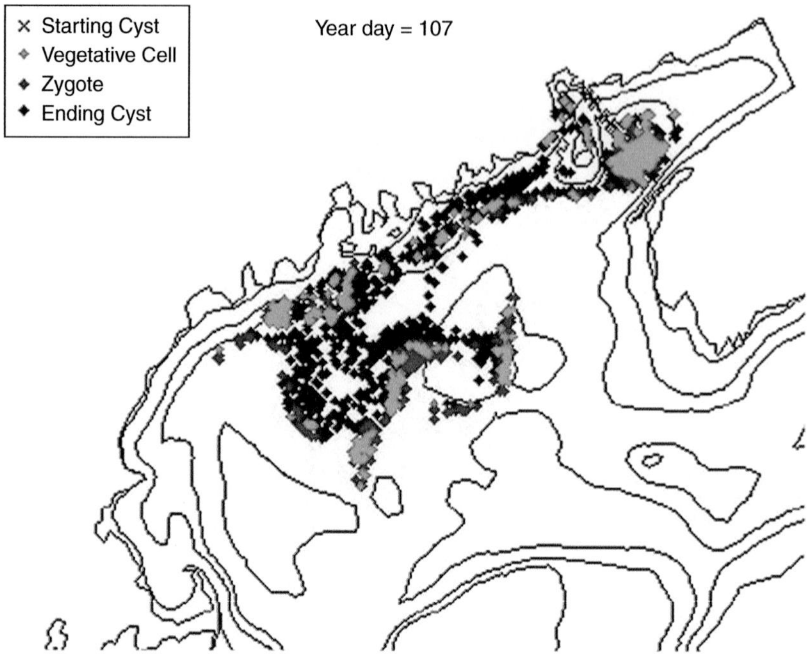

Figure 11.38. *Alexandrium* IBM in the Gulf of Maine. The benthic cysts were seeded in the Bay of Fundy (the northeastern end of the domain) during winter. Excystment was triggered by an endogenous clock. Vegetative cells were transported along coast near the surface, dividing along the way; encountered a major branch point directing part of the flow south; and generally reached nutrient limitation based on climatological records, which triggered the encystment process. Major life stages are indicated by colors. For a color version of this figure, please see the color plate section.

- 2 = Vegetative cell
- 3 = Gamete (ignored here)
- 4 = Zygote
- 5 = Ending cyst

Notice that the numbering key used in this figure differs from the governing stage numbering convention, Figure 11.35. This proves necessary to implement all possible transitions, illustrating the subtle distinctions needed between biological stage and the biological state vector needed to represent all possible state transitions.

- **Particle identity** reflects the details of the organism, and the type of interactions with the environment enabled.
  - All particles have a unique identifier throughout any simulation.
  - The identifier persists through all individual state transitions.
  - Vegetative growth occurs by cell subdivision: one particle becomes two.
  - When subdivision occurs, one particle retains the original identity, and the clone is initialized with the identical inheritance.

- **Output files:** There are three types of files resulting from such a simulation:
  - A Lagrangian file (.3dp) contain the time histories of a selected set of particles. A detailed description is contained here.
  - A set of Eulerian files, containing the location and state of all particles at one point in time.
  - A state transition file containing the time, location and nature of all biological events during a simulation. Transitions are numbered as follows:

    1: State 1 → 2 (excystment or germination)
    2: State 2 → 4 (zygote formation)
    3: State 2 → 2 (vegetative growth)
    4: State 4 → 5 (encystment)
    5: State 1 → 0 (death)
    6: State 2 → 0 (death)
    7: State 4 → 0 (death)
    8: State 5 → 0 (death)

These standards were found adequate to implement this IBM (Figure 11.38 and the supporting animations, NML-03-19 [273]). They are not unique, but cover the states, transitions, and general organism features necessary to function in a simulated environment.

11.11.4 Summary – Plankton IBMs

Here we have given some example applications from the literature, covering planktonic IBMs and their use. Included are phytoplankton, zooplankton, ichthyoplankton and larval fish, and shellfish applications. In addition to the scientific merit of these studies, they are conceived in the context of important public management issues relative to natural resources and the environment including

- Understanding the stock-recruitment relation and the role of planktonic states in generating variable recruitment
- Overfishing, habitat degradation, and fisheries effort management
- Benthic aggregation and retention of shellfish populations
- Transboundary larval transport separating spawners and recruitment
- Zooplankton population dynamics: persistence, importance of early life stages in providing food for larval fish
- Phytoplankton toxicity (red tide) – occurrence and persistence

This is a diverse set of problems, all addressing complex issues combining physical and biological dynamics, very dependent on site-specific details. The examples illustrate, but do not constrain, future possibilities with this important approach.

We note that there are freely available software packages that readers may find suitable for their own IBM studies. One such package is the Larval TRANSport Lagrangian model (LTRANS: North et al. [336], Schlag and North [404]), an offline particle-tracking model that runs with the stored predictions of a 3-D hydrodynamic model, specifically the Regional Ocean Modeling System (ROMS). Written in Fortran 90, LTRANS v.2 is based on a 4th order Runge–Kutta scheme for particle advection and includes a random displacement model for vertical turbulent particle motion. For more information see http://northweb.hpl.umces.edu/LTRANS.htm.

Generally, particle tracking routines and codes are summarized on the publisher's web site.

11.12 Population Connectivity

The issue of population connectivity is intrinsic to many applications that lend themselves to the IBM approach. Key questions include: What is the fate of planktonic organisms originating from a particular location? What was the origin of planktonic organisms currently occupying a particular location? To what degree are subpopulations connected? Under what conditions is a given population self-sustaining? Kritzer and Sale [233] describe a conceptual framework for metapopulations, ranging from networks of closed populations to patchy subpopulations (Figure 11.39). Increasing connectivity among the subpopulations increases resilience to disturbance.

To quantify the connectivity among subpopulations (Cowen et al. [87, 88]; Cowen and Sponaugle [89]), we adopt the formalism provided in Mitarai et al. [317]. The discrete Lagrangian probability density function (PDF) for a set of N particles with initial locations $\mathbf{a}$ for advection time τ given trajectories $\mathbf{X}_n$ is denoted $f'_{\mathbf{X}}(\xi;\tau,\mathbf{a})$:

$$f'_{\mathbf{X}}(\xi;\tau,\mathbf{a}) = \frac{1}{N}\sum_{n=1}^{N}\delta(\mathbf{X}_n(\tau,\mathbf{a})-\xi) \tag{11.79}$$

where ξ is the sample space variable for $\mathbf{X}$. Here bold denotes vector quantities. It is assumed that the discrete Lagrangian PDFs from starting locations $\mathbf{a}$ are representative of starting *areas* of radius R centered on $\mathbf{a}$:

$$f'_{\mathbf{X}}(\xi;\tau,\mathbf{a}) \approx \frac{1}{\pi R^2}\int_{|\mathbf{r}|\leq R} f'_{\mathbf{X}}(\xi;\tau,\mathbf{a}+\mathbf{r})\,d\mathbf{r} \tag{11.80}$$

A continuous PDF, $f_{\mathbf{X}}(\xi;\tau,\mathbf{a})$ is given by spatially filtering the discrete PDF in the sample space:

$$f_{\mathbf{X}}(\xi;\tau,\mathbf{a}) \approx \int_{-\infty}^{\infty} G(\mathbf{x}-\xi) f'_{\mathbf{X}}(\xi;\tau,\mathbf{a})\,d\mathbf{x} \tag{11.81}$$

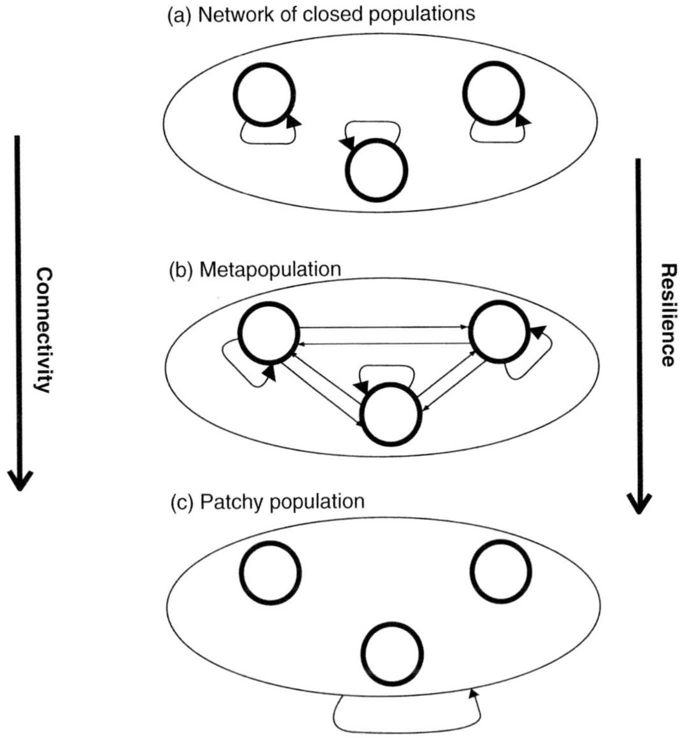

Figure 11.39. Connectivity in metapopulations (Adapted from Kritzer and Sale [233]. Reproduced with permission).

$G(\mathbf{x}-\xi)$ is a filtering function, taken in this case to be Gaussian with an e-folding scale similar to the site radius R. Connectivity is defined as the probability a water parcel leaving a source site j arrives a destination site i over a time interval τ, and is denoted here as $C_{ji}(\tau)$. Values of $C_{ji}(\tau)$ are evaluated from the Lagrangian PDF for a source location ($\mathbf{x}_j$) and a destination location ($\mathbf{x}_i$) as

$$C_{ji}(\tau) = f_{\mathbf{X}}(\xi = \mathbf{x}_i; \tau, \mathbf{a} = \mathbf{x}_j) \cdot \pi R^2 \tag{11.82}$$

where the normalization by πR^2 converts probability densities to probabilities.

One can exploit the connectivity matrix to compute ecologically meaningful diagnostics, such as the destination attractiveness $D_i(\tau)$ and source strength $S_j(\tau)$:

$$D_i(\tau) = \sum_{j \subseteq J} C_{ij}(\tau) \qquad J = j_1, j_2, j_3, \ldots j_N \tag{11.83}$$

$$S_j(\tau) = \sum_{i \subseteq I} C_{ji}(\tau) \qquad I = i_1, i_2, i_3, \ldots i_M \tag{11.84}$$

Values of $D_i(\tau)$ reflect the relative "attractiveness" of site i for all of the Lagrangian particles released in the domain for a release time of τ. Similarly, the source strength $S_j(\tau)$ reflects the relative success that particles from site j encounter another site within the time scale τ.

These techniques have been applied to questions relevant to fisheries management and pollutant transport in the Southern California Bight (Figure 11.40). Hydrodynamic simulations portray highly variable transport pathways, owing to the energetic eddy field characteristic of this region (Figure 11.41). The particle tracking results provide a basis for evaluating a connectivity matrix for this system (Figure 11.42). For shorter advection times, self-seeding and connectivity among neighboring sites are prevalent. In general, the three major subregions are not well connected with each other – although there are some exceptions such as a transport from the mainland to the North Islands. As the advection time increases, connectivity among sites becomes more homogeneous, yet the imprint of the northward mean flow persists in the connectivity along the mainland and its connection with the North Islands.

Assessment of source strength (Figure 11.43) reveals that all three subregions have distinct source locations. Water parcels arriving at the Northern Islands originate mostly from the northern portion of the mainland coast and the northern shore of islands themselves. In contrast, particles arriving at the Southern Islands come

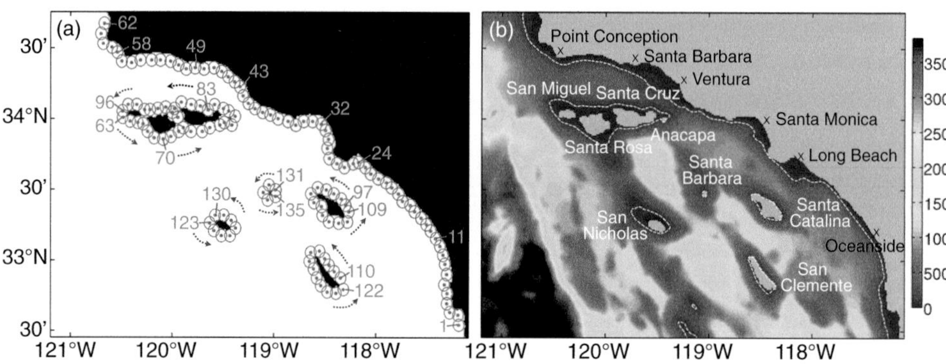

Figure 11.40. (a) Site locations used for the assessment of Lagrangian PDFs. Red dots indicate the center location of each site, with blue circles showing the area covered by each site (with a 5-km radius). The sites are categorized into three regional groups, that is, along the mainland (sites 1–62; from south to north), Northern Islands (sites 63–96), and Southern Islands (sites 97–135). (b) Model bathymetry (color bar in meters). The white dashed line indicates the 100 m isobath. (c) Sample Lagrangian particle trajectories from a site near San Nicholas Island (yellow circle). The blue lines indicate simulated 30-day trajectories, with red dots indicating the endpoints. (From Mitarai et al. [317], with permission. Source courtesy S. Mitarai (pers. comm)). For a color version of this figure, please see the color plate section.

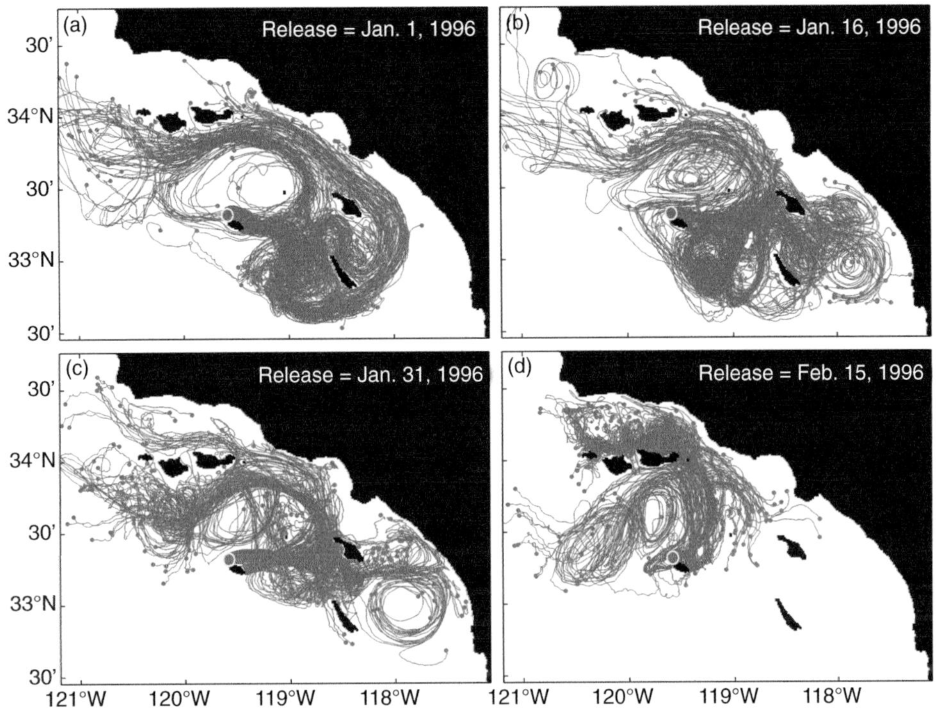

Figure 11.41. Sample Lagrangian particle trajectories from a site near San Nicholas Island (yellow circle). The blue lines indicate simulated 30-day trajectories, with red dots indicating the endpoints. (From Mitarai et al. [317], with permission. Source courtesy S. Mitarai (pers. comm)). For a color version of this figure, please see the color plate section.

from all around those islands as well as the mainland. Water parcels arriving at the mainland sites come mainly from the mainland and not from the offshore Islands.

Destination attractiveness distributions (Figure 11.44) illustrate the tendency for retention of particles originating at the Northern and Southern Islands, with relatively more transport to the mainland. The mainland primarily reseeds itself, with the south-to-north gradient reflecting the directionality of the mean flow. There is modest transport from the mainland to the Northern Islands, whereas the connectivity between the mainland and the Southern Islands is substantially less.

Growing appreciation for the utility of IBM-based studies of connectivity is leading to their application in a variety of coastal regimes. Examples are presented in Li et al. [251] (coastal Maine, toxic algae), Edwards et al. [127] (South Atlantic bight, larval fish dispersal) and Aretxabaleta et al. [10] (Gulf of Maine, benthic cysts).

11.13 Summary – Individual-Based Modeling

Cohorts of One. In the classic Beverton–Holt form, one simulates entire cohorts of *identical* individuals, by simulating the individual's weight $W(t)$ and the number

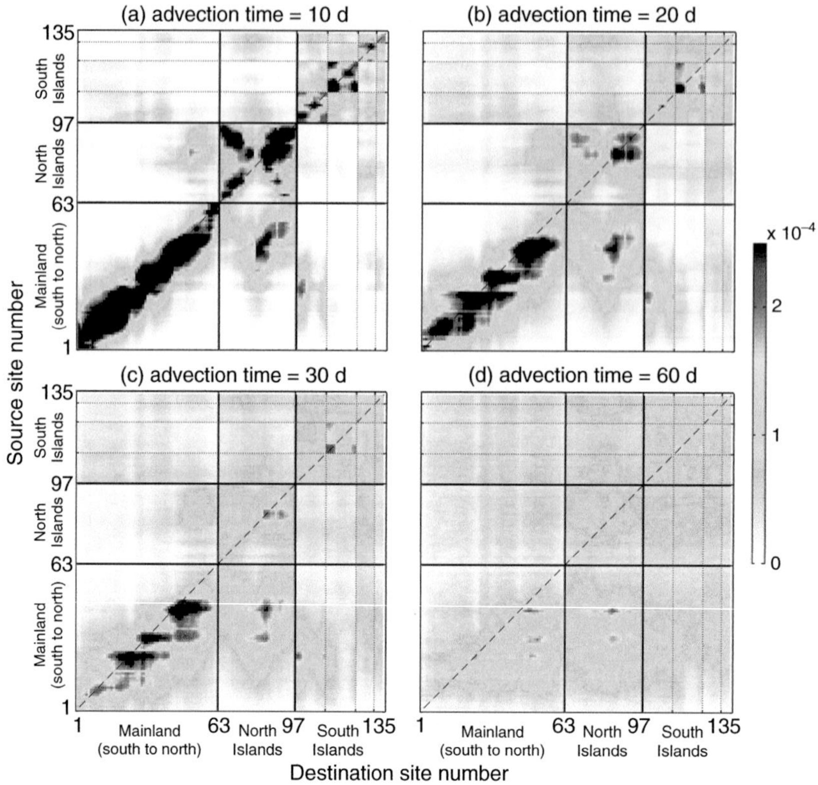

Figure 11.42. Mean coastal connectivity for the advection time of (a) 10 days, (b) 20 days, (c) 30 days, and (d) 60 days. The mean connectivity for a given source site *j* (horizontal axis) and destination site *i* (vertical axis) is determined by evaluating Lagrangian PDFs. Connectivity values are normalized so that the summation of the mean connectivity over all possible combinations is one. The black dashed line indicates self connectivity, that is, the source and destination sites are identical. The solid lines define three regions: the mainland (sites 1–62), Northern Islands (sites 63–96), and Southern Islands (sites 97–135). (From Mitarai et al. [317], with permission. Source courtesy S. Mitarai (pers. comm)). For a color version of this figure, please see the color plate section.

or abundance $N(t)$ of them is a cohort. Vital rates concerning growth, health, and maturation are simulated as continuous variables applicable to all living individuals in a cohort; recruitment, stage progression, and natural and fishing mortality are treated as sources and sinks of N. Variability exists between cohorts, but not within one.

The natural extension is individual-based, recognize diversity in vital rates and behavior *within* cohorts. One simulates an ensemble of individuals, whether initially present, imported across boundaries, or spawned within. It is the *outcomes* that are aggregated.

Early general work is represented in Sammarco and Heron [398], Grimm and Railsback [188], Deangelis and Gross [107]. Werner et al. provide a useful review of early

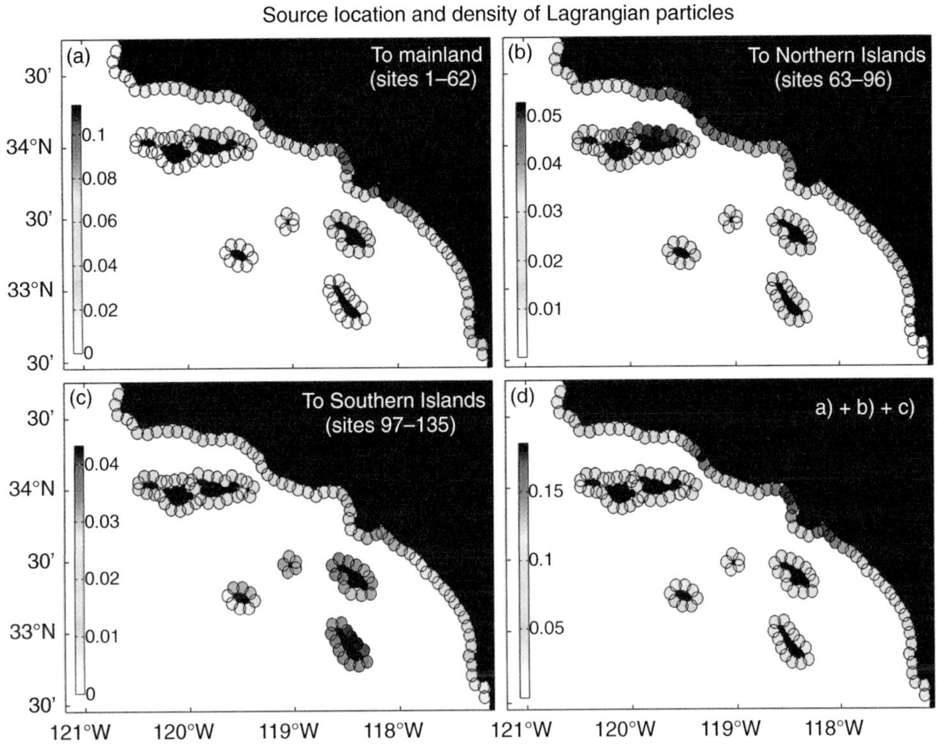

Figure 11.43. The source location and density (source strength) of particles transported to (a) the mainland (sites 1–62), (b) the Northern Islands (sites 63–96), (c) the Southern Islands (sites 97–135), and (d) all sites (sites 1–135) for an advection time of 30 days. For example, a value of 0.1 indicates that 10% of Lagrangian particles that were released from this site were found somewhere along the mainland after 30 days. Note that the color scales differ among the four panels. (From Mitarai et al. [317], with permission. Source courtesy S. Mitarai (pers. comm)). For a color version of this figure, please see the color plate section.

research [474]. A recent review of IBM research and prospects appears in Deangelis and Mooij [108]. The important influence of diffusion in realistic environments is explored by Okubo and Levin [343]. Clark and Mangel [84] provide a detailed look at behavioral development (a rework of the earlier work [297]). A good basic reference on probability models is Ross [389].

Simulation. In simulation one reduces all processes to the simplest describable units. In the IBM case, this amounts to describing a single individual evolving over one time step. The simulation takes over from there, generating a cohort with individual outcomes reflective of its initial variability, the nonlinearities and stochastic influences distributed across the population, and each individual's motion with a heterogeneous environment. In essence, one simulates the individual parts, then aggregates the outcomes.

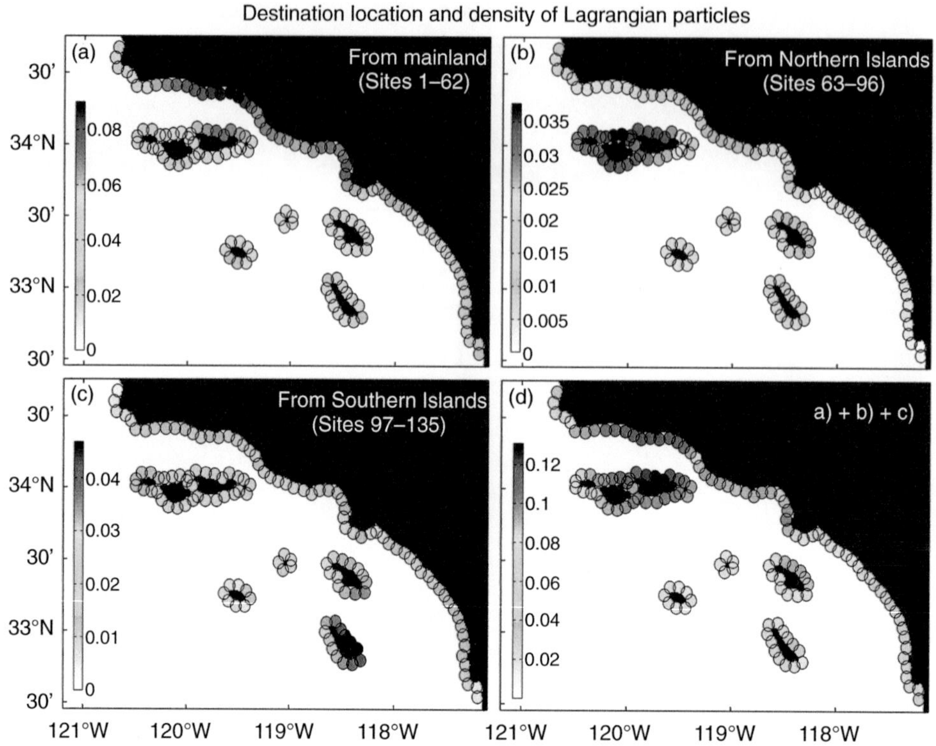

Figure 11.44. The destination location and density (destination attractiveness) of particles transported from sites (a–d) as in Figure 11.43. Advection time 30 days. (From Mitarai et al. [317], with permission. Source courtesy S. Mitarai (pers. comm)). For a color version of this figure, please see the color plate section.

Some cohort processes are readily modeled as continuous processes – for example, individual weight, maturity, and location. Individually, these retain their character as continuous, real variables requiring integration over time. Individuals vary due to parameter variation; and as in the case of random walk, by stochastic forcing. But the canonical problem is the integration of conventional differential equations for weight or location over a single time step.

There are other processes that are binary at the individual level. Included are mortality, stage transition, and reproduction. In these cases, the procedure for an individual would be to (a) form the probability of an event during the coming Δt; (b) generate the uniform deviate $\mathcal{U}$; and (c) accept or reject the event based on the comparison.

When the vital rate parameters – for example, growth rates, mortalities, reproduction rates – depend on $\mathbf{X}$, then we have "physical-biological" coupling with between motion and growth.

A simple example individual might have the following state variables:

- Real variables
  - Weight
  - "Condition" or health (possibly a convolution of history)
  - Time since entering stage
  - Position (a pointer to the environment)
- Integer variables:
  - Alive or dead (binary)
  - Male or female (binary)
  - Stage
  - Number of prior reproductive events
  - Family lineage

The Simulator as Experimentalist. In using IBMs, it is useful to adopt the ecologist's viewpoint. Especially relative to space-time patterns of abundance and distribution, the design of experiments, and parameter estimation, this field is still in its infancy. Among the main issues is that in simulation one is dealing with controlled experiments in which *all* variables and individuals are observable with complete precision. This accompanies of course the *imprecision* of a finite model substituting for nature.

Some good resources that combine ecological data and phenomena with a probabilist's view include Hilborn and Mangel [195] and the Gotelli primer [174].[11]

Coupling Physics and Biology. In an IBM, a specific model environment – heat, light, nutrients, predators, motion – is inhabited by a model cohort. In many practical cases, the knowledge of physics supports detailed site-specific simulations that are not affected by the biology – this has been the goal of circulation modeling for a long time, and is now quite mature. On the other hand, field-scale biology is still rapidly evolving. This is most apparent when one is concerned with manipulation of harvesting regimes or protecting habitat, at the regional scale. There is much value in detailed, spatially explicit simulation of model cohorts in realistic environments in this case.

The coupling between physics and biology is critical when the environment varies spatially and temporally, on scales comparable to that observed in the population itself. Recent examples using a common physical model focused on specific biological population dynamics and management include populations of larval fish, shellfish, zooplankton, and phytoplankton, as illustrated in Section 11.11. These and similar

[11] More fundamental aspects of mathematical biology generally are reviewed by, for example, Britton [55] and Murray [322, 323], although these lack a specific focus on field ecology.

simulation projects have the possibility of achieving regional-scale population abundance and distribution, constrained by variations in current, temperature, wind, and so forth and their effect on aggregation, dispersal, and vital rates. Quantification of population connectivity (Section 11.12) is a straightforward application of the IBM approach.

Part IV

Appendices

Appendix A

Series, Sums, Limits

Series (any θ)

$$e^{\theta} = \left[1 + \theta + \frac{\theta^2}{2!} + \frac{\theta^3}{3!} + \frac{\theta^4}{4!} \dots\right] = 1 + \sum_{n=1}^{\infty} \left(\frac{\theta^n}{n!}\right) \tag{A.1}$$

$$= \frac{1}{\left[1 - \theta + \frac{\theta^2}{2!} - \frac{\theta^3}{3!} + \frac{\theta^4}{4!} \dots\right]} \tag{A.2}$$

$$\theta e^{\theta} = \sum_{n=1}^{\infty} n \left(\frac{\theta^n}{n!}\right) \tag{A.3}$$

$$\left[1 - \frac{\theta}{2}\right] e^{\theta} = 1 + \sum_{n=1}^{\infty} \left(1 - \frac{n}{2}\right) \left(\frac{\theta^n}{n!}\right) \tag{A.4}$$

$$= \left[1 + \frac{\theta}{2}\right] + \sum_{n=3}^{\infty} \left(1 - \frac{n}{2}\right) \left(\frac{\theta^n}{n!}\right) \tag{A.5}$$

$$e^{\theta} = \left[\frac{1 + \frac{\theta}{2}}{1 - \frac{\theta}{2}}\right] + \sum_{n=3}^{\infty} \left[\frac{1 - \frac{n}{2}}{1 - \frac{\theta}{2}}\right] \left(\frac{\theta^n}{n!}\right) \tag{A.6}$$

$$= \left[\frac{1 + \frac{\theta}{2}}{1 - \frac{\theta}{2}}\right] + \mathcal{O}(\theta^3) \tag{A.7}$$

$$1 - e^{-\theta} = \left[\theta - \frac{\theta^2}{2!} + \frac{\theta^3}{3!} - \frac{\theta^4}{4!} \dots\right] \tag{A.8}$$

$$\frac{1 - e^{-\theta}}{\theta} = \left[1 - \frac{\theta}{2!} + \frac{\theta^2}{3!} - \frac{\theta^3}{4!} \dots\right] \tag{A.9}$$

$$\sin x = x - \frac{x^3}{3!} + \frac{x^5}{5!} - \frac{x^7}{7!} + \ldots \tag{A.10}$$

$$\cos x = 1 - \frac{x^2}{2!} + \frac{x^4}{4!} - \frac{x^6}{6!} + \ldots \tag{A.11}$$

$$\tan x = x + \frac{x^3}{3} + \frac{2x^5}{15} + \frac{17x^7}{315} + \ldots \qquad (\,|x| < \frac{\pi}{2}\,) \tag{A.12}$$

(A.13)

$$\tan^{-1} x = x - \frac{x^3}{3} + \frac{x^5}{5} - \frac{x^7}{7} + \ldots \qquad (\,x^2 < 1\,) \tag{A.14}$$

$$\ln(1+x) = x - \frac{x^2}{2} + \frac{x^3}{3} - \frac{x^4}{4} + \ldots$$

$$= -\sum_{n=1}^{\infty} (-1)^n \frac{x^n}{n} \quad (-1 < x < 1) \tag{A.15}$$

Sums ($|\rho| < 1$; any z)

Infinite Sums

$$\sum_{i=0}^{\infty} \frac{z^i}{i!} = e^z \tag{A.16}$$

$$\sum_{i=0}^{\infty} i \frac{z^i}{i!} = z e^z \tag{A.17}$$

$$\sum_{i=0}^{\infty} i^2 \frac{z^i}{i!} = (z + z^2) e^z \tag{A.18}$$

$$\sum_{i=0}^{\infty} \rho^i = \frac{1}{1-\rho} \tag{A.19}$$

$$\sum_{i=0}^{\infty} (i+1) \rho^i = \frac{1}{[1-\rho]^2} \tag{A.20}$$

$$\sum_{i=1}^{\infty} \rho^i = \frac{\rho}{1-\rho} \tag{A.21}$$

$$\sum_{i=1}^{\infty} i \rho^i = \frac{\rho}{(1-\rho)^2} \tag{A.22}$$

$$\sum_{i=1}^{\infty} i^2 \rho^i = \frac{\rho(1+\rho)}{(1-\rho)^3} \tag{A.23}$$

Partial Sums

$$\sum_{i=0}^{n} z^i = \frac{1-z^{n+1}}{1-z} \tag{A.24}$$

$$\sum_{i=1}^{n} z^i = z\left[\frac{1-z^n}{1-z}\right] \tag{A.25}$$

$$\sum_{i=1}^{n} iz^i = z\left[\frac{1-(n+1)z^n+nz^{n+1}}{(1-z)^2}\right] \tag{A.26}$$

$$\sum_{i=1}^{n} i^2 z^i = z\left[\frac{1+z-(n+1)^2 z^n+(2n^2+2n-1)z^{n+1}-n^2 z^{n+2}}{(1-z)^3}\right] \tag{A.27}$$

Limits (any z)

$$\lim_{n\to\infty}\left[1+\frac{z}{n}\right]^n = e^z \tag{A.28}$$

Appendix B
Complex Numbers

There are many excellent expositions of complex numbers and their applications; readers are referred to, for example, Hildebrand [140] or Boas [41]. Here we record some elementary features for convenience.

Let the basic imaginary unit be j:

$$\mathrm{j} = \sqrt{-1} \tag{B.1}$$

Z is complex; $(a, b, \theta, t, \omega, \phi)$ are real (Figure B.1).

The basic representation is in terms of real and imaginary parts:

$$Z = a + \mathrm{j}b \tag{B.2}$$

The complex conjugate is indicated by the asterisk:

$$Z^* = a - \mathrm{j}b \tag{B.3}$$

The magnitude $|Z|$ is real, positive:

$$|Z|^2 = a^2 + b^2 = ZZ^* \tag{B.4}$$

The polar representation is in terms of magnitude $|Z|$ and angle θ in the complex plane:

$$Z = |Z|e^{\mathrm{j}\theta} \qquad Z^* = |Z|e^{-\mathrm{j}\theta} \tag{B.5}$$

for magnitude $|Z|$ and real angle θ:

$$\tan\theta = \frac{b}{a} \tag{B.6}$$

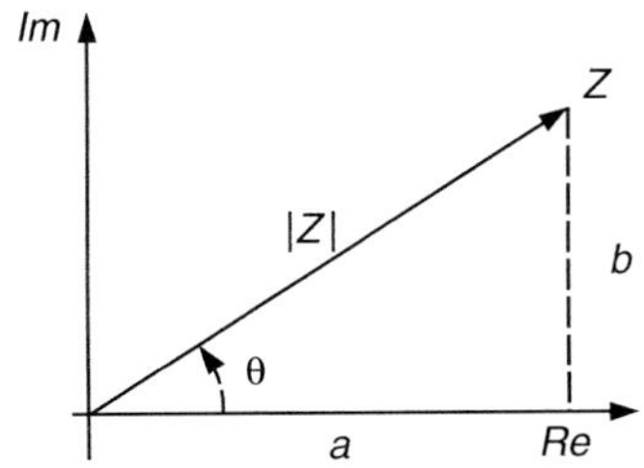

Figure B.1. Complex Z on real and imaginary axes.

The complex exponential:

$$e^{j\theta} = \cos\theta + j\sin\theta \tag{B.7}$$

$$|e^{j\theta}| = 1 \tag{B.8}$$

$$e^{\pm j2\pi} = 1 \tag{B.9}$$

$$e^{\pm j\pi} = -1 \tag{B.10}$$

$$e^{\pm j\pi/2} = \pm j \tag{B.11}$$

Periodic motions: for complex amplitude A, real frequency ω:

$$Z(t) = Ae^{j\omega t} \tag{B.12}$$

$A = |A|e^{j\phi}$ is the complex amplitude; $|A|$ is its size; ϕ is its real phase or angle in the complex plane.

$$Z(t) = |A|e^{j\phi}e^{j\omega t} \tag{B.13}$$

$$= |A|e^{j(\omega t+\phi)} \tag{B.14}$$

The trigonometric forms:

$$Z(t) = |A|\left[\cos(\omega t+\phi) + j\sin(\omega t+\phi)\right] \tag{B.15}$$

or in terms of the real and imaginary parts of A:

$$Z(t) = [Re(A)\cos\omega t - Im(A)\sin\omega t] \tag{B.16}$$

$$+ j[Im(A)\cos\omega t + Re(A)\sin\omega t] \tag{B.17}$$

The real part of $Z(t)$:

$$Re(Z(t)) = |A|\cos(\omega t+\phi) \tag{B.18}$$

$$= Re(A)\cos\omega t - Im(A)\sin\omega t \tag{B.19}$$

Some miscellaneous items :

$$\sqrt{j} = \pm\frac{1+j}{\sqrt{2}} \tag{B.20}$$

$$\frac{1}{Z} = \frac{Z^*}{|Z|^2} \qquad \left(\frac{1}{Z}\right)^* = \frac{Z}{|Z|^2} \tag{B.21}$$

(B.22)

$$Re(Z) = \frac{Z+Z^*}{2} \qquad Im(Z) = \frac{Z-Z^*}{2j} \tag{B.23}$$

Appendix C
Wiener Integrals

Increment and Integral. The Wiener *increment* is $d\mathcal{W}$. It is a random variable, Gaussian, with zero mean and variance accumulated over the infinitesimal interval dt:

$$d\mathcal{W} \sim \sqrt{dt}\,\mathcal{G}(0,1) \tag{C.1}$$

By definition, $d\mathcal{W}$ is independent of t and has zero autocorrelation. It is neither continuous nor differentiable; but it is integrable.

$\Delta\mathcal{W}$ is the Wiener *integral* over the finite interval τ:

$$\Delta\mathcal{W}_\tau = \int_t^{t+\tau} d\mathcal{W} \tag{C.2}$$

$\Delta\mathcal{W}$ is a Gaussian with zero mean and variance τ:

$$\Delta\mathcal{W}_\tau \sim \sqrt{\tau}\,G(0,1) \tag{C.3}$$

There is no time dependence nor is there any autocorrelation. $\Delta\mathcal{W}_\tau$ is the noise accumulated over the interval τ. The distribution $\Delta\mathcal{W}$ depends uniquely on the interval τ; it is the same regardless of the starting time t.

For nonoverlapping τ, the integrals add:

$$\Delta\mathcal{W}_{\tau+\tau} = \Delta\mathcal{W}_\tau + \Delta\mathcal{W}_\tau \tag{C.4}$$

$$\Delta\mathcal{W}_{2\tau} \sim \sqrt{\tau}\,\mathcal{G}_1(0,1) + \sqrt{\tau}\,\mathcal{G}_2(0,1) \tag{C.5}$$

$$\sim \sqrt{2\tau}\,\mathcal{G}(0,1) \tag{C.6}$$

Scaling. $\Delta\mathcal{W}$ integrates *unit* randomness. The result scales with the size S of the randomness:

$$Sd\mathcal{W} \sim S\sqrt{dt} \cdot \mathcal{G}(0,1) \tag{C.7}$$

$$\int_t^{t+\tau} Sd\mathcal{W} = S\Delta\mathcal{W}_\tau \tag{C.8}$$

$$\sim S\sqrt{\tau}\,\mathcal{G}(0,1) \tag{C.9}$$

Convolution. If we convolve a deterministic function $f(t)$ with $d\mathcal{W}$, the result scales with S_f, the size of f:

$$\int_{t'=0}^{\tau} f(\tau - t')d\mathcal{W} \sim S_f \Delta\mathcal{W}_\tau \tag{C.10}$$

$$\sim S_f\sqrt{\tau}\,G(0,1) \tag{C.11}$$

The convolution has zero mean and variance τS_f^2. S_f is the RMS value of $f(t)$ on the integration interval:

$$S_f^2 = \overline{f^2} = \frac{1}{\tau}\int_{t=0}^{\tau} f(t)^2 dt \tag{C.12}$$

Appendix D
Rates and Rate Limiters

Below we summarize some common expressions in ecological modeling. The base case is the widely used linear case. Several nonlinear functional forms are in common use. The Belehradek form indicates a general class of power-law dependencies [23]. Some more specific forms appear below. Fekedulegn et al. [142] discuss parameter estimation for several of these functions. An excellent exposition of ecological applications is Gotelli [174].

Base Case

$$\frac{dB}{dt} = -r(B-K) \tag{D.1}$$

This is the common linear "first-order" rate expression. There is a single parameter r. The solutions have a single equilibrium at $B = K$, stable for $r > 0$. In the form introduced in Section 1.2.2, let $B^{'}$ be the disequilibrium $B-K$:

$$\frac{1}{B^{'}}\frac{dB^{'}}{dt} = -r \tag{D.2}$$

The transient approach to this equilibrium is the exponential e^{-rt}; if B_0 is the initial condition,

$$B(t)-K = (B_0-K)e^{-rt} \tag{D.3}$$

Logistic

$$\frac{dB}{dt} = rB\left[\left(1-\frac{B}{K}\right)\right] \qquad \frac{1}{B}\frac{dB}{dt} = r\left[\left(1-\frac{B}{K}\right)\right] \tag{D.4}$$

This form has a stable equilibrium at $B = K$. For small B there is growth at the exponential rate e^{rt}. At high B the equilibrium is stable; departures from this equilibrium relax as e^{-rt}. There are two parameters, r and K. K is commonly referred to as "carrying capacity"; and r, the "growth rate constant". The rate limiter

$$L = \left[\left(1 - \frac{B}{K}\right)\right] \tag{D.5}$$

modulates between these extremes. If B_0 is the initial condition, then the approach to equilibrium is, in closed form,

$$\frac{B(t)}{K} = \frac{1}{1 + \left(\frac{K - B_0}{B_0}\right) e^{-rt}} \tag{D.6}$$

An equivalent expression is in terms of the disequilibrium $B' = K - B$:

$$\frac{dB'}{dt} = -rB'\left[\left(1 - \frac{B'}{K}\right)\right] \qquad \frac{1}{B'}\frac{dB'}{dt} = -r\left[\left(1 - \frac{B'}{K}\right)\right] \tag{D.7}$$

Generalized Logistic

Here an additional parameter ν is introduced into the basic logistic form,

$$\frac{1}{B}\frac{dB}{dt} = r\left[1 - \left(\frac{B}{K}\right)^{\nu}\right] \tag{D.8}$$

When $\nu = 1$ we recover the basic logistic case. The transient starting at B_0:

$$\frac{B(t)}{K} = \left[\frac{1}{1 + Qe^{-r\nu t}}\right]^{\frac{1}{\nu}} \tag{D.9}$$

with the parameter combination

$$Q \equiv -1 + \left[\frac{K}{B_0}\right]^{\nu} \tag{D.10}$$

(Figure D.1). This is sometimes referred to as the "Richards' Function." There are variants.

Double Logistic

This function has two sigmoid growth curves bonded at the origin; there is a positive and negative equilibrium, $B = \pm K$, with the origin unstable.

$$\frac{1}{B}\frac{dB}{dt} = r\left[\frac{1 - e^{-\left(\frac{B}{K}\right)^2}}{|B|}\right] \tag{D.11}$$

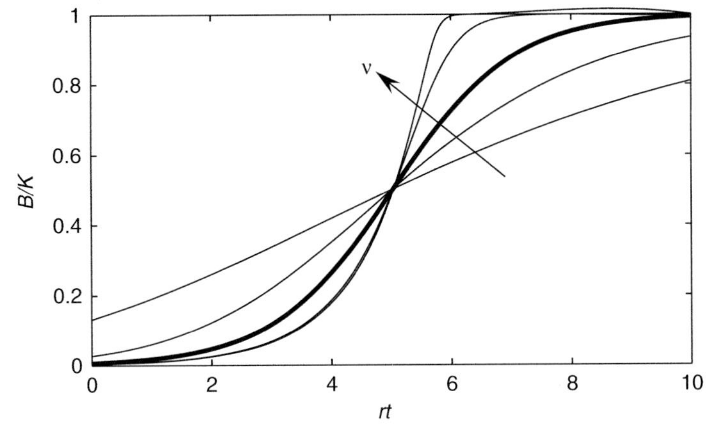

Figure D.1. Generalized logistic growth, Equation D.8, initiated to reach 50% of carrying capacity at $rt = 5$. The parameter $\nu = [0.25, 0.5, 1, 3, 10]$. $\nu = 1$ is the basic logistic case.

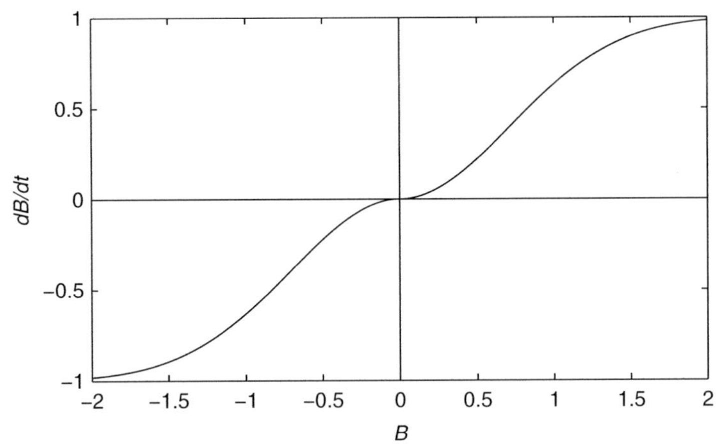

Figure D.2. Double logistic growth. $r = K = 1$.

Gompertz

The Gompertz form is the limit of the generalized logistic form, as $\nu \to \infty$

$$\frac{1}{B}\frac{dB}{dt} = r\left[\ln\left(\frac{K}{B}\right)\right] \tag{D.12}$$

There are the familiar two parameters (r, K). The transient response is

$$\frac{B(t)}{K} = e^{be^{-rt}} = \left(\frac{B_0}{K}\right)^{e^{-rt}} \tag{D.13}$$

with initial condition B_0 and parameter combination $e^b = \frac{B_0}{K}$

Bertalanffy

This form is actually the same as the base case, linear with exponential rise to carrying capacity K:

$$\frac{dB}{dt} = r[K - B] \tag{D.14}$$

$$= rK\left[1 - \frac{B}{K}\right] \tag{D.15}$$

It can be thought of as the "second part of logistic growth" with K taking the place of B in the formulation. But fundamentally, it is an instance of the base case with relaxation of initial B_0:

$$B(t) - K = (B_0 - K)e^{-rt} \tag{D.16}$$

Geometric Saturation

$$\frac{dB}{dt} = rB\left[\frac{1}{1+\frac{B}{B_h}}\right] \qquad \frac{1}{B}\frac{dB}{dt} = r\left[\frac{1}{1+\frac{B}{B_h}}\right] \tag{D.17}$$

there are two parameters, the rate constant r and the "half-saturation constant" B_h. At low B, $\frac{dB}{dt} = rB$ just as in the base case. At intermediate B: $\frac{dB}{dt}$ is lower, halved when $B = B_h$. At high B the rate saturates, $\frac{dB}{dt} = rB_h$, a constant. This form is the same as the popular Michaelis–Menten and Monod forms.

Exponential Saturation

$$\frac{dB}{dt} = R_0\left[1 - e^{-\frac{B}{K}}\right] \qquad \frac{1}{B}\frac{dB}{dt} = R_0\left[\frac{1 - e^{-\frac{B}{K}}}{B}\right] \tag{D.18}$$

this from is similar to the geometric saturation form. At low B asymptote we recover the base case, $\frac{dB}{dt} = \left(\frac{R_0}{K}\right)B$ The high B asymptote is a constant rate, fully saturated: $\frac{dB}{dt} = R_0$. This form is also referred to as the Ivlev form [159]

Mazard–Poulet Saturation

$$\frac{dB}{dt} = R_0 \left[\frac{B}{K}\right] \left[1 - e^{-\frac{B}{K}}\right] \qquad \frac{1}{B}\frac{dB}{dt} = R_0 \left[\frac{1 - e^{-\frac{B}{K}}}{K}\right] \tag{D.19}$$

This is another form of exponential saturation, but modulates between the base case asymptote at high B, $\frac{dB}{dt} = R_0(\frac{B}{K})$, and second-order growth at the low B asymptote, $\frac{dB}{dt} = R_0(\frac{B}{K})^2$ [159].

Appendix E
Diffusion Solutions

For solutions of the diffusion equation are heavily studied. Among the classic sources are Carslaw and Jaeger [69], Crank [90], and Van Genuchten and Alves [458]. The extension to the special case of moving boundaries (e.g., in phase change problems, ablation or freezing) are treated by Crank [91] and Finlayson [151].

The computational context presents many special problems, notably when diffusion is accompanied by advection. Problems of resolution and advection-dominance (high Peclet number) are covered in Baptista et al. (the Convective-Diffusive forum) [16] and Gresho and Sani [184].

Some simple solutions for point sources at the origin are reproduced below. These are diffusion-only solutions, useful for testing random walk models in highly idealized contexts. The diffusion coefficient is constant; the sources are concentrated at the origin only; the boundaries are at $\pm\infty$. The source nomenclature: M = Mass input at origin, an impulsive loading at $t=0$; $\dot{M}$ = Mass input rate at origin, beginning at $t=0$ and constant thereafter.

Summary: STEP and IMPULSE FUNCTIONS FOR DIFFUSION

1-D

$$C(x,t) = \frac{M}{\sqrt{4\pi Dt}} \exp\left(\frac{-x^2}{4Dt}\right) \qquad \text{(impulse)}$$

$$C(x,t) = \dot{M}\left[\sqrt{\frac{t}{\pi D}}\, exp\left(-\frac{x^2}{4Dt}\right) - \frac{|x|}{2D}\, erfc\left(\frac{|x|}{\sqrt{4Dt}}\right)\right] \qquad \text{(step)}$$

$$C(x,t) = C_0\, erfc\left(\frac{x}{\sqrt{4Dt}}\right) \qquad \text{(step in } C \text{ at origin)}$$

2-D

$$C(R,t) = \frac{M}{4\pi Dt} exp\left(\frac{-R^2}{4Dt}\right) \qquad \text{(impulse)}$$

$$C(R,t) = \frac{\dot{M}}{4\pi D}\left[-Ei\left(\frac{-R^2}{4Dt}\right)\right] = \frac{\dot{M}}{4\pi D} W\left(\frac{R^2}{4Dt}\right) \qquad \text{(step)}$$

3-D

$$C(r,t) = \frac{M}{(4\pi Dt)^{3/2}} exp\left(\frac{-r^2}{4Dt}\right) \qquad \text{(impulse)}$$

$$C(r,t) = \frac{\dot{M}}{4\pi rD} erfc\left(\frac{r}{\sqrt{4Dt}}\right) \qquad \text{(step)}$$

Appendix F

Covariance Matrix for Shear and Convergence

For a 2-D sheared and convergent Eulerian flow, the time-dependent covariance of particle positions $\widehat{X}, \widehat{Y}$ is developed in [289]. For the Eulerian velocity (u, v) we have a gradient in the x direction:

$$u = U_o - bx \qquad v = V_o + ax \tag{F.1}$$

Everything is uniform in the y direction. The mean motion satisfies

$$\frac{d\overline{X}}{dt} = -b\overline{X} + U_o \qquad \frac{d\overline{Y}}{dt} = a\overline{X} + V_o \tag{F.2}$$

The mean trajectory converges on constant $\overline{X}$ and $\overline{V}$

$$\overline{X} = \frac{U_o}{b} \qquad \frac{d\overline{Y}}{dt} = V_o + \frac{aU_o}{b} \tag{F.3}$$

Superposed on this mean motion are independent AR0 walks in x and in y.

$$\Delta\widehat{X} = -b\Delta t\widehat{X} + \mathcal{R}_x \qquad \Delta\widehat{Y} = a\Delta t\widehat{X} + \mathcal{R}_y \tag{F.4}$$

with independent and uncorrelated disturbances $\mathcal{R}$:

$$\mathcal{R}_x \sim \sqrt{2D_x\Delta t} \cdot \mathcal{G}(0,1) \qquad \mathcal{R}_y \sim \sqrt{2D_y\Delta t} \cdot \mathcal{G}(0,1) \tag{F.5}$$

Because of the underlying shear and/or convergence, covariance develops between $\widehat{X}$ and $\widehat{Y}$ – in other words, the off-diagonal terms in the covariance matrix grow.

Two dimensionless times emerge, $T = at$ and $\tau = bt$. $\mathcal{D}$ is the ratio of diffusivities $\frac{D_y}{D_x}$. For an isotropic random walk, $\mathcal{D} = 1$. β is b/a.

There are five useful forms.

Form 1: General:

$$Cov_{\widehat{X}\widehat{Y}} = 2D_x t \begin{bmatrix} \left[\frac{1-e^{-2\tau}}{2\tau}\right] & \frac{T}{2}\left[\frac{1+e^{-2\tau}-2e^{-\tau}}{\tau^2}\right] \\ \frac{T}{2}\left[\frac{1+e^{-2\tau}-2e^{-\tau}}{\tau^2}\right] & T^2\left[\frac{\tau+\frac{1}{2}(1-e^{-2\tau})-2(1-e^{-\tau})}{\tau^3}\right]+\mathcal{D} \end{bmatrix}$$

Form 2: General, with $b \neq 0$

$$Cov_{\widehat{X}\widehat{Y}} = \frac{D_x}{b}\begin{bmatrix} \left[1-e^{-2\tau}\right] & \frac{1}{\beta}\left[1+e^{-2\tau}-2e^{-\tau}\right] \\ \frac{1}{\beta}\left[1+e^{-2\tau}-2e^{-\tau}\right] & 2\left\{(\frac{1}{\beta})^2\left[\tau+\frac{1}{2}(1-e^{-2\tau})-2(1-e^{-\tau})\right]+\mathcal{D}\tau\right\} \end{bmatrix}$$

Form 3: In the small τ limit (small t or small b; $\tau = 0$ is the no-divergence case, $b=0$):

$$Cov_{\widehat{X}\widehat{Y}} = 2D_x t \begin{bmatrix} \left[1-\tau+\frac{2}{3}\tau^2-\frac{1}{3}\tau^3\ldots\right] & \frac{T}{2}\left[1-\tau+\frac{7}{12}\tau^2\ldots\right] \\ \frac{T}{2}\left[1-\tau+\frac{7}{12}\tau^2\ldots\right] & \frac{T^2}{3}\left[1-\frac{3}{4}\tau+\frac{7}{20}\tau^2\ldots\right]+\mathcal{D} \end{bmatrix}$$

Form 4: In the limit of large τ:

$$Cov_{\widehat{X}\widehat{Y}} = \begin{bmatrix} \frac{D_x}{b} & \frac{aD_x}{b^2} \\ \frac{aD_x}{b^2} & 2D_x t\frac{a^2}{b^2}\left[1-\frac{3}{2\tau}\right]+2D_y t \end{bmatrix}$$

Form 5: In the absence of shear and convergence, the familiar diagonal form is recovered:

$$Cov_{\widehat{X}\widehat{Y}} = 2D_x t \begin{bmatrix} 1 & 0 \\ 0 & \mathcal{D} \end{bmatrix}$$

Appendix G
Summary – ARn Recursions

Here is a summary of the general recursions, expressed in AR2 form; AR1 is the special case $\rho_2 = 0$.

Definitions:

$$\Delta\widehat{X}^k = \widehat{X}^k - \widehat{X}^{k-1} = \Delta t \widehat{V}^k \tag{G.1}$$

$$\widehat{X}^k = \sum_1^k \Delta\widehat{X}^k = \Delta t \sum_1^k \widehat{V}^k \tag{G.2}$$

Initialization:

$$0 = \widehat{X}^k = \widehat{V}^k = \mathcal{R}_1^k \qquad k \leq 0 \tag{G.3}$$

Randomness:

$$\mathcal{R}_1 \sim S_1 \sqrt{\Delta t} \cdot \mathcal{G}(0,1) \tag{G.4}$$

Recursion:

$$\widehat{V}^k = \rho_1 \widehat{V}^{k-1} + \rho_2 \widehat{V}^{k-2} + \mathcal{R}_1^k \tag{G.5}$$

Velocity Convolution:

$$\widehat{V}^N = \sum_{i=1}^{i=N} \mathcal{R}_1^i Z_{N-i} \tag{G.6}$$

$$Z_0 = 1 \tag{G.7}$$

$$Z_1 = \rho_1 \tag{G.8}$$

$$Z_j = \rho_1 Z_{j-1} + \rho_2 Z_{j-2} \tag{G.9}$$

Displacement Accumulation:

$$\widehat{X}^N = \Delta t \sum_{i=1}^{i=N} \mathcal{R}_1^i \; B^{N-i} \tag{G.10}$$

$$B^j = \sum_{k=0}^{k=j} Z_k \tag{G.11}$$

$$B^j = B^{j-1} + Z_j \tag{G.12}$$

$$B^j = 1 + \rho_1 B^{j-1} + \rho_2 B^{j-2} \tag{G.13}$$

$$B^0 = 1 \tag{G.14}$$

$$B^1 = 1 + \rho_1 \tag{G.15}$$

$$B^\infty = \frac{1}{[1 - \rho_1 - \rho_2]} \tag{G.16}$$

Variance:

$$\sigma_{\widehat{V}}^2 = \left[\frac{1}{1 - \rho_1^2 \left(\frac{1+\rho_2}{1-\rho_2} \right) - \rho_2^2} \right] \sigma_{\mathcal{R}_1}^2 \tag{G.17}$$

$$\Delta \sigma_{\widehat{X}}^{2^{k+1}} = \Delta t^2 \sigma_{\mathcal{R}_1}^2 \left[B^k \right]^2 \tag{G.18}$$

Velocity Covariance:

$$C_i \equiv E \left[\widehat{V}^k \widehat{V}^{k-i} \right] \tag{G.19}$$

$$C_0 = \sigma_{\widehat{V}}^2 \tag{G.20}$$

$$C_1 = \left[\frac{\rho_1}{1 - \rho_2} \right] C_0 \tag{G.21}$$

$$C_n = \rho_1 C_{n-1} + \rho_2 C_{n-2} \tag{G.22}$$

Appendix H

Distribution Properties for Linear Triangles

We consider one triangle at a time; everything is local. Continuity among elements is guaranteed.

Bases and Their Integrals

Let $\phi_j(\mathbf{x})$ denote the usual local linear bases, $j = 1 \rightarrow 3$, for an individual element, area A. Each varies smoothly (linearly) over the element between 0 and 1; the sum of the three bases is constrained. $< \;\; >$ indicates integration over a single element.

$$0 \leq \phi_j \leq 1 \tag{H.1}$$

$$\phi_1 + \phi_2 + \phi_3 = 1 \tag{H.2}$$

$$\tag{H.3}$$

$$< 1 > = A \tag{H.4}$$

$$< \phi_j > = A/3 \tag{H.5}$$

$$< \phi_j \phi_i > = A/6 \qquad (i = j) \tag{H.6}$$

$$= A/12 \qquad (i \neq j) \tag{H.7}$$

$$\tag{H.8}$$

$$< \phi_1^l \phi_2^m \phi_3^n > = 2A \frac{l!\, m!\, n!}{(l+m+n+2)!} \tag{H.9}$$

See for example [279]. Any two of the ϕ_j constitute a local coordinate system, suitable for mapping from local to global coordinates, and for routine linear interpolation. If $\mathbf{x}_j$ is the global location of node j, then the mapping within the element is

$$\mathbf{x} = \sum_1^3 \mathbf{x}_j \phi_j \tag{H.10}$$

This is a continuous, 1:1 mapping; every point in the local ϕ space has a unique partner in the global space.

For interpolation: if s_j is the value of a scalar field at node j, and $\mathbf{v}_j$ the nodal value of a vector field, then the interpolation on the element is

$$s = \sum_1^3 s_j \phi_j \qquad \mathbf{v} = \sum_1^3 \mathbf{v}_j \phi_j \tag{H.11}$$

All interpolants will be continuous across element boundaries. Higher order bases can be expressed in terms of these linear ϕ.

Particle k's location in local coordinates is $(\phi_{1k}, \phi_{2k}, \phi_{3k})$.[1] The corresponding global location is

$$\mathbf{X}_k = \sum_{j=1}^3 \mathbf{x}_j \phi_{jk} \tag{H.12}$$

Moments

The moments of a distribution of particles on a single triangle are easy to acquire, using the standard integration results above. They appear below, for the uniform and linear distributions.

Some definitions: The mean or expectation of $a(\mathbf{x})$ is the weighted average

$$\bar{a} = E(a) \equiv < a\, \nu > / < \nu > \tag{H.13}$$

$\nu(\mathbf{x})$ is the intensity function for the occurrence of a particle. (Normalized, it is the probability density function.) It is constant for the uniform distribution; and interpolated among the nodal values ν_i in the linear case.

$\tilde{a}$ is the departure from expectation and the familiar relations hold:

$$a = \bar{a} + \tilde{a} \tag{H.14}$$

$$E(\tilde{a}) = 0 \tag{H.15}$$

$$E(a\, b) = E\left(\tilde{a}\, \tilde{b}\right) + \bar{a}\, \bar{b} \tag{H.16}$$

$$E(a\, b\, c) = E\left(\tilde{a}\, \tilde{b}\, \tilde{c}\right) + \bar{a}\, \bar{b}\, \bar{c} + \bar{a}\, E\left(\tilde{b}\, \tilde{c}\right) + \bar{b}\, E(\tilde{c}\, \tilde{a}) + \bar{c}\, E\left(\tilde{a}\, \tilde{b}\right) \tag{H.17}$$

The location of a particular particle k by given by its three local coordinates ϕ_{jk}. For an ensemble of particles, the moments of all particle locations are given the familiar

[1] ϕ_{jk} is "the value of basis function ϕ_j where particle k lies."

Table H.1. *Moments of Particle Distributions on a Triangle*

1	1	1	ν_j	1	0	0
$\frac{1}{3}$	$\frac{1}{3}$	$\frac{1}{3}$	μ_j	$\frac{1}{2}$	$\frac{1}{4}$	$\frac{1}{4}$
$\frac{1}{18}$	$\frac{1}{18}$	$\frac{1}{18}$	σ_j^2	$\frac{4}{80}$	$\frac{3}{80}$	$\frac{3}{80}$
$\frac{1}{135}$	$\frac{1}{135}$	$\frac{1}{135}$	$\overline{\widetilde{\phi}_j^3}$	0	$\frac{1}{160}$	$\frac{1}{160}$
$\frac{2\sqrt{2}}{5}$	$\frac{2\sqrt{2}}{5}$	$\frac{2\sqrt{2}}{5}$	sk_j	0	$\sqrt{\frac{20}{27}}$	$\sqrt{\frac{20}{27}}$
1	$-\frac{1}{2}$	$-\frac{1}{2}$		1	$-\frac{1}{\sqrt{3}}$	$-\frac{1}{\sqrt{3}}$
$-\frac{1}{2}$	1	$-\frac{1}{2}$	$[r]$	$-\frac{1}{\sqrt{3}}$	1	$-\frac{1}{3}$
$-\frac{1}{2}$	$-\frac{1}{2}$	1		$-\frac{1}{\sqrt{3}}$	$-\frac{1}{3}$	1

Note: Left: Uniform Distribution; Right: Unit linear distribution. $\nu(\mathbf{x}) = \sum \nu_j \phi_k(\mathbf{x})$ is the unnormalized probability density function.

statistical names:

$$\mu_j = E(\phi_j) \tag{H.18}$$

$$\sigma_j^2 = E\left(\widetilde{\phi}_j^2\right) \tag{H.19}$$

$$sk_j = \frac{E\left(\widetilde{\phi}_j^3\right)}{\sigma_j^3} \tag{H.20}$$

$$Cov_{ij} = E\left(\widetilde{\phi}_i \widetilde{\phi}_j\right) \tag{H.21}$$

$$r_{ij} = \frac{E\left(\widetilde{\phi}_i \widetilde{\phi}_j\right)}{\sigma_i \sigma_j} \tag{H.22}$$

sk_j is the skew coefficient; $[Cov]$ is the covariance matrix; $[r]$ is its normalization. Indices i and j run from 1 to 3.

Results: Routine integration over a triangle produces the distribution moments shown in Table H.1, for two particular cases: Uniform ($\nu_j = [1,1,1]$) and a "unit" linear distribution ($\nu_j = [1,0,0]$). For these cases, all coordinate distributions have non-negative skew. The negative correlation reflects the nature of the constraints on position coordinates (Equations H.1, H.2).

For general linear distributions of particles, the intensity is a weighted sum of "unit" distributions as computed above: $\nu(\mathbf{x}) = \sum_{j=1}^{3} \nu_j \phi_j(\mathbf{x})$. The moments will depend on the specifics of the ν_j:

$$E(\phi_j) = \frac{1}{12}\left(3 + \frac{\nu_j}{\nu_e}\right) = \mu_j \tag{H.23}$$

$$E(\phi_j^2) = \frac{2}{60}\left(3 + \frac{2\nu_j}{\nu_e}\right) = \sigma_j^2 + \mu_j^2 \tag{H.24}$$

$$E(\phi_j \phi_i) = \frac{1}{60}\left(3 + \frac{\nu_j + \nu_i}{\nu_e}\right) \qquad i \neq j \tag{H.25}$$

where $\nu_e = (\nu_1 + \nu_2 + \nu_3)/3$ is the average ν on the element. Generally for all possible nodal values $\nu_j \geq 0$, we have the range

$$0 \leq \frac{\nu_j}{\nu_e} \leq 3 \tag{H.26}$$

The lower limit is when $\nu_j = 0$; the upper limit is when ν_j is huge compared with neighbors. For uniform ν, the ratio is unity. (The range of this ratio is one metric of resolution.)

Bibliography

[1] A.D. Abrahams and P. Gao. A bed-load transport model for rough turbulent open-channel flows on plane beds. *Earth Surface Processes and Landforms*, 31(7):910–928, 2006.

[2] M. Abramowitz and I.A. Stegun. *Handbook of Mathematical Functions: with Formulas, Graphs, and Mathematical Tables*. Dover, 1965.

[3] J.P. Ahrens. A fall-velocity equation. *Journal of Waterway, Port, Coastal, and Ocean Engineering*, 126(2):99–102, 2000.

[4] D.M. Anderson, B.A. Keafer, D.J. McGillicuddy Jr., M.J. Mickelson, K.E. Keay, P.S. Libby, J.P. Manning, C.A. Mayo, D.K. Whittaker, J.M. Hickey, R. He, D.R. Lynch, and K.W. Smith. Initial observations of the 2005 Alexandrium fundyense bloom in southern New England: General patterns and mechanisms. *Deep-Sea Research, Part II*, 52:2856–2876, 2005.

[5] D.M. Anderson, D.W. Townsend, D.J. McGillicuddy Jr., and J.T. Turner. The ecology and oceanography of toxic *Alexandrium fundyense* blooms in the Gulf of Maine. *Deep-Sea Research, Part II*, 52(19–21), 2005. Special Issue.

[6] D.M. Anderson. Bloom dynamics of toxic Alexandrium species in the northeastern US. *Limnology and Oceanography*, 42(5):1009–1022, 1997.

[7] B.K. Arbic, P. St-Laurent, G. Sutherland, and C. Garrett. On the resonance and influence of the tides in Ungava Bay and Hudson Strait. *Geophysical Research Letters*, 34(17):L17606, 2007.

[8] A. Aretxabaleta, J. Manning, F.E. Werner, K. Smith, B.O. Blanton, and D.R. Lynch. Data assimilative hindcast on the southern flank of Georges Bank during May 1999: Frontal circulation and implications. *Continental Shelf Research*, 25(7–8):849–874, 2005.

[9] A.L. Aretxabaleta, D.J. McGillicuddy Jr., K.W. Smith, and D.R. Lynch. Model simulations of the Bay of Fundy gyre: 1. Climatological results. *Journal of Geophysical Research*, 113(C10027), 2008.

[10] A.L. Aretxabaleta, B. Butman, R.P. Signell, P.S. Dalyander, C.R. Sherwood, V.A. Sheremet, and D.J. McGillicuddy Jr. Near-bottom circulation and dispersion of sediment containing alexandrium fundyense cysts in the Gulf of Maine during 2010–2011. *Deep Sea Research Part II: Topical Studies in Oceanography*, 103:96–111, 2014.

[11] A.L. Aretxabaleta, D.J. McGillicuddy, K.W. Smith, J.P. Manning, and D.R. Lynch. Model simulations of the Bay of Fundy gyre: 2. hindcasts for 2005–2007 reveal interannual variability in retentiveness. *Journal of Geophysical Research: Oceans (1978–2012)*, 114(C9), 2009.

[12] P. Armstrong. *Copper, Zinc, Chromium, Lead, and Cadmium in the Unconsolidated Sediments of Great Bay Estuary, New Hampshire*. PhD thesis, University of New Hampshire, 1974.

[13] P.B. Armstrong, G.M. Hanson, and H.E. Gaudette. Minor elements in sediments of Great Bay Estuary, New Hampshire. *Environmental Geology*, 1(4):207–214, 1976.

[14] Task Committee on Modeling of Oil Spills ASCE. State-of-the-art review of modeling transport and fate of oil spills. *Journal of Hydraulic Engineering*, 122(11):594–609, 1996.

[15] T.E. Baldock, M.R. Tomkins, P. Nielsen, and M.G. Hughes. Settling velocity of sediments at high concentrations. *Coastal Engineering*, 51(1):91–100, 2004.

[16] A.M. Baptista, E.E. Adams, and P. Gresho. Benchmarks for the transport equation: The convection-diffusion forum and beyond. In D.R. Lynch and A.M. Davies, ed., *Quantitative Skill Assessment for Coastal Ocean Models*, pp. 241–268. Coastal and Estuarine Studies, No. 47. pages 241–268. American Geophysical Union, 1995.

[17] C.N. Barron, L.F. Smedstad, J.M. Dastugue, and O.M. Smedstad. Evaluation of ocean models using observed and simulated drifter trajectories: Impact of sea surface height on synthetic profiles for data assimilation. *Journal of Geophysical Research: Oceans*, 112(C7), 2007.

[18] H.P. Batchelder. Forward-in-time-/backward-in-time-trajectory (fitt/bitt) modeling of particles and organisms in the coastal ocean. *Journal of Atmospheric and Oceanic Technology*, 23(5):727–741, 2006.

[19] H.P. Batchelder and C.B. Miller. Life history and population dynamics of *Metridia pacifica*: Results from simulation modelling. *Ecological Modeling*, 113–136, 1989.

[20] H.P. Batchelder and R. Williams. Individual-based modelling of the population dynamics of *Metridia lucens* in the North Atlantic. *ICES Journal of Marine Science*, 469–482, 1995.

[21] E.J. Baum. *Chemical Property Estimation. Theory and Application*. CRC Press, 1998.

[22] H. Baumert, J. Simpson, and J. Sundermann, eds. *Marine Turbulence: Theories, Observations and Methods Results of the CARTUM Project*. Cambridge University Press, 2005.

[23] J. Belehradek. Physiological aspects of heat and cold. *Annual Review of Physiology*, 59–82, 1957.

[24] A. Bennett. *Lagrangian Fluid Mechanics*. Cambridge University Press, 2006.

[25] P.S. Berloff and J.C. McWilliams. Material transport in oceanic gyres. Part II: Hierarchy of stochastic models. *Journal of Physical Oceanography*, 32:797–830, 2002.

[26] A. Berry, T. Dabrowski, and K. Lyons. The oil spill model OILTRANS and its application to the Celtic Sea. *Marine Pollution Bulletin*, 64(11):2489–2501, 2012.

[27] F. Betancourt, A. Palacio, and A. Rodriguez. Effects of the mass transfer process in oil spill. *American Journal of Applied Sciences*, 2(5):939–946, 2005.

[28] R.J.H. Beverton and S.J. Holt. On the dynamics of exploited fish populations. Fisheries Investigation Series II, Volume XIX, Ministry of Agriculture, Fisheries and Food, 1957.

[29] M.O. Bezerra, M. Diez, C. Medeiros, A. Rodriguez, E. Bahia, A. Sanchez-Arcilla, and J.M. Redondo. Study on the influence of waves on coastal diffusion using image analysis. *Applied Scientific Research*, 59(2–3):191–204, 1997.

[30] H.B. Bigelow. *Physical Oceanography of the Gulf of Maine*. US Government Printing Office, 1927.

[31] A. Bilgili, J.A. Proehl, D.R. Lynch, K.W. Smith, and M.R. Swift. Estuary/ocean exchange and tidal mixing in a Gulf of Maine Estuary: A Lagrangian modeling study. *Estuarine, Coastal, and Shelf Science*, 65:607–624, 2005.

[32] A. Bilgili, K.W. Smith, and D.R. Lynch. BatTri: A two-dimensional bathymetry-based unstructured triangular grid generator for finite element circulation modeling. *Computers and Geosciences*, 32:632–642, 2006.
[33] A. Bilgili, M.R. Swift, D.R. Lynch, and J.T.C Ip. Modeling bed-load transport of coarse sediments in the Great Bay Estuary, New Hampshire. *Estuarine, Coastal, and Shelf Science*, 58:937–950, 2003.
[34] A. Bilgili, M.R. Swift, and B. Celikkol. Shoal formation in the Piscataqua River, New Hampshire. *Estuaries*, 19(3):518–525, 1996.
[35] A.K. Blackadar. The vertical distribution of wind and turbulence exchange in a neutral atmosphere. *Journal of Geophysical Research*, 67:3095–3102, 1962.
[36] P.C. Blokker. Spreading and evaporation of petroleum products on water. In *Proceedings of the 4th Harbor Conference*, Antwerp, pp. 913–919, 1964.
[37] A.F. Blumberg, B. Galperin, and D.J. O'Connor. Modeling vertical structure of open-channel flows. *ASCE Journal of Hydraulic Engineering*, 118:1119–1134, 1992.
[38] A.F. Blumberg. Numerical modeling of estuarine circulation. *Journal of Hydraulics Division ASCE*, 103:295–310, 1977.
[39] A.F. Blumberg. Turbulent mixing processes in lakes, reservoirs, and impoundments. In W.G. Gray, ed. *Physics-Based Modeling of Lakes, Reservoirs, and Impoundments*, pp. 79–104. ASCE, 1986.
[40] A.F. Blumberg and G.L. Mellor. A description of a three-dimensional coastal ocean circulation model. In N.S. Heaps, ed, *Three-Dimensional Coastal Models*, Vol. 4 of Coastal and Estuarine Series, pp. 1–16. American Geophysical Union, 1987.
[41] M.L. Boas. *Mathematical Methods in the Physical Sciences*, 3rd edn. Wiley, 2005.
[42] J. Bogardi. *Sediment Transport in Alluvial Streams*. Akademiai Kiado Budapest, 1974.
[43] W.C. Boicourt, S.-Y. Chao, H.W. Ducklow, P.M. Glibert, T.C. Malone, M.R. Roman, L.P. Sanford, J.A. Fuhrman, C. Garside, and R.W. Garvine. Physics and microbial ecology of a buoyant estuarine plume on the continental shelf. *Eos, Transactions American Geophysical Union*, 68(31):666–668, 1987.
[44] L. Bonoli and P.A. Witherspoon. Diffusion of aromatic and cycloparaffin hydrocarbons in water from 2 to 60.deg. *The Journal of Physical Chemistry*, 72(7):2532–2534, 1968.
[45] M.S. Borgas, T.K. Flesch, and B.L. Sawford. Turbulent dispersion with broken reflectional symmetry. *Journal of Fluid Mechanics*, 332:141–156, 1997.
[46] M.S. Borgas and B.L. Sawford. A family of stochastic models for two-particle dispersion in isotropic homogeneous stationary turbulence. *Journal of Fluid Mechanics*, 279:66–99, 1994.
[47] K.F. Bowden. Turbulent process estuaries. In C.B. Officer, ed. *Estuaries, Geophysics, and the Environment*, Studies in Geophysics, pp. 46–56. National Academy of Sciences, 1977.
[48] K.F. Bowden and P. Hamilton. Some experiments with a numerical model of circulation and mixing in a tidal estuary. *Estuarine and Coastal Marine Science*, 3:281–301, 1975.
[49] G.E.P. Box, G.M. Jenkins, and G.C. Reinsel. *Time Series Analysis: Forecasting and Control* , 4th ed. John Wiley & Sons, 2008.
[50] P. Bradshaw and G.P. Huang. The law of the wall in turbulent flow. *Proceedings of the Mathematical and Physical Sciences*, 451(1941):65–188, 1995. Osborne Reynolds Centenary Volume (Oct. 9, 1995).
[51] F.P. Bretherton, R.E. Davis, and C.B. Fandry. A technique for objective analysis and design of oceanographic experiments applied to MODE-73. *Deep Sea Research*, pp. 559–582, 1976.

[52] D. Brickman and P.C. Smith. Lagrangian stochastic modeling in coastal oceanography. *Journal of Atmospheric and Oceanic Technology*, 19:83–99, 2002.

[53] D. Brickman, B. Adlandsvik, U.H. Thygesen, C. Parada, K. Rose, A.J. Hermann, and K. Edwards. Particle tracking. In E. Watkins North, A. Gallego, and P. Petitgas, editors, *Manual of Recommended Practices for Modelling Physical-Biological Interactions during Fish Early Life*, number 295 in ICES Cooperative Research Reports, pp. 111. International Council for the Exploration of the Sea, No. 295. Copenhagen, Denmark, May 2009.

[54] K.H. Brink and A.R. Robinson, eds. *The Global Coastal Ocean – Processes and Methods*. The Sea, No. 10. John Wiley & Sons, 1998.

[55] N.F. Britton. *Essential Mathematical Biology*. Springer, 2003.

[56] D.A. Brooks. A model study of the buoyancy-driven circulation in the Gulf of Maine. *Journal of Physical Oceanography*, 24(11):2387–2412, 1994.

[57] D.A. Brooks and D.W. Townsend. Variability of the coastal current and nutrient pathways in the eastern Gulf of Maine. *Journal of Marine Research*, 47(2):303–321, 1989.

[58] I. Buchanan and N. Hurford. Methods for predicting the physical changes in oil spilt at sea. *Oil and Chemical Pollution*, 4(4):311–328, 1988.

[59] D.F. Bumpus. A description of the circulation on the continental shelf of the East Coast of the United States. *Progress in Oceanography*, 6:111–157, 1973.

[60] D.F. Bumpus and L.M. Lauzier. Surface circulation on the continental shelf off eastern North America between Newfoundland and Florida. In W. Webster, ed., *Serial Atlas of the Marine Environment*, volume folio 7, p. 4. American Geographic Society, 1965.

[61] H. Burchard, K. Bolding, and M.R. Villarreal. GOTM – A General Ocean Turbulence Model: Theory, Implementation, and Test Cases. Technical Report EUR 18745 EN, European Commission, 1999.

[62] H. Burchard, K. Boulding, M. Ruiz Villarreal, T. Rippeth, N. Fisher, and A. Stips. The GOTM modeling system. In *Marine Turbulence: Theories, Observations and Methods Results of the CARTUM Project*, pp. 213–224. Cambridge University Press, 2005.

[63] H. Burchard and K. Bolding. Comparative analysis of four second-moment turbulence closure models for the oceanic mixed layer. *Journal of Physical Oceanography*, 31(8):1943–1968, 2001.

[64] L.G. Burns. *Tracking Trash: Flotsam, Jetsam, and the Science of Ocean Motion*. Scientists in the Field. Houghton Mifflin, 2007.

[65] B. Camenen. Simple and general formula for the settling velocity of particles. *Journal of Hydraulic Engineering*, 133(2):229–233, 2007.

[66] R. Camilli, C.M. Reddy, D.R. Yoerger, B.A.S. Van Mooy, M.V. Jakuba, J.C. Kinsey, C.P. McIntyre, S.P. Sylva, and J.V. Maloney. Tracking Hydrocarbon Plume Transport and Biodegradation at Deepwater Horizon. *Science*, 330(6001):201–204, 2010.

[67] G.F. Carey. Mesh Generation, a posteriori error estimation, and mesh refinement. In D.R. Lynch and A.M. Davies, eds., *Quantitative Skill Assessment for Coastal Ocean Models*. Coastal and Estuarine Studies, No. 47, pp. 15–29. American Geophysical Union, 1995.

[68] F. Carlotti and G. Radach. Seasonal dynamics of phytoplankton and Calanus finmarchicus in the North Sea as revealed by a coupled one-dimensional model. *Limnology and Oceanography*, 41(3):522–539, 1996.

[69] H.S. Carslaw and J.C. Jaeger. *Conduction of Heat in Solids*. Oxford University Press, 1946.

[70] V. Casulli. A high-resolution wetting and drying algorithm for free-surface hydrodynamics. *International Journal for Numerical Methods in Fluids*, 60(4):391–408, 2009.

[71] H.M. Cekirge and S.L. Palmer. Mathematical modeling of oil spilled into marine waters. In C.A. Brebbia, ed., *Oil Spill Modelling and Processes*, pp. 1–21. WIT Press, 2001.
[72] H.M. Cekirge, M. Koch, C. Long, C.P. Giammona, K. Binkley, R. Engelhardt, and R. Jamail. State-of-the-art techniques in oil spill modeling. In *International Oil Spill Conference*, Vol. 1, pp. 67–72. American Petroleum Institute, 1995.
[73] H. Chamley. *Sedimentology*. Springer-Verlag, 1990.
[74] C. Chan-Braun, M. Garcia-Villalba, and M. Uhlmann. Direct numerical simulation of sediment transport in turbulent open channel flow. In *High Performance Computing in Science and Engineering'10*, pp. 295–306. Springer, 2011.
[75] Y. Chang, D. Hammond, A.C. Haza, P. Hogan, H.S. Huntley, A.D. Kirwan Jr., B.L. Lipphardt Jr., V. Taillandier, A. Griffa, and T.M. Ozgokmen. Enhanced estimation of sonobuoy trajectories by velocity reconstruction with near-surface drifters. *Ocean Modelling*, 36(3-4):179–197, 2011.
[76] X. Chao, N.J. Shankar, and H.F. Cheong. Two- and three-dimensional oil spill model for coastal waters. *Ocean Engineering*, 28(12):1557–1573, 2001.
[77] D.C. Chapman and S.J. Lentz. Trapping of a coastal density front by the bottom boundary layer. *Journal of Physical Oceanography*, 24(7):1464–1479, 1994.
[78] S.C. Chapra. *Surface Water-Quality Modeling*. McGraw-Hill, 1997.
[79] C. Chatfield. *The Analysis of Time Series: An Introduction*, 6th ed. Chapman & Hall/CRC, 2003.
[80] F. Chen and P. Yapa. Estimating the oil droplet size distributions in deepwater oil spills. *Journal of Hydraulic Engineering*, 133(2):197–207, 2007.
[81] N.-S. Cheng. Simplified settling velocity formula for sediment particle. *Journal of Hydraulic Engineering*, 123(2):149–152, 1997.
[82] N.-S. Cheng and Y.M. Chiew. Pickup probability for sediment entrainment. *Journal of Hydraulic Engineering*, 124(2):232–235, 1998.
[83] C.W. Clark. *Mathematical Bioeconomics: The Optimal Management of Renewable Resources*. Wiley, 2nd edn, 1990.
[84] C.W. Clark and M. Mangel. *Dynamic State Variable Models in Ecology: Methods and Applications*. Oxford University Press, 2000.
[85] A.K. Coker. *Ludwig's Applied Process Design for Chemical and Petrochemical Plants*, Vol. 1 of *Chemical, Petrochemical & Process*, 4th ed. Gulf Professional, Publications, 2007.
[86] C. Coulliette, F. Lekien, J.D. Paduan, G. Haller, and J.E. Marsden. Optimal pollution mitigation in Monterey Bay Based on coastal radar data and nonlinear dynamics. *Environmental Science & Technology*, 41(18):6562–6572, 2007.
[87] R.K. Cowen, C.B. Paris, and A. Srinivasan. Scaling of connectivity in marine populations. *Science*, 311(5760):522–527, 2006.
[88] R.K. Cowen, G. Gawarkiewicz, J. Pineda, S.R. Thorrold, and F.E. Werner. Population connectivity in marine systems: An overview. *Oceanography*, 20(3):14–21, 2007.
[89] R.K. Cowen and S. Sponaugle. Larval dispersal and marine population connectivity. *Annual Review of Marine Science*, 1(1):443–466, 2009.
[90] J. Crank. *The Mathematics of Diffusion*. Oxford University Press, 1956.
[91] J. Crank. *Free and Moving Boundary Problems*. Oxford University Press, 1984.
[92] N.A.C. Cressie. *Statistics for Spatial Data*. John Wiley & Sons, 1991.
[93] G.T. Csanady. The arrested topographic wave. *Journal of Physical Oceanography*, 8(1):47–62, 1978.
[94] G.T. Csanady. *Turbulent Diffusion in the Environment*. Geophysics and Astrophysics Monographs 3. D. Reidel, 1973.
[95] G.T. Csanady. *Circulation in the Coastal Ocean*. D. Reidel, 1982.

[96] G.T. Csanady. *Air-Sea Interaction – Laws and Mechanisms*. Cambridge University Press, 2001.

[97] L.K. Dasanayaka and P.D. Yapa. Role of plume dynamics phase in a deepwater oil and gas release model. *Journal of Hydro-environment Research*, 2(4):243–253, 2009.

[98] E.A. D'Asaro. Performance of Autonomous Lagrangian floats. *Journal of Atmospheric and Oceanic Technology*, 20(6):896–911, 2003.

[99] A.M. Davies. Spectral models in continental shelf sea oceanography. In N.S. Heaps, ed., *Three-Dimensional Coastal Ocean Models*, pp. 71–106. Coastal and Estuarine Series 4. American Geophysical Union, 1987.

[100] A.M. Davies and G.K. Furnes. Observed and computed M2 tidal currents in the North Sea. *Journal of Physical Oceanography*, 10:237–257, 1980.

[101] A.M. Davies and J. Xing. An intercomparison and validation of a range of turbulence closure schemes used in three dimensional tidal models. In D.R. Lynch and A.M. Davies, eds. *Quantitative Skill Assessment for Coastal Ocean Models*. Coastal and Estuarine Studies, No. 47, pp. 71–95. American Geophysical Union, 1995.

[102] M. Davies. Psed 4.3, a Lagrangian sediment transport model, technical documentation. Document release 1.3 published as an annex to the Sedsim reference manual, Canadian Hydraulics Center, National Research Council of Canada, Ottawa, ON, Canada, February 2006.

[103] R. Davis. Drifter observations of coastal surface currents during code: The method and descriptive view. *Journal of Geophysical Research*, 90:4741–4755, 1985.

[104] R.E. Davis. Oceanic property transport, Lagrangian particle statistics, and their prediction. *Journal of Marine Research*, 41(1):163–194, 1983.

[105] R.E. Davis, C.E. Eriksen, and C.P. Jones. Autonomous buoyancy-driven underwater vehicle gliders. In G. Griffiths, editor, *Technology and Applications of Autonomous Underwater Vehicles*, pp. 37–62. Taylor & Francis, 2003.

[106] R.G. Dean and R.A. Dalrymple. *Coastal Processes with Engineering Applications*. Cambridge University Press, 2004.

[107] D. DeAngelis and L. Gross, ed. *Individual-Based Models and Approaches in Ecology*. Coastal and Estuarine Studies. Chapman and Hall, 1992.

[108] D.L. DeAngelis and W.M. Mooij. Individual-based modeling of ecological and evolutionary processes. *Annual Review of Ecology, Evolution, and Systematics*, 36:147–168, 2005.

[109] A. Defant. *Physical Oceanography*, Vol. 1. Pergamon, 1961.

[110] E. Deleersnijder and P. Luyten. On the practical advantages of the quasi-equilibrium version of the mellor and yamada level 2.5 turbulence closure applied to marine modeling. *Applied Mathematical Modeling*, 18:281–287, 1994.

[111] E.J.M. Delhez and E. Deleersnijder. Overshootings and spurious oscillations caused by biharmonic mixing. *Ocean Modelling*, 17(3):183–198, 2007.

[112] G.A.L. Delvigne and C.E. Sweeney. Natural dispersion of oil. *Oil and Chemical Pollution*, 4(4):281–310, 1988.

[113] G.A.L. Delvigne. Natural dispersion of oil by different sources of turbulence. In *International Oil Spill Conference*, pp. 415–419. American Petroleum Institute, March 1993.

[114] J.W. Demmel. *Applied Numerical Linear Algebra*. SIAM, 1997.

[115] L. Devroye. *Non-Uniform Random Variate Generation*. Springer-Verlag, 1986. Out of print; available at http://luc.devroye.org/rnbookindex.html.

[116] W.E. Dietrich. Settling velocity of natural particles. *Water Resources Research*, 18(6):1615–1626, 1982.

[117] K.N. Dimou and E.E. Adams. A random-walk, particle tracking model for well-mixed estuaries and coastal waters. *Estuarine, Coastal, and Shelf Science*, 37:99–110, 1993.

[118] F.T. Dodge, J.T. Park, J.C. Buckingham, and R.J. Magott. Revision and experimental verification of the hazard assessment computer system models for spreading, movement, dissolution, and dissipation of insoluble chemicals spilled onto water. volume 1. Contractor report CG-D-35-83, DTIC Document, June 1983.

[119] DoE. Crude oil analysis database. Website http://www.netl.doe.gov/technologies/oil-gas/Software/database.html, 11 2013. retrieved 11/22/2001.

[120] B. Dunnewind, M.A. Bos, and W. Koops. Entrainment of oil from oil spills into the water column: A new theory. In *International Oil Spill Conference*, pp. 1059–1066. American Petroleum Institute, 2003.

[121] F. Dupont, C.G. Hannah, and D.A. Greenberg. Modeling the sea level in the Upper Bay of Fundy. *Atmosphere-Ocean*, 43(1):33–47, 2005.

[122] O. Duran, B. Andreotti, and P. Claudin. Numerical simulation of turbulent sediment transport, from bed load to saltation. *Physics of Fluids*, 24(10):103306, 2012.

[123] G. Dyson. *Turing's Cathedral: The Origins of the Digital Universe*. Vintage, 2012.

[124] C. Ebbesmeyer and E. Scigliano. *Flotsametrics and the Floating World*. HarperCollins, 2009.

[125] Environment Canada EC. Oil properties. Website http://www.etc-cte.ec.gc.ca/databases/oilproperties/, 11 2013. Retrieved 11/22/2013.

[126] K.P. Edwards, J.A. Hare, and F.E. Werner. Dispersal of Black Sea bass (Centropristis striata) larvae on the southeast U.S. continental shelf: Results of a coupled vertical larval behavior – 3D circulation model. *Fish. Oceanogr.*, 17(4):299–315, 2008.

[127] K.P. Edwards, J.A. Hare, F.E. Werner, and H. Seim. Using 2-dimensional dispersal kernels to identify the dominant influences on larval dispersal on continental shelves. *Marine Ecology Progress Series*, 352:77–87, 2007.

[128] K.P. Edwards, F.E. Werner, and B.O. Blanton. Comparison of observed and modeled drifter trajectories in coastal regions: An improvement through adjustments for observed drifter slip and errors in wind fields. *Journal of Atmospheric and Oceanic Technology*, 23:1614–1620, 2006.

[129] Energy Institute EI. Oil transport information. Website http://www.oil-transport.info/crudedata/crudeoildata/crudeoildata.html, 11 2013. Retrieved 11/22/2013.

[130] K. Eiane and D. Parisi. Towards a robust concept for modelling zooplankton migration. *SARSIA*, pp. 465–475, 2001.

[131] H.A. Einstein. The bed-load function for sediment transportation in open channel flows. Technical Bulletin 1026, US Department of Agriculture, 1950.

[132] D.D. Eley, M.J. Hey, and J.D. Symonds. Emulsions of water in asphaltene-containing oils 1. Droplet size distribution and emulsification rates. *Colloids and Surfaces*, 32:87–101, 1988.

[133] A.J. Elliott, N. Hurford, and C.J. Penn. Shear diffusion and the spreading of oil slicks. *Marine Pollution Bulletin*, 17(7):308–313, 1986.

[134] A.J. Elliott. Shear diffusion and the spread of oil in the surface layers of the North Sea. *Deutsche Hydrografische Zeitschrift*, 39(3):113–137, 1986.

[135] A.J. Elliott and D.C. Wallace. Dispersion of surface plumes in the southern North Sea. *Deutsche Hydrografische Zeitschrift*, 42(1):1–16, 1989.

[136] F. Engelund and J. Fredsoe. A sediment transport model for straight alluvial channels. *Nordic Hydrology*, 7(5):293–306, 1976.

[137] J.A. Fay. The spread of oil slicks on a calm sea. In D.P. Hoult, ed., *Oil on the Sea, Proceedings of a symposium on the scientific and engineering aspects of oil pollution of the sea*, Ocean Technology, pp. 53–64, New York, Held at Cambridge, Massachusetts, May 16 1969. Massachusetts Institute of Technology and Woods Hole Oceanographic Institution, Plenum.

[138] J.A. Fay. Physical processes in the spread of oil on a water surface. In *International Oil Spill Conference*, pp. 463–467. American Petroleum Institute, 1971.
[139] J.A. Fay. *Introduction to Fluid Mechanics*. MIT Press, 1994.
[140] F.B. Hildebrand. *Advanced Calculus for Applications*. Prentice Hall, 1962.
[141] R.P. Feinman. *Surely You're Joking, Mr. Feinman*. W.W. Norton, 1997 (originally published 1984).
[142] D. Fekedulegn, M.P. MacSiurtain, and J.J. Colbert. Parameter estimation of nonlinear growth models in forestry. *Silva Fennica*, 33(4):327–336, 1999.
[143] R.I. Ferguson and M. Church. A simple universal equation for grain settling velocity. *Journal of Sedimentary Research*, 74(6):933–937, 2004.
[144] R. Fernandez Luque. *Erosion and Transport of Bed Load Sediment*. PhD thesis, Technische Hoogeschool Delft, Kris Repro BV, Netherlands, December 1974.
[145] R. Fernandez Luque and R. Van Beek. Erosion and transport of bed-load sediment. *Journal of Hydraulic Research*, 14(2):127–144, 1976.
[146] M.B. Fiering. *Streamflow Synthesis*. Harvard University Press, 1967.
[147] M.B. Fiering and B.B. Jackson. *Synthetic Streamflows*. American Geophysical Union, 1971.
[148] M. Fingas, B. Fieldhouse, and J. Mullin. Water-in-oil emulsions results of formation studies and applicability to oil spill modelling. *Spill Science & Technology Bulletin*, 5(1):81–91, 1999.
[149] M.F. Fingas. Modeling oil and petroleum evaporation. *Journal of Petroleum Science Research*, 2(3), 2013.
[150] M. Fingas. Introduction to spill modeling. In M. Fingas, ed., *Oil Spill Science and Technology*, pp. 187–199. Elsevier Science, 2011.
[151] B. Finlayson. *Numerical Methods for Problems with Moving Fronts*. Ravenna Park Publishing, 1992.
[152] H.B. Fischer. Mixing and dispersion in estuaries. *Annual Reviews of Fluid Mechanics*, 8:107–133, 1976.
[153] H.B. Fischer, E.J. List, R.C.Y. Koh, J. Imberger, and N.H. Brooks. *Mixing in Inland and Coastal Waters*. Academic Press, 1979.
[154] Fisheries Research Services Scottish Government. Zooplankton and climate change - the calanus story. Website http://www.scotland.gov.uk/Uploads/Documents/ME07Ocean.pdf, Nov 2005.
[155] R.A. Flather and N.S. Heaps. Tidal computations for Morecambe Bay. *Geophysical Journal International*, 42(2):489–517, 1975.
[156] S.W. Fong and N.S. Heaps. Note on quarter-wave tidal resonance in the Bristol Channel. I.O.S. Report, Unpublished Manuscript 63, Institute of Oceanographic Sciences, 1978.
[157] C. Forbes, M. Evans, N. Hastings, and B. Peacock. *Statistical Distributions*, 4th ed. John Wiley & Sons, 2011.
[158] J.R.D Francis. Experiments on the motion of solitary grains along the bed of a water-stream. *Proceedings of the Royal Society of London. A. Mathematical and Physical Sciences*, 332(1591):443–471, 1973.
[159] P.J.S. Franks, J.S. Wroblewski, and G.R. Flierl. Behavior of simple plankton model with food-level acclimation by herbivores. *Marine Biology*, 121–129, 1986.
[160] D.M. Fratantoni. North Atlantic surface circulation during the 1990's observed with satellite-tracked drifters. *Journal of Geophysical Research: Oceans*, 106(C10):22067–22093, 2001.
[161] J. Fredsoe and R. Deigaard. *Mechanics of Coastal Sediment Transport*. Advanced Series on Ocean Engineering, Vol. 3. World Scientific, 1992.

[162] D. French-McCay. State-of-the-art and research needs for oil spill impact assessment modeling. In *Proceedings of the 32nd AMOP Technical Seminar on Environmental Contamination and Response*, pp. 601–653, Ottawa, ON, Canada, 2009. Emergencies Science Division, Environment Canada.

[163] D.P. French-McCay. Oil spill impact modeling: Development and validation. *Environmental Toxicology and Chemistry*, 23(10):2441–2456, 2004.

[164] A. Gallego, ed. Special Issue: Advances in modelling physical-biological interactions in fish early life history. *Marine Ecology Progress Series*, 121–206, 2007.

[165] A. Gallego, E.W. North, P. Petitgas, and H.I. Browman. Advances in modelling physical-biological interactions in fish early life history. Idea and coordination. *Marine Ecology Progress Series*, 121–126, 2007.

[166] B. Galperin, L.H. Kantha, S. Hassid, and A. Rosati. A quasi-equilibrium turbulent energy model for geophysical flows. *Journal of Atmospheric Sciences*, 45:55–62, 1988.

[167] J.A. Galt. The integration of trajectory models and analysis into spill response information systems. *Spill Science & Technology Bulletin*, 4(2):123–129, 1997. IMO Special Issue-Part III.

[168] P.R. Gent and J.C. Mcwilliams. Isopycnal mixing in ocean circulation models. *Journal of Physical Oceanography*, 20(1):150–155, 1990.

[169] M. Germano, U. Piomelli, P. Moin, and W.H. Cabot. A dynamic subgrid-scale eddy viscosity model. *Physics of Fluids A: Fluid Dynamics*, 3:1760, 1991.

[170] W.R. Geyer. Field calibration of mixed-layer drifters. *Journal of Atmospheric and Oceanic Technology*, 6(2):333–342, 1989.

[171] K. Ghoshal and D. Pal. Grain-size distribution in suspension over a sand-gravel bed in an open channel flow. *International Journal of Sediment Research*, Accepted 2013.

[172] D.M. Glover, W.J. Jenkins, and S.C. Doney. *Modeling Methods for Marine Science*. Cambridge University Press, 2011.

[173] G. Godin. On tidal resonance. *Continental Shelf Research*, 13(1):89–107, 1993.

[174] N.J. Gotelli. *A Primer of Ecology*. Sinauer Associates, 4 ed., 2008.

[175] N.J. Gotelli and A.M. Ellison. *A Primer of Ecological Statistics*. Sinauer Associates, 2004.

[176] D.I. Graham and R.A. Moyeed. How many particles for my Lagrangian simulations? *Powder Technology*, 125:179–186, 2002.

[177] D. Greenberg, J. Loder, Y. Shen, D. Lynch, and C. Naimie. Spatial and temporal structure of the barotropic response of the Scotian Shelf and Gulf of Maine to surface wind stres: A model-based study. *Journal of Geophysic Research*, 102(C9):20897–20915, 1997.

[178] D.A. Greenberg, F. Dupont, F.H. Lyard, D.R. Lynch, and F.E. Werner. Resolution issues in numerical models of oceanic and coastal circulation. *Continental Shelf Research*, 27:1317–1343, 2007.

[179] D.A. Greenberg, J.A. Shore, F.H. Page, and M. Dowd. A finite element circulation model for embayments with drying intertidal areas and its application to the Quoddy region of the Bay of Fundy. *Ocean Modeling*, 10:211–231, 2005.

[180] D.A. Greenberg, F.E. Werner, and D.R. Lynch. A diagnostic finite-element ocean circulation model in spherical-polar coordinates. *Journal of Atmospheric and Oceanic Technology*, 15(C9):942–958, 1998.

[181] D.A. Greenberg. A numerical model investigation of tidal phenomena in the Bay of Fundy and Gulf of Maine. *Marine Geodesy*, 2(2):161–187, 1979.

[182] D.A. Greenberg. Modelling the mean barotropic circulation in the Bay of Fundy and Gulf of Maine. *JPO*, 13(5):886–904, 1983.

[183] D.A. Greenberg. *Mathematical Studies of Tidal Behaviour in the Bay of Fundy*. PhD thesis, University of Liverpool, Liverpool, UK, 1975.

[184] P.M. Gresho and R.L. Sani. *Incompressible Flow and the Finite Element Method*, Vol. 1, *Advection-Diffusion*. John Wiley & Sons, 2000.

[185] A. Griffa. Applications of stochastic particle models to oceanographic problems. In R.J. Adler, P. Muller, and R.B. Rozovskii, ed., *Stochastic Modelling in Physical Oceanography*, pp. 113–140. Birkhauser, 1996.

[186] A. Griffa, A.D. Kirwan Jr., A.J. Mariano, T. Ozgokmen, and H.T. Rossby, ed. *Lagrangian Analysis and Prediction of Coastal and Ocean Dynamics*. Cambridge University Press, 2007. 500 pp.

[187] S.M. Griffies and R.W. Hallberg. Biharmonic friction with a smagorinsky-like viscosity for use in large-scale eddy-permitting ocean models. *Monthly Weather Review*, 128(8):2935–2946, 2000.

[188] V. Grimm and S. Railsback. *Individual-Based Modeling and Ecology*. Princeton Studies in Theoretical and Computational Biology. Princeton University Press, 2005.

[189] R.J Hallermeier. Terminal settling velocity of commonly occurring sand grains. *Sedimentology*, 28(6):859–865, 1981.

[190] C.G. Hannah and F. Peters. The workshop on future directions in modelling physical-biological interactions. *Journal of Marine Systems*, 61(3/4):115–117, 2006.

[191] T. Hayashi, David A. Greenberg, and C. J.R. Garrett. Open boundary conditions for numerical models of shelf sea circulation. *Continental Shelf Research*, 5(4):487–497, 1986.

[192] A.C. Haza, A. Griffa, P. Martin, A. Molcard, T.M. Ozgokmen, A.C. Poje, R. Barbanti, J.W. Book, P.M. Poulain, M. Rixen, and P. Zanasca. Model-based directed drifter launches in the Adriatic Sea: Results from the dart experiment. *Geophysical Research Letters*, 34(10), 2007.

[193] R. He, D.J. McGillicuddy Jr., D.R. Lynch, K.W. Smith, C.A. Stock, and J.P. Manning. Data assimilative hindcast of the Gulf of Maine coastal circulation. *Journal of Geophysical Research*, 110(C10011), 2005.

[194] R. He and R.H. Weisberg. Tides on the west Florida shelf. *Journal of Physical Oceanography*, 32(12):3455–3473, 2002.

[195] R. Hilborn and M. Mangel. *The Ecological Detective: Confronting Models with Data*. Princeton University Press, 1997. pp. 315.

[196] F. Hillier and G. Lieberman. *Introduction to Operations Research*, 8th ed., McGraw-Hill, 2005.

[197] D.T. Ho, J.R. Ledwell, and W.M. Smethie. Use of Sf5Cf3 for ocean tracer release experiments. *Geophysical Research Letters*, 35(4), 2008.

[198] A. Hochstaedter and D. Sullivan. Oceanography and Google Earth: Observing ocean processes with time animations and student-built ocean drifters. *Geological Society of America Special Papers*, 492:441–451, 2012.

[199] D. Hohn. *Moby-Duck: The True Story of 28,800 Bath Toys Lost at Sea and of the Beachcombers, Oceanographers, Environmentalists, and Fools, Including the Author, Who Went in Search of Them*. Viking Adult, March 2011.

[200] D.P. Hoult. Oil spreading on the sea. *Annual Review of Fluid Mechanics*, 4(1):341–368, 1972.

[201] H.L. Hunt, D.C. Fugate, and R.J. Chant. Modeling bedload transport of juvenile bivalves: Predicted changes in distribution and scale of postlarval dispersal. *Estuaries and Coasts*, 32(6):1090–1102, 2009.

[202] J.R. Hunter, P.D. Craig, and H.E. Phillips. On the use of random walk models with spatially variable diffusivity. *Journal of Computational Physics*, 106:366–376, 1993.

[203] H.S. Huntley, B.L. Lipphardt Jr., and A.D. Kirwan Jr. Lagrangian predictability assessed in the East China Sea. *Ocean Modelling*, 36(1–2):163–178, 2011.

[204] J.M. Huthnance. Tidal current asymmetries over the Norfolk Sandbanks. *Estuarine and Coastal Marine Science*, 1(1):89–99, 1973.

[205] L.S. Incze and C.E. Naimie. Modeling the transport of lobster (*Homarus americanus*) larvae and postlarvae in the Gulf of Maine. *Fisheries Oceanography*, 9:99–113, 2000.

[206] T. Isaji, M. Spaulding, and A. Allen. Stochastic particle trajectory modeling techniques for spill and search and rescue models. In Malcolm L. Spaulding, ed., *Estuarine and Coastal Modeling: Proceedings of the Ninth International Conference*, October 31–November 2, 2005, Charleston, South Carolina, pp. 537–547. ASCE Publications, 2005.

[207] A. Isobe, S. Kako, P.H. Chang, and T. Matsuno. Two-way particle-tracking model for specifying sources of drifting objects. Application to the East China Sea shelf. *Journal of Atmospheric and Oceanic Technology*, 26(8):1672–1682, 2009.

[208] J. Jaffe and C. Schurgers. Sensor networks of freely drifting autonomous underwater explorers. In *Proceedings of the 1st ACM international workshop on Underwater networks*, WUWNet '06, pp. 93–96. ACM, 2006.

[209] C. Ji, A. Munjiza, E. Avital, J. Ma, and J.J.R. Williams. Direct numerical simulation of sediment entrainment in turbulent channel flow. *Physics of Fluids*, 25(5):056601, 2013.

[210] J.A. Jimenez and O.S. Madsen. A simple formula to estimate settling velocity of natural sediments. *Journal of Waterway, Port, Coastal, and Ocean Engineering*, 129(2):70–78, 2003.

[211] H.E. Jobson and W.W. Sayre. Vertical transfer in open channel flow. *Journal of the Hydraulics Division*, 96(3):703–724, 1970.

[212] O. Johansen. Deepblow – a Lagrangian plume model for deep water blowouts. *Spill Science & Technology Bulletin*, 6(2):103–111, 2000.

[213] O. Johansen. Development and verification of deep-water blowout models. *Marine Pollution Bulletin*, 47(9):360–368, 2003.

[214] O. Johansen, H. Rye, and C. Cooper. Deepspill–field study of a simulated oil and gas blowout in deep water. *Spill Science & Technology Bulletin*, 8(5-6):433–443, 2003.

[215] D. Johnson, R. Stocker, R. Head, J. Imberger, and C. Pattiaratchi. A compact, low-cost GPS drifter for use in the oceanic nearshore zone, lakes, and estuaries. *Journal of Atmospheric and Oceanic Technology*, 20(12):1880–1884, 2003.

[216] W.P. Jones and B.E. Launder. The prediction of laminarization with a two-equation model of turbulence. *International Journal of Heat and Mass Transfer*, 15(2):301–314, 1972.

[217] P.Y. Julien and B. Bounvilay. Velocity of rolling bed load particles. *Journal of Hydraulic Engineering*, 139:177–186, February 2013.

[218] P.A. Jumars, A.R.M. Nowell, and R.F.L. Self. A simple model of flow–sediment–organism interaction. *Marine Geology*, 42(1):155–172, 1981.

[219] L. Kantha. Computational ocean modeling. In A.B. Tucker Jr., ed., *The Computer Science and Engineering Handbook*, 40, pp. 934–958. CRC Press, 1997.

[220] L. Kantha and S. Carniel. Comments on "A generic two-equation model for geophysical flows." *Journal of Marine Research*, 61:693–702, 2003.

[221] L.H. Kantha. The length scale equation in turbulence models. *Nonlinear Processes in Geophysics*, 11:83–97, 2004. Special Issue Dedicated to Prof. A.D. Kirwan, eds. S. Wiggins, P. Haynes, and A. Osborne.

[222] L.H. Kantha and C.A. Clayson. An improved mixed layer model for geophysical applications. *Journal of Geophysical Research*, 99(C12):25235–25266, 1994.

[223] L.H. Kantha and C.A. Clayson. *Numerical Models of Oceans & Oceanic Processes*. Academic Press, 2000.

[224] L.H. Kantha and C.A. Clayson. *Small Scale Processes in Geophysical Fluid Flows*. Academic Press, 2000.

[225] R. Keeley, M. Pazos, and B. Bradshaw. Data management system for surface drifters. Proceedings OceanObs'09: *Sustained Ocean Observations and Information for Society 2*, 2010.

[226] J.F. Kenney and E.S. Keeping. *Mathematics of Statistics, Part 2*. Van Nostrand, 1951.

[227] R.E. Kent and D.W. Pritchard. A test of mixing-length theories in a coastal-plain estuary. *Journal of Marine Research*, 18:62–72, 1959.

[228] P.K. Kitanidis. Particle-tracking equations for the solution of the advection-dispersion equation with variable coefficients. *Water Resources Research*, 30(11):3225–3227, 1994.

[229] P.E. Kloeden and E. Platen. *Numerical Solution of Stochastic Differential Equations*. Stochastic Modelling and Applied Probability, No. 23. Springer, 2010.

[230] P.E. Kloeden, E. Platen, and H. Schurz. *Numerical Solution of SDE Through Computer Experiments*. Universitext, Volume XIV. Springer, 2007.

[231] J.A. Knauss. *Introduction to Physical Oceanography*. Waveland Press, 2005.

[232] B.R. Knight, R. Zyngfogel, and B. Forrest. PartTracker–a fate analysis tool for marine particles. In *Proceedings of Australasian Coasts and Ports Conference 2009*, p. 8. New Zealand Coastal Society, 2009.

[233] J.P. Kritzer and P.F. Sale. *Marine Metapopulations*. Elsevier Academic Press, 2006.

[234] W.C. Krumbein and L.L. Sloss. *Stratigraphy and Sedimentation*. Freeman San Francisco, 1963.

[235] L. Kuznetsov, K. Ide, and C.K.R.T. Jones. A method for assimilation of Lagrangian data. *Monthly Weather Review*, 131(10):2247–2260, 2003.

[236] J.H. LaCasce. Statistics from Lagrangian observations. *Progress in Oceanography*, 77(1):1–29, 2008.

[237] R. Lange. A 100 tons experimental oil spill at Halten bank, off Norway. In *International Oil Spill Conference*, pp. 503–505. American Petroleum Institute, 1985.

[238] P. Langevin. Sur la théorie du mouvement brownien. *Comptes Rendus de l'Académie des Sciences* (Paris), 146(530–533), 1908.

[239] B.E. Launder and B.I. Sharma. Application of the energy-dissipation model of turbulence to the calculation of flow near a spinning disc. *Letters in Heat and Mass Transfer*, 1(2):131–137, 1974.

[240] K. Lavender Law, S. Moret-Ferguson, N.A. Maximenko, G. Proskurowski, E.E. Peacock, J. Hafner, and C.M. Reddy. Plastic accumulation in the North Atlantic subtropical gyre. *Science*, 329(5996):1185–1188, 2010.

[241] C. Le Provost, M.-L. Genco, and F. Lyard. Modeling and predicting tides over the world ocean. In D.R. Lynch and A.M. Davies, eds, *Quantitative Skill Assessment for Coastal Ocean Models*. American Geophysical Union, 1995.

[242] W. Lehr, R. Jones, M. Evans, D. Simecek-Beatty, and R. Overstreet. Revisions of the adios oil spill model. *Environmental Modelling & Software*, 17(2):189–197, 2002.

[243] W.J. Lehr. Review of modeling procedures for oil spill weathering behavior. In C.A. Brebbia, ed., *Oil Spill Modelling and Processes*, pp. 51–90. WIT Press, 2001.

[244] W.J. Lehr, R.J. Fraga, M.S. Belen, and H.M. Cekirge. A new technique to estimate initial spill size using a modified Fay-type spreading formula. *Marine Pollution Bulletin*, 15(9):326–329, 1984.

[245] D.S. Lemons and A. Gythiel. Paul Langevin's 1908 paper "on the theory of Brownian motion" ["sur la théorie du mouvement brownien," *Comptes Rendus de l'Académie des Sciences* (Paris) 146, 530–533 (1908)]. *American Journal of Physics*, 65(11):1079, 1997.

[246] S. Lentz. Buoyant coastal currents. In E. Chassignet, C. Cenedese, and J. Verron, ed., *Buoyancy-Driven Flows*, pp. 164–202. Cambridge University Press, New York, 2012.

[247] S.J. Lentz and M.R. Fewings. The wind- and wave-driven inner-shelf circulation. *Annual Review of Marine Science*, 4(1):317–343, 2012.

[248] C. Lett, P.V., C. Mullon, C. Parada, T. Brochier, P. Penven, and B. Blanke. A Lagrangian tool for modelling ichthyoplankton dynamics. *Environmental Modelling & Software*, 23(9):1210–1214, 2008.

[249] C.V.W. Lewis, D.R. Lynch, M.J. Fogarty, and D. Mountain. Effect of area closures on Georges Bank bivalves: Larval transport and population dynamics. Numerical Methods Laboratory Report NML-00-2, Dartmouth College, Hanover NH USA. [http://www-nml.dartmouth.edu/ Publications/internal_reports/NML-00-2] , 2000.

[250] M. Li and C. Garrett. The relationship between oil droplet size and upper ocean turbulence. *Marine Pollution Bulletin*, 36(12):961–970, 1998.

[251] Y. Li, R. He, and J.P. Manning. Coastal connectivity in the Gulf of Maine in spring and summer of 2004-2009. *Deep-Sea Research Part II*, 103: 199–209, May 2014.

[252] J. Lin, H.V. Wang, J.-H. Oh, K. Park, S.C. Kim, J. Shen, and A.Y. Kuo. A new approach to model sediment resuspension in tidal estuaries. *Journal of Coastal Research*, 19(1):76–88, 2003.

[253] B.L. Lipphardt, D. Small, A.D. Kirwan, S. Wiggins, K. Ide, C.E. Grosch, and J.D. Paduan. Synoptic Lagrangian maps: Application to surface transport in Monterey Bay. *Journal of Marine Research*, 64(2):221–247, 2006.

[254] Y. Liu, A. MacFadyen, Z.G. Ji, and R. Weisberg, ed. *Monitoring and Modeling the Deepwater Horizon Oil Spill: A Record-Breaking Enterprise. Geophysical Monograph Series*, Vol. 195. American Geophysical Union, 2011.

[255] Y.Y. Liu, R.H.R.H. Weisberg, C.C. Hu, and L. L. Zheng. Trajectory forecast as a rapid response to the Deepwater Horizon oil spill. In Y. Liu, A. MacFadyen, Z.-G. Ji, and Robert Weisberg, eds. *Monitoring and Modeling the Deepwater Horizon Oil Spill: A Record-Breaking Enterprise*, pp. 153–165. American Geophysical Union, 2011.

[256] J. Loder, Y. Shen, and H. Ridderinkhof. Characterization of three-dimensional Lagrangian circulation associated with tidal rectification over a submarine bank. *Journal of Physical Oceanography*, 27:1729–1742, 1997.

[257] J.W. Loder. Topographic rectification of tidal currents on the sides of Georges Bank. *Journal of Physical Oceanography*, 10:1399–1416, 1980.

[258] S.A. Lonin. Lagrangian model for oil spill diffusion at sea. *Spill Science & Technology Bulletin*, 5(5–6):331–336, 1999.

[259] D.P. Loucks, E. VanBeek, J.R. Stedinger, J.P.M. Dijkman, and M.T. Villars. *Water Resource Systems Planning and Management*. UNESCO, 2005.

[260] R.G. Lough, W.G. Smith, F.E. Werner, J.W. Loder, F.H. Page, C.G. Hannah, C.E. Naimie, R.I. Perry, M. Sinclair, and D.R. Lynch. Influence of wind driven advection on the interannual variability in cod egg and larval distributions on Georges Bank: 1982 vs 1985. *ICES Marine Science Symposium*, 198:356–378, 1994.

[261] R.G. Lough, E.A. Broughton, L.J. Buckley, L.S. Incze, K. Pehrson Edwards, R. Converse, A. Aretxabaleta, and F.E. Werner. Modeling growth of Atlantic cod larvae on the southern flank of Georges Bank in the tidal-front circulation during May 1999. *Deep-Sea Research Part II*, 53:2771–2788, 2006.

[262] R.G. Lough, L.J. Buckley, F.E. Werner, J.A. Quinlan, and K.P. Edwards. A general biophysical model of larval cod (Gadus morhua) growth applied to populations on Georges Bank. *Fisheries Oceanography*, 14(4), 2005.

[263] R.G. Lough, E.M. Caldarone, T.K. Rotunno, E.A. Broughton, B.R. Burns, and L.J. Buckley. Vertical distribution of cod and haddock eggs and larvae, feeding and condition in stratified and mixed waters on southern Georges Bank, May 1992. *Deep-Sea Research Part II*, 43(7–8), 1996.

[264] R.G. Lough and J.P. Manning. Tidal-front entrainment and retention of fish larvae on the southern flank of Georges Bank. *Deep-Sea Research Part II*, 631–644, 2001.

[265] R.G. Lough and D.G. Mountain. Effect of small-scale turbulence on feeding rates of larval cod and haddock in stratified water on Georges Bank. *Deep-Sea Research Part II*, 43(7-8):1745–1772, 1996.

[266] R.G. Lough and L. O'Brien. Life-stage recruitment models for Atlantic cod (*Gadus morhua*) and haddock (*Melanogrammus aeglefinus*) on Georges Bank. *Fish. Bull.*, pages 123–140, 2011.

[267] R.A. Luettich and J.J. Westerink. Continental shelf scale convergence studies with a barotropic tidal model. In D.R. Lynch and A.M. Davies, eds., *Quantitative Skill Assessment for Coastal Ocean Models*, pages 349–372. Coastal and Estuarine Series 47, American Geophysical Union, Washington, DC, 1995.

[268] R. Lumpkin and M. Pazos. Measuring surface currents with surface velocity program drifters: The instrument, its data, and some recent results. In A. Griffa, A.D. Kirwan, A.J. Mariano, T. Ozgokmen, and T. Rossby, editors, *Lagrangian Analysis and Prediction of Coastal and Ocean Dynamics (LAPCOD)*, pp. 39–67 Cambridge University Press, 2006.

[269] R. Lumpkin and S. Garzoli. Interannual to decadal changes in the western South Atlantic's surface circulation. *Journal of Geophysical Research: Oceans*, 116(C1), 2011.

[270] R. Lumpkin, S.A. Grodsky, L. Centurioni, M.H. Rio, J. A Carton, and D. Lee. Removing spurious low-frequency variability in drifter velocities. *Journal of Atmospheric and Oceanic Technology*, 30(2):353–360, 2013.

[271] R. Lumpkin, N. Maximenko, and M. Pazos. Evaluating where and why drifters die. *Journal of Atmospheric and Oceanic Technology*, 29(2):300–308, 2012.

[272] P.J. Luyten, J.E. Jones, R. Proctor, A. Tabor, P. Tett, and K. Wild-Allen. COHERENS – A Coupled Hydrodynamical-Ecological Model for Regional and Shelf Seas: User Documentation. Technical Report, Management Unit of the Mathematical Models of the North Sea, Brussels, 1999.

[273] D. Lynch, K. Smith, and D.J. McGillicuddy Jr. Individual-Based Model for Alexandrium fundyense. Numerical Methods Laboratory Report NML-03-19, Dartmouth College, Hanover NH USA. [http:// www-nml.dartmouth.edu/ Publications/internal_reports/NML-03-19] , 2003.

[274] D. Lynch, K. Smith, and D.J. McGillicuddy Jr. Bay of Fundy Recirculation. Numerical Methods Laboratory Report NML-05-2, Dartmouth College, Hanover NH USA. [http:// www-nml.dartmouth.edu/ Publications/internal_reports/NML-05-2] , 2005.

[275] D.R. Lynch, J.T.C. Ip, C.E. Naimie, and F.E. Werner. Convergence studies of tidally-rectified circulation on Georges Bank. In D.R. Lynch and A.M. Davies, eds., *Quantitative Skill Assessment for Coastal Ocean Models*, pages 153–174. Coastal and Estuarine Series 47, American Geophysical Union, Washington, DC, 1995.

[276] D.R. Lynch and F.E. Werner. Three-dimensional hydrodynamics on finite elements. Part II: Nonlinear time-stepping model. *Int. J. Num. Methods Fluids*, 12:507–533, 1991.

[277] D.R. Lynch, C.E. Naimie, J.T. Ip, C.V. Lewis, F.E. Werner, R.A. Luettich, B.O. Blanton, J. Quinlan, D.J. McGillicuddy, J.R. Ledwell, et al. Real-time data assimilative modeling on Georges Bank. *Oceanography*, 14:65–77, 2001.

[278] D.R. Lynch, F.E. Werner, D.A. Greenberg, and J.W. Loder. Diagnostic model for baroclinic, wind-driven and tidal circulation in shallow seas. *Continental Shelf Research*, 12(1):37–64, 1992.

[279] D.R. Lynch. *Numerical Partial Differential Equations for Environmental Scientists and Engineers*. Springer, 2005.

[280] D.R. Lynch. *Sustainable Natural Resource Management for Scientists and Engineers*. Cambridge University Press, 2009.

[281] D.R. Lynch, M.J. Holboke, and C.E. Naimie. The Maine Coastal Current: spring climatological circulation. *Continental Shelf Research*, 17(6):605–634, 1997.

[282] D.R. Lynch, J.T.C. Ip, C.E. Naimie, and F.E. Werner. Comprehensive coastal circulation model with application to the Gulf of Maine. *Continental Shelf Research*, 16(7):875–906, 1996.

[283] D.R. Lynch, J.T.C. Ip, F.E. Werner, and E.M. Wolff. Environmental Hydrodynamics: Comprehensive model of the Gulf of Maine. In G.F. Carey, ed., *Finite Element Modeling of Environmental Problems*, pp. 17–38. John Wiley & Sons, 1995.

[284] D.R. Lynch, C.V.L. Lewis, and F.E. Werner. Can Georges Bank cod survive on a calanoid diet? *Deep-Sea Research Part II*, 48(2):609–630, 2001.

[285] D.R. Lynch and D.J. McGillicuddy Jr. Objective analysis for coastal regimes. *Continental Shelf Research*, 21:1299–1315, 2001.

[286] D.R. Lynch and C.E. Naimie. The M_2 tide and its residual on the outer banks of the Gulf of Maine. *Journal of Physical Oceanography*, 23:2222–2253, 1993.

[287] D.R. Lynch, K. Smith, B. Blanton, R. Luettich, and F. Werner. Forecasting the coastal ocean: Resolution, tides, and operational data in the South Atlantic Bight. *Journal of Atmospheric and Oceanic Technology*, 21(7):1074–1085, 2004.

[288] D.R. Lynch, K. Smith, and B. Cahill. Seasonal mean circulation on the Irish Shelf – A model-generated climatology. *Continental Shelf Research*, 24(18), 2004.

[289] D.R. Lynch and K.W. Smith. Lagrangian dispersion in sheared flow. *Continental Shelf Research*, 30:2092–2105, 2010.

[290] J. Ma, D. Xu, Y. Bai, and J.J.R. Williams. Probability model for gravel sediment entrainment in turbulent flows. *Journal of Hydro-environment Research*, 7(3): 154–160, September 2013.

[291] N.J. MacDonald, M.H. Davies, A.K. Zundel, J.D. Howlett, Z. Demirbilek, J.Z. Gailani, T.C. Lackey, and J. Smith. PTM: Particle tracking model. report 1: Model theory, implementation, and example applications. Technical Report ERDC/CHL TR-06-20, U.S. Army Corps of Engineers, U.S. Army Engineer Research and Development Center, 3909 Halls Ferry Road, Vicksburg, MS 39180-6199, September 2006.

[292] A. Macfadyen, G.Y. Watabayashi, C.H. Barker, and C.J. Beegle-Krause. Tactical modeling of surface oil transport during the Deepwater Horizon spill response. In Y. Liu, A. Macfadyen, Z.-G. Ji, and R.H. Weisberg, editors, *Monitoring and Modeling the Deepwater Horizon Oil Spill: A Record-Breaking Enterprise*, pp. 167–178. American Geophysical Union, 2013.

[293] D. Mackay, I. Buist, K. Mascarennas, and S. Paterson. Oil spill processes and models. Contractor report, Environment Protection Service, Environment Canada, Ottawa Ontario, December 1980.

[294] D. Mackay and P.J. Leinonen. Mathematical model of the behaviour of oil spills on water with natural and chemical dispersion. Economic and Technical Review Report EPS-3-EC-77-19, Environment Protection Service, Department of Fisheries and the Environment, Ottawa, Canada, October 1977.

[295] D. Mackay. Formation and stability of water-in-oil emulsions. Unpublished Report EE-93, Environment Protection Directorate, Environment Canada, Ottawa, December 1987.

[296] M. Mangel. *The Theoretical Biologist's Toolbox*. Cambridge University Press, 2006.

[297] M. Mangel and C.W. Clark. *Dynamic Modeling in Behavioral Ecology*. Princeton University Press, 1988.

[298] T.O. Manley. Drifters, drogues, and circulation. *Oceanography*, 23(4):165, 2010.

[299] J. Manning, D.J. McGillicuddy Jr., N. Pettigrew, J. Churchill, and L. Incze. Drifter observations of Gulf of Maine coastal current, continental shelf. *Continental Shelf Research*, 29:835–845, 2009.

[300] J.P. Manning. Student-built, fishermen-deployed, satellite-tracked drifters, community white paper for US IOOS Summit 2012. Website https://docs.google.com/file/d/0BwW7dLnWT2C0dTVwTTQtYXJwM2s/edit, 2012.

[301] J.P. Manning and J.H. Churchill. Estimates of dispersion from clustered-drifter deployments on the southern flank of Georges Bank. *Deep-Sea Res. II*, 53(23–24):2501–2519, 2006.

[302] A.J. Mariano and E.H. Ryan. Lagrangian analysis and prediction of coastal and ocean dynamics (LAPCOD). In A. Griffa, A.D. Kirwan Jr., A.J. Mariano, T. Özgökmen and H.T. Rossby, eds. *Lagrangian Analysis and Prediction of Coastal and Ocean Dynamics*, pp. 423–479. Cambridge University Press, 2007.

[303] J.L. Martin and S.C. McCutcheon. *Hydrodynamics and Transport for Water Quality Modeling*. Lewis, 1999.

[304] J.L. McClean, P.-M. Poulain, J.W. Pelton, and M.E. Maltrud. Eulerian and Lagrangian statistics from surface drifters and a high-resolution pop simulation in the north Atlantic. *Journal of Physical Oceanography*, 32(9):2472–2491, 2002.

[305] D.J. McGillicuddy Jr., D.M. Anderson, C.A. Stock, D.R. Lynch, and D.W. Townsend. Modelling blooms of *Alexandrium fundyense* in the Gulf of Maine. In M. Babin, C.S. Roesler, and J.J. Cullen, eds., *Real-time Coastal Observing Systems for Marine Ecosystem Dynamics and Harmful Algal Blooms: Theory, Instrumentation and Modelling*, pp. 599–626. UNESCO, 2008.

[306] D.J. McGillicuddy Jr., D.M. Anderson, D.R. Lynch, and D.W. Townsend. Mechanisms regulating large-scale seasonal fluctuations in *alexandrium fundyense* populations in the Gulf of Maine: Results from a physical–biological model. *Deep Sea Research Part II*, 52(19):2698–2714, 2005.

[307] J.W. McLaughlin, A. Bilgili, and D.R. Lynch. Numerical modeling of tides in the Great Bay Estuarine System: dynamical balance and spring-neap residual modulation. *Estuarine, Coastal, and Shelf Science*, 57:283–296, 2003.

[308] G.J. McNally, W.C. Patzert, A.D. Kirwan, and A.C. Vastano. The near-surface circulation of the North Pacific using satellite tracked drifting buoys. *Journal of Geophysical Research: Oceans*, 88(C12):7507–7518, 1983.

[309] C.J. Meise and J.E. OReilly. Spatial and seasonal patterns in abundance and age-composition of *Calanus finmarchicus* in the Gulf of Maine and on Georges Bank: 1977–1987. *Deep-Sea Research Part II*, 43(7–8):1473–1501, 1996.

[310] C.J. Meise-Munns, J. Green, M. Ingham, and D. Mountain. Interannual variability in the copepod populations of Georges Bank and the western Gulf of Maine. *Marine Ecology Progress Series*, 225–232, 1990.

[311] G.L. Mellor and T. Yamada. Development of a turbulence closure model for geophysical fluid problems. *Reviews of Geophysics and Space Physics*, 20:851–875, 1982.

[312] G.L. Mellor and H.J. Herring. A survey of mean turbulent field closure models. *AIAA Journal*, 11:590–599, 1973.

[313] G.L. Mellor and T. Yamada. A hierarchy of turbulence closure models for planetary boundary layers. *Journal of Atmospheric Science*, 31:1791–1806, 1974.

[314] C.B. Miller, D.R. Lynch, F. Carlotti, W.C. Gentleman, and C.V.W. Lewis. Coupling of an individual-based population dynamic model of *Calanus finmarchicus* to a circulation model for the Georges Bank region . Numerical Methods Laboratory Report NML-98-7, Dartmouth College, Hanover NH USA. [http:// www-nml.dartmouth.edu/ Publications/internal_reports/NML-98-7] , 1998.

[315] C.B. Miller, D.R. Lynch, F. Carlotti, W. Gentleman, and C. Lewis. Coupling of an individual-based population dynamics model for stocks of *Calanus finmarchicus* with a circulation model for the Georges Bank region. *Fisheries Oceanography*, 7(3-4):219–234, 1998.

[316] C.B. Miller and K.S. Tande. Stage duration estimation of *Calanus* populations, a modelling study. *Marine Ecology Progress Series*, 15–34, 1993.

[317] S. Mitarai, D.A. Siegel, J.R. Watson, C. Dong, and J.C. McWilliams. Quantifying connectivity in the coastal ocean with application to the Southern California Bight. *Journal of Geophysical Research: Oceans*, 114(C10), 2009.

[318] A. Molcard, P.M. Poulain, P. Forget, A. Griffa, Y. Barbin, J. Gaggelli, J.C. De Maistre, and M. Rixen. Comparison between VHF radar observations and data from drifter clusters in the Gulf of La Spezia (Mediterranean Sea). *Journal of Marine Systems*, 78, Supplement, Coastal Processes: Challenges for Monitoring and Prediction: S79–S89, 2009.

[319] A. Molcard, A.C. Poje, and T.M. Ozgokmen. Directed drifter launch strategies for Lagrangian data assimilation using hyperbolic trajectories. *Ocean Modelling*, 12(3-4):268–289, 2006.

[320] W. Morse and C. Meise. *Calanus finmarchicus* – THE MOVIE. Numerical Methods Laboratory Report NML-98-13, Dartmouth College, Hanover NH USA. [http:// www-nml.dartmouth.edu/ Publications/internal_reports/NML-98-13] , 1998.

[321] W.H. Munk and E.R. Anderson. Notes on a theory of the thermocline. *Journal of Marine Research* , 1:276–295, 1948.

[322] J.D. Murray. *Mathematical Biology I. An Introduction*, 3rd ed. Interdisciplinary Applied Mathematics, Vol. 17 Springer, 2001.

[323] J.D. Murray. *Mathematical Biology II. Spatial Models and Biomedical Applications*, 3rd ed. Interdisciplinary Applied Mathematics, Vol. 18 Springer, 2003.

[324] C.E. Naimie. A turbulent boundary layer model for the linearized shallow water equations NUBBLE USER'S MANUAL (Release 1.1). Numerical Methods Laboratory Report NML-00-2, Dartmouth College, Hanover NH USA. [http:// www-nml.dartmouth.edu/ Publications/internal_reports/NML-96-1/nubble.html] , 1996.

[325] C.E. Naimie, C.A. Blain, and D.R. Lynch. Seasonal mean circulation in the Yellow Sea – A model-generated climatology. *Continental Shelf Research*, 21:667–695, 2001.

[326] C.E. Naimie, J.W. Loder, and D.R. Lynch. Seasonal variation of the three-dimensional residual circulation on Georges Bank. *Journal of Geophysical Research*, 99:15,967–15,989, 1994.

[327] C.E. Naimie. Georges Bank residual circulation during weak and strong stratification periods: Prognostic numerical model results. *Journal of Geophysical Research*, 101(C3):6469–6486, 1996.

[328] P. Nielsen. *Coastal Bottom Boundary Layers and Sediment Transport*. Advanced Series on Ocean Engineering, Vol. 4. World Scientific, 1992.

[329] P. Niiler, P.A. Sybrandy, K. Bi, P. Poulan, and D. Bitterman. Measurements of the water-following capability of holey-sock and tristar drifters. *Deep-Sea Research*, 42(11–12), 1995.

[330] J.A.U. Nilsson, S. Dobricic, N. Pinardi, V. Taillandier, and P.M. Poulain. On the assessment of Argo float trajectory assimilation in the Mediterranean forecasting system. *Ocean Dynamics*, 61(10):1475–1490, 2011.

[331] Y. Nino and M. Garcia. Using Lagrangian particle saltation observations for bedload sediment transport modelling. *Hydrological Processes*, 12(8):1197–1218, 1998.

[332] National Ocean Service NOAA. Adios (automated data inquiry for oil spills). Website http://response.restoration.noaa.gov/adios, 2013. Site visited 2013/12/31.

[333] E. North, A. Gallego, and P. Pettigas, eds. *Manual of Recommended Practices for Modelling Physical-biological Interactions during Early Fish Life*. ICES Cooperative Research Report No. 295, 2009. 111 pp.

[334] E.W. North, E.E. Adams, Z. Schlag, C.R. Sherwood, R. He, K.H. Hyun, and S.A. Socolofsky. Simulating oil droplet dispersal from the Deepwater Horizon spill with a Lagrangian approach. In Y. Liu, A. Macfadyen, Z.-G. Ji, and R.H. Weisberg, eds., *Monitoring and Modeling the Deepwater Horizon Oil Spill: A Record-Breaking Enterprise*, Vol. 195, pp. 217–226. American Geophysical Union, 2011.

[335] E.W. North, R.R. Hood, S.-Y. Chao, and L.P. Sanford. Using a random displacement model to simulate turbulent particle motion in a baroclinic frontal zone: A new implementation scheme and model performance tests. *Journal of Marine Systems*, pp. 365–380, 2006.

[336] E.W. North, Z. Schlag, R.R. Hood, M. Li, L. Zhong, T. Gross, and V.S. Kennedy. Vertical swimming behavior influences the dispersal of simulated oyster larvae in a coupled particle-tracking and hydrodynamic model of Chesapeake Bay. *Marine Ecology Progress Series*, 99–115, 2008.

[337] NRC. Oil in the sea III: Inputs, fates, and effects. Technical report, National Research Council Committee on Oil in the Sea, 2003.

[338] NRC. Oil spill dispersants: Efficacy and effects. Technical report, Committee on Understanding Oil Spill Dispersants: Efficacy and Effects, National Research Council, The National Academic Press, Washington DC, 2005.

[339] V. Oeresland and M. Ulmestrand. European lobster subpopulations from limited adult movements and larval retention. *ICES Journal of Marine Science: Journal du Conseil*, 70(3):532–539, 2013.

[340] C.B. Officer. *Physical Oceanography of Estuaries (and Associated Coastal Waters)*. John Wiley & Sons, 1976.

[341] B. Oksendal. *Stochastic Differential Equations: An Introduction with Applications*, 6th ed., Universitext. Springer Verlag, New York, 2010.

[342] A. Okubo. Dynamical aspects of animial grouping: Swarms, schools, flocks, and herds. *Advances in Biophysics*, 22:1–94, 1986.

[343] A. Okubo and S. Levin, eds. *Diffusion and Ecological Problems: Modern Perspectives*. Interdisciplinary Applied Mathematics, No. 14. Springer, 2001.

[344] A. Okubo. Oceanic diffusion diagrams. *Deep-Sea Research and Oceanographic Abstracts*, 18(8):789–802, 1971.

[345] M.J. Olascoaga. Isolation on the West Florida shelf with implications for red tides and pollutant dispersal in the Gulf of Mexico. *Nonlinear Processes in Geophysics*, 17(6):685, 2010.

[346] P.B. Ortner, L.B. Crowder, and D.E. Hoss. The South Atlantic Bight Recruitment Experiment: Introduction and overview. *Fisheries Oceanography* (Suppl. 2), 1–6, 1999.

[347] T.M. Ozgokmen, A. Molcard, T.M. Chin, L.I. Piterbarg, and A. Griffa. Assimilation of drifter observations in primitive equation models of midlatitude ocean circulation. *Journal of Geophysical Research: Oceans*, 108(C7), 2003.

[348] N. Paldor, Y. Dvorkin, A.J. Mariano, T. Ozgokmen, and E.H. Ryan. A practical, hybrid model for predicting the trajectories of near-surface ocean drifters. *Journal of Atmospheric and Oceanic Technology*, 21:1246–1258, August 2004.

[349] A.N. Papanicolaou, P. Diplas, N. Evaggelopoulos, and S. Fotopoulos. Stochastic incipient motion criterion for spheres under various bed packing conditions. *Journal of Hydraulic Engineering*, 128(4):369–380, 2002.

[350] B.B. Parker. *Tidal Analysis and Prediction.* NOAA Special Publication NOS-CO-OPS 3, NOAA, Silver, 2007.

[351] C. Pasquero, A. Bracco, A. Provenzale, and J.B. Weiss. Particle motion in a sea of eddies. In A. Griffa, A.D. Kirwan Jr, A.J. Mariano, T. Özgökmen, and H.T. Rossby, eds. *Lagrangian Analysis and Prediction of Coastal and Ocean Dynamics*, p. 89. Cambridge University Press, 2007.

[352] S.E. Pazan and P.P. Niiler. Recovery of near-surface velocity from undrogued drifters. *Journal of Atmospheric and Oceanic Technology*, 18:476–489, 2001.

[353] R. Perianez and F. Caravaca. A set of rapid-response models for pollutant dispersion assessments in southern Spain coastal waters. *Marine Pollution Bulletin*, 60(9):1412–1422, 2010.

[354] R. Perianez and A. Pascual-Granged. Modelling surface radioactive, chemical and oil spills in the Strait of Gibraltar. *Computers & Geosciences*, 34(2):163–180, 2008.

[355] A. Perlin, J.N. Moum, J.M. Klymak, M.D. Levine, T. Boyd, and P.M. Kosro. A modified law-of-the-wall applied to oceanic bottom boundary layers. *Journal of Geophysical Research*, 110(C10S10), 2005.

[356] F. Peters and C.G. Hannah. Special issue devoted to the Workshop on Future Directions in Modelling Physical - Biological Interactions. *Journal of Marine Systems*, 61(3/4):115–274, 2006.

[357] G.W. Platzman. Normal modes of the world ocean I - Design of a finite-element barotropic model. *Journal of Physical Oceanography*, 8, 1978.

[358] A.H. Pollard. Random mortality fluctuations and the binomial hypothesis. *Journal of the Institute of Actuaries*, 96(2):251–264, 1970.

[359] P. Poulain. Drifter observations of surface circulation in the Adriatic Sea between December 1994 and March 1996. *Journal of Marine Systems*, 20, 1999.

[360] M.D. Prater and T. Rossby. Observations of the Faroe Bank Channel overflow using bottom-following RAFOS floats. *Deep-Sea Research Part II*, 52(3–4):481–494, 2005. Direct observations of oceanic flow: A tribute to Walter Zenk.

[361] R.D. Present. *Kinetic Theory of Gases*. McGraw-Hill, 1958.

[362] W.H. Press, B.P. Flannery, S.A. Teukolsky, and W.T. Vetterling. *Numerical Recipes: The Art of Scientific Computing*. Cambridge University Press, 1986.

[363] J.M. Price, M. Reed, M.K. Howard, W.R. Johnson, Z.-G. Ji, C.F. Marshall, N. L. Guinasso Jr., and G.B. Rainey. Preliminary assessment of an oil-spill trajectory model using satellite-tracked, oil-spill-simulating drifters. *Environmental Modelling & Software*, 21(2):258–270, 2006.

[364] D.W. Pritchard. The movement and mixing of contaminants in tidal estuaries. In E.A. Pearson, ed., *Proceedings of the First International Conference on Waste Disposed in the Marine Environment*, pp. 512–525, Pergamon Press, 1960.

[365] J.A. Proehl and D.R. Lynch. Experiments in Lagrangian diffusion. Numerical Methods Laboratory Report NML-01-5, Dartmouth College, Hanover, NH USA. [http:// www-nml.dartmouth.edu/Publications/internal_reports/NML-01-5], 2001.

[366] J.A. Proehl, D.R. Lynch, D.J. McGillicuddy Jr., and J.R. Ledwell. Modeling turbulent dispersion on the north flank of Georges Bank using Lagrangian particle methods. *Continental Shelf Research*, 25(7-8):875–900, 2005.

[367] D.T. Pugh. *Tides, Surges and Mean Sea-level* (reprinted with corrections). John Wiley & Sons, 1996.

[368] R.E. Rathbun and V.C. Kennedy. Transport and dispersion of fluorescent tracer particles for the dune-bed condition, Atrisco Feeder Canal near Bernalillo, New Mexico: An experimental study of the rates of transport and dispersion of sediment particles of various diameters and specific gravities for a dune-bed condition of alluvial-channel flow. Geological Survey professional paper 1037, United States Department of the Interior, Geological Survey, U.S. Govt. Print. Off., 1978.

[369] M. Reed, C. Turner, A. Odulo, T. Isaji, S.E. Soerstroem, J.P. Mathisen, and J.P. Mathisen. Field evaluation of satellite-tracked surface drifting buoys in simulating the movement of spilled oil in the marine environment. Technical Report PB95-192241, Bureau of Land Management New Orleans Louisiana, Narragansett, RI, July 1990. Final report.

[370] M. Reed. The physical fates component of the natural resource damage assessment model system. *Oil and Chemical Pollution*, 5(2-3):99–123, 1989. Resource Damage Assessment in the Marine Environment.

[371] M. Reed, O. Johansen, P.J. Brandvik, P. Daling, A. Lewis, R. Fiocco, D. Mackay, and R. Prentki. Oil spill modeling towards the close of the 20th century: Overview of the state of the art. *Spill Science & Technology Bulletin*, 5(1):3–16, 1999.

[372] M. Reed, C. Turner, and A. Odulo. The role of wind and emulsification in modelling oil spill and surface drifter trajectories. *Spill Science & Technology Bulletin*, 1(2):143–157, 1994.

[373] A.M. Reynolds. On Lagrangian stochastic modeling of material transport in oceanic gyres. *Physica D*, 172:124–138, 2002.

[374] A.M. Reynolds and M. Veneziani. Rotational dynamics of turbulence and tsallis statistics. *Physics Letters A*, 327(1):9–14, 2004.

[375] J.F. Richardson and W.N. Zaki. Sedimentation and fluidization: Part I. *Trans. Inst. Chem. Eng*, 32:35–53, 1954.

[376] L.F. Richardson. Atmospheric diffusion shown on a distance-neighbor graph. *Proceedings of the Royal Society*, A110:709–737, 1926.

[377] P.L. Richardson. Drifters and floats. In J.H. Steele, S.A. Thorpe, and K.K. Turekian, eds., *Measurement Techniques, Sensors and Platforms-A Derivative of Encyclopedia of Ocean Sciences*, (2nd ed.). Elsevier Science, 2010.

[378] H. Ridderinkhof. Lagrangian flows in complex Eulerian current fields. In D.R. Lynch and A.M. Davies, editors, *Quantitative Skill Assessment for Coastal Ocean Models*, pp. 31–48. Coastal and Estuarine Series 47, American Geophysical Union, 1995.

[379] H. Ridderinkhof and J. Loder. Lagrangian characterization of circulation over submarine banks with application to the outer Gulf of Maine. *Journal of Physical Oceanography*, 24(6):1184–1200, 1994.

[380] H. Ridderinkhof and J.T.F. Zimmerman. Chaotic stirring in a tidal system. *Science*, 258:1107–1111, 1992.

[381] H. Ridderinkhof, J.T.F. Zimmerman, and M.E. Philippart. Tidal exchange between the North Sea and Dutch Wadden Sea and mixing time scales of the tidal basins. *Netherlands Journal of Sea Research*, 25(3):331–350, 1990.

[382] A.M. Riddle. The specification of mixing in random walk models for dispersion in the sea. *Continental Shelf Research*, 18:441–456, 1998.

[383] M. Roberts and D. Marshall. Do we require adiabatic dissipation schemes in eddy-resolving ocean models? *Journal of Physical Oceanography*, 28(10):2050–2063, 1998.

[384] A.R. Robinson and K.H. Brink, eds. *The Global Coastal Ocean – Regional Studies and Syntheses*. The Sea, No. 11. John Wiley & Sons, 1998.

[385] H.C. Rodean. *Stochastic Lagrangian Models of Turbulent Diffusion*. Meteorological Monographs No. 48. American Meteorological Society, 1996.

[386] W. Rodi. *Turbulence Models and Their Application in Hydraulics*. International Association for Hydraulic Research, Delft, NE, 1980. (2nd revised ed., 1984).

[387] W. Rodi. Examples of calculation methods for flow and mixing in stratified fluids. *Journal of Geophysical Research*, 92(C5):5305–5328, 1987.

[388] O.N. Ross and J. Sharples. Recipe for 1-D Lagrangian particle tracking models in space-varying diffusivity. *Limnology and Oceanography: Methods*, 2:289–302, 2004.

[389] S.M. Ross. *Introduction to Probability Models*. Academic Press, 9 ed., 2007.

[390] H.T. Rossby, S.C. Riser, and A.J. Mariano. The Western North Atlantic – a Lagrangian viewpoint. In A.R. Robinson, ed., *Eddies in Marine Science*, Topics in Atmospheric and Oceanographic Sciences, pp. 66–91. Springer-Verlag, 1983.

[391] T. Rossby. Evolution of Lagrangian methods in oceanography. In A. Griffa, ed., *Lagrangian Analysis and Prediction of Coastal and Ocean Dynamics*. pp 1–38 Cambridge University Press, 2007.

[392] H. Rouse. Experiments on the mechanics of sediment suspension. In *Proceedings of the 5th International Congress on Applied Mechanicss*, Cambridge, MA, pp. 550–554, 1939.

[393] V. Rupolo, B.L. Hua, A. Provenzale, and V. Artale. Lagrangian velocity spectra at 700m in the Western North Atlantic. *Journal of Physical Oceanography*, 26:1591–1607, 1996.

[394] H. Rye, P.J. Brandvik, and T. Strom. Subsurface blowouts: Results from field experiments. *Spill Science & Technology Bulletin*, 4(4):239–256, 1997.

[395] I.I. Rypina, A. Kirincich, R. Limeburner, and I.A. Udovydchenkov. Results from a mass drifter deployment in the coastal ocean near Martha's Vineyard, MA: Testing the correspondence to the high-frequency radar surface currents. *Journal of Geophysical Research*, 2014. Submitted.

[396] R.H. Sabersky, A.J. Acosta, and E.G. Hauptmann. *Fluid Flow: A First Course in Fluid Mechanics*, 2nd ed. Macmillan, 1971.

[397] H. Salman, L. Kuznetsov, C.K.R.T. Jones, and K. Ide. A method for assimilating Lagrangian data into a shallow-water-equation ocean model. *Monthly Weather Review*, 134(4):1081–1101, 2006.

[398] P. Sammarco and M. Heron, eds. *The Bio-Physics of Marine Larval Dispersal*. Number 45 in Coastal and Estuarine Studies. AGU, 1994. 303pp.

[399] C. Sammari, V.G. Koutitonsky, and M. Moussa. Sea level variability and tidal resonance in the Gulf of Gabes, Tunisia. *Continental Shelf Research*, 26(3):338–350, 2006.

[400] W.B. Samuels, N.E. Huang, and D.E. Amstutz. An oilspill trajectory analysis model with a variable wind deflection angle. *Ocean Engineering*, 9(4):347–360, 1982.

[401] L.P. Sanford. New sedimentation, resuspension, and burial. *Limnology and Oceanography*, 37:1164–1164, 1992.

[402] B.L. Sawford. Rotation of trajectories in Lagrangian stochastic models of turbulent dispersion. *Boundary-Layer Meteorology*, 93(3):411–424, 1999.

[403] R.L. Scheaffer and L.J. Young. *Introduction to Probability and Its Applications*. Brooks/Cole, 2010.

[404] Z.R. Schlag and E.W. North. Lagrangian TRANSport model (LTRANS) v.2 User's Guide. Technical report, University of Maryland Center for Environmental Science, Cambridge, MD., January 2012.

[405] J.L. Schnoor. *Environmental Modeling: Fate and Transport of Pollutants in Water, Air, and Soil*. John Wiley & Sons, 1996.

[406] K. Schroeder, A.C. Haza, A. Griffa, T.M. Oezgoekmen, P.M. Poulain, R. Gerin, G. Peggion, and M. Rixen. Relative dispersion in the Liguro-Provençal basin: From sub-mesoscale to mesoscale. *Deep-Sea Research Part I*, 58(3):209–228, 2011.

[407] R.J. Seymour and R.A. Geyer. Fates and effects of oil spills. *Annual Review of Energy and the Environment*, 17(1):261–283, 1992.

[408] S.C. Shadden, F. Lekien, and J.E. Marsden. Definition and properties of Lagrangian coherent structures from finite-time Lyapunov exponents in two-dimensional aperiodic flows. *Physica D: Nonlinear Phenomena*, 212(3–4):271–304, 2005.

[409] P.T. Shaw and G.T. Csanady. Self-advection of density perturbations on a sloping continental shelf. *Journal of Physical Oceanography*, 13(5):769–782, 1983.

[410] H.T. Shen, P.D. Yapa, D.S. Wang, and X.Q. Yang. A mathematical model for oil slick transport and mixing in rivers. CRREL Special Report 93-21 AD-A269929, US Army Core of Engineers, Cold Regions Research and Engineering Laboratory, August 1993.

[411] J.R. Shewchuck. Triangle: Engineering a 2D quality mesh generator and Delaunay triangulator. *Proceedings of the First Workshop on Applied Computational Geometry; Association for Computing Machinery*, Philadelphia, pp. 124–133, 1996.

[412] J.R. Shewchuck. *Delaunay Refinement Mesh Generation*. PhD thesis, Carnegie Mellon University, 1997.

[413] J.R. Shewchuck. Delaunay refinement algorithms for triangular mesh generation. *Computational Geometry:Theory and Applications*, 22(1–3):21–74, 2002.

[414] A Shields. Anwendung der Aehnlichkeit-Mechanik und der Turbulenz-forschung auf die Geschiebebewegung [Application of similarity mechanics and turbulence research on shear flow]. *Preussische Versuchsanstalt fur Wasserbau und Schiffbau*, pp. 524–526, 1936.

[415] W.Y. Shiu, M. Bobra, A.M. Bobra, A. Maijanen, L. Suntio, and D. Mackay. The water solubility of crude oils and petroleum products. *Oil and Chemical Pollution*, 7(1):57–84, 1990.

[416] B.W. Silverman. *Density Estimation for Statistics and Data Analysis*, 1st ed. Monographs on Statistics & Applied Probability, No. 26. Chapman & Hall/CRC, 1986.

[417] R.D. Simons, D.A. Siegel, and K.S. Brown. Model sensitivity and robustness in the estimation of larval transport: A study of particle tracking parameters. *Journal of Marine Systems*, 119–120:19–29, 2013.

[418] J.H. Simpson and J.R. Hunter. Fronts in the Irish Sea. *Nature*, 250:404–406, 1974.

[419] J.H. Simpson and J. Sharples. *Intro to the Physical and Biological Oceanography of Shelf Seas*. Cambridge University Press, 2012.

[420] J.G. Skellam. Random dispersal in theoretical populations. *Biometrika*, 38:196–218, 1951.

[421] J. Smagorinsky. General circulation experiments with the primitive equations I. The basic experiment. *Monthly Weather Review*, 91:99–164, 1963.

[422] R.D. Smith and P.R. Gent. Anisotropic Gent-McWilliams parameterization for ocean models. *Journal of Physical Oceanography*, 34(11):2541–2564, 2004.

[423] R.D. Smith and J.C. McWilliams. Anisotropic horizontal viscosity for ocean models. *Ocean Modelling*, 5(2):129–156, 2003.

[424] S. Socolofsky and E. Adams. Liquid volume fluxes in stratified multiphase plumes. *Journal of Hydraulic Engineering*, 129(11):905–914, 2003.

[425] S. Socolofsky and E. Adams. Role of slip velocity in the behavior of stratified multiphase plumes. *Journal of Hydraulic Engineering*, 131(4):273–282, 2005.

[426] S.A. Socolofsky, E. Eric Adams, and C.R. Sherwood. Formation dynamics of subsurface hydrocarbon intrusions following the Deepwater Horizon blowout. *Geophysical Research Letters*, 38(9), 2011.

[427] R. Soulsby. *Dynamics of Marine Sands: A Manual for Practical Applications*. Thomas Telford, 1997.

[428] R.L. Soulsby, C.T. Mead, and B.R. Wild. A model for simulating the dispersal tracks of sand grains in coastal areas: 'sandtrack'. *Geological Society, London, Special Publications*, 274(1):65–72, 2007.

[429] R.L. Soulsby, C.T. Mead, B.R. Wild, and M.J. Wood. Lagrangian model for simulating the dispersal of sand-sized particles in coastal waters. *Journal of Waterway, Port, Coastal, and Ocean Engineering*, 137(3):123–131, 2011.

[430] R.L. Soulsby and R.J.S.W. Whitehouse. Threshold of sediment motion in coastal environments. Paper presented at pacific coasts and Ports' 97 Conference. In *Pacific Coasts and Ports' 97 Conference*, p. 1, University of Canterbury, Christchurch, New Zealand, 1997.

[431] M. Spaulding. Drift current under the action of wind and waves. In S.G. Sajjadi, N.H. Thomas, and J.C.R. Hunt, eds., *Wind-Over-Wave Couplings*, pp. 243–256. OUP/Institute of Mathematics and its Applications, 1999.

[432] M.L. Spaulding. A state-of-the-art review of oil spill trajectory and fate modeling. *Oil and Chemical Pollution*, 4(1):39–55, 1988.

[433] M.L. Spaulding, Tatsu Isaji, Paul Hall, and Arthur A. Allen. A hierarchy of stochastic particle models for search and rescue (SAR): Application to predict surface drifter trajectories using HF radar current forcing. *Journal of Marine Environmental Engineering*, 8(3):181, 2006.

[434] M.L. Spaulding, P.R. Bishnoi, E. Anderson, and T. Isaji. An integrated model for prediction of oil transport from a deep water blowout. In *Arctic and Marine Oilspill Program Technical Seminar*, Vol. 2, pp. 611–636. Environment Canada; 1999, 2000.

[435] D.C. Spears, W.S.C. Gurney, M.R. Heath, W. Horbelt, S.N. Wood, and B.A. deCuevas. Ocean-scale Modelling of the distribution, abundance, and seasonal dynamics of the copepod *Calanus finmarchicus*. *Marine Ecology Progress Series*, 313:173–192, 2006.

[436] D.C. Spears, W.S.C. Gurney, M.R. Heath, and S.N. Wood. Modelling the basin-scale demography of *Calanus finmarchicus* in the North-East Atlantic. *Fisheries Oceanography*, 14(5):333–358, 2005.

[437] P.M. Stegman, J.A. Quinlan, F.E. Werner, and B.O. Blanton, and P. Berrien. Atlantic Menhaden recruitment to a southern estuary: Defining potential spawning regions. *Fisheries Oceanography* (Suppl. 2), 8:111–123, 1999.

[438] P.L. Stewart and S.H. Arnold. Environmental requirements of the sea scallop (*Placopecten magellanicus*) in eastern Canada and its response to human impacts. Canadian Technical Report on Fisheries and Aquatic Science, 1994:1–36, 1994.

[439] W. Stiver and D. Mackay. Evaporation rate of spills of hydrocarbons and petroleum mixtures. *Environmental Science & Technology*, 18(11):834–840, 1984.

[440] A. Stohl. Computation, accuracy and applications of trajectories – a review and bibliography. *Atmospheric Environment*, 32(6):947–966, 1998.

[441] U. Svensson. PROBE: Program for boundary layers in the environment. System Description and Manual. Technical Report 24, Swedish Meteorological and Hydrological Institute, 1998.

[442] V. Taillandier, S. Dobricic, P. Testor, N. Pinardi, A. Griffa, L. Mortier, and G.P. Gasparini. Integration of Argo trajectories in the Mediterranean forecasting system and impact on the regional analysis of the western Mediterranean circulation. *Journal of Geophysical Research: Oceans*, 115(C3), 2010.

[443] V. Taillandier, A. Griffa, P.-M. Poulain, and K. Beranger. Assimilation of Argo float positions in the north western Mediterranean Sea and impact on ocean circulation simulations. *Geophysical Research Letters*, 33(11), 2006.

[444] V. Taillandier, A. Griffa, P.M. Poulain, R. Signell, J. Chiggiato, and S. Carniel. Variational analysis of drifter positions and model outputs for the reconstruction of surface currents in the central Adriatic during fall 2002. *Journal of Geophysical Research: Oceans*, 113(C4), 2008.

[445] G.I. Taylor. Diffusion by continuous movements. *Proceedings of the London Mathematical Society*, 20:196–212, 1921.

[446] R.V. Thomann and J.A. Mueller. *Principles of Surface Water Quality Modeling and Control*. Harper Collins, 1987.

[447] K. Thompson, M. Dowd, Y. Shen, and D.A. Greenberg. Probabalistic characterization of tidal mixing in a coastal embayment: A Markov Chain approach. *Continental Shelf Research*, 22:1603–1614, 2002.

[448] C.W. Thomson. *The Voyage of the "Challenger": The Atlantic; a Preliminary Account of the General Results of the Exploring Voyage of H. M. S. "Challenger" during the Year 1873 and the Early Part of the Year 1876*. Number v. 1 in The Voyage of the "Challenger." Macmillan and Company, 1878.

[449] D.J. Thomson. Criteria for the selection of stochastic models of particle trajectories in turbulent flows. *Journal of Fluid Mechanics*, 180:529–556, 1987.

[450] P. Tkalich and E.S. Chan. Vertical mixing of oil droplets by breaking waves. *Marine Pollution Bulletin*, 44(11):1219–1229, 2002.

[451] M. Toner, A.D. Kirwan, L.H. Kantha, and J.K. Choi. Can general circulation models be assessed and their output enhanced with drifter data? *Journal of Geophysical Research: Oceans*, 106(C9):19563–19579, 2001.

[452] D.W. Townsend. *Oceanography and Marine Biology*. Sinauer, 2012.

[453] L.N. Trefethen and D. Bau. *Numerical Linear Algebra*. SIAM, 1997.

[454] M.J. Tremblay, J.W. Loder, F.E. Werner, C.E. Naimie, F.H. Page, and M.M. Sinclair. Drift of sea scallop larvae *Placopecten magellanicus* on Georges Bank: A model study of the roles of mean advection, larval behaviour, and larval origin. *Deep-Sea Research Part II*, 41(1):7–49, 1994.

[455] L. Umlauf and H. Burchard. A generic two-equation model for geophysical flows. *Journal of Marine Research*, 61:235–265, 2003.

[456] L. Umlauf, H. Burchard, and K. Hutter. Extending the k–ω turbulence model towards oceanic applications. *Ocean Modelling*, 5(3):195–218, 2003.

[457] G. van Belle. *Statistical Rules of Thumb*, 2nd ed. John Wiley & Sons, 2008.

[458] M.Th. van Genuchten and W.J. Alves. *Analytical Solutions of the One-Dimensional Convective-Dispersive Solute Transport Equation*. Technical Bulletin, No. 1661. U.S. Deptartment of Agriculture, 1982.

[459] L.C. van Rijn. Sediment pick-up functions. *Journal of Hydraulic Engineering*, 110(10):1494–1502, 1984.

[460] L.C. van Rijn. Sediment transport, Part I: Bed load transport. *Journal of Hydraulic Engineering*, 110(10):1431–1456, 1984.

[461] V.A. Vanoni. Fifty years of sedimentation. *Journal of Hydraulic Engineering*, 110(8):1021–1057, 1984.

[462] M. Veneziani, A. Griffa, A.M. Reynolds, and A.J. Mariano. Oceanic turbulence and stochastic models from Subsurface Lagrangian Data for the Northwest Atlantic Ocean. *Journal of Physical Oceanography*, 34:1884–1906, 2004.

[463] M. Veneziani, A. Griffa, Z.D Garraffo, and E.P Chassignet. Lagrangian spin parameter and coherent structures from trajectories released in a high-resolution ocean model. *Journal of Marine Research*, 63(4):753–788, 2005.

[464] A.W. Visser. Using random walk models to simulate the vertical distribution of particles in a turbulent water column. *Marine Ecology Progress Series*, 158:275–281, 1997.
[465] R.J. Vos. Comparison of 5 oil weathering models. Werkdocument RIKZ.ZD/2005.011w, Rijkswaterstaat Dienst Noordzee, Ministerie van Verkeer en Waterstaat, July 2005.
[466] M.P. Wand and M.C. Jones. *Kernal Smoothing*. Chapman and Hall, 1994.
[467] S.-D. Wang, Yong-Ming Shen, Ya-Kun Guo, and Jun Tang. Three-dimensional numerical simulation for transport of oil spills in seas. *Ocean Engineering*, 35(5-6):503–510, 2008.
[468] L.G. Ward. Sedimentology of the lower Great Bay/Piscataqua River Estuary. NCCOSC RDTE division report, Department of the Navy, San Diego, CA, 1995.
[469] J.C. Warner, C.R. Sherwood, H.G. Arango, and R.P. Signell. Performance of four turbulence closure methods implemented using a generic length scale method. *Ocean Modeling*, 8:81–113, 2005.
[470] D.J. Webb. Tidal resonance in the Coral Sea. *Nature*, 243:511, 1973.
[471] R.H. Weisberg, L.L. Zheng, and Y.Y. Liu. Tracking subsurface oil in the aftermath of the Deepwater Horizon well blowout. In Y. Liu, A. Macfadyen, Z.-G. Ji, and R.H. Weisberg, eds., *Monitoring and Modeling the Deepwater Horizon Oil Spill: A Record-Breaking Enterprise*, pp. 205–215. American Geophysical Union, 2011.
[472] E.W. Weisstein. *CRC Concise Encyclopedia of Mathematics*, 2nd ed. CRC Press, 2002.
[473] E.W. Weisstein. Sample variance distribution. Website http://mathworld.wolfram.com/SampleVarianceDistribution.html, 2005.
[474] F. Werner, J. Quinlan, G. Lough, and D. Lynch. Spatially-explicit individual based modeling of marine populations: A review of the advances in the 1990's. *Sarsia*, 86(6):411–421, 2001.
[475] F.E. Werner, B.R. MacKenzie, R.I. Perry, R.G. Lough, C.E. Naimie, B.O. Blanton, and J.A. Quinlan. Larval trophodynamics, turbulence, and drift on Georges Bank: a sensitivity analysis of cod and haddock. *Scientia Marina*, 65(Suppl.1):99–115, 1996.
[476] F.E. Werner, R.I. Perry, R.G. Lough, and C.E. Naimie. Trophodynamic and advective influences on Georges Bank larval cod and haddock. *Deep-Sea Research Part II*, 43:1793–1822, 1996.
[477] F.E. Werner, J.A. Quinlan, B.O. Blanton, and R.A. Luettich. The role of hydrodynamics in explaining fish populations. *Journal of Sea Research*, 37:195–212, 1997.
[478] F.E. Werner, F.H. Page, D.R. Lynch, J.W. Loder, R.G. Lough, R.I. Perry, D.A. Greenberg, and Michael M. Sinclair. Influences of mean advection and simple behavior on the distribution of cod and haddock early life stages on Georges Bank. *Fish. Oceanogr.*, 2:43–64, 1993.
[479] J.J. Westerink, R.A. Luettich Jr., and J. Muccio. Modeling tides in the Western North Atlantic using unstructured graded grids. *Tellus*, 46A:178–199, 1994.
[480] J.J. Westerink, R.A. Luettich, J.C. Feyen, J.H. Atkinson, C. Dawson, H.J. Roberts, M.D. Powell, J.P. Dunion, E.J. Kubatko, and H. Pourtaheri. A basin-to channel-scale unstructured grid hurricane storm surge model applied to southern Louisiana. *Monthly Weather Review*, 136(3):833–864, March 2008.
[481] R. Wheeler, K. Smith, and D. Lynch. File standards for individual-based models of *Alexandrium*. Numerical Methods Laboratory Report NML-05-6, Dartmouth College, Hanover NH USA. [http:// www-nml.dartmouth.edu/Publications/internal_reports/NML-05-6] , 2005.
[482] F. White. *Viscous Fluid Flow*, 3rd ed. Mechanical Engineering. McGraw-Hill, 2005.

[483] D.C. Wilcox. Reassessment of the scale-determining equation for advanced turbulence models. *AIAA Journal*, 26(11):1299–1310, 1988.

[484] E. Wolanski. Facts and numerical artefacts in modelling the dispersal of crown-of-thorns starfish larvae in the Great Barrier Reef. *Marine and Freshwater Research*, 44(3):427–436, 1993.

[485] P.M.J. Woodhead and A.J. Lee. A new instrument for measuring residual currents near the sea-bed. *Conseil Internation poor Exploration de la Mer, Hydrographic Communications*, 12:6, 1960.

[486] D.G. Wright. On the alongshelf evolution of an idealized density front. *Journal of Physical Oceanography*, 19(4):532–541, 1989.

[487] D.G. Wright, D.A. Greenberg, J.W. Loder, and P.C. Smith. The steady state barotropic response of the Gulf of Maine and adjacent regions to surface wind stress. *Journal of Physical Oceanography*, 16:947–966, 1986.

[488] F.-C. Wu and Y.-J. Chou. Rolling and lifting probabilities for sediment entrainment. *Journal of Hydraulic Engineering*, 129(2):110–119, 2003.

[489] C. Wunsch. *The Ocean Circulation Inverse Problem*. Cambridge University Press, 1996.

[490] H. Xie, P.D. Yapa, and K. Nakata. Modeling emulsification after an oil spill in the sea. *Journal of Marine Systems*, 68(3-4):489–506, 2007.

[491] D. Xu, H. Xue, and D.A. Greenberg. A numerical study of the circulation and drifter trajectories in Cobscook Bay. In M.L. Spaulding, ed., *Estuarine and Coastal Modeling 2005*, pp. 176–195. American Society of Civil Engineers, 2006.

[492] D. Xu. *Lagrangian Study of Particle Transport Processes in the Coastal Gulf of Maine*. PhD thesis, University of Maine, 2008.

[493] Q. Yang, B. Parvin, and A. Mariano. Detection of vortices and saddle points in SST data. *Geophysical Research Letters*, 28(2):331–334, 2001.

[494] P. Yapa, L. Zheng, and K. Nakata. Modeling underwater oil/gas jets and plumes. *Journal of Hydraulic Engineering*, 125(5):481–491, 1999.

[495] P.D. Yapa. Modeling oil spills to mitigate coastal pollution. In H.J. Fernando, ed., *Handbook of Environmental Fluid Dynamics: Systems, Pollution, Modeling, and Measurements*, Vol. 2. pp. 243–255. CRC Press, 2013.

[496] P.D. Yapa, M.R. Wimalaratne, A.L. Dissanayake, and J.A. DeGraff Jr. How does oil and gas behave when released in deepwater? *Journal of Hydro-environment Research*, 6(4):275–285, 2012.

[497] P.D. Yapa, L. Zheng, and F. Chen. A model for deepwater oil/gas blowouts. *Marine Pollution Bulletin*, 43(7–12):234–241, 2001. Marine Environmental Modelling.

[498] K.Y. Yick, C.R. Torres, T. Peacock, and R. Stocker. Enhanced drag of a sphere settling in a stratified fluid at small Reynolds numbers. *Journal of Fluid Mechanics*, 632:49, 2009.

[499] M.M. Youssef. The behavior of the near ocean surface under the combined action of waves and currents in shallow water. Dissertations and master's theses, University of Rhode Island, 1993. Paper AAI9332472.

[500] D.F. Zhang, A.K. Easton, and J.M. Steiner. Simulation of coastal oil spills using the random walk particle method with Gaussian kernel weighting. *Spill Science & Technology Bulletin*, 4(2):71–88, 1997. IMO Special Issue-Part III.

[501] L. Zheng and P. Yapa. Buoyant velocity of spherical and nonspherical bubbles/droplets. *Journal of Hydraulic Engineering*, 126(11):852–854, 2000.

[502] S. Zhiyao, W. Tingting, X. Fumin, and L. Ruijie. A simple formula for predicting settling velocity of sediment particles. *Water Science and Engineering*, 1(1):37–43, 2008.

[503] D. Zhong, G. Wang, and Y. Ding. Bed sediment entrainment function based on kinetic theory. *Journal of Hydraulic Engineering*, 137(2):222–233, Feb 2011.

[504] J.T.F. Zimmerman. Circulation and water exchange near a tidal watershed in the Dutch Wadden Sea. *Netherlands Journal of Sea Research*, 8(2–3):126–138, 1974.

[505] J.T.F. Zimmerman. Mixing and flushing of tidal embayments in the western Dutch Wadden Sea. Part I: Distribution of salinity and calculation of mixing time scales. *Netherlands Journals of Sea Research*, 10(2):149–191, 1976.

[506] J.T.F. Zimmerman. Mixing and flushing of tidal embayments in the western Dutch Wadden Sea. Part II: Analysis of Mixing Processes. *Netherlands Journal of Sea Research*, 10(4):397–439, 1976.

[507] J.T.F. Zimmerman. The tidal whirlpool: A review of horizontal dispersion by tidal and residual currents. *Netherlands Journal of Sea Research*, 20(2/3):133–154, 1986.

Index

Printed in the United Kingdom by TJ Clays Ltd.